4th International Conference on Wireless Communications and Smart Grid (ICWCSG 2022)

AA007221

Journal of Physics: Conference Series Volume 2384

Hulun Buir, China
19-21 August 2022

ISBN: 978-1-7138-6490-5
ISSN: 1742-6588

Printed from e-media with permission by:

Curran Associates, Inc.
57 Morehouse Lane
Red Hook, NY 12571

Some format issues inherent in the e-media version may also appear in this print version.

Institute of Physics
Dirac House, Temple Back
Bristol BS1 6BE UK

Phone: 44 1 17 929 7481
Fax: 44 1 17 920 0979

techtracking@iop.org

Additional copies of this publication are available from:

Curran Associates, Inc.
57 Morehouse Lane
Red Hook, NY 12571 USA
Phone: 845-758-0400
Fax: 845-758-2633
Email: curran@proceedings.com
Web: www.proceedings.com

TABLE OF CONTENTS

Journal of Physics: Conference Series **2384** (2022) 011001 doi:10.1088/1742-6596/2384/1/011001

Preface

The 2022 4th International Conference on Wireless Communications and Smart Grid (ICWCSG 2022) was successfully held on August 19-21, 2022 in Hulunbuir, China in the form of online conference. ICWCSG 2022 promoted research and developmental activities in wireless communications and smart grid as well as scientific information interchange between researchers, engineers, and practitioners working all around the world.

We were delighted and honored to invite both Prof. Jizhong Zhu from South China University of Technology, China and Prof. Ting Yang from Tianjin University, China to serve as our General Conference Chairs. There were 90 individuals and enterprises attending the conference. Divided into three parts, the conference agenda covered keynote speeches, oral presentations, and online Q&A discussion. Firstly, keynote speakers were each allocated 30-45 minutes to hold their speeches. Then in the oral presentations, the excellent papers we had selected were presented by their authors by turns.

During the conference, seven sophisticated professors were invited to address keynote speeches. Among them, Prof. Tao Zhang from North China University of Technology, China delivered a speech on the topic: Application of Urban Brain in Data Center. In this report, he explained that urban brain is an artificial intelligence development, innovation and operation platform based on cloud computing, big data, artificial intelligence and the new generation of information technology of the Internet of things. Additionally, Prof. Chuanming Liu from National Taipei University of Technology (NTUT), China gave a report on the Uncertain Data Management over IoT Data Streams in Edge Computing. The wireless sensor networks to the Internet of Things (IOT) with cloud computing is one of the examples that modern computing environments tend to be decentralized and autonomous. In this presentation, he introduced the trends of emerging computing environments and the promising edge cloud systems with IoT applications. The wonderful and dramatic speeches of each keynote speaker had triggered heated discussion. Moreover, every participant praised this conference for disseminating useful and insightful knowledge.

After months of well preparation and hard work, the proceedings of ICWCSG 2022 covering a bunch of papers is smoothly published. These papers feature but are not limited to the following areas: Signal Processing & Applications, Optical Communications and Networking, Microwave Engineering, Communications Networks and Systems, etc. All the papers have been checked through rigorous review and processes to meet the requirements of publication.

On behalf of the conference organizing committee, we would like to express our heartfelt appreciation to all the keynote speakers, peer reviewers, and all the participants. In particular, we would like to acknowledge the Journal of Physics: Conference Series, for the endeavor of all the colleagues in publishing this paper volume. We firmly believe that all the attendees have had fruitful discussions and gained valuable knowledge, and will enjoy the opportunity for future collaborations.

Committee of ICWCSG 2022

Journal of Physics: Conference Series **2384** (2022) 011001 doi:10.1088/1742-6596/2384/1/011001

Committee member

General Conference Chair
Professor Jizhong Zhu, South China University of Technology, China
Professor Ting Yang, School of Electrical and Information Engineering, Tianjin University, China

Publication Chair
Assoc. Prof. Marina Yusoff, Universiti Teknologi MARA, Malaysia

International Technical Program Chairs
Prof. Yang Yue, Xi'an Jiaotong University, China
Assoc. Prof. Pavel Loskot, Zhejiang University-University of Illinois at Urbana-Champaign Institute (ZJUI), China
Assoc. Prof. Dong Li, Faculty of Information Technology Macau University of Science and Technology Avenida Wai Long, Taipa, Macau SAR, China
Dr. Wan Nor Shuhadah Wan Nik, Universiti Sultan Zainal Abidin, Terengganu, Malaysia

Program Committee
Prof. Panos M. Pardalos, Department of Industrial & Systems Engineering, University of Florida, USA
Prof. Zhengxiang Xie, Medical informatics college of Chongqing University, China
Prof. Zhibin Lv, School of Data Science and Software Engineering, Qingdao University, China
Prof. Carlos Borrego, Departament Enginyeria of Information Comunicacions Autonomous Universityof Barcelo na, Spain
Prof. Jiang Wu, University of Electronic Science and Technology of China, China
Prof. Steven Sheng-Uei Guan, Department of Computer Science & Software Engineering, Xi'an Jiaotong-Liverpool University, China
Prof. Tao Wang, SUN YAT-SEN UNIVERSITY, China
Doctor Xuan Huang, Computer Academy of Guangdong, China
Engineer. Xiai Yang, Computer Academy of Guangdong, China
Secretary. Ling Long, Computer Academy of Guangdong, China
Assoc. Prof. N. Rajathi, Department of Information Technology, Kumaraguru College of Technology, India
Assoc. Prof. RAHUL VISHWANATH DANDAGE, Rajendra Mane College of Engineering & Technology, Ambav (Devrukh)
Dr. Uttam Ghosh, IEEE Senior Member. Vanderbilt University, India
Dr. Nirmalya Thakur, IEEE Member, Department of Electrical Engineering & Computer Science, University of Cincinnati, USA.
Dr. Ayush Dogra, CSIR NPDF at CSIR-CSIO(Research Lab-Government of India), Chandigarh, India
Dr. Hamid R. Ghaeini, CISPA - Helmholtz Center for Information Security, Saarbrücken, Germany

Peer Review Statement

All papers published in this volume have been reviewed through processes administered by the Editors. Reviews were conducted by expert referees to the professional and scientific standards expected of a proceedings journal published by IOP Publishing Publishing.

- **Type of peer review:** Single Anonymous
- **Conference submission management system:** Morressier
- **Number of submissions received:** 106
- **Number of submissions sent for review:** 98
- **Number of submissions accepted:** 58
- **Acceptance Rate (Submissions Accepted / Submissions Received × 100):** 54.7
- **Average number of reviews per paper:** 2
- **Total number of reviewers involved:** 30
- **Contact person for queries:**
  Name: Xuexia Ye
  Email: xx.ye@keoaeic.org
  Affiliation: AEIC Academic Exchange Information Centre

Simulation Verification and Testing of Dual Circular Polarization Diversity

Haocong Ji[1,a], Ning Chen[1,b], Peng Zhao[1,c], Guoshun Tang[1,d]

[1]School of Information and Electronics Beijing Institute of Technology, Beijing, China

[a] jihaocong_bit@163.com, [b] chenning@bit.edu.cn,
[c]1005805904@qq.com[d]1036832863@qq.com

Abstract—This paper presents a test method of satellite communication based on dual circular polarization diversity. A polarized channel model is established based on two validated models to obtain the bit error rate curve of MIMO system. Moreover, the specific antennas are chosen through the simulation result of the cross-polar discrimination (XPD). Finally, a hardware platform with USRPE310 and polarized-antenna is established to verify the simulation result.

1. Introduction

In order to get the diversity gain obtained by MIMO (Multi-Input Multi-Output) compared with SISO (Single-Input Single -Output,) in satellite communication, it is necessary to establish an accurate channel model. The satellite-ground signal path modeling which consists of Loo and two-state Markov are used to imitate the real signal path. Then, as the changing of environment and XPD, the diversity gain acquired by MIMO can be contrasted with SISO. Moreover, the suitable antenna is selected through the simulation result of XPD. Finally, the system is accomplished on a hardware system which consist of USRPE310 and dual-circularly-polarized antennas.

2. Satellite channel model analysis

2.1 C.LOO model

The C.Loo model is a signal path model which apply to Land-Mobile Satellite systems, including shadowing fading called large-scale decline and multipath effects called small-scale decline [1].

The large-scale fading's contour defers the distribution of log-normal:

$$f(z) = \frac{1}{z\sqrt{2\pi d_0}} e^{-\frac{(lnz-\mu)^2}{2d_0}}, z \geq 0 \quad (1)$$

where μ is the mean of lnz and d_0 is the variance of lnz,

The small-scale fading's contour defers the distribution of Rayleigh:

$$f(m) = \frac{m}{b_0} e^{-\frac{m^2}{2b_0}}, m \geq 0 \quad (2)$$

So the channel parameter H considering both shadowing and multipath can be expressed as:

$$H = \overline{H} + \widetilde{H} \quad (3)$$

where $\overline{H}$ represents the influence caused by shadowing, and $\widetilde{H}$ represents the influence caused by small-scale fading.

So, the probability function of H is as followed:[1]

$$f(r)=\frac{r}{b_0\sqrt{2\pi d_0}}\int_0^{\infty}\frac{1}{z}e^{-\frac{(lnz-\mu)^2}{2d_0}-\frac{r^2+z^2}{2b_0}}I_0(\frac{rz}{b_0})dz \qquad (4)$$

where $\alpha=20log_{10}(e^{\mu}), \varphi=20\sqrt{d_0}log_{10}(e), MP=10log_{10}(2b_0)$.

2.2 Markov probabilistic transfersion model

Markov model with two state is shown in Figure 1, the condition of the channel signal path is divided into superior state and inferior state[2]. " Superior state" means that from the satellite transmitter to the ground mobile receiver. " Inferior state" channel represents that the satellite-to-ground channel is effected by shadowing and multipath heavily.

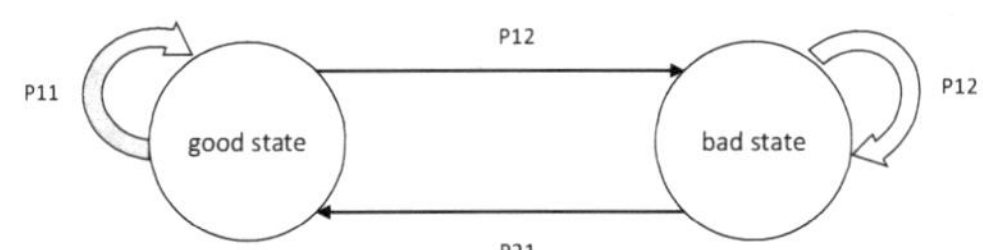

Figure 1. Markov probabilistic transfersion model

Define the state probability vector $w=[p_{11}\ p_{22}]$.

$$wp=w \qquad (5)$$

In [3], the numerical value of α, φ, MP and P in different angles between satellite and ground are provided in [3].

2.3 The effect of polarization on communication channels

Two different polarized signals can exist in the same beam, and there is no spatial correlation between different polarized waves in ideal conditions. But unsatisfactory antenna polarization characteristics and environmental factors will lead to cross-polarization also called polarization signal leakage. Therefore, the dual-polarization satellite MIMO signal power needs to consider the cross-polar discrimination (XPD) of the antenna and cross-polar coupling (XPC) of the environment. The effects of cross-polar discrimination and cross-polar coupling on large-scale and small-scale fading are shown in (6) and (7):

$$\begin{bmatrix}E\{|\overline{H_{11}}|^2\} & E\{|\overline{H_{12}}|^2\}\\ E\{|\overline{H_{21}}|^2\} & E\{|\overline{H_{22}}|^2\}\end{bmatrix}=(\alpha^2+\varphi^2)\begin{bmatrix}1-\beta & \beta\\ \beta & 1-\beta\end{bmatrix} \qquad (6)$$

where $\beta=\frac{1}{1+XPD}$

$$\begin{bmatrix}E\left\{\left|\widetilde{H_{11}}\right|^2\right\} & E\left\{\left|\widetilde{H_{12}}\right|^2\right\}\\ E\left\{\left|\widetilde{H_{21}}\right|^2\right\} & E\left\{\left|\widetilde{H_{22}}\right|^2\right\}\end{bmatrix}=(\alpha^2+\varphi^2)\begin{bmatrix}1-\beta & \beta\\ \beta & 1-\beta\end{bmatrix} \qquad (7)$$

where $\gamma=\frac{XPD+XPC}{(1+XPD)(1+XPC)}$

β and γ represent two kinds of fading polarization factors.

3. Simulation result of channel model

3.1 Satellite polarization channel simulation

The establishment steps of signal path model is shown in Figure 2,

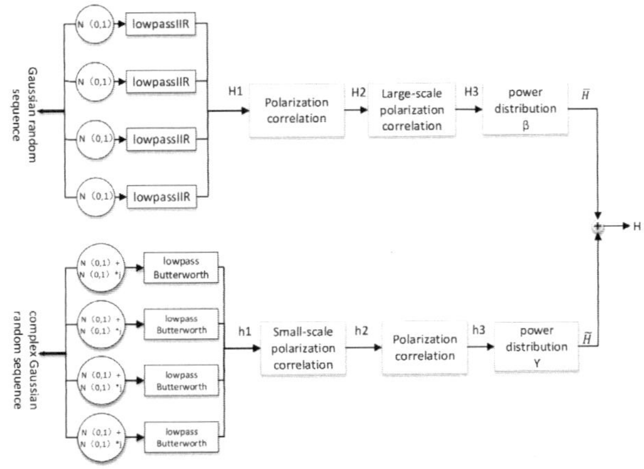

Figure 2. Large-scale and Small-scale fading establishment process

Large-scale fading establishment steps:

1) Generating a Gaussian random sequence x_n with mean 0 and variance 1, and then these random sequences x_n are introduced with temporal correlation through a low-pass IIR filter, which can be expressed as $y_n = x_n + Ay_{n-1}$, where $A = \exp\,(-VT/L_{cor})$.

2) The data is multiplied by the spatial correlation matrix C to introduce spatial correlation.

$$\mathrm{H2} = \mathrm{C}^{1/2} \times \mathrm{H1} \tag{8}$$

3) Combining H2 with the statistical parameters (α, ψ) yields the large-scale fading matrix.

$$\mathrm{H3} = 10^{\mathrm{H2}\times\frac{\varphi}{20}+\frac{\alpha}{20}} \tag{9}$$

4) The XPD is considered into the model to distribute the power

$$\overline{\mathrm{H}} = \begin{bmatrix} \sqrt{1-\beta}\cdot \mathrm{H3}_{11} & \sqrt{\beta}\cdot \mathrm{H3}_{12} \\ \sqrt{\beta}\cdot \mathrm{H3}_{21} & \sqrt{1-\beta}\cdot \mathrm{H3}_{22} \end{bmatrix} \tag{10}$$

Small-scale fading establishment steps:

1) A four-way complex Gaussian random sequence is passed through a low-pass Butterworth filter to simulate the Doppler effect.

2) Combining h1 with the statistical parameters MP yields the small-scale fading matrix.

$$\mathrm{h2} = (0.5 * 10^{\frac{MP}{10}})^{1/2} \times \mathrm{h1} \tag{11}$$

3) Using spatial correlation matrix C to introduce the polarization correlation.

$$\mathrm{h3} = \mathrm{C}^{1/2} \times \mathrm{h2} \tag{12}$$

4) The XPD is considered into the model to distribute the power

$$\widetilde{H} = \begin{bmatrix} \sqrt{1-\chi}\cdot \mathrm{h3}_{11} & \sqrt{\chi}\cdot \mathrm{h3}_{12} \\ \sqrt{\chi}\cdot \mathrm{h3}_{21} & \sqrt{1-\chi}\cdot \mathrm{h3}_{22} \end{bmatrix} \tag{13}$$

Figure 3 below represents the whole building procedure of the channel model with both C.Loo model and Markov transition model.

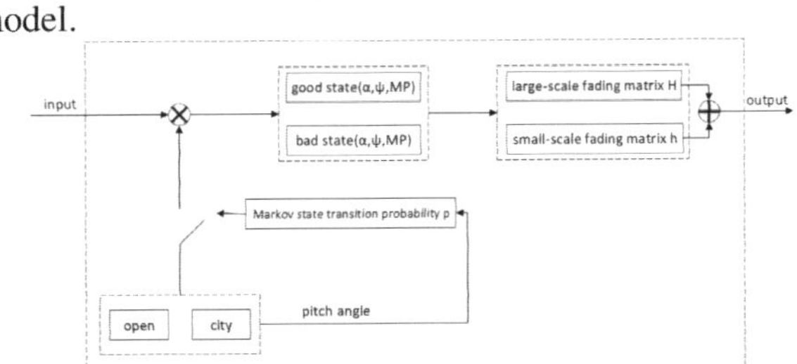

Figure 3. Frame of satellite polarization channel model

3.2 Simulation results of satellite polarization channel model

In line with the dual-polarization channel model establishment procedure, simulations result with different parameters are obtained. Spatial correlation matrix C in different environment can be found in [5].

As shown in Figure 4, superior states continue longer than inferior states and the difference between H11 and H12 is about 10dB.

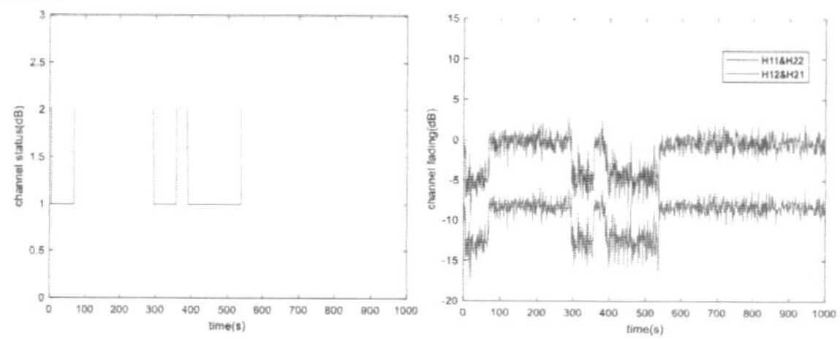

Figure 4. 2*2 MIMO channel status and gain in open area

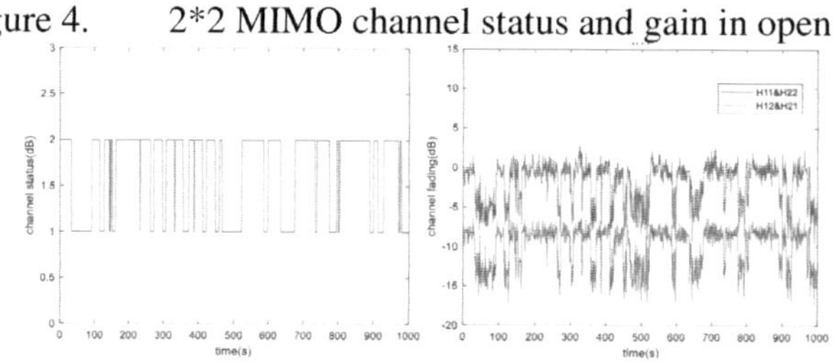

Figure 5. 2*2 MIMO channel status and gain in urban area

Then the communication system simulation results of MIMO and SISO are obtained. The BER simulation result is shown in Figure 6:

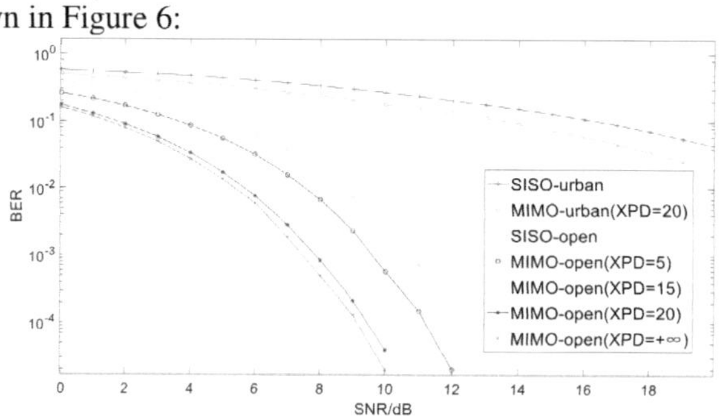

Figure 6. Satellite dual polarization bit error rate curve in different environment and XPD

From Figure can see that in open environment MIMO can reach 3.2dB gain by diversity. In urban environment, MIMO can get 2.9dB diversity gain. And from Figure 6 can also see that when XPD=20dB, the BER curve is close to the curve when XPD=+∞ which is the ideal state. So the MIMO system has a good performance of BER when in open environment and XPD=20dB.

4. Polarized Antenna Design and Test Results

4.1 Description of outdoor test environment

In order to simulate the actual scenario of the application of satellite dual circular polarization diversity technology, it is necessary to design a corresponding field experiment plan. The most important step in the experimental scheme is the selection of the polarized antenna. The design of outdoor test scene is as followed.

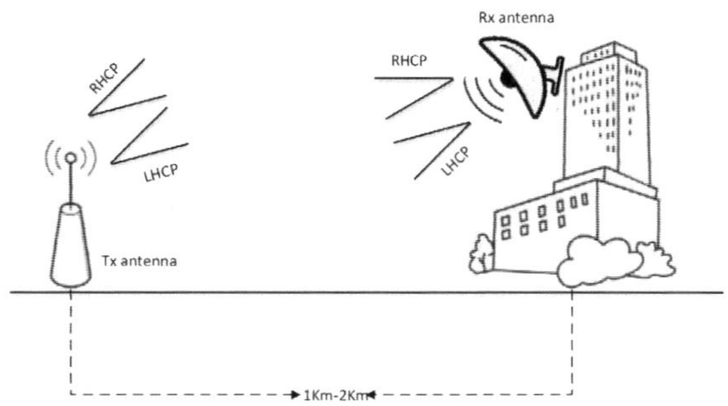

Figure 7. Field test simulation diagram

4.2 Transmit antenna

In order to simulate the transmitting antenna of a personal terminal, the transmitting antenna needs to be miniaturized and omnidirectional. The transmitting antenna adopts a 3dB bridge as the feeding method, and the signal realizes LHCP (left-hand circular polarization) and RHCP (right-hand circular polarization) through the 3dB bridge. Through this structure, the antenna can simultaneously generate dual circularly polarized waves. The transmitting antenna parameters are shown in Table 1,

TABLE 1 TRANSMIT ANTENNA PARAMETER

Parameter	Unit	Value
VSWR		≤1.5
Polarization mode		RHCP/LHCP
Transmit gain(normal direction)	dB	≥3
Axial Ratio(normal direction)	dB	≤2
Port isolation	dB	≥18
passband range	GHz	1.98~2.4

The S21 parameter of the antenna measured by vector network analyzer shown in Figure 8. For omnidirectional antenna, S21 can be acted as XPD which is about -19dB.

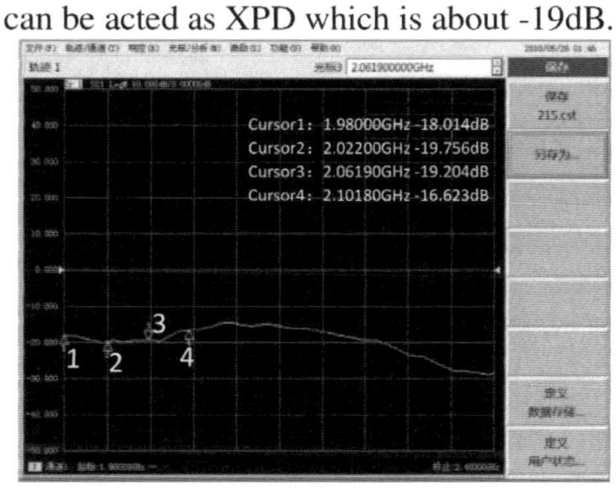

Figure 8. Omnidirectional Antenna S21 Parameter test result

4.3 Receive antenna

In actual satellite communication, the receiving antenna possess the features of high gain and directional. Therefore, the directional horn antenna is used as the receiving antenna. The receiving antenna parameters are shown in Table 2,

TABLE 2 TRANSMIT ANTENNA PARAMETER

Parameter	Unit	Value
VSWR		≤1.5
Polarization mode		RHCP/LHCP
Transmit gain(normal direction)	dB	≥16

Parameter	Unit	Value
Axial Ratio(normal direction)	dB	≤2
Port isolation	dB	≥18
passband range	GHz	1.98~2.4

The feed port of the horn directional antenna is divided into port 1 and port 2. The port isolation of receiving antenna is measured by the vector network analyzer, and XPD of the antenna in the bandwidth range is at least -22dB.

From Figure 6 can see that when in XPD=20dB and BER $=10^{-4}$, the SNR is about 8.3dB, indicating that the MIMO system with Alamouti space-time encoding has a good performance of BER when the antenna polarization isolation is 20dB. Therefore, the selection of current antenna can well verify the polarization gain obtained by MIMO compared with SISO in the satellite communication system

4.4 Test result

The BER curve is obtained through multiple tests shown in Figure 10. Compared with SISO, MIMO can get 3.1dB improvement of SNR compared to SISO.

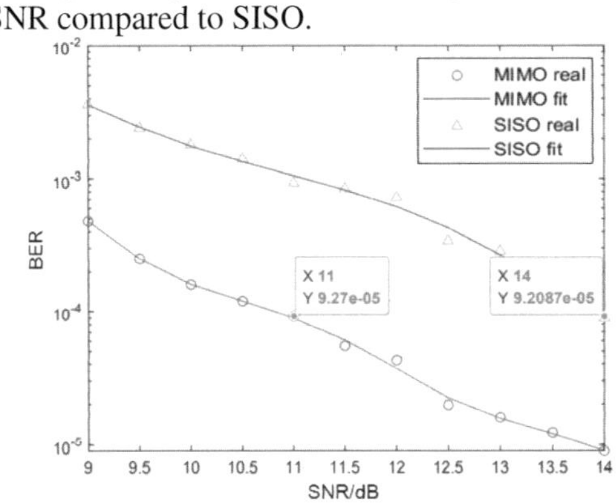

Figure 9. SISO and MIMO bit error rate measurement result

The experimental results indicate that in the identical BER situation, 2×2 MIMO with dual circular polarization diversity obviously declines the needed SNR compared to SISO.

References

[1] Konstantinos Liolis, Ilias Andrikopoulos, Panayotis Cottis. On Statistical Modeling and Performance Evaluation of SIMO Land Mobile Satellite Channels. IEEE Advanced Satellite Mobile Systems. 2008, 4: 76-81

[2] Byman A, Hulkkonen A, Arapoglou P D, et al. MIMO for mobile satellite digital broadcasting : from theory to practice[J]. IEEE Transactions on Vehicular Technology, 2016, 65(7) : 4839-4853

[3] Fassoi S, Michailidis E T, Kanatas A G. Performance evaluation of MIMO satellite multiple-relay multi-user fading channels[C]. In : Proc. of the 9th European Conference on Antennas and Propagation (EuCAP). Lisbon, 2015, 1-5

[4] Liolis K P, Gómez-Vilardebó J, Casini E, et al. Statistical modeling of dual-polarized MIMO land mobile satellite channels[J]. IEEE Transactions on Communications, 2010, 58(11): 3077-3083.

[5] King P R, Stavrou S. Capacity improvement for a land mobile single satellite MIMO system[J]. IEEE Antennas and Wireless Propagation Letters, 2006, 5: 98-100.

Journal of Physics: Conference Series **2384** (2022) 012002 doi:10.1088/1742-6596/2384/1/012002

C/X/Ku triple-band all-metal slot antenna

Xiaoyu Cheng[1], Shaofeng Bi[1], Yunfei Sun[1] and Chengwei Yuan[1,2]

[1] College of Advanced Interdisciplinary Studies, National University of Defense Technology, Changsha 410073, China

[2] Corresponding Author, cwyuan@nudt.edu.cn

Abstract. Most multi-band all-metal slot antennas are dual-band. Their design concept is not feasible for triple-band antennas. A triple-band all-metal slot antenna is designed using an electromagnetic band gap structure (EBG). The antenna can work at C/X/Ku-band. Simulated results are presented to demonstrate the feasibility of the design. The compact antenna expands the working frequency of an all-metal slot antenna, which is especially appealing for high-power antennas.

1. Introduction

With the development of communication technology, the demand for multi-band antennas has been growing. Microstrip is widely used in multi-band antenna [1-4]. However, microstrip antennas are not suitable for high frequency. Due to low loss, high structural strength, and high power capacity, multi-band all-metal waveguide slot antennas have been studied [5-8]. At present, most multi-band slot antennas are dual-band antennas. In this article, a triple-band slot antenna is proposed. By placing the EBG structure in the ridge waveguide, a slot waveguide antenna can work at the C/X-band, and the Ku-band antenna is placed above the C/X-band antenna. The proposed antenna is a standing wave antenna with three bands. The central frequencies are 4.3 GHz (C-band), 8.5 GHz (X-band), and 12.5GHz (Ku-band).

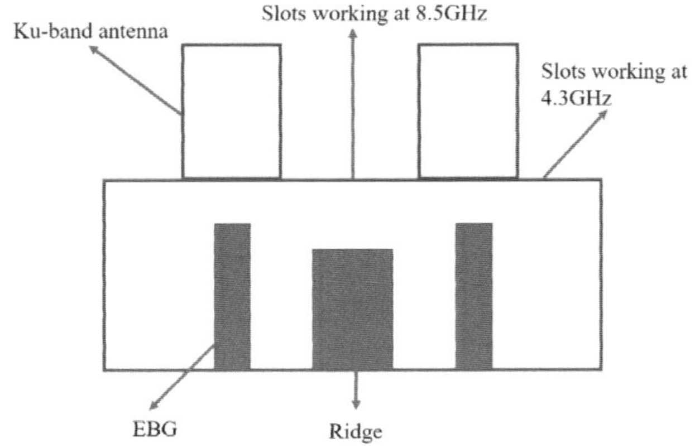

Figure 1. Diagram of the triple-band antenna

2. Antenna configuration and design

The antenna diagram is shown in Figure 1. EBG structure can stop the propagation of electromagnetic waves at the cut-off band. Placing a nailed EBG structure with suitable dimensions on both sides of the ridge can block electromagnetic waves at 8.5 GHz and pass electromagnetic waves at 4.3 GHz. The electromagnetic wave at 8.5 GHz can only propagate in the central area of the waveguide, while the electromagnetic wave at 4.3 GHz can propagate throughout the waveguide. There are X-band slots near the middle of the waveguide, and C-band slots are close to the edge of the waveguide. Since electromagnetic waves at 8.5 GHz cannot pass through the EBG structure, they can only radiate through slots near the middle. As for electromagnetic waves at 4.3GHz, the middle slots are too close to the center, so little energy radiates through them.

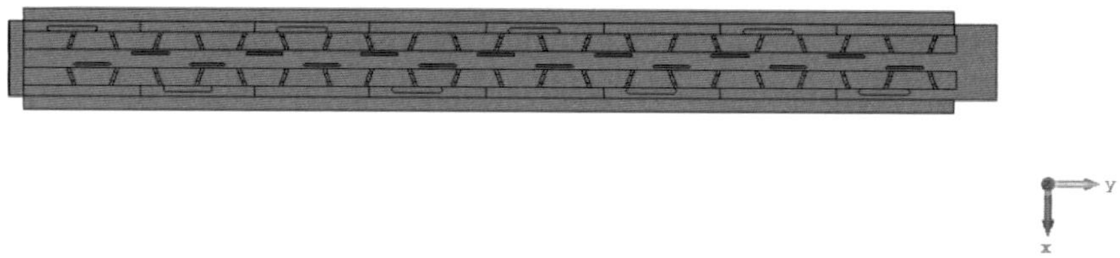

(a)

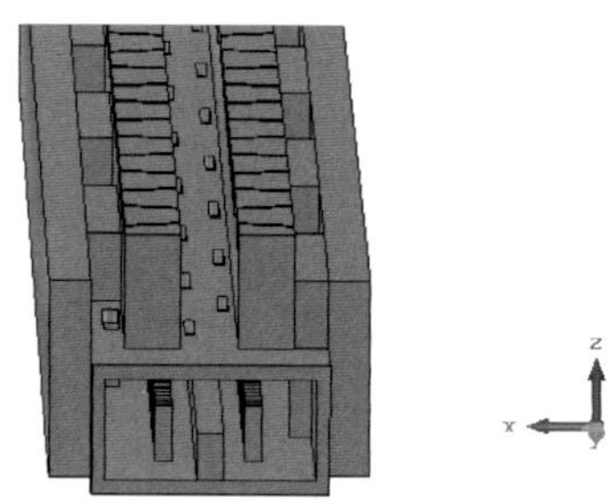

(b)

Figure 2. Configuration of the proposed antenna array. (a) Top view. (b) Front view

However, due to the limited isolation of single-row EBG structures, branches are placed beside the C-band slots to reduce the length of the slot. Adjusting the propagation constant makes the C-band slots located at the position where the wall current at 8.5GHz is low, further reducing the interference. Step structure is placed in the ridge waveguide to offset reflection at C-band. The C/X-band antenna is extended at the end to increase the bandwidth. The antenna working at the Ku band is a narrow side inclined slot antenna placed between the slots. The structure of the antenna is shown in Figure 2.

3. Results and discussion

The antenna shown above has 8 C-band slots, 15 X-band slots, and 2 * 21 Ku-band slots. Its radiation pattern is shown in Figure 3, and its reflection is shown in Figure 4. Side lobes are -12.7dB at X-band and below -14dB at C-band. The bandwidth with a reflection below -15dB is 124MHz at C-band and 125MHz at X-band. That demonstrates a slot antenna proposed above can work at C/X-band.

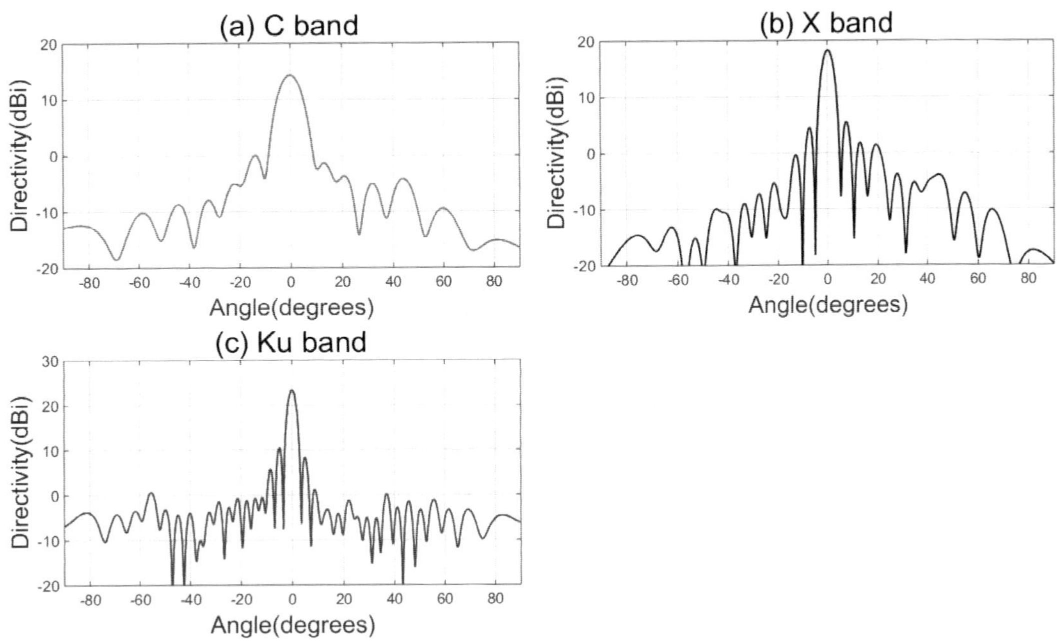

Figure 3. Radiation patterns (Y-Z plane) at the C/X/Ku band

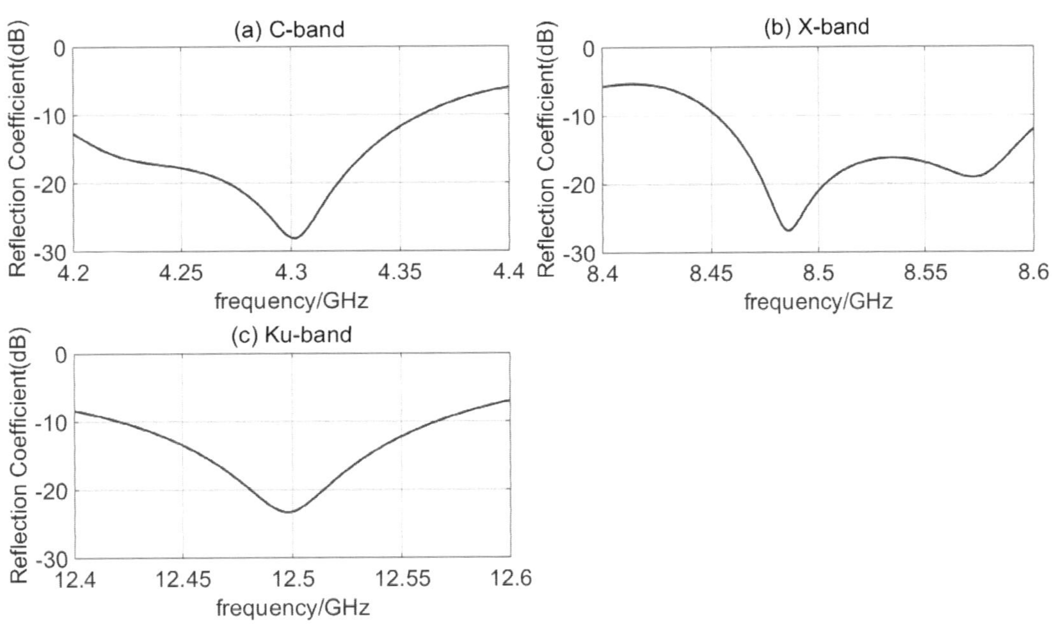

Figure 4. Reflection coefficient at the C/X/Ku band

To demonstrate the feasibility of composing array at X/Ku-band, a small antenna array model is established, as shown in Figure 5. Its radiation pattern is shown in Figure 6. There is a very low grating lobe in the figure, which demonstrates the size of the antenna meets the needs of the array.

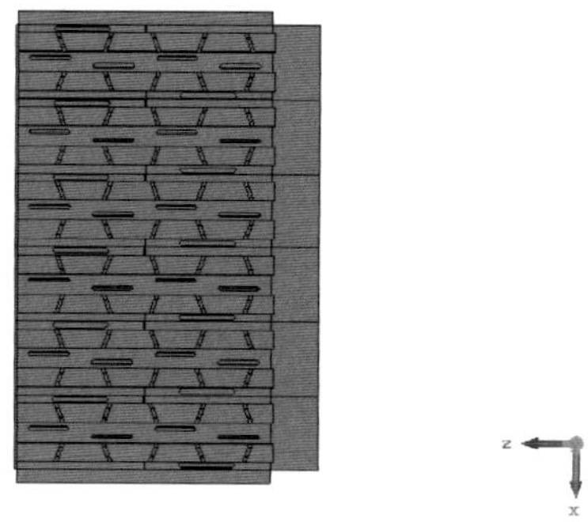

Figure 5. Top view of antenna array

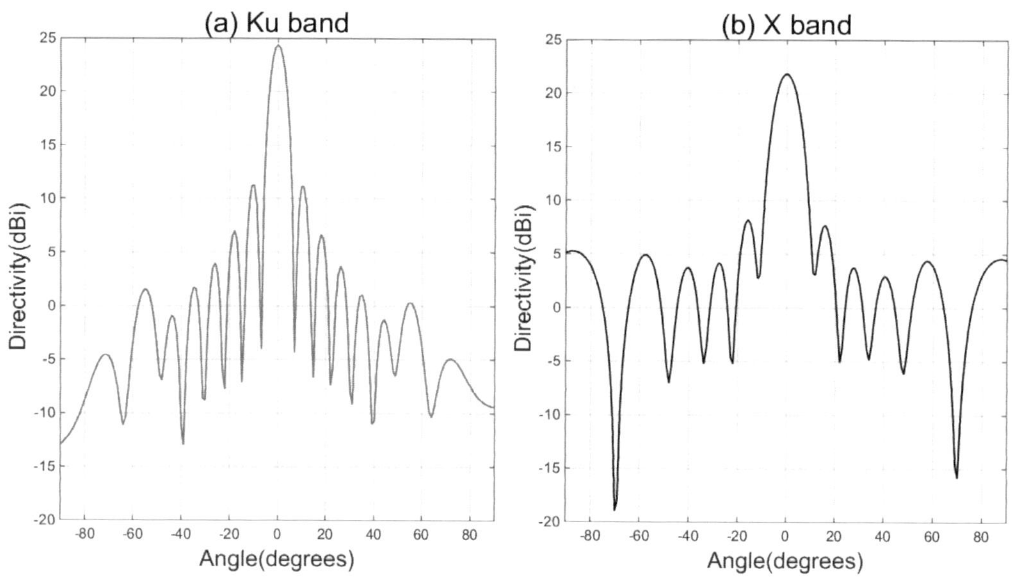

Figure 6. Radiation patterns (X-Y plane) at the X/Ku band

4. Conclusion

The triple-band all-metal slot waveguide antenna at the C/X/Ku band is designed and simulated. Placing EBG in the ridge waveguide, placing branches, and adjusting the position of slots make one slot antenna can work at the C/X band without sacrificing radiation features. The Ku-band antenna was placed above the C/X band antenna. The simulation result demonstrates the feasibility of the design. Compared with traditional all-metal multi-band slot antennas, the most attractive point of the proposed antenna is it can work in three bands.

References

[1] Chen Y, Jiao Y, Gang Z, Fan Z, Liao Z and Yu T. (2011) Dual-band dual-sense circularly polarized slot antenna with a c-shaped grounded strip. *IEEE Antennas & Wireless Propagation Letters*, **10**: 915-918.

[2] Mao C, Gao S, Wang Y, Luo Q and Chu Q. (2017) A shared-aperture dual-band dual-polarized filtering-antenna array with improved frequency response. *IEEE Transactions on Antennas and Propagation*, **65**: 1836-1844.

[3] Vazquez-Roy J L, Krozer V, and Dall J (2012). Wideband dual-polarization microstrip patch antenna array for airborne ice sounder. *IEEE Antennas and Propagation Magazine*, **54**: 98-107.

[4] Zhong S, Sun Z, Kong L, Gao C, Wang W, and Jin M. (2012) Tri-band dual-polarization shared-aperture microstrip array for SAR applications. *IEEE Transactions on Antennas and Propagation*, **60**: 4157–4165.

[5] Chen M, Fang X, Wang W, Zhang H, and Huang G L. (2020). Dual-band dual-polarized waveguide slot antenna array for SAR applications. *IEEE Antennas and Wireless Propagation Letters*, **19**: 1719-1723.

[6] Li T, Meng H, and Dou W. (2014) Design and implementation of dual-frequency dual-polarization slotted waveguide antenna array for Ka-band application. *IEEE Antennas & Wireless Propagation Letters*, **13**: 1317-1320.

[7] Park S, Okajima Y, Hirokawa J, and Ando M. (2005) A slotted post-wall waveguide array with the interdigital structure for 45° linear and dual polarization. *IEEE Transaction Antennas and Propagation*, **53**: 2865–2871.

[8] Chen Y, and Vaughan R G. (2018) Dual-polarized L-band and single-polarized X-band shared-aperture SAR array. *IEEE Transaction Antennas and Propagation*, **66**:3391-3400.

Reference Map Generation for Synthetic Aperture Radar Based on CycleGAN

Lingyu Meng[1], Zhiqiang Xing*[1](Corresponding author), Dongning Fu[2]

[1]School of Information science and technology, North China University of Technology, Beijing, China

[2]Hwa Create Technology-NCUT Joint Laboratory for RF Simulation, Hwa Create Technology Cop., Ltd, Beijing, China

e-mail: speech@ncut.edu.cn*, 864377922@qq.com, fudnmail@vip.sohu.com

Abstract. The real-time closed-loop simulator of Synthetic Aperture Radar (SAR) needs to input the SAR radar reference map as the simulation target scene information in the process of simulation. However, it is difficult to obtain the scene map of SAR radar in the detection range that is realistic and consistent with the task target, and the data sets are limited. In this article, Cycle-Consistent Adversarial Networks (CycleGAN) is applied to realize the migration learning of SAR radar reference map and solve the task required of a directional generation of SAR images.

1. Introduction

In recent years, the rapid development of deep learning, especially the generative adversarial network, has been used to generate images or transfer images' styles by virtue of the unique game characteristics between the generative network and the adversarial network, promotes the rapid development of image processing. This method is used to solve the problems of difficult acquisition of SAR scene maps, insufficient data, and serious noise interference.[1]

Literature is a new method of generating high-quality visible images using synthetic aperture radar (SAR) images based on deep convolution generative adversarial network (GAN) architecture proposed by Wang et al. in 2018 [2]. It is mainly based on cascaded convolutional neural networks (CNNs) for denoising and image coloring. A SAR image data enhancement method based on a generative adversarial network (GAN) was proposed by GUI et al. in 2019 [3]. WGAN with gradient penalty is used to increase the number of training data samples, avoid the randomness of data enhancement and improve data quality.

At present, the SAR image generation based on CycleGAN is still in its infancy. Because of the above problems, this paper proposes a method of using the CycleGAN network, which uses its strong generation ability, stable training, and fast convergence to generate and change SAR scene images, to meet the experimental requirements and ensure certain image quality.

2. Models

2.1 Cycle-consistent adversarial networks

CycleGAN network is a cycle structure that defines two groups of generators and discriminators. The network structure and network implementation logic are shown in Figure 1. G and F are generators, and

Dx and Dy are discriminators. x is an image of domain X, y is an image of domain Y, x' is the return image of domain X, and y' is the return image of domain Y. The loss function of generators and discriminators is defined as the same as the traditional GAN network. However, due to the poor image quality and unstable training of the traditional GAN network, CycleGAN introduces the least-squares GAN (LSGAN) objective function to replace the objective function of the traditional GAN, which uses the square difference as the loss function rather than the original log-likelihood as the loss function. The mapping is as follows:

$$\begin{aligned} L_{LSGAN}(G,D_Y,X,Y) &= E_{y\sim P_{data}(y)}[(D_Y(y)-1)^2]+E_{x\sim P_{data}(x)}[D_Y(G(x))^2] \\ L_{LSGAN}(F,D_X,Y,X) &= E_{x\sim P_{data}(x)}[(D_X(x)-1)^2]+E_{y\sim P_{data}(y)}[D_X(G(y))^2] \end{aligned} \quad (1)$$

The cycle consistency loss function is introduced into the CycleGAN network to ensure that the learning function maps the input image to the desired output. The loss function is as follows:

$$L_{cyc}(G,F) = E_{x\sim P_{data}(x)}[\|F(G(x)-x)\|_1]+E_{y\sim P_{data}(y)}[\|G(F(y)-y)\|_1] \quad (2)$$

The final objective function of CycleGAN is as follows:

$$L(G,F,D_X,D_Y)=L_{LSGAN}(G,D_Y,X,Y)+L_{LSGAN}(F,D_X,Y,X)+\lambda L_{cyc}(G,F) \quad (3)$$

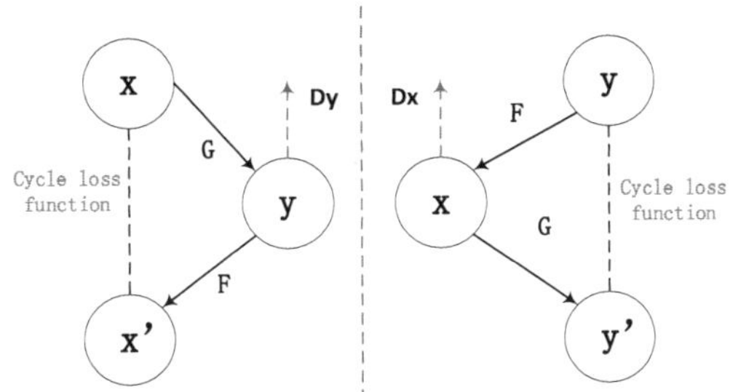

Figure 1 Structure of double cycles

2.2 Network model structure of CycleGAN

The generation network of CycleGAN consists of an encoder, converter, and decoder, as shown in Figure 2(a).

Encoder: It consists of three convolution layers. The image will be filled with a mirror image to obtain better convolution results. The convolution layer carries out the feature extraction of the image, and then the normalization processing is carried out to accelerate the convergence of the model, and the independence of the image instances is maintained.

Converter: Using the improved residual network to enhance learning ability. When the input layer image is 256 * 256 pixels, nine residual blocks are used. The use of jump connections in residual networks can alleviate the problem of gradient disappearance in networks.

Decoder: It is composed of two-layer deconvolution and one-layer convolution, and the Tanh activation function is used for the convolution layer to obtain better output.

The discriminant network of CycleGAN improves the discriminant model of the GAN network. PatchGAN[4] of 70 * 70 receptive field is used to design the full convolution network with a total of five layers, as shown in Figure 2(b).

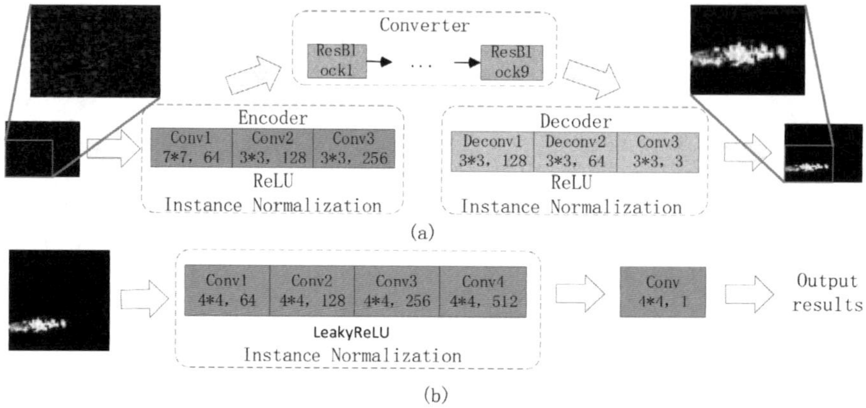

Figure 2 Network Structure of Generator and Discriminator

3. Experiment and result analysis

3.1 Preparation before experiment

The deep learning open-source network structure Pytorch used in this experiment was run on the Linux server, and training uses a dataset of 256 * 256 pixels created by myself: 309 noisy sea images, 479 complex sea scene images, 1043 ship images, 506 ship images at all angles.

3.2 Analysis of experimental results

For the generation of simple background SAR images, CGAN (Conditional Generative Adversarial Networks) and DCGAN (Deep Convolutional Generative Adversarial Networks) are used to train and generate the results[5]. As shown in Figure 3, the left, middle and right SAR images are the SAR scene data images generated by CGAN, DCGAN, and CycleGAN networks, respectively.

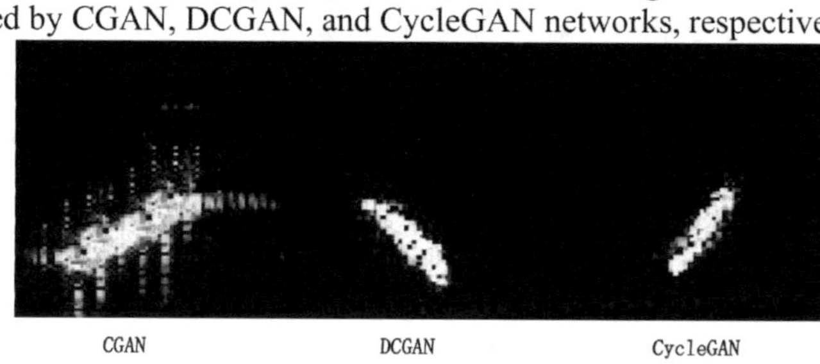

Figure 3. Results of various SAR generation

Table 1. Scores of various generating results

	CGAN	DCGAN	CycleGAN
SSIM	0.3768	0.7746	0.7972
Histogram similarity	0.5188	0.7806	0.8202

By comparing the image display of different generation methods and the original data set, the generation results of CGAN are obviously disturbed by noise, and the generation effect is not good. The scores of DCGAN and CycleGAN are also significantly higher than those of CGAN, and the generation effect is better. The performance of CycleGAN is better than that of DCGAN.

SAR image target generation and SAR information modification using complex scene information for CycleGAN. Figure 4 (a) and Figure 4 (b), from left to right, are the original images of input and the image generated after the information of output is modified.

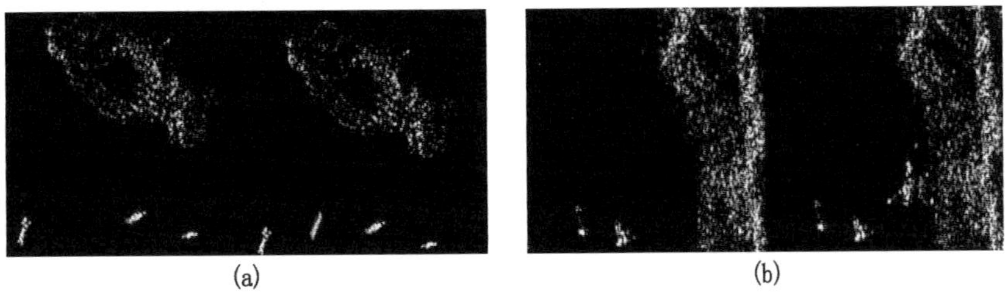

Figure 4 . Complex scene image generation

Table 2 Evaluation of SAR Images

	Complex scenes a	Complex scenes b
SSIM	0.6654	0.6846
Histogram similarity	0.6964	0.7195

The SSIM of the generated (a) graph reaches 66 %, and the SSIM of the (b) graph reaches 68 %. The migrated generated image has reached the same background as the original image. The histogram similarity of the (a) graph reaches 69 %, and the histogram similarity of the (b) graph reaches 71 %, which can ensure the authenticity of the migrated generated image.

4. Conclusions

Aiming at the problem of insufficient scene data of synthetic aperture radar echo simulator reference map, this paper proposes a method of generating a network by cyclic consistency confrontation. The available reference image data for SAR radar echo simulation are added to realize the migration generation and partial information directional modification of the directional scene target of the existing SAR image and ensure the image quality of the migrated SAR.

References

[1] Huang Qiong male, Zhu Weigang, Li Yonggang. Overview of SAR Data Expansion Based on GAN [J]. Journal of Weapon Equipment Engineering, 202,42 (11): 31-38.

[2] P. Wang and V. M. Patel, "Generating high-quality visible images from SAR images using CNNs," 2018 IEEE Radar Conference (RadarConf18), 2018, pp. 0570-0575, doi: 10.1109/RADAR.2018.8378622.

[3] Z. Cui, M. Zhang, Z. Cao, and C. Cao, "Image Data Augmentation for SAR Sensor via Generative Adversarial Nets," in IEEE Access, vol. 7, pp. 42255-42268, 2019, doi: 10.1109/ACCESS.2019.2907728.

[4] Zhu J Y, Park T, Isola P, et al. Unpaired Image-to-Image Translation using Cycle-Consistent Adversarial Networks[J]. 2017 IEEE International Conference on Computer Vision (ICCV), 2017.

[5] Lu Q L, Ye W, Overview of the application of generative adversarial network for SAR image processing [J]. Telecommunication Technology, 2020, 60 (1) : 121 – 128. DOI : 10.3969 / j. issn. 1001 – 893x. 2020.01.020.

Research on defect detection and sorting device of curved label based on multi-template matching

Si Chen*, Yihang Peng, Ming Yi

School of Intelligent Manufacturing, Foshan Polytechnic, Foshan, Guangdong Province, 528137, China

*Corresponding author: 1513945755@qq.com

ABSTRACT: Label defects will affect word-of-mouth and sales of goods and bring consumers a bad consumption experience, so it is of great significance to detect label defects. In order to quickly and accurately detect the defects of the curved label and sort the defective products with label defects, the mechanical mechanism and control circuit of curved label defect detection and intelligent sorting device and the label defect method based on visual detection are studied. In the aspect of label defect detection, taking the label defect of a curved bottle as the detection target, using the method of multi-template matching, image acquisition is carried out on the curved label from multiple angles, multiple templates are made, the shape of the image is matched and compared with the established multiple templates in turn, and the highest matching score is taken as the final matching score. The statistical results of 600 surface label defect detection experiments show that this method's accuracy is 96.5%. Using this method to detect the defects of curved bottle labels has a good effect, and the detection accuracy of curved labels with dense characters and patterns is very high.

1. Introduction

Almost all goods will be labeled with information about goods. Label defects will inevitably affect the reputation and sales of goods and bring consumers a bad consumption experience. Therefore, before goods are shipped, label defects need to be tested. Traditional production lines mostly use manual detection of label defects, which is not only costly, but also prone to missed and wrong detection due to worker fatigue and mental negligence. Such detection efficiency and effect are not ideal. [1,2] In the modern manufacturing industry, label defect detection technology based on machine vision is widely used in automated production lines, which has the advantages of high detection efficiency and high detection accuracy. Literature [3] proposed a method to detect the defects of printed labels by combining machine vision detection technology and deep learning technology, which can directly input the original image, and use the principal component analysis method to reduce the dimensionality disaster in the feature learning process and improve the efficiency of feature learning. Literature [4] proposed a label defect detection method based on Caffe's deep learning framework, which pays attention to the extraction of effective features of images. Literature [5] uses visual detection technology to detect defects in the character label area and uses template matching technology to extract three feature quantities of area, the center of gravity, and direction from the non-character label area, which are matched with the standard template data, and the detection accuracy reaches 95%. Literature [6] established a BP neural network model, combined with machine vision detection technology, to detect defects such as missing stickers, offsets, tilts, overlays, missing, offset, tilts, multiple sprays of production date characters, and date printing errors in the production of paper labels for cosmetics. The above cases mainly focus on flat labels and use visual inspection technology to detect the defects of flat labels. In actual production activities, curved labels need to be detected on many occasions, such as cosmetic labels, wine bottle labels, etc. Literature [7] studies the cylindrical surface defect detection method based on machine vision, uses nonlinear weights to fuse into a complete cylindrical label expansion map and uses double threshold and morphological processing to judge whether the beverage label is qualified. It has achieved certain results on cylindrical label defects, but the accuracy rate is only 95%. Literature [8] uses multiple cameras to collect and splice images of cylindrical labels and then uses the template matching method to detect label defects. This method has a high detection accuracy, but it requires multiple cameras to take photos of labels in multiple directions.

Because a large number of curved labels need to be detected in real production, machine vision detection technology, PLC application technology, sensor detection technology, and human-computer interaction technology are comprehensively applied to study the mechanical structure, control circuit, and defect detection method of curved label defect detection and sorting device. In the aspect of the label defect detection method based on visual inspection technology, the method of multi-template matching is used to collect the image of the curved label from multiple angles and make multiple templates. In the process of detection, the collected image is matched with multiple templates.

2. Working principle of curved label defect detection device

The defect detection device of curved labels comprehensively applies machine vision detection technology, PLC application technology, sensor detection technology, and human-computer interaction technology. Among them, machine vision detection technology is used for non-contact detection of curved labels on the mainstream waterline. Siemens PLC is used as the main control device to receive sensor information and output signals to control the movement of the manipulator. Using optical fiber sensors to detect objects, on the one hand, trigger the visual detection system and trigger the manipulator to work, on the other hand, count the total number of objects and the number of defective products. Human-computer interaction technology is applied to monitor the operation status of the system and display the total quantity of items and the number of defective products. When the first optical fiber sensor detects an object and triggers the visual inspection system to detect a defect in the label, when the defective product runs to the second optical fiber sensor, the manipulator clamps the defective product and places it on the defective product assembly line.

3. Mechanical structure design of curved label defect detection device

The mechanical structure diagram of the curved label defect detection device is shown in Figure 1. In the figure, the light shield is used to reduce external light interference, install the optical fiber sensor, ring light source, and camera on the light shield, and install the second optical fiber sensor 10 cm away from the light shield. The manipulator is composed of three cylinders, one of which is used to complete the lifting movement of the manipulator, one of which is used to complete the rotation movement of the manipulator, and the other is used to complete the clamping and loosening movement of the fixture.

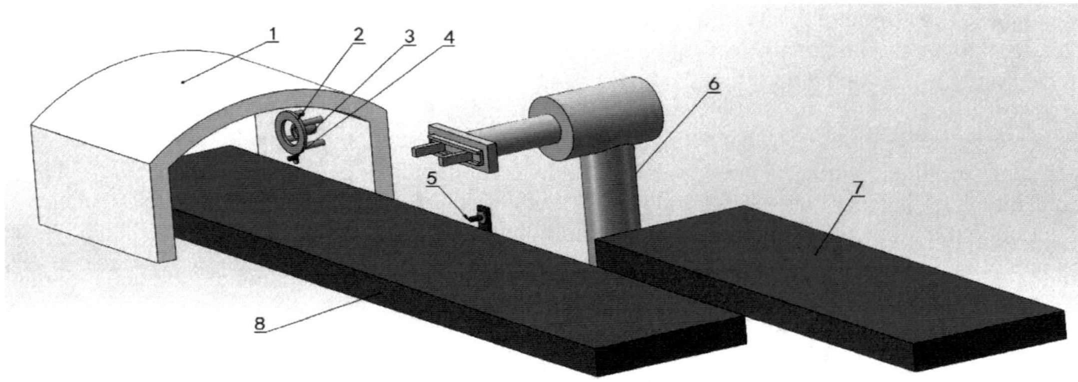

Fig. 1 Schematic diagram of mechanical structure
1-hood, 2-light source, 3-camera, 4-first optical fiber sensor, 5-second optical fiber sensor, 6-manipulator, 7-defective assembly line, 8-mainstream water line

4. Control circuit design of curved label defect detection device

The defect detection device of curved label uses Siemens PLC with the model of CPU 1214 dc/dc/dc as the control device. The first and second optical fiber sensors pass through the digital input interfaces ia.0 and ia.1 and connect to PLC; select the Siemens touch screen with the model of TPC7062Ti as the human-computer interaction equipment, connect to PLC through Ethernet interface, and communicate with PLC based on TPC/IP protocol; digital output interface Qa.1 connect with the relay to control the opening and closing of the ring light source through the relay; PLC controls the lifting, rotation, tightening and loosening of the manipulator by controlling three solenoid valves.

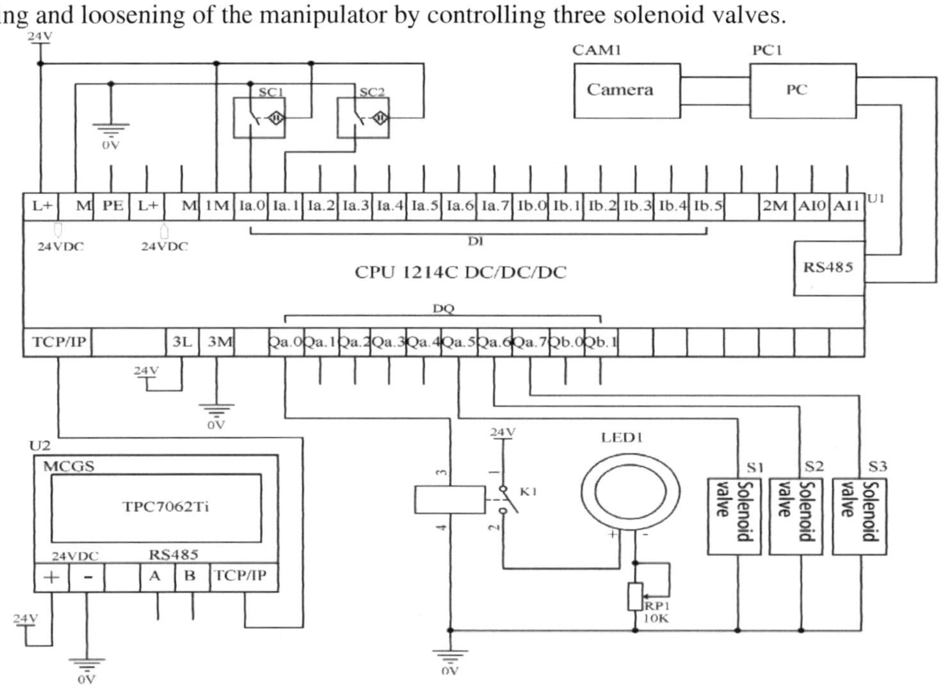

Fig.2 Schematic diagram of the control circuit

5. Human-computer interaction interface design

The Siemens touch screen with the model of TPC7062Ti is selected as the human-computer interaction equipment of the curved label defect detection device, and the interface design is shown in Figure 3. The start button, stop button, and system reset button are designed on the touch screen to control the start, stop and reset of the system, and red and green indicator lights are designed to indicate the running state of the system. The total number of the items display boxes, the total number of defective products display boxes are designed, and the total number of items reset button and the number of defective products reset button are set. Using the animation display function of the touch screen, the motion of the manipulator is simulated by associating the output signal of the digital output interface Qa.5-Qa.7 of PLC.

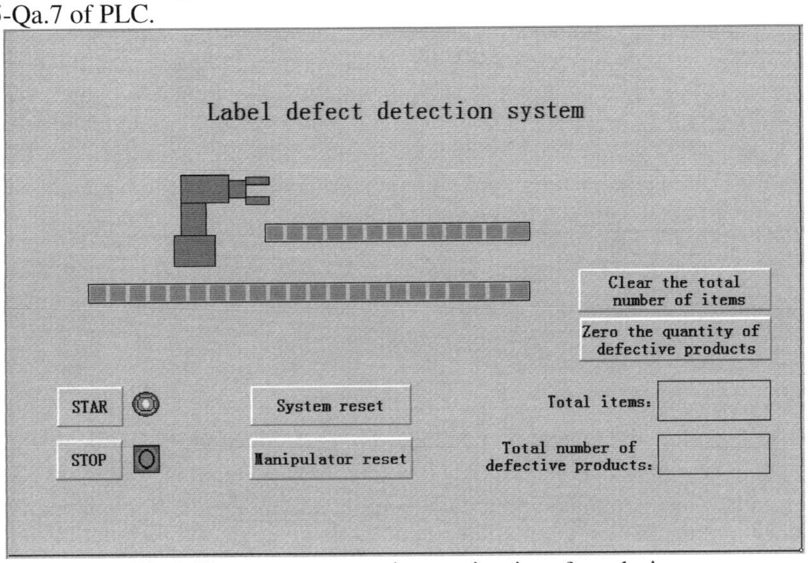

Fig.3 Human-computer interaction interface design

6. Surface label defect detection algorithm based on multi-template matching

Because it is difficult to keep the same camera angle when taking photos of labels on the moving production line, it is necessary to collect images from different angles, make multiple templates, match the images with all templates in turn during detection, and select the highest matching score as the final matching score, which can reduce the detection error caused by inconsistent camera angles to a certain extent.

6.1 Feature-based template matching

Template matching methods are mainly divided into the pixel gray-scale matching method and the feature-based matching method. [9] The matching method based on pixel gray is to establish the corresponding relationship between all pixels of the input image and all pixels of the template image. It has the advantages of high detection accuracy in the field of defect detection, but it also has the disadvantages of computational complexity and slow detection speed. Feature matching is the process of establishing a corresponding relationship between the feature points of the input image and the feature points of the template image. Because the pixels of the feature points are much less than the total number of pixels, the amount of calculation is much less. [10,11] Common feature matching methods include the moment invariant matching method, distance variation matching method, and minimum mean square error matching method.

The minimum mean square error matching method is to calculate the transformation parameters between images by solving the transformation equation of feature points, construct matrix A with the feature points in the template, input the feature points of the image to construct matrix B, and then solve the transformation matrix from matrix A to matrix B. the position with the minimum mean square error is the best matching position. [12,13]

Based on the HALCON platform, take the image whose abscissa of the label center coincides with the abscissa of the image center as the reference angle zero, and make ten standard templates within the range of label rotation $\pm 1°$. Some standard templates are shown in Figure 4. The operator provided by the Halcon platform is used to collect the image, draw the rectangular area surrounding the label, cut the label image using the rectangular area, and use the template matching assistant to extract feature points from the cut label image to generate a standard template.

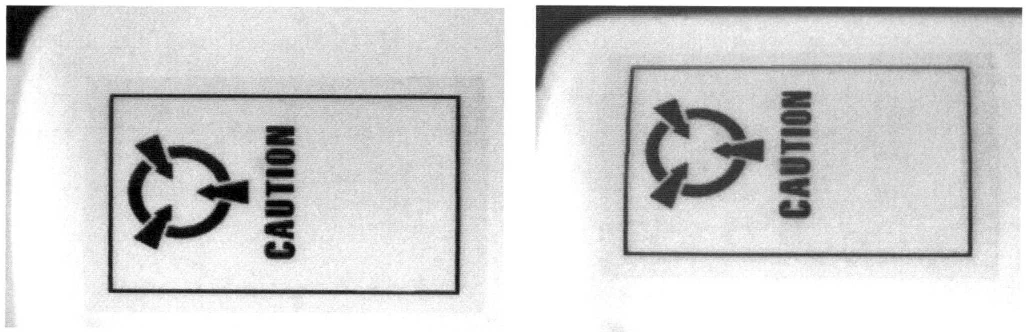

(a) (b)

Fig.4 Partial template image

6.2 Experimental results

On the HALCON platform, after making ten standard templates, conduct 600 detection experiments, set the rotation angle of the template to ± 0.5 °, set the matching score threshold to 0.95, and the detection accuracy reaches 96.5%. The detection results of some defect-free labels are shown in Figure 5, and the detection results of some label defects are shown in Figure 6. Since it is set to find the image matching the template within ± 0.5 ° rotation, when the label is skewed more than ± 0.5 °, the image matching the template will not be found, and it will be judged as a defective label.

The experiment found that if the label defects affect the characters or patterns of the label, the detection accuracy is very high. When the defects appear in the blank of the label, it is easy to cause false detection and missed detection. Some false detection results are shown in Figure 7. Therefore, the feature-based matching method is more suitable for detecting tags with dense characters and patterns. For labels with more blank areas, the matching method based on the image gray value is suitable, but the amount of calculation is relatively large.

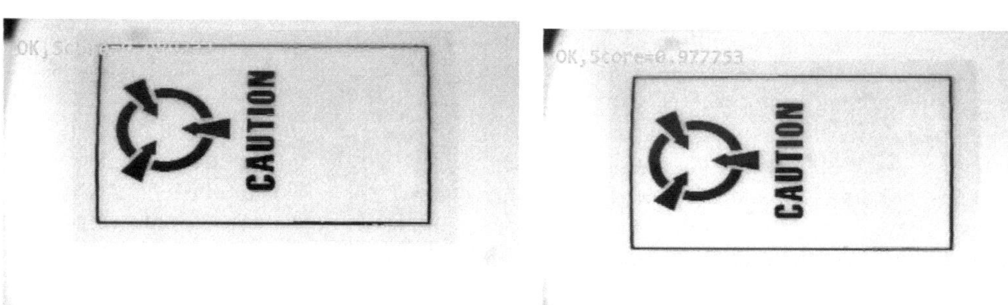

(a) (b)

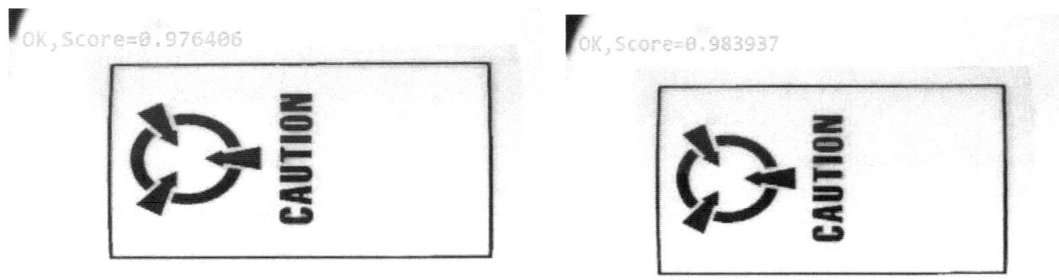

(c) (d)

Fig. 5 Test results of some defect-free labels

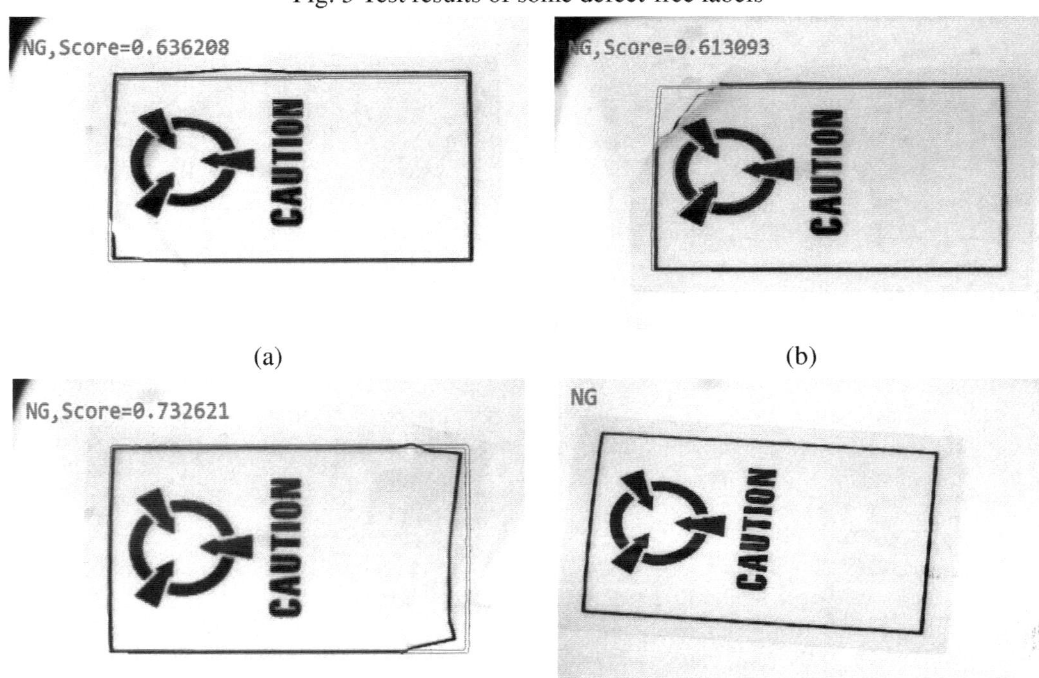

(a) (b)

(c) (d)

Fig. 6 Defect detection results of some labels

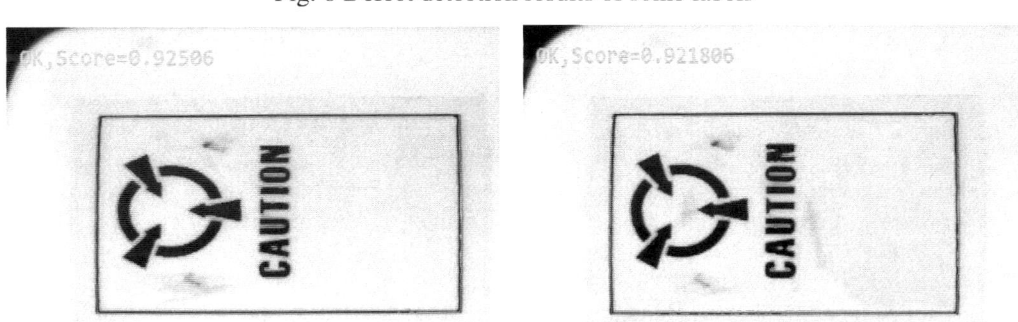

(a) (b)

Fig.7 Example of false detection

7. Conclusion

Combined with the actual situation of the current automatic production line, the defect detection and sorting device of the curved label is studied and designed. This device has high feasibility and is expected to be applied to production activities. In surface label defect detection, the method of multi-template matching is used, which has the characteristics of fast detection speed. The tag defect detection method based on the deep learning method needs to collect many tag defect images to make a training set, and the collected defect images are difficult to cover all kinds of defect types. The matching method based on pixel gray needs to establish a corresponding relationship between all pixels of the input image and all pixels of the template image, and the amount of calculation is also large. The feature-based matching method only needs to establish a corresponding relationship between the input image's feature points and the template image's feature points. The amount of calculation is relatively small, the detection speed is fast, and the detection accuracy of labels with dense characters and patterns is very high.

Projects

Characteristic Innovation Project of Colleges and Universities in Guangdong Province, China(2020KQNCX214); 2019 Characteristic Innovation Project of Colleges and Universities in Guangdong Province, China (2019GKTSCX121); 2021 Research and Practice on Teaching Reform of the Teaching Steering Committee of Automation Major in Guangdong Higher Vocational Colleges (GDZDHJZW202106); 2021 Scientific Research Project of Foshan Polytechnic (KY2021Y06).

About the authors

Si Chen (1986 --), born in Zhanjiang, Guangdong province, is a master and lecturer. Her research interests include machine vision technology, pattern recognition, and intelligent systems; Yihang Peng (1985--), born in Heze, Shandong province, is a master and lecturer. His research interest is intelligent control technology; Ming Yi (1973.09--), born in Xiangxiang, Hunan province, is a master and associate professor. His research interest is automatic control. Tel.: 18928511426; E-mail: 1513945755@qq.com

References:

[1] Fang Chuncheng, Chen Gengxin. Discussion on defect detection of printed labels based on machine vision [J]. China Equipment Engineering, 2022 (07): 180-182

[2] Hu Zongliang. Design and development of white wine bottle label detection system based on machine vision [J]. Electronic Technology and Software Engineering, 2020 (13): 77-78

[3] Li Zhijin, Li Peixiu, Zhu Chao. Application of label defect detection system based on deep learning [J]. Modern Electronic Technology, 2019,42 (07): 153-156.doi:10.16652/j.issn.1004-373x.2019.07

[4] Li Peixiu, Li Zhijin, Han Ke, Zhu Chao. Applied research on tag defect detection based on Caffe deep learning framework [J]. Journal of China Academy of Electronic Sciences, 2019,14 (02): 118-122

[5] Tan Zhiren, Lu Jun. label defect detection method based on image sensor [J]. Modular Machine Tool and Automatic Processing Technology, 2014 (03): 127-130. Doi:10.13462/j.cnki.mmtamt.2014.03.034

[6] Liu Peng, Dai Wenzhan. Defect detection method of cosmetic paper labels based on machine vision [j]. Journal of Zhejiang University of Science and Technology (NATURAL SCIENCE EDITION), 2019,41 (02): 231-238

[7] Liang Qiaokang, Xie Bingbing, Guo Dongni, Sun Wei, Wang Yaonan. Cylindrical label defect detection method based on machine vision [j]. Unmanned System Technology, 2020,3 (02): 43-48

[8] Yang Haiming, Gou Yanyan. Research on 360°omni-directional detection system of beverage bottle labels [j]. Packaging and Food Machinery, 2017,35 (03): 41-45

[9] Yu Qian, Chang Jiang, Gong Xue, Ding Changyu. Printing quality inspection scheme of beer boxes based on template matching [J]. Packaging Engineering, 2022,43 (03): 290-296. Doi:10.19554/j.cnki.1001-3563.2022.03
[10] Fu xingyong, Ren Dejun, Yan zhajie, Chen Ruxia, Gao yongxun. Transparent box defect detection based on template matching [J]. Software Guide, 2019,18 (03): 187-190
[11] Liu Wei, Han Zhenyu. SMD character defect detection based on machine vision [j]. Science and Technology and Innovation, 2018 (09): 114-115. Doi:10.15913/j.cnki.kjycx.2018.09.114
[12] Song Baobao, Du Wenhua, Duan Nengquan, Li Xian. Study on human eye pupil localization algorithm of non-contact tonometer [j]. Electronic Measurement Technology, 2022,45 (03): 112-117. Doi:10.1965/j.cnki.emt.2108272
[13] Ren Haohao Research on target recognition method of synthetic aperture radar image [d]. University of Electronic Science and Technology, 2021.doi:10.27005/d.cnki.gdzku.2021.005028

A method for locating abnormal sound in transmission line channels based on time delay estimation and equipment coordinates

Xiaobin Sun[1], Fudong Cai[2*], Jie Yang[3], Guoxin Guo[2], Huanyun Liu[2], Xiaoqing Liu[2,4], Zhiqiang Kong[2] and Changfeng Lv[2]

[1]State Grid Shandong Electric Power Co., Ltd., No. 150 Jinger Road, Jinan, Shandong, China

[2]Shandong Senter Electronic Co., No. 18, Liuyishan Road, Zibo, Shandong, China

[3]Jinan Power Supply Co., Ltd., State Grid Shandong Electric Power Co., Ltd., No. 238 Luoyuan Street, Jinan, Shandong, China

*Corresponding author's e-mail: caifd@senter.com.cn

Abstract. To get rid of the current transmission line visual monitoring device which only relies on a single photo-monitoring method, this paper proposes a method based on time-delay estimation and equipment coordinates to locate abnormal sound in the transmission line channel. Firstly, the audio-visual equipment is selected to collect the abnormal sound information of any transmission line as the reference warning record, and the relative time difference between other warning records and the reference warning record is determined. Then, the corresponding audio-visual equipment of the reference alarm record is used as a reference point to set up a Cartesian coordinate system, obtain the GPS information of other audio-visual equipment, unify the GPS information, and get the coordinates of other audio-visual equipment; finally, based on the reference point coordinates, other audio-visual equipment coordinates, and the relative time difference, the transmission channel abnormal sound source coordinates are obtained. Through the experiment of firecracker sound, we can get the firecracker sound within 2km in the open environment, which proves that the location method can effectively enhance the monitoring effect of the transmission channel.

1. Introduction

The transmission line is an important piece of power equipment, that needs to carry out a regular inspection to ensure its safe operation [1-3]. At present, there are some common inspection methods, such as manual inspection, unmanned aerial vehicle inspection, and visual monitoring, among which visual monitoring is one of the most widely used inspection methods, however, this method is subject to many factors such as the power of the monitoring device, the network, and the platform processing capacity, and the current monitoring device takes pictures at regular intervals, such as 30 minutes, 60 minutes, etc., so that it is difficult to meet the needs of transmission line fault monitoring [4].

Using sound to monitor the environment of the transmission channel in real-time all day long, taking pictures of the scene with the visual equipment around the transmission channel can get rid of the transmission channel monitoring mode of taking pictures at a single time. It is of great significance to the research of visual intelligent transmission operation and maintenance mode. The existing technology usually uses the lightning location system to carry on the sound real-time monitoring of the transmission line environment, through the substation traveling wave distance measuring device to give the fault pole tower location and the fault time, the location of the tower is judged by combining the site fault trace of the tower and whether there are any mines around the tower in the Lightning location system [5-6]. At present, the existing audio monitoring device for the transmission channel does not form a linkage mechanism with the surrounding visual monitoring equipment group, and can not realize real-time multi-angle location and tracking of hidden trouble sources.

To solve these problems, this paper proposes a novel method based on time delay estimation and equipment coordinates to locate abnormal sounds in transmission lines. According to the audio-visual equipment alarm sound information, it is important to select any alarm sound as a reference alarm record and correlate the two alarm sounds to obtain the delay error; by using the device as a reference point, it is needed to set up a Cartesian coordinate system, use the monitoring platform to get the GPS information of the rest of the device and unify the values; based on the coordinate position of the reference point in the Cartesian coordinate system, the coordinate position of other audio-visual equipment in the Cartesian coordinate system, and the relative time difference between the other warning records and the reference warning records, respectively, the source coordinates of abnormal fault sound in transmission channel are obtained.

2. Sound localization method based on time delay estimation and device coordinates

2.1. Cross-correlation-based method for calculating the sound difference of environmental monitoring in transmission channels

TDOA is generally carried out in two steps. First, the time difference of the sound arrival is estimated, and the sound delay between the elements of the sensor array is obtained, the position of the sound source is further determined by combining the spatial position of the known array. The alarm sound record collection is obtained by automatic clustering based on the Spatio-temporal domain. It needs to select the equipment to which one of the records belongs as the reference point and calculate the relative time difference between the other warning records and the reference records when the target sounds (such as thunder, bird calls, mechanical construction, etc.) occur. The mathematical expression is:

$$r_1(t) = s(t) + n_1(t) \tag{1}$$

$$r_2(t) = s(t - D) + n_2(t) \tag{2}$$

Among them, the sound collected for the first and second equipment, which represents the target sound signal, is the noise collected by the two audio-visual equipment, and D is the time difference (TDOA) of the sound signal between the two audio-visual equipment, or Time Lag. Cross-correlation is used to calculate the time delay, which is the relative time difference of the target sound.

$$R_{1,2}(\tau) = E\{r_1(t) \cdot r_2(t + \tau)\} \tag{3}$$

$$\widehat{D} = \arg\max_{\tau} R_{1,2}(\tau) \tag{4}$$

2.2. GPS to the Cartesian coordinate system

The reference point is used as the origin to establish the space Cartesian coordinate system, and the GPS information corresponding to other audio-visual devices is obtained. The GPS information includes the latitude, longitude, altitude, and curvature radius of other audio-visual equipment. The altitude and the radius of curvature are unified respectively. The coordinates of other audio-visual equipment in the Cartesian coordinate system are determined based on the GPS information after numerical unification.

The GPS data collection of the acquisition device is:

$$\{(\mathrm{lati}, \mathrm{loni}, \mathrm{alti}) | i \in N+\} \tag{5}$$

Among them, Lati, Loni, and Alti are the latitude, longitude, and altitude of the first device, respectively; Ni is the radius of curvature of the ellipsoid at the location of the first device.

The distance measuring device used in this method is located in a plain area, and the distance between the positioning devices is less than 5 km. The N value of each installation point of the audio-visual equipment can be regarded as the same and the installation height as the same height:

X direction:

$$\mathbf{x_i = (N + alt) * (cos(lat_i) * lon_i/\pi - cos(lat_k) * lon_k/\pi)} \tag{6}$$

Y direction:

$$\mathbf{y_i = (N + alt) * (lat_i/\pi - lat_k/\pi)} \tag{7}$$

Z direction:

$$\mathbf{z_i = alt_i} \tag{8}$$

Among them,

$$N=\sum_{i=1}^{n} N_i /n, \quad alt=\sum_{i=1}^{n} alt_i /n \tag{9}$$

2.3. The method of acoustic localization based on variable elimination

According to the latitude and longitude of the reference point, the radius of curvature after unification, and the altitude after unification, the coordinate position of the sound source in the Cartesian coordinate system is inversely transformed to obtain the GPS position information corresponding to the sound source. Taking one of the audio-visual devices (defined as device 0) as the reference point, the time when the target sound is captured by the device is zero, the time when the sound source produces sound is t, and the coordinates are (x, y, z). In other audio-visual acquisition equipment, I receive a sound at Ti at coordinates (xi, Yi, Zi) and K at the speed of sound (340 m/s).

Distance from the sound source to the reference point:

$$x^2 + y^2 + z^2 = k^2 t^2 \tag{10}$$

The distance from the sound source to the device I is:

$$(x-x_i)^2 + (y-y_i)^2 + (z-z_i)^2 = k^2(t-t_i)^2 \tag{11}$$

The above formula is simplified as follows:

$$xi*x + yi*y + zi*z - k2*ti*t = (xi2+yi2+zi2 - k2*ti2)/2 \tag{12}$$

Namely:

$$\begin{cases} x_1 * x + y_1 * y + z_1 * z - k^2 * t_1 * t = (x_1^2 + y_1^2 + z_1^2 - k^2 * t_1^2)/2 \\ x_2 * x + y_2 * y + z_2 * z - k^2 * t_2 * t = (x_2^2 + y_2^2 + z_2^2 - k^2 * t_2^2)/2 \\ \ldots\ldots\ldots\ldots \\ x_{n-1} * x + y_{n-1} * y + z_{n-1} * z - k^2 * t_{n-1} * t = (x_{n-1}^2 + y_{n-1}^2 + z_{n-1}^2 - k^2 * t_{n-1}^2)/2 \\ x_n * x + y_n * y + z_n * z - k^2 * t_n * t = (x_n^2 + y_n^2 + z_n^2 - k^2 * t_n^2)/2 \end{cases} \tag{13}$$

$$\begin{bmatrix} x_1 & y_1 & z_1 & -k^2 t_1 \\ x_2 & y_2 & z_2 & -k^2 t_2 \\ \ldots\ldots & & & \\ x_n & y_n & z_n & -k^2 t_n \end{bmatrix} \begin{bmatrix} x \\ y \\ z \\ t \end{bmatrix} = \begin{bmatrix} (x_1^2 + y_1^2 + z_1^2 - k^2 t_1^2)/2 \\ (x_2^2 + y_2^2 + z_2^2 - k^2 t_2^2)/2 \\ \ldots\ldots \\ (x_n^2 + y_n^2 + z_n^2 - k^2 t_n^2)/2 \end{bmatrix} \tag{14}$$

Order

$$A = \begin{bmatrix} x_1 & y_1 & z_1 & -k^2 t_1 \\ x_1 & y_1 & z_1 & -k^2 t_1 \\ x_1 & y_1 & z_1 & -k^2 t_1 \end{bmatrix} X = \begin{bmatrix} x \\ y \\ z \\ t \end{bmatrix} B = \begin{bmatrix} (x_1^2 + y_1^2 + z_1^2 - k^2 t_1^2)/2 \\ (x_2^2 + y_2^2 + z_2^2 - k^2 t_2^2)/2 \\ \ldots\ldots \\ (x_n^2 + y_n^2 + z_n^2 - k^2 t_n^2)/2 \end{bmatrix} \tag{15}$$

That is AX=B. Using the least squares of the system of linear equations:

$$X = (A^T A)^{-1} A^T B \tag{16}$$

Find the coordinates of the sound source(x,y,z). The GPS positions of the Cartesian coordinate system sources were obtained by inverse computation of the coordinates of the sound positions in the source:

Longitude:

$$lat = lat_k + y * \pi/(N + alt) \tag{17}$$

Latitude:

$$lon = (x/(N + alt) + \cos(lat_k) * lon_k/\pi) * \pi/\cos(lat) \tag{18}$$

Altitude:

$$\text{alt=z} \tag{19}$$

3. Experimental verification

Through the firecracker experiment, the collected samples were tested, through the actual verification in the open environment, the device can be collected within 2km of the firecracker sound. Three sets of data were collected at 15:01, 16:17, and 17:10. The relevant location data records are shown in table 1.

Table 1. Locate the data record table

	Corresponding point latitude and longitude	Device ID	About kilometers
Firing time 15:01	Latitude=36.75041865,Longitude=117.36621722	625637900	A few hundred meters
	Latitude=36.75140321, Longitude=117.37127083	626131200	Less than half a kilometer
	Latitude=36.74943471, Longitude=117.36092884	625717900	0.5 km
	Latitude=36.74777433, Longitude=117.35262735	625883900	About 1 km
	Latitude=36.76349647, Longitude=117.37305248	625749200	Less than 1.5 km
	Latitude=36.76029994, Longitude=117.35389548	627441400	About 1.5 km
Firing time 16:17	Latitude=36.75041865, Longitude=117.36621722	625637900	A few hundred meters
	Latitude=36.75140321, Longitude=117.37127083	626131200	Less than half a kilometer
	Latitude=36.74943471, Longitude=117.36092884	625717900	0.5 km
	Latitude=36.76349647, Longitude=117.37305248	625749200	Less than 1.5 km
	Latitude=36.7628666, Longitude=117.3633298	626419100	About 1.5 km
Firing time 17:10	Latitude=36.75302259, Longitude=117.38017581	210809500	About 2 km
	Latitude=36.76333638, Longitude=117.38033265	625847400	Less than 3 km
	Latitude=36.74943471,	625717900	Less than 3

Longitude=117.36092884	km

The sound source location map is as follows:

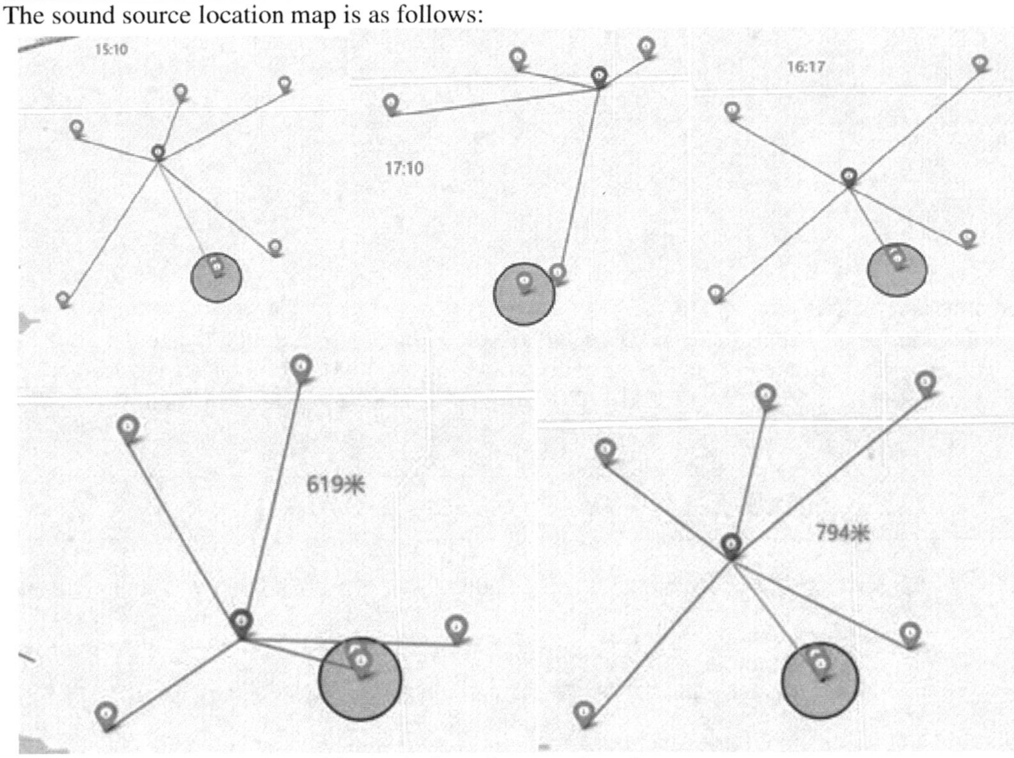

Figure 1. Sound source location map

4. Conclusion

To solve the problem that the current transmission visual monitoring system only relies on a single way to monitor the transmission channel, this paper proposes a method based on time delay estimation and equipment coordinates to locate the abnormal fault sound in the transmission channel. Firstly, a warning record sent by audio-visual equipment is chosen as the first start time of abnormal sound, and the correlation algorithm is used to correlate other warning records and reference warning records in pairs, the relative time difference of alarm recording is determined, and then coordinate transformation is carried out, and the sound source of alarm sound is located by cross-correlation time delay estimation. The validity and feasibility of the algorithm are verified by the location record of the firecracker sound.

References

[1] Yang CK, Kong ZZ, Xie QN and Du JC. 2021 J. Electric Power Engineering Technology. Image recognition method for transmission line based on the DeepLab v3+ deep conv olutional network 40 189.

[2] Zhou YX, Chen JN, Zhang L, Zhang YX, Teng CY and Huang X. 2021 J. High Voltage Engineering. Opportunity for Developing Ultra High Voltage Transmission Technology Under the Emission Peak, Carbon Neutrality and New Infrastructure 47 2396.

[3] Sui Y, Ning PF, Niu PJ, Wang CY, Zhao D, Zhang WL, Han SZ, Liang LJ, Xue GJ and Cui YJ. 2021 J. Power System Technology. Review on Mounted UAV for Transmissi on Line Inspection 45 3636.

[4] Liu ZY, Miao XR, Chen J and Jiang H. 2020 J. Power System Technology. Review of V isible Image Intelligent Processing for Transmission Line Inspection 44 1057.

[5] Wu Q, Ji L, Yu Y, Wang YB and Gong QW. 2018 J. Engineering Journal of Wuhan Uni versity. Identification of direct lightning fault based on PSO improved atomic decomp osition 51 620.
[6] Xiao JF, Wang JF, He ZM, Wan H, Tang XF and Mi Y. 2018 J. Electrical Measurement & Instrumentation. Reviews of lightning current online monitoring based on Rogowski coil and differential loop 55 1.

Journal of Physics: Conference Series **2384** (2022) 012006 doi:10.1088/1742-6596/2384/1/012006

Adaptive Not-direct-path Identification in UWB Localization

Ziyu Lu[1a], Tao Zuo[2b*]

[1]Wuhan University of Science and Technology, Wuhan, China.

[2]Wuhan University of Science and Technology, Wuhan, China.

[a] luziyu1999@163.com.

[b] corresponding author: zuomu666@163.com.

Abstract. In specific ultra-wideband (UWB) localization scenarios, conventional non-line-of-sight (NLOS) identification algorithm cannot detect other not-direct-path conditions. We proposed an adaptive not-direct-path identification method based on machine learning and artificial neural networks. Compared to the artificial division of not-direct-path data sets, we verified the advantages of data set partitioning by the method. The performance of NLOS recognition using different methods in different scenes is also analyzed, and the average identification accuracy in different scenarios can reach 92% or more.

1. Introduction

In many positioning systems, including UWB positioning, the distance between the mobile station (MS) and the base station (BS) is estimated by measuring the product of the speed of light and the time, and the time comes from the time of flight(Tof) between the MS and the BS. Under direct-path conditions, the signal transmission is almost unaffected, and the positioning accuracy performance is excellent. But the signal transmission is extremely susceptible to not-direct-path interference such as NLOS and multi-path effects conditions[13]. Even though UWB positioning has such powerful positioning performance, it still produces errors that are several times higher than its highest accuracy when facing interference with not-direct-path conditions[2]. The direct path is the line-of-sight(LOS) transmission scenario, as shown in **Figure 1**(a). The various complications called not-direct-path that affect signal transmission are shown in **Figure 1**(b), including NLOS, multi-path effects, and so on[1].

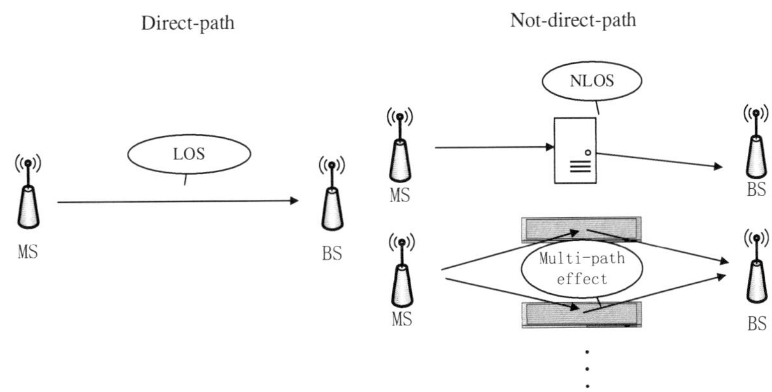

Figure 1. Scenarios of direct-path conditions and not-direct-path conditions. (a)Direct-path condition scenario; (b)Not-direct-path conditions scenarios.

Being aimed at the problems caused by not-direct-path conditions, many scholars have given a large number of studies[3]. A practical NLOS identification (PNI) technique is proposed in [4] for reducing the serious effect and a set of appropriate features for improving the speed of NLOS identification. And based on the identification results, the corresponding NLOS error mitigation method is given as well; In [5], machine learning (ML) techniques are used to analyze several sets of real-world UWB measurements captured in different scenarios, which attempt to identify measurements that face non-line-of-sight propagation conditions.

The above study barely researched NLOS identification, but the new studies show that in some scenarios, except NLOS, the multi-path effect and other not-direct-path conditions have a great impact on the wireless ranging and positioning system as well. Sometimes, wireless measurement work was implemented in a narrow corridor, which confirmed that the error caused by the multi-path effect on the final result cannot be ignored. Literature[12] takes advantage of the characteristics of the signal in the signal waveform, the signal is not only divided into NLOS and LOS but also further divided into Direct-Path NLOS(NP-NLOS) and Not-Direct-Path NLOS(NDP-NLOS).

However, in specific scenarios, we cannot ascertain what the not-direct-path information contains through prior knowledge. Carrying out NLOS identification directly like the literature above cannot detect other potential not-direct-path conditions. Not-direct-path conditions neglected that have a great impact on signal transmission will inevitably cause greater errors in ranging or positioning results[6]. In addition, in the case of simple scene environment information, not-direct-path information is relatively single. Therefore, a variety of not-direct-path conditions identification (such as NLOS and multi-path effect identification) will affect the identification efficiency in the positioning process, because in the process of the identification algorithm training must waste more resources; Moreover, in the process of identifying NLOS and multi-path effects, the data sets collection stage needs to artificially distinguish, and it is hard to ensure the rationality of the data sets division between NLOS and multi-path effects by using such a classification method[7]. The reliability of the samples in the training stage is thereby get reduced, which will affect the final identification accuracy.

In view of the above problem that existed in the previous literature, we proposed the adaptive not-direct-path identification algorithm in the specific scenario[8]. Based on this algorithm, the not-direct-path conditions in the scene are determined, and different not-direct-path conditions are classified. In addition, the deep feed-forward artificial neural network is used for not-direct-path identification to improve the identification accuracy.

2. Algorithm framework and data sets preparation

2.1. Algorithm Framework

In the specific wireless ranging and positioning systems, the solution of not-direct-path conditions includes two stages of identification and mitigation, this paper mainly focuses on the identification stage.

The proposed algorithm framework in this paper is shown in **Figure 2**, the algorithm framework is mainly divided into two major stages: the determination of not-direct-path conditions and not-direct-path identification. The determination of not-direct-path conditions is mainly accomplished by the Gap Statistic algorithm, which consists of the following steps. 1)The total feature data sets of the received signal are processed to get the number of types of not-direct-path conditions in the scene. 2)For the case of multi-not-direct-path(the types of not-direct-path conditions are more than one) conditions, K-Means is performed on the not-direct-path data sets based on the number of types obtained to get a variety of not-direct-path data sets. 3)According to the proportion of not-direct-path samples after clustering and the prior information to determine the not-direct-path conditions. In the not-direct-path identification stage for the scene containing only NLOS, a deep-feed forward artificial neural network is used to train LOS and NLOS data sets. However, for the case of multi-not-direct-path conditions, the direct path data set and the not-direct-path data sets are integrated as the training sets and trained by a deep feed-forward artificial neural network, then the algorithm model is obtained to identify not-direct-path conditions.

2.2. Gap statistic not-direct-path conditions determination

Gap statistic[10] is an algorithm proposed by Professor Robert and his colleagues at Stanford University for the K-means algorithm for autonomous judgment of the classification number. The algorithm framework proposed in this paper is mainly used to determine the number of types of not-direct-path conditions.

The basic idea of the Gap Statistic algorithm is to introduce a reference measurement value, which can be obtained by the Monte Carlo sampling method, the Gap statistic can be defined as:

$$Gap(k) = E_n\{\log D_k\} - \log D_k \tag{1}$$

$$E_n\{\log D_k\} = \frac{1}{B}\sum_{b=1}^{B}\log(D_{kb}^{*}) \tag{2}$$

Where D_k is the mean value, $E_n\{\log D_k\}$ is the expectation of $\log D_k$, and B is the times of sampling, as to correct the error brought by the Monte Carlo sampling.

After the formulation of the formula above, the smallest value k that satisfies $Gap_k \geq Gap_{k+1} - s_{k+1}$ is selected as the optimal number of clusters.

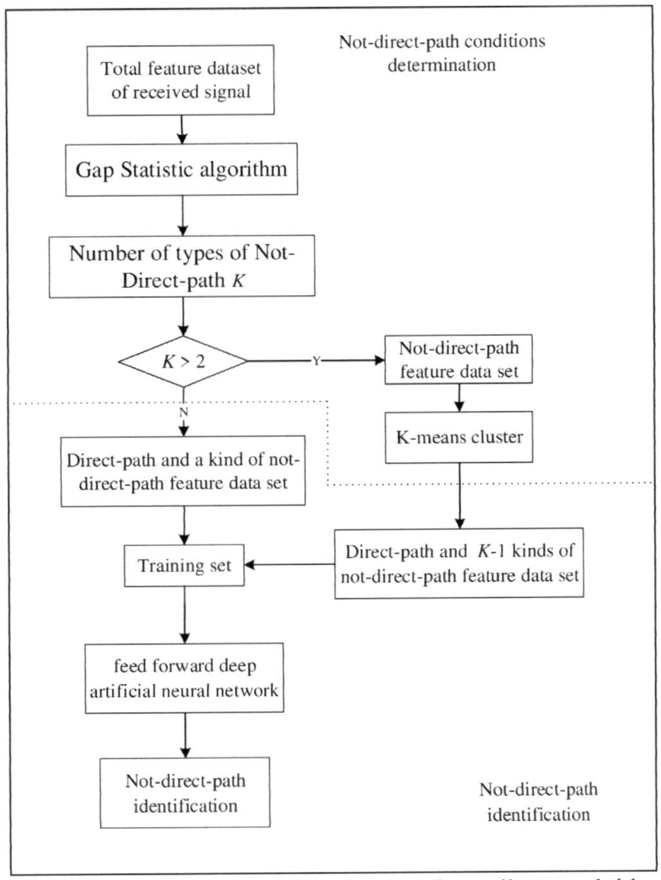

Figure 2. The system architecture of the proposed not-direct-path identification.

2.3. K-means not-direct-path conditions cluster

In the framework of the algorithm proposed in this paper, the classification of different clusters can be carried out by the clustering algorithm based on the similar relationship of the features in different not-direct-path conditions.

The K-means algorithm is followed by:1) Select the initialized k samples as the initial cluster $a = a_1, a_2, \cdots a_k$ centroid; 2)For each sample x_i, calculate its distance from the centroid of clusters in the data sets and divide it into the clusters corresponding to the k centroid of the cluster with the smallest distance; 3) For each cluster a_j, recalculate its cluster center $a_j = \frac{1}{|c_i|}\sum_{x \in c_i} x$ (i.e. the centroid of all samples belonging to that cluster); 4) Repeat the above two steps 2)and 3) until a certain termination condition is reached (e.g. number of iterations, minimum error change, etc).

2.4. Deep feed-forward artificial neural network

Artificial Neural Networks (ANN) is a mathematical model that simulates biological neural networks for information processing, and the not-direct-path identification stage in this paper is based on artificial neural networks.

The goal of a feed-forward network is to approximate a function f^*. For example, for a classifier, the function $y = f^*(x)$ maps the input x to a category y. The feed-forward network defines a map $y = f(x;\theta)$ and learns the values of the parameters θ so that it can get the best approximation of the function. Inside the feed-forward neural network, the parameters propagate unidirectionally from the input layer to the output layer, which is different from the recurrent neural network, which does not form a directed ring inside. This one is called the feed-forward model, because the function through which information flows, flows through x and the intermediate computation process is used for f definition, and finally reaches the output y. There is no feedback connection between the output of the model and the model itself. When feedforward neural networks are extended to one that includes feedback connections, they are called recurrent neural networks[11].

3. Data sets preparation

There are the distribution of FP(i.e. First-path Power) and CIR(i.e. Channel Impulse Response) in direct-path conditions and different not-direct-path conditions shown in **Figure 3**. The feature distribution has a relatively obvious distribution threshold, so FP and CIR are used as the basis for the determination of not-direct-path conditions[9].

In order to obtain a higher identification accuracy, some or all of the following signal eigenvalues are selected as the feature values used by the identification algorithm in reference to many previous studies on NLOS identification and NLOS and multi-path effect detection:

(1) Estimated distance ; (2) The compound amplitudes of multiple harmonics in the FP signal; (3) The amplitude of the first harmonic in the FP signal; (4) The amplitude of the second harmonic in the FP signal; (5) The amplitude of the third harmonic in the FP signal; (6) The amplitude of the channel impulse response (CIR); (7) Pulse repetition frequency rate in MHz; (8) Bit rate; (9) Standard deviation of noise; (10) Maximum value of noise; (11) Length of frame ; (12) Preamble length; (13) Received RX preamble symbols.

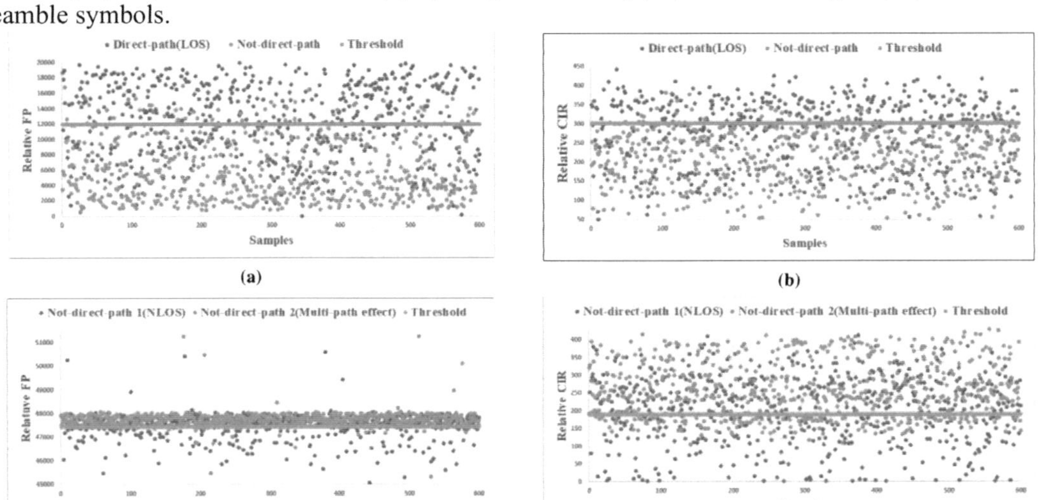

Figure 3. Sample feature distribution of direct-path and not-direct-path conditions. (a)Relative FP distribution of samples between direct-path and not-direct-path; (b)Relative CIR distribution of samples between direct-path and not-direct-path; (c) Relative FP distribution of samples between not-direct-path conditions; (d) Relative CIR distribution of samples between not-direct-path conditions.

The data sets used in this paper are from two scenarios, as shown in **Figure 4**(a) is a schematic diagram of the environment of Scenario 1, which is defined in a complex experimental building consisting of multiple rooms including a corridor, which is known to exist two not-direct-path conditions (NLOS and multi-path effect) in this environment, while Scenario 2 (shown in **Figure 4**(b).) is a simple office environment where the not-direct-path conditions are not distinguished during data sets acquisition. And some of the data sets come from [9].

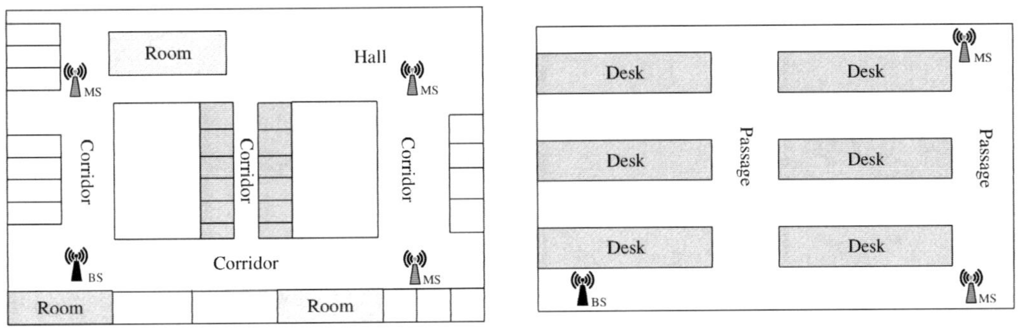

(a) (b)

Figure 4. Data sets provided scenario.

4. Experiment and result analysis

In the experimental verification, the experimental results are analyzed in the two stages of not-direct-path conditions determination and not-direct-path identification.

In the experimental verification stage of determination of not-direct-path conditions in Scenario 1, in which we got direct-path, NLOS, and multi-path effects. We use the Gap Statistic algorithm to determine the number of not-direct-path types, the Elbow method, the Canopy algorithm, and the Silhouette Coefficient method are used to determine the types respectively. The results of the four methods are shown in **Table 1**, the Gap Statistic got the best performance.

Table 1. Experimental results of the not-direct-path conditions determination algorithm.

Method	Outcome	Procedure time-consuming	Conclusion
Gap Statistic	3	624717ms	The number of classifications is correct and takes less time
Elbow method	/	959453ms	In artificial observation, the inflection point is uncertain
Canopy	3	90700ms	The initial distance threshold is difficult to determine
Silhouette	2	16 minutes	It takes a long time and the number of classifications is incorrect

Not-direct-path data sets with K-means clustering are classified based on the number of result classifications obtained by the Gap Statistic algorithm. In the clustering process, the value of the feature is selected as FP and CIR, while the clustered result preserves all the features of each sample. Compared with the manual acquisition of the not-direct-path conditions data sets in **Figure 5**(a), the feature distribution curve of the maximum noise classified by our algorithm in **Figure 5**(b) has a more obvious threshold. So the features of the not-direct-path conditions classified by the method proposed in this paper, as the features of not-direct-path identification, can obtain higher training reliability.

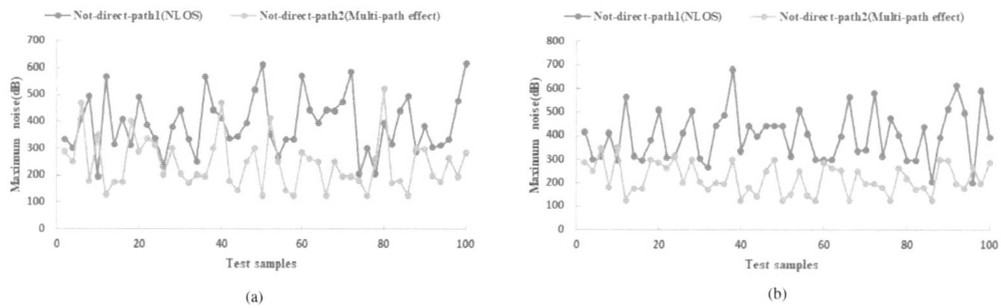

Figure 5. Comparison of sample feature distribution curve between manual classification and the method proposed. (a)Maximum noise of samples distribution in manual classification; (b)Maximum noise of samples distribution that uses the method proposed.

In the experimental verification stage of not-direct-path identification in Scenario 1, a deep feed-forward artificial neural network is used as the identification algorithm. There is one input layer, three hidden layers(100 neurons in each hidden layer), and one output layer. In the algorithm, the Relu function is used as the activation function of the hidden layer and the Softmax function is used as the activation function of the output layer. In addition, for configuring the neural network, we choose Adaptive Moment Estimation(Adam) as the optimization algorithm.

In order to show the advantages of the artificial neural network algorithm, the support vector machine and convolutional neural network are selected for comparison.

We have verified the identification results of the test sets. All the identification results of the not-direct-path identification in Scenario 1 are shown in **Figure 6**, in which sample points under different conditions were plotted in the figure by referring to the distance between each point and the base station. According to the proportion of test sets in different conditions (direct-path, not-direct path 1 and 2), the size of sample points existing in different intervals is different, among which the sample points incorrectly identified are also represented with a symbol.

The identification results of the not-direct-path identification using the deep feed-forward artificial neural network are shown in **Figure 6**(a) and **Figure 6**(b), in which the training sets used in the former come from the data sets classified by the method proposed in this paper and the one in the latter comes from manual classification. As the figures showing out, compared to the conventional not-direct-path classification(i.e manual classification), the algorithm proposed in this paper is more reliable for providing the training sets for not-direct-path identification.

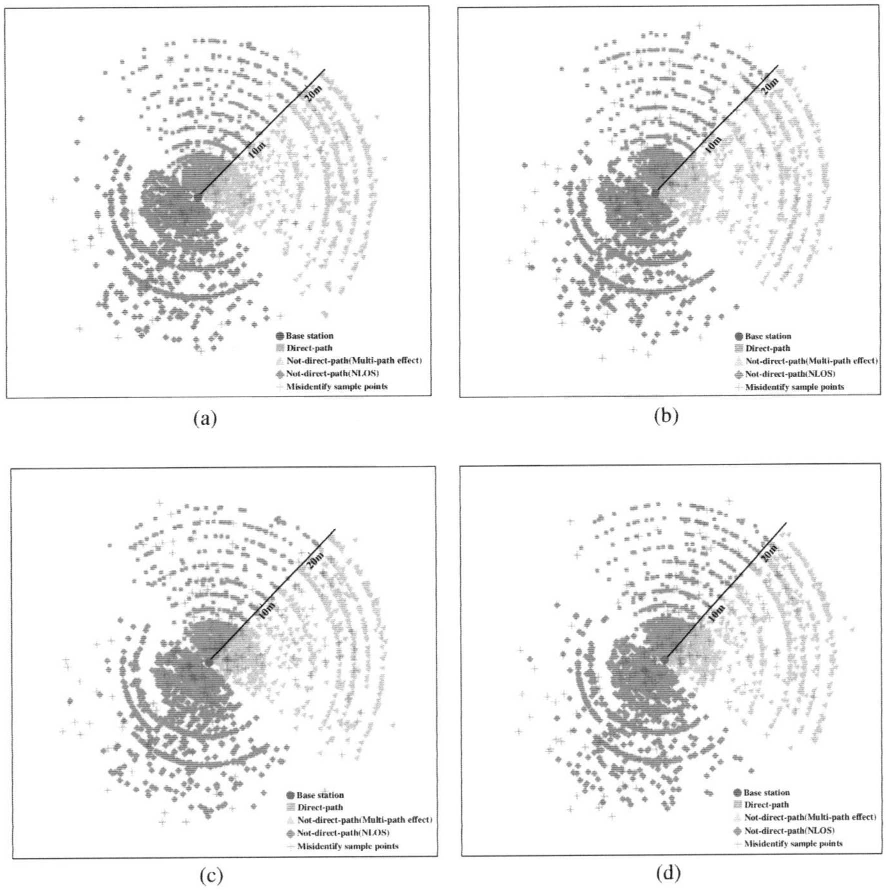

Figure 6. Not-direct-path identification validation results. (a)Identification result using the method proposed; (b) Identification result where the data set comes from manual classification; (c) Identification result using convolutional neural network; (d) Identification result using support vector machine.

In addition, two features, three features, and twelve features are selected in the training sets obtained by the classification of the algorithm proposed in this paper for training verification. **Figure 6**(c) and **Figure 6**(d) are the identification results obtained by using the convolutional neural network and support vector machine, and the identification accuracy is lower. **Figure 7** shows the comparison of the accuracy of not-direct-path identification of the three algorithms. Using more features as the training basis for not-direct-path identification can improve the identification accuracy rate well, and the deep feed-forward artificial neural network algorithm has obvious identification accuracy advantages in not-direct-path identification.

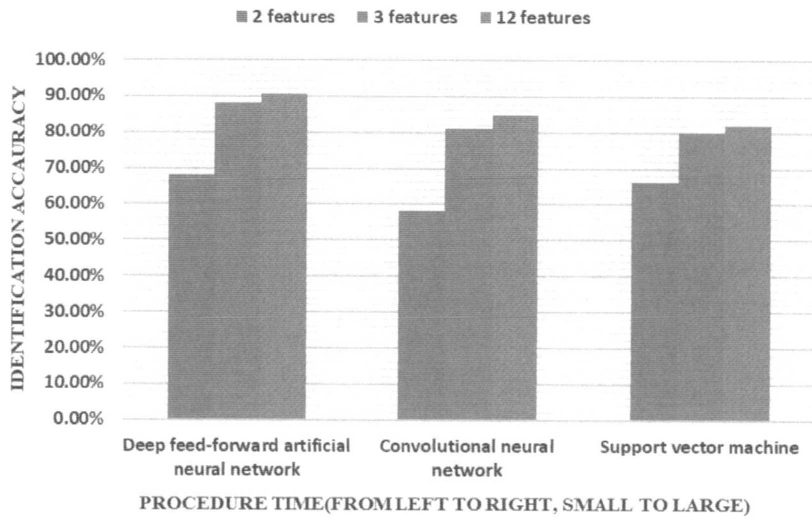

Figure 7. Comparison of not-direct-path identification accuracy.

For the classification information of scenario 2, there is only one not-direct-path condition of NLOS, and not-direct-path identification is only the conventional NLOS identification, which uses three algorithms of deep feed-forward artificial neural network, support vector machine, and convolutional neural network for NLOS identification, and the final identification accuracy rate using three features as the training basis is 91.54%, 83.63%, and 84.37%.

5. Conclusions

Aiming at the situation of the not-direct-path that neglects the multi-path effect in the conventional NLOS identification that causes greater ranging and positioning error, the adaptive identification algorithm is proposed in this paper. For the multi-not-direct-path conditions, the method proposed provides more reliable classification data sets for training in not-direct-path identification. The deep feed-forward artificial neural network algorithm can obtain higher identification accuracy based on the classification results of the former in the identification process. And the identification accuracy can be obtained higher than that of other algorithms for simple NLOS identification as well.

Data Availability

This publication is supported by multiple datasets, which are openly available at locations cited in the reference section.

Conflicts of Interests

The authors declare that they have no known competing financial interests or personal relationships that could have appeared to influence the work reported in this paper.

References

[1] Musa, A. , Nugraha, G. D. , Han, H. , Choi, D. , Seo, S. , & Kim, J. . (2019). A decision tree-based nlos detection method for the uwb indoor location tracking accuracy improvement. International Journal of Communication Systems, 32(13): e3997.1-e3997.13.

[2] Yan, L., Lu, Y., & Zhang, Y. (2017). An improved nlos identification and mitigation approach for target tracking in wireless sensor networks. IEEE Access, 5: 2798-2807.

[3] Wang, W. , Jost, T. , & Raulefs , R. (2017). A semi-deterministic path loss model for in-harbor los and nlos environment. IEEE Transactions on Antennas & Propagation, 65(12 pt.2): 7399-7404.

[4] Tomic, S. , Beko, M. , Rui, D. , & Montezuma, P. . (2017). A robust bisection-based estimator for toa-based target localization in nlos environments. IEEE Communications Letters, 21(11):

2488-2491.

[5] Liu, M., Lou, X., Jin, X., Jiang, R., Ye, K., & Wang, S. (2021). NLOS Identification for Localization Based on the Application of UWB. Wireless personal communications: An International Journal, 119(4): 3651-3670.

[6] Barral, V., Escudero, C. J., JA García-Naya, & Maneiro-Catoira, R. (2019). Nlos identification and mitigation using low-cost uwb devices. Sensors, 19(16): 3464.

[7] Khodjaev, J., Park, Y., & Malik, A. S. (2010). Survey of nlos identification and error mitigation problems in uwb-based positioning algorithms for dense environments. annals of telecommunications - annales des télécommunications, 65(5): 301-311.

[8] Huang, C., Molisch, A. F., He, R., Wang, R., & Zhong, Z. (2020). Machine learning-enabled los/nlos identification for mimo systems in dynamic environments. IEEE Transactions on Wireless Communications, 19(6), 3643-3657.

[9] Geng, C., Yuan, X., & Huang, H. (2019). Exploiting channel correlations for nlos toa localization with multivariate gaussian mixture models. IEEE Wireless Communication Letters, 9(1): 70-73.

[10] Ilarri, S., Mena, E., & Illarramendi, A. (2006). Location-dependent queries in mobile contexts: distributed processing using mobile agents. Mobile Computing IEEE Transactions on, 5(8); 1029-1043.

[11] Sang, C, L., Steinhagen, B., Homburg, J, D., Adams, M., Hesse, M., & Rückert, U. (2020). Identification of NLOS and Multi-Path Conditions in UWB Localization Using Machine Learning Methods. Applied Sciences, 10(11): 3980.

[12] Shavakand, M. Y., & Ahmadi-Shokouh, J. (2019). Compact uwb filter with narrow notched band based on grounded circular patch resonator. International Journal of Ultra Wideband Communications and Systems, 4(1): 16-21.

[13] Tibshirani, R., & Hastie, W. T. (2001). Estimating the number of clusters in a data set via the gap statistic. Journal of the Royal Statistical Society B, 63(2): 411-423.

Research on the construction scheme of the emergency communication link in electric power system without public network signals

Zijian Meng, Ming Hu, Shuang Fang, Zhigang Wang, Qing Wu, Xunwei Zhao and Di Liu

State Grid Information & Telecommunication Group Co. Ltd

mengzijian96@qq.com

Abstract. Recently, natural disasters have occurred frequently which usually cause great damage to communication links. Emergency communication technology is the key technology to connecting the emergency site and the command center under the condition of communication system collapse, and it plays an extremely important role in the process of emergency rescue. Based on the disaster scene of the electric power system, this paper fully considers the extreme conditions of power outages, network disconnection, and traffic interruption at the emergency site, and proposes an emergency communication link construction scheme. After the pilot test, the scheme proposed in this paper can quickly restore the network signal coverage of the emergency site, and establish the communication link between the command center and the emergency site which can provide a reliable communication guarantee for power emergency rescue operations.

1. Introduction

In recent years, under the combined influence of climate change and environmental factors, various natural disasters have occurred frequently all around the world. Electric power facilities and wireless communication base stations are damaged to varying degrees which would result in power outages and network disconnection at the disaster site. Communication among rescue personnel is blocked. The efficiency of information transmission is extraordinarily low. And emergency rescue forces are in a state of "blind scheduling" for a long time. The problems mentioned above may seriously affect the efficiency of command decisions and on-site rescue which could cause great difficulties in the emergency rescue of the electric power system. In view of these problems, the construction of an emergency communication system is particularly important[1-2].

Nowadays, a certain amount of research on emergency communication has been done. In [3], the construction and operation of the emergency communication system in the railway system have been proposed. In view of the situation that the on-site public network is destroyed, [4] proposes a scheme to restore the on-site communication network by using the satellite ground station carried by the emergency communication vehicle. [5] also describes the use of emergency communication vehicles in China Unicom's emergency communication scenarios. However, for large-scale disaster-stricken power system emergency scenarios, such as the heavy rainstorm disaster in Henan Province on July 20, the emergency site is of power failure, network disconnection, and traffic interruption. The emergency communication vehicle cannot reach the emergency site immediately. Traditional emergency communication methods based on emergency communication vehicles cannot be used properly in an electric power system.

In view of the electric power system emergency scene, this paper fully considers the extreme situation of "three interruptions" consisting of power failure, network disconnection, and traffic interruption at the scene of large-scale disaster emergency rescue. An emergency communication link construction scheme of electric power system based on ultra-small portable satellite ground station, portable dispatching all-in-one machine, and wireless broadband Mesh ad-hoc network equipment is proposed. All the equipment in this scheme can be manually carried to the emergency site, featuring lightweight, quick deployment, and easy operation. After pilot validation and testing, the scheme in this paper can meet the communication requirements between the command center and the emergency site.

2. Equipment of the scheme

2.1. Ultra-small portable satellite ground station

In the field of emergency communication, satellite communication is an integral part. The satellite communication system consists of satellites and ground stations[6]. Communication between two ground stations is achieved through the relay satellite. So satellite communication can not be affected by terrain factors and distance restrictions, and can still work normally in areas without public network signal coverage[7].

Figure 1. The structure of the satellite Ground station.

With the continuous development of communication means, emergency communication is no longer limited to the transmission of voice. Real-time video and images can enable commanders to understand the site more clearly, which is helpful for unified scheduling. However, due to the limited bandwidth of traditional satellite communication technology, its links are difficult to carry high-traffic real-time services such as high-definition videos and images. For the emergency scene of the electric power system, the scheme proposed in this paper adopts a kind of high-throughput ultra-small portable satellite ground station. Compared with traditional satellite stations, high-throughput satellites use multi-beam antennas and frequency multiplexing technology, which could greatly improve bandwidth efficiency. So it effectively alleviates the communication bottleneck caused by bandwidth limitations [8-9].

The power emergency communication link construction scheme proposed in this paper adopts a 0.6 m high-throughput feedforward ultra-small portable satellite ground station. In order to be convenient for rescue workers to carry, the whole machine adopts a split design that the satellite antenna plate, and the host machine are detachable. The structure of the station is shown in Figure 1. According to the requirements of emergency rescue, the portable satellite ground station is powered by a detachable battery, which can work continuously for more than 6 hours with fully charged. The station adopts

Journal of Physics: Conference Series **2384** (2022) 012007 doi:10.1088/1742-6596/2384/1/012007

dual-frequency leaky-wave antennas whose receiving gain is higher than 36 dBi and transmits gain is higher than 32.6 dBi. The high-throughput ultra-small portable satellite station operates in the Ku-band, and its downlink transmission rate can reach up to 60 Mbps. The forward transmission channel adopts LDPC (Low-Density Parity Check Code) and BCH encoding, and the return channel adopts Turbo encoding, which further improves the channel transmission efficiency[10]. This satellite station is equipped with a Gigabit Ethernet interface and a WIFI module, which facilitates the device to access the satellite network in various ways. In addition, in order to facilitate the operation of emergency rescue, the station is equipped with a satellite auto-aiming function, which can complete aiming the satellite within five minutes after being powered on. Its process can be described as follows

Step 1: GPS module collects current geographic location information.

Step 2: Calculate the theoretical value of antenna azimuth and pitch angle according to geographic location information.

Step 3: Magnetic declination correction to the theoretical value.

Step 4: Calculate the difference between the corrected theoretical value and the current azimuth and elevation angles of the antenna.

Step 5: Stepper motor rotation to achieve aiming the satellite roughly.

Step 6: Control the stepper motor to scan in a small area and monitor the satellite signal reception strength in real time until satellite signal reception strength reaches maximum.

2.2. Portable integrated scheduling all-in-one machine

In the large-scale disaster scene of "three interruptions", there are often multiple simultaneous emergency repairs at the power system rescue site. A large amount of data information from multiple sites needs to be sent back in time. In response to this problem, considering that the emergency communication command vehicle cannot reach the emergency site, this paper proposes to use the portable integrated scheduling machine as the front data collection node. The portable integrated scheduling machine mainly has the following five functions

1) Unified management. The intelligent individual equipment of each emergency spot can be connected to the management system of the all-in-one machine through the deployed Mesh network.

2) GIS scheduling. It supports GIS map inquiring real-timely. The all-in-one machine can display the geographic location, working status, and collected data information of each registered device simultaneously.

3) Visualization. The rescue worker can use the all-in-one machine to check the video surveillance of the registered individual equipment at the emergency site at any time, and grasp the emergency repair progress in real time.

4) Intelligent routing. The all-in-one machine has the function of automatic routing configuration, which can extend the satellite network signal to the emergency site through the Mesh ad-hoc network to realize the network signal coverage of the emergency operation site.

5) Three-way linkage. After the all-in-one machine is connected to the satellite network through wired or wireless means, it can participate in or initiate a three-party video conference with the command center and the emergency site, which can realize the integrated scheduling negotiation. The data of each individual piece of equipment at the emergency scene is transferred to the command center by the all-in-one machine through the satellite network.

The portable integrated scheduling all-in-one machine used in the electric power system emergency communication link construction scheme proposed in this paper adopts a lightweight design. The weight of the whole machine is less than 12 kg, which is easy to carry manually. Taking into account the factor that the bandwidth of the satellite system will be limited by bad weather conditions, the all-in-one machine is also equipped with a video combining function. When the command center obtains the real-time video images of multiple on-site individual pieces of equipment at the same time, the all-in-one machine can combine multiple video streams into one stream to reduce the bandwidth required for transmission and improve the stability of the communication link.

2.3. Wireless broadband Mesh ad-hoc network equipment

In the case of power outages and network interruptions at the disaster site, restoring the network signal coverage at the emergency site is a necessary condition for the unified scheduling of the command center. This paper proposes a network construction scheme at the emergency site based on Mesh ad-hoc network equipment. By deploying multiple wireless Mesh devices in the field, the multi-hop ad-hoc network composed of each node can extend the satellite network signal to the rescue site, and realize the coverage of the on-site network signal.

The Mesh ad-hoc network has the characteristics of rapid deployment and flexible networking without a central node. After each node is powered on, it can automatically form a network according to the channel conditions between nodes. Mesh ad-hoc network supports the features of multi-hop and routing management. And the network can automatically allocate routes according to the real-time communication load of each node, which can effectively reduce network congestion. In addition, the Mesh ad-hoc network has high robustness. When a node in the ad-hoc network fails, the network will update the current topology, and the data packets will be routed to the current optimal path for transmission, so the overall network connectivity will not be affected.

The Mesh equipment used in the scheme proposed in this paper adopts a backpack buckle design, which is convenient for rescue workers to carry. It is equipped with a wireless WIFI module, so personal smartphones and intelligent individual equipment carried by workers can access the Mesh network through WIFI hotspots. Moreover, this Mesh equipment supports the function of working frequency and bandwidth configuration. Under the condition of the LOS (Line of Sight) channel, the coverage distance of a single hop is more than 2km, and the maximum number of hops is not less than 8 hops. It adopts the COFDM (coded orthogonal frequency division multiplexing) coding method, which effectively suppresses channel selective fading in a multipath channel environment. And the peak transmission rate can reach up to 80 Mbps, which can meet the Network signal coverage and transmission rate requirements at the power emergency site.

3. Networking Scheme

The network topology of the emergency communication link construction scheme in the power system proposed in this paper is shown in Figure 2. Emergency rescue workers carried wireless broadband Mesh equipment, personal smartphones, and individual intelligent equipment to the emergency site to carry out rescue work. The individual equipment that emergency rescue workers can carry includes but is not limited to a smart helmet, smart bracelet, gas detector, and action recorder, all of which can be connected to the Mesh network by connecting to a WIFI hotspot.

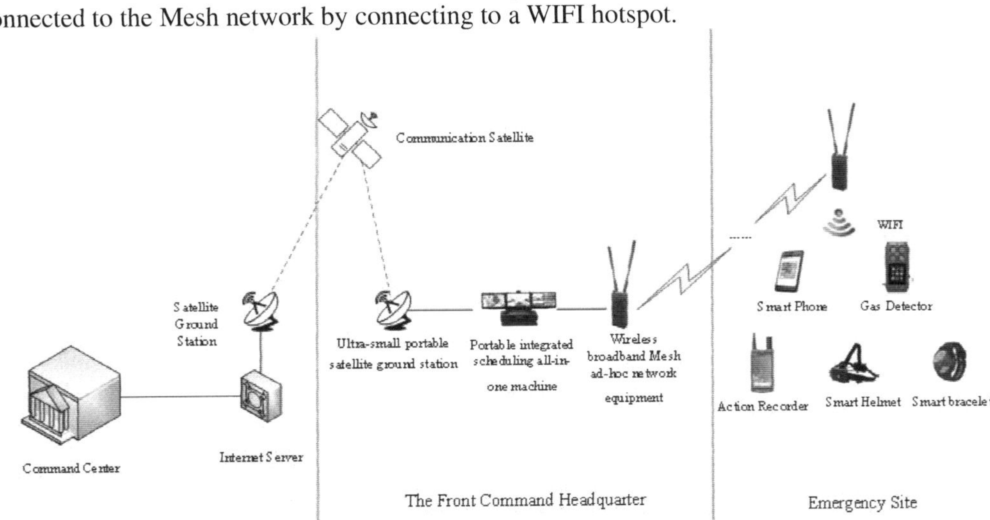

Figure 2. The Structure of the Networking Topology.

The forward command center deploys a portable integrated scheduling all-in-one machine and a portable ultra-small satellite ground station. The all-in-one machine is connected to the Mesh ad-hoc network and the satellite network at the same time through wired or wireless means. After all kinds of intelligent individual equipment at the emergency site are registered to the management system of the all-in-one machine through the Mesh network, they can send back information such as voice, video, and data at the emergency site. The all-in-one machine uploads the data collected at the emergency site to the command center management platform to achieve tripartite interconnection through the satellite network. The server equipped with the management platform can be deployed in the cloud, and the command center can complete access to the management platform through any computer connected to the Internet. The platform adopts a B/S(Browser and Server) architecture, so there is no need to install additional software on the computer, which is convenient for quick access and can reduce the cost of system maintenance and upgrades. In addition, through the intelligent routing function of the all-in-one machine, the satellite network can realize the network coverage of the emergency site through the mesh network, and the emergency rescue workers on the emergency site can connect to the Mesh WIFI hotspot to access the Internet through the personal smartphone.

4. Pilot Test

In order to verify the effectiveness and practicability of the power emergency communication link construction scheme proposed in this paper under the condition of "three interruptions", the testers in this paper conducted a pilot test in Mangshan Forest Park, Changping, Beijing. The test environment is shown in Figure 3.

Figure 3. Field Test Environment.

The front command headquarters and emergency site are located along the transmission line in the northern mountainous area of Mangshan Forest Park. The command center is deployed in the office of the industrial company in the urban area, and the management platform is set up in the public network cloud service center of the industrial company. The test site has no power supply, no public network signal coverage, and vehicles cannot enter, which simulates the "three interruptions" of emergency scenarios. The equipment brought to the test site is shown in Table 1. After all kinds of equipment arrive at the scene, the principle of first building a communication link and then connecting to each individual piece of equipment should be followed. The deployment process is shown in Figure 4.

Table 1. Equipment List.

Equipment Name	Type	Amount
Portable integrated scheduling all-in-one machine		1
Ultra-small portable satellite ground station	High-throughput	1
Wireless broadband Mesh ad-hoc network	600 MHz	4

Journal of Physics: Conference Series **2384** (2022) 012007 doi:10.1088/1742-6596/2384/1/012007

equipment		
Wireless broadband Mesh ad-hoc network equipment	1.4 GHz	4
Individual Equipment	Smart Helmet	1
Individual Equipment	Smart Bracket	1
Individual Equipment	Gas Detector	1
Individual Equipment	Action Recorder	1

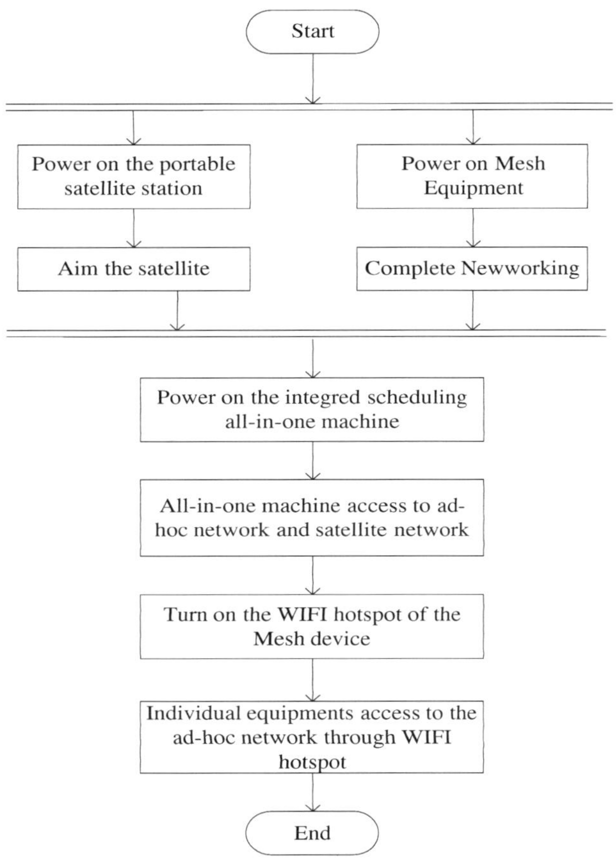

Figure 4. The process of deployment.

Considering the influence of mountainous terrain on the propagation of different working frequency signals of Mesh, the working frequency of Mesh is selected to test the low-frequency band of 600MHz and the high-frequency band of 1.4 GHz respectively. The test results are shown in Table 2.

Table 2. Mesh multi-hop transmission rate test.

Working Frequency	One-hop Transmission rate	Two-hops Transmission Rate
1.4 GHz	6.6 Mbps	845 Kbps
600 MHz	9.0 Mbps	2.7 Mbps

The relationship between Mesh coverage distance and rate is shown in Figure 5.

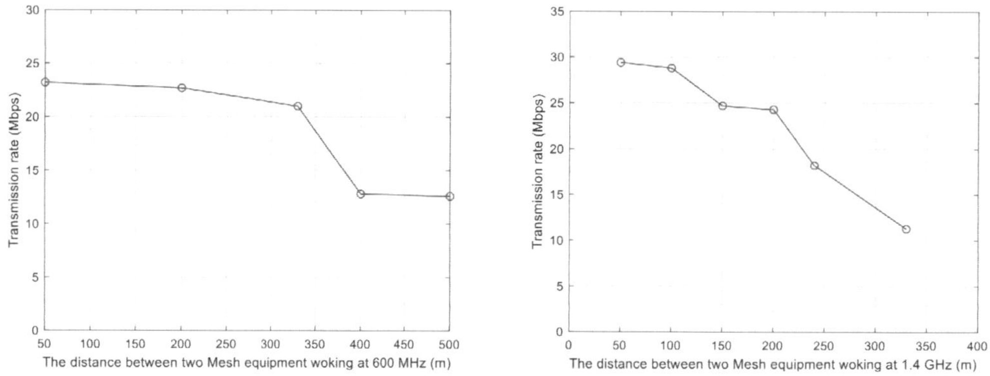

Figure 5. The coverage of Mesh signal.

According to Table 2 and Figure 5, under the condition which is blocked by mountain obstacles, the coverage distance and data transmission rate of Mesh that works in a low-frequency state are obviously better than those of high frequency. Therefore, in an environment with obstacles such as mountainous areas, in order to improve the reliability of network transmission, Mesh can preferentially select the low-frequency working state for networking.

Each intelligent individual piece of equipment on site is connected to the ad-hoc network through WiFi. According to the measured data, the WIFI coverage of Mesh equipment can reach more than 80 m, and the average video transmission of each channel is 0.5 Mbps. The on-site ad-hoc network can meet the needs of real-time video and data transmission of individual equipment.

The ultra-small portable satellite ground station can access the Internet after aiming the satellite successfully. The specific configuration parameters and on-site test results are shown in Table 3.

Table 3. The specific configuration and test results of the satellite ground station

Error Correcting Codes	Code Efficiency	Modulation	Bandwidth	Demodulation Threshold（dB）
LDPC+BCH	5/6	8PSK	UL：9.6 DL：14.7	10.5
Protocol	Throughput（Mbps）	Packet Loss Rate	Average Delay（ms）	Average Jitter（ms）
UDP	UL：26.71 DL：6.22	0%	711	253

After the satellite network is configured by the intelligent routing of the all-in-one machine, the network signal coverage of the emergency site can be realized through the Mesh ad-hoc network, and the emergency rescue workers can connect to the wireless hotspot of the Mesh equipment through the personal smartphone to access the Internet. The interface of the command center is shown in Figure 6. Information such as video images and position data of each individual soldier at the test site can be accessed in real time. The video screen is free of lag, and the geographic location information is accurate. The command center can conduct real-time audio and video calls with the front command headquarter and rescue team at the emergency site through the satellite network to realize unified scheduling.

Figure 6. The real-time screen of the command center

5. Conclusion

With the continuous advancement of the construction of the new power system, the power grid is striding toward a new era of digitalization and informatization. As an important support force in the new electric power system, emergency communication technology plays an important role in the safe production of the power grid and the process of disaster relief. Based on the extreme scenario of "three interruptions" at the electric power emergency site, this paper proposes an emergency communication link construction scheme using ultra-small portable satellite ground stations, Portable integrated scheduling all-in-one machines, and Mesh ad-hoc network equipment. After field tests, the scheme proposed in this paper can quickly restore the network signal coverage of the emergency site and open up the front and rear communication links. The scheme realizes unified command and integrated scheduling at the emergency site and provides a rich theoretical basis and practical experience for electric power emergency rescue work, which has wide application prospects.

References

[1] Tian S M, Chen X, Zhu C Y and Xie Y J. 2007 Theory of Electric Power Emergency Management and Its Technological Countermeasures. *Power System Technology*, 24: 22-27

[2] Wang H T. 2011 Analysis on Current Status of Emergency Communication and Its Technical Means. *Telecommunications for Electric Power System*, 5: 1-6

[3] Du Y, Zhong H and Li M C. 2021 Research on railway emergency command communication system based on high- throughput satellite. *China Safety Science Journal*, 31: 187-191

[4] Wang S H and Yang Z W. 2004 Design of a New Type of Wireless Communication Commanding Vehicle for Emergency. *Jiangxu Communication*, 20: 26-29

[5] Zou Y, Luan S, Ma H and Yang J C. 2012 Research on China Unicom's large and medium-sized emergency communication vehicle scheme. *Telecommunications and network technology* 7: 6-11

[6] Zhang H T, Wang M and Cui W Z. *2018 Satellite Communication.* (Beijing: Beijing Institute of Technology) p 23

[7] Yi K C, Li Y, Sun C H and Nan C G. 2015 Recent development and its prospect of satellite communications. *Journal on Communications*. 36: 161-176

[8] Fenech H, Amos S, Tomatis A, et al. 2015 High throughput satellite systems: An analytical approach. *IEEE Transactions on Aerospace & Electronic Systems*, 51: 192-202

[9] Vidal O, Verelst J, Lacan J, E. Alberty, J. Radzik and M. Bousquet. 2012. Next generation High Throughput Satellite system. **In:** *2012 IEEE First AESS European Conference on Satellite Telecommunications*, Rome. pp 1-7

[10] Berrou C and Glavieux A. 1996 Near optimum error correcting coding and decoding: turbo-codes. *IEEE Transactions on Communications*, 44: 1261-71.

Journal of Physics: Conference Series **2384** (2022) 012008 doi:10.1088/1742-6596/2384/1/012008

A low insertion loss RF switch using a 130 nm SOI process

Caozheng Zhang*[a], Libo Zheng[b], Jinghao Yu[b], Pengfei Ye[b], Yi Liu†[b]

[a] State Grid Electric Power Research Institute, Nanjing, 210000, China;

[b] Nanjing University of Posts and Telecommunications, Nanjing, 210000, China

* Corresponding author: zhangcaozheng@sgepri.sgcc.com.cn

† Corresponding author: liuyi@njupt.edu.cn

ABSTRACT: To realize a better radio frequency (RF) communication system, the RF switch, as an important module, is used to cooperate with other devices to form an RF front end for optimizing communication performances. This paper designs a single-pole multi-throw (SPMT) RF switch, which exhibits a low insertion loss of 0.57 dB at 1.9 GHz under the voltage of 2.2 to 4.7 V. The RF switch is fabricated using a 130 nm silicon-on-insulator (SOI) complementary metal-oxide-semiconductor (CMOS) process. It can be used in multi-frequency and multi-mode applications such as 4G/5G. The proposed RF switch has competitive performance and can be combined with other modules to achieve better wireless communication and promote the development of ultra-wideband and multi-mode wireless communication systems.

1. INTRODUCTION

With the rapid development of the Internet of Things (IoT), the number of end-side devices has surged, and the demand for radio frequency (RF) communication in the entire society has increased rapidly. Working as most front-end parts of an RF communication system, the performance of the RF switch directly affects the performance of the entire communication system. A poorly functional switch can limit power amplifier operation and reduce receiver sensitivity[1]. Past studies have shown that switches based on GaAs/GaN are less integrated and more expensive[2,3]. SiGe-based switches have the disadvantage of large insertion loss[4,5]. So, new technologies are desired to fabricate high-quality RF switches with the advantages of low insertion loss, high isolation, wide bandwidth, and a high power handling capability.

With the continuous development of semiconductor processing technology in recent years, more circuit systems can be integrated into a single chip. Thanks to this, each component of the RF communication system can also be compatible with CMOS technology, thereby achieving higher integration and lower cost[6]. The silicon-on-insulator (SOI) complementary metal-oxide-semiconductor CMOS technology developed based on traditional CMOS can bring lower parasitic capacitance to the switch so that it has a higher communication bandwidth while having a higher integration level[7,8]. Although the SOI process will bring about many problems, such as increasing the insertion loss, the problem can still be avoided by rationally designing the structure of the switch, and the improvement of the integration level brought about by it still has high research value[9,10]. Using CMOS technology to manufacture RF switches to improve the integration and cost-effectiveness of communication chips, and avoid their inherent shortcomings through reasonable design, will be the focus of research for a long time in the future. Compared with the traditional standard CMOS technology, the SOI CMOS technology offers superior performance, including a lower parasitic capacitance, a higher frequency, and so on, because the transistor is isolated from the silicon substrate.

Therefore, this paper presents a low insertion loss single-pole multi-throw (SPMT) RF switch based on SOI CMOS technology using the 130 nm SOI process compatible with CMOS technology. The article mainly introduces the details of the design and fabrication of the switch and analyzes its input and output performances at various frequencies.

2. DESIGN OF THE SWITCH

After selecting a suitable process, to facilitate analysis and design, it is still necessary to select a suitable model. This design is based on a PSP-SOI-PD model, where PSP is physical surface potential, and PD is partially depleted. The PSP-SOI-PD model is a physical surface potential-based model based on the PSP model. The extraction flow of it is similar to that of the PSP model and can cover PD SOI devices. This model has better g_m/g_{ds} (This symbol represents the transistor intrinsic gain) fitting capability and better harmonics simulation capability. Contrary to the PSP-SOI-PD model, the commonly used B4-SOI model, a V_{th}-based model based on BSIM3/4, has some shortcomings in describing the analog characteristics and the continuity of the current high-order derivatives and cannot accurately describe the trend of each harmonic[11].

A simplified diagram of the model used for the SOI RF switch application is shown in Fig. 1. Each field effect transistor (FET) of this switch is modeled by the PSP model. A MOS varactor connects the body node of the device and the substrate. The substrate network should be generated with an electromagnetic (EM) simulation according to the layout because the most critical factor in determining the voltage imbalance of a switch made of many (>10) transistors stacked in series in the OFF state is the total loss of the substrate. The substrate loss consists of two components: the lateral loss into the adjacent structures and the vertical loss into the substrate. S-parameter characterization shows that the lateral loss can be more significant than the vertical loss as the spacing between the two unrelated active structures decreases. Therefore, a layout-dependent EM approach is adopted to design the substrate network.

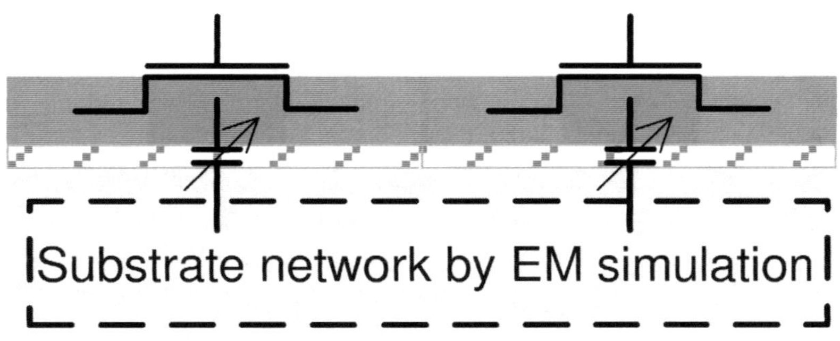

Figure 1. A simplified diagram of the SOI RF switch model.

Based on the full simulation analysis of the process model, this single-pole sixteen-throw switch was designed. The whole switch contains 40 FETs and has a total area of 2150 μm × 2085 μm, with a power consumption of 54 μW. Besides the 16 select one function, the decoder also has a level shift which is used to hold the 2.5 V and -2.5 V voltage. When designing, the switch is divided into a channel part and a control part. The channel part supplies power and provides channels, and the control part realizes the control function. The transistors in the channel part are arranged symmetrically as a whole, which aims to use the symmetry to reduce the insertion loss of the switch and the internal crosstalk. The control part adopts a low-power design to reduce the overall switching loss. Such a design shows good performance in theoretical analysis and simulation.

By design and tape-out, the switch has been fabricated using a 130 nm SOI process, which is shown in Fig. 2. The chip is a square with a side length of 3 mm. Switch arrays and analog circuits are spatially distributed to the left and right parts of the chip. The test tape-out only integrates the switch and the corresponding control circuit on one chip, and other parts of the receiver will be integrated on this chip in the later work, which further improves the integration degree.

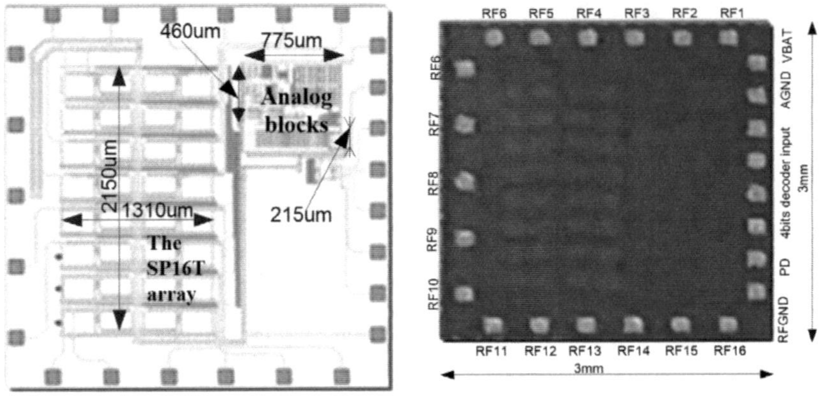

Figure 2. The chip photo of implemented SPMT RF switch.

3. MEASUREMENT DATA

The performance test of the switch is completed with the assistance of the supporting circuit, spectrum analyzer, digital oscilloscope, and other equipment used in the test. Two sets of measurements were performed, the first is the performance of the switch's linearity, and the second is the input and output power of the switch at different operating frequencies. Under the working condition of 900 MHz, the insertion loss (IL) of the switch is 0.5 dB, and under the working condition of 1.9 GHz, the IL is 0.57 dB. Compared with other RF switches, its performance is better[12].

This switch is designed mainly for communication applications in GSM mode. For the Global System for Mobile Communications (GSM) mode, linearity is an important parameter to consider, and these switches have very good linearity. In Fig. 3, the P1dB of the switch is up to 29.48 dBm at 900 MHz. The 2nd and 3rd harmonics are also measured with the results of -63.26 dBm and -56.93 dBm at 900 MHz. The switch also has low insertion loss. The value is 0.5 dB at 0.9 GHz and 0.57 dB at 1.9 GHz. Its return loss value is also excellent, and its value is -25 dB at 0.9 GHz and -19.5 dB at 1.9 GHz. Compared to other SPMT switches, this switch does have advantages in IL, return loss (RL), and linearity.

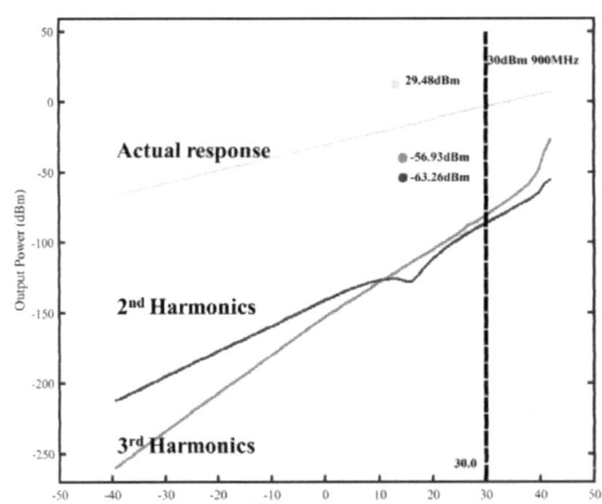

Figure 3. Linearity of the SPMT switch.

For RF switches, their power characteristics are also very important. The greater the output power, the stronger the communication ability. The switch is also capable of outputting at a suitable power which is enough for GSM mode (See Fig. 4 for specific data). When the switch is on, if the input is 0 dBm, the output at 900MHz is -0.52 dBm, and the output at the 2nd and 3rd harmonics are -123.7 dBm and -137.0 dBm, respectively. When the switch is off, the output at 900MHz is -65.5 dBm, and the output at the 2nd and 3rd harmonics are -167.5 dBm and -182.5 dBm, respectively. Similar to this, When the switch is on, if the input is 35 dBm, the output at 900MHz is 34.5 dBm, and the output at 2nd and 3rd harmonics are -51.73 dBm and -42.11 dBm, respectively. When the switch is off, the output at 900MHz is -30.4 dBm, and the output at the 2nd and 3rd harmonics are -100.2 dBm and -92.61 dBm, respectively. In general, the switch achieves lower insertion loss and higher linearity under the condition of high output and input power performance at different operating frequencies and meets the design requirements.

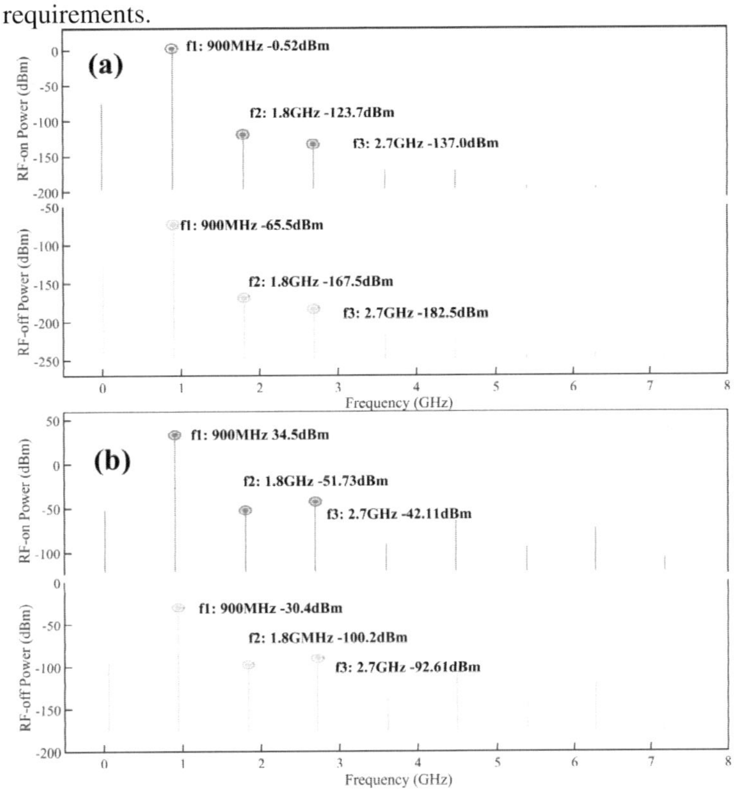

Figure 4. RF off-port and on-port with 900 MHz a) at 0 dBm input, b) at 35 dBm input.

4. CONCLUSION

This article describes an RF switch fabricated using the 130 nm SOI CMOS process. This article describes the advantages and disadvantages of this process and the corresponding solutions. Various models of the SOI CMOS process are analyzed in the paper, and the switch design and fabrication are carried out according to the model analysis results. After testing, the switch has better IL, RL, and linearity parameters. The test results show that the switch avoids some disadvantages of the process through reasonable circuit design and further expands its advantages. The switch also integrates the corresponding analog control circuit, which provides a basis for further integration. Future work will continue to improve the structure of the switch to serve the development of wireless communication systems better.

5. ACKNOWLEDGMENTS

The authors also wish to thank Qiuliang Li for guidance on paper writing, Rocket Liu for packaging and measurement, Kiry Shen for layout contribution, and Hiter Hou for his modeling suggestion and Keysight modeling extraction.

REFERENCES

[1] Ciocoveanu, R., Weigel, R., Hagelauer, A. and Issakov, V., "A Low Insertion-Loss 10-110 GHz Digitally Tunable SPST Switch in 22 nm FD-SOI CMOS," 2018 IEEE BiCMOS and Compound Semiconductor Integrated Circuits and Technology Symposium (BCICTS), 1-4 (2018).

[2] Polli, G., Longhi, P. E., Colangeli, S., Fenu, S., Costanzo, F., Ciccognani, W. and Limiti, E., "GaN/Si Ka-band SPDT for observation payloads," 2019 IEEE Asia-Pacific Microwave Conference (APMC), 288-290 (2019).

[3] Fletcher, A. S. A., Nirmal, D., Arivazhagan, L., Ajayan, J., Raj, M. G., Hamza, K. H., Murugapandiyan, P. and Natarajan, R., "A 28-GHz Low-Loss AlGaN/GaN HEMT for TX/RX Switches in 5G Base Stations," J. Electron. Mater. 51(3), 1215-1225 (2022).

[4] Cheon, C. D., Cho, M.-K., Rao, S. G., Cardoso, A. S., Connor, J. D. and Cressler, J. D., "A New Wideband, Low Insertion Loss, High Linearity SiGe RF Switch," IEEE Microwave and Wireless Components Letters 30(10), 985-988 (2020).

[5] Kazemi Esfeh, B., Rack, M., Makovejev, S., Allibert, F. and Raskin, J.-P., "A SPDT RF Switch Small- and Large-Signal Characteristics on TR-HR SOI Substrates," IEEE Journal of the Electron Devices Society 6, 543-550 (2018).

[6] Mansour, R. R., "RF MEMS-CMOS device integration: An overview of the potential for RF researchers," IEEE Microwave Magazine, 14(1), 39-56 (2013).

[7] Li, C., Ustundag, B., Kumar, A., Boenke, M., Kodak, U. and Rebeiz, G., "< 0.8dB IL 46dBm OIP3 Ka-band SPDT for 5G communication," 2018 IEEE 18th Topical Meeting on Silicon Monolithic Integrated Circuits in RF Systems (SiRF), 1-3 (2018).

[8] Parlak, M. and Buckwalter, J. F., "A 2.5-dB Insertion Loss, DC-60 GHz CMOS SPDT Switch in 45-nm SOI," 2011 IEEE Compound Semiconductor Integrated Circuit Symposium (CSICS), 1-4 (2011).

[9] Adabi, E. and Niknejad, A. M., "An mm-wave transformer based transmit/receive switch in 90nm CMOS technology," 2009 European Microwave Conference (EuMC), 389-392 (2009).

[10] Tomkins, A., Garcia, P., and Voinigescu, S. P., "A 94GHz SPST Switch in 65nm Bulk CMOS," 2008 IEEE Compound Semiconductor Integrated Circuits Symposium, 1-4 (2008).

[11] Grujić, D. N., Božović, M. and Savić, M., "BSIM4 to PSP Model Conversion: A Case Study," J CIRCUIT SYST COMP 25(03), 1-17 (2016).

[12] Joshi, A., Lee, T. and Chen, Y., "Recent advances and future trends in SOI for RF applications," Subthreshold Microelectronics Technology Unified Conference, 1-3 (2014).

OFDM Spectrum Sensing Analysis and Implementation Method

Hao Liu[1,†], Yufan Xia[2,†], Haonan Yu[3,*,†]

[1]Hao Liu, School of electronic engineering, Xidian University, Shaanxi Province, 710126, China

[2]Yufan Xia, Electronic communication engineering, Changzhou University, Jiangsu Province, 213164, China

[3]Haonan Yu, Electronic communication engineering, Nanjing Tech University, Jiangsu Province, 211899, China

*Email:202021139033@njtech.edu.cn

†These authors contributed equally to this work

Abstract: OFDM is a carrier modulation technology, which realizes the parallel transmission of high-speed serial data through frequency division multiplexing. OFDM has an excellent ability to resist multipath fading and is commonly used in wireless communication transmission technology. Spectrum sensing technology is a technology that allows cognitive users to accurately obtain the authorized spectrum usage in wireless networks through signal detection and other processing methods. It is the basis for effectively using different idle channels in different time domains. This paper mainly discusses the OFDM technology, deduces the algorithm of the implementation principle and mathematical characteristics of OFDM, analyses its spectrum characteristics and performance advantages, focuses on the principle and method of OFDM spectrum sensing based on spectrum sensing technology, combs the characteristics of several existing commonly used OFDM spectrum sensing methods, and analyses the implementation cost, hardware requirements, transmission performance The performances of interference factors and other aspects are sorted out, the advantages and disadvantages of several existing OFDM spectrum sensing methods are compared, their respective applicable environments are compared, and their possible problems in the future development prospect of OFDM technology and the direction that needs further research and improvement are put forward.

1. Introduction

Orthogonal frequency division multiplexing (OFDM) is a communication technology developed based on multi-carrier technology for high-speed transmission under wireless conditions. The main idea of OFDM technology is to divide a given channel into many orthogonal subchannels in the frequency domain, use a subcarrier for modulation on each subchannel, and transmit each subcarrier in parallel. Due to the channel characteristics of OFDM, the mutual interference between each subchannel is very small, and inter-symbol crosstalk can be eliminated, to improve the spectral efficiency of the system and transmit signals under clutter interference [1]. Therefore, OFDM technology is often used in transmission media that are vulnerable to interference or have the poor anti-interference ability.

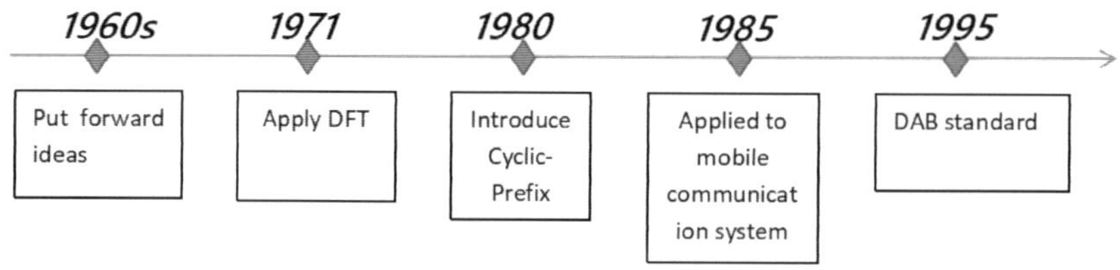

Figure 1 OFDM development timeline

Figure 1 is a timeline diagram, briefly describing the development of OFDM technology from its introduction to its application. In the 1960s, the idea of orthogonal frequency division multiplexing was put forward[2]. It is a multi-carrier technology that uses high-speed serial data to realize parallel transmission through frequency division multiplexing. R. W. Chang published a paper on multi-channel transmission synthesis of band-limited signals [3]. Saltzberg completes the performance analysis. However, because the parallel transmission system requires a baseband-shaped carrier array, sine wave carrier generator array, and coherent demodulation array, the traditional simulation method is very complex and expensive, so it has not been put into practical application in the early stage. In 1970, the OFDM patent was published. In 1971, Weinstein and Ebert proposed using discrete Fourier transform (DFT) to realize multi-carrier modulation [4]. People began to study the digital implementation method of parallel transmission multi-carrier systems. DFT was applied to OFDM modulation and demodulation, which laid a foundation for the practicality of OFDM, simplified the implementation of multi-carrier technology, and made theoretical preparations for the full digital scheme of OFDM. However, due to the limitations of digital signal processing technology at that time, It was still not widely used. In the 1980s, people conducted in-depth research on applying multi-carrier modulation in high-speed modems and digital mobile communications. In 1980, Peled and Ruiz solved the orthogonality problem of OFDM by introducing the concept of cyclic prefix (CP) [4]. In 1985, l.j.cimini first analyzed the problems existing in the application of OFDM Technology in mobile communication and proposed corresponding solutions, which simplified the implementation scheme of OFDM and made rapid progress in the field of wireless mobile communication [5]. In 1995, the DAB standard was proposed by the European Telecommunications Standards Association and became the first standard to adopt OFDM. In 2001, the Radio Management Bureau of the Ministry of Information Industry approved China Netcom to test the OFDM fixed wireless access system celerflex, which enabled the system to be tested and put into operation in China.

With the development over time, modern communication technology is becoming more and more mature, which is enough to support the wide application of OFDM Technology in modern society. At present, OFDM technology is essential in civil and military communications. However, with the rapid development of wireless communication technology and mobile terminals, spectrum congestion has become more serious, which can not be ignored. Using spectrum sensing technology can improve spectrum utilization and effectively solve the problem of frequency resource shortage.[6]. Spectrum sensing is generally considered the cornerstone of cognitive radio technology. Its purpose is to accurately and quickly find the free spectrum in the complex wireless communication environment's time, frequency, and space domain, detect spectrum vulnerabilities, and provide a reasonable way for unauthorized users to access the idle authorized channel. At present, OFDM signal spectrum sensing technology has attracted the attention of many researchers. However, the actual electromagnetic environment is complex and changeable, putting high requirements for spectrum sensing technology and basic sensing ability. It also requires cognitive knowledge in multiple fields, foresight, timeliness, and adaptability under extreme conditions.

This paper focuses on the application of OFDM Technology in spectrum sensing, aims to explore the significance and importance of OFDM Technology in the field of spectrum sensing, and compares their shortcomings and deficiencies. This paper starts with a brief background introduction and development of OFDM technology. Secondly, the mathematical principle of OFDM technology is discussed in detail through mathematical algorithm and principle formula, and the spectrum sensing technology is explained. Then, through the existing experimental methods and conclusions, this paper summarizes five spectrum sensing methods and compares their advantages and disadvantages, summarizes the most outstanding performance of each method, and puts forward the outlook of OFDM Technology in the field of spectrum sensing based on these five methods.

2. Properties of OFDM

2.1 Mathematical expression of OFDM

OFDM originates from frequency division multiplexing (FDM). The corresponding carrier frequency formula is shown below.

$$f_n = f_0 + (\mathrm{n}-1)\,\Delta f \tag{1}$$

where f_n is in accordance with the signal carrier frequency,Δf is the interval of the signal carrier interval.

Suppose that the OFDM signal transmitted by the transmitter contains n subcarriers, and the transmitter is fully synchronized. Each OFDM symbol comprises N signals after PSK modulation or QAM modulation. The OFDM symbol is represented as $X(0), X(N-1)$. After the serial-parallel transform and inverse fast Fourier transform of this OFDM symbol, the time-domain baseband signal of the nth subcarrier is:

$$X(k) = \frac{1}{\sqrt{N}}\sum_1^N x(n)exp(j2\pi f_n k), k = 0,\ldots,N-1 \tag{2}$$

where N is the number of subcarriers of the OFDM signal and the length of IFFT [7].

2.2 The modulation and demodulation of the OFDM system

A single OFDM signal is composed of multiple subcarriers. At the transmitter of the system, for OFDM digital signals, T is the length of time required for data symbols, and the modulation symbol of the nth carrier is x(n). N represents the total number of subcarriers, T represents the width of OFDM, $di\ (i = 0,1,\ldots,N-1)$ represents the data symbol allocated to each subchannel, and f_i represents the frequency of the ith subcarrier. $rect\ (t) = 1,\ |\,t\,| < T/2$ represents a rectangular function, and the OFDM symbol starting from $t = t_s$ can be expressed as:

$$S(t) = Re\{\sum_{i=0}^{N-1} d_i\, rect(t - t_s - \tfrac{T}{2})\, exp[j2\pi(f_c + \tfrac{i}{T})(t - t_s)]\}, t_s < t < t_s + T \tag{3}$$

Then the bit data is allocated to each subcarrier. After modulation, it will be mapped to the amplitude and phase of the subcarrier. Usually, the output signal of OFDM is expressed as:

$$S(t) = \sum_{i=0}^{N-1} d_i\, rect(t - t_s - \tfrac{T}{2})\, exp[j2\pi \tfrac{i}{T}(t - t_s)], t_s < t < t_s + T \tag{4}$$

The in-phase component of the OFDM symbol corresponds to $s\ (T)$ real part, and the orthogonal component corresponds to $s\ (T)$ imaginary part. In the actual system, the components are represented by the real and imaginary parts of $S\ (T)$ are multiplied by the cos and sin components of the corresponding subcarriers to form OFDM symbols.

Then we demodulate the jth subcarrier and perform the integral operation within the time length T.

$$\begin{aligned} d_j &= \frac{1}{T}\int_{t_s}^{t_s+T} exp[-j2\pi\tfrac{j}{T}(t-t_s)]\sum_{i=0}^{N-1} d_i\, \mathrm{exp}[\mathrm{j}2\pi\tfrac{i}{T}(t\text{-}t_s)]d_t \\ &= \frac{1}{T}\sum_{i=0}^{N-1} d_i \int_{t_s}^{t_s+\mathrm{T}} \mathrm{exp}[\mathrm{j}2\pi\tfrac{j-i}{T}(t-t_s)]\, d_t \end{aligned} \tag{5}$$

The formula shows that the original data can be recovered after demodulating the jth subcarrier. For the remaining subcarriers, an integer multiple of the period can be generated by controlling the value of (J-I) /t within the integration interval.

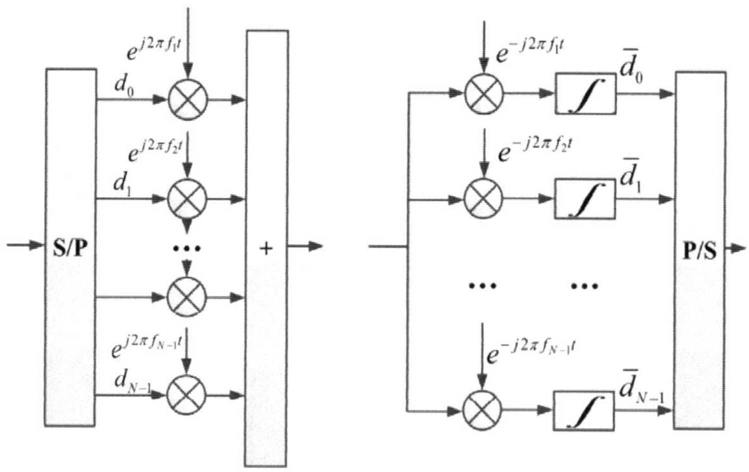

Figure 2 Modulation and demodulation diagram of the system

2.3 Spectrum characteristics of OFDM signals

In the OFDM system, within each symbol period T, the integration of each carrier meets the orthogonal condition, as shown in the following formula:

$$R = \frac{1}{T}\int_0^T e^{j2\pi t f_l} e^{j2\pi t f_k} dt = \begin{cases} 1, l = k \\ 0, l \neq k \end{cases} \tag{6}$$

OFDM is divided by the orthogonal method in the frequency domain, and Sa (x) function represents each carrier spectrum. The carrier spectrum of OFDM can be obtained. [9]

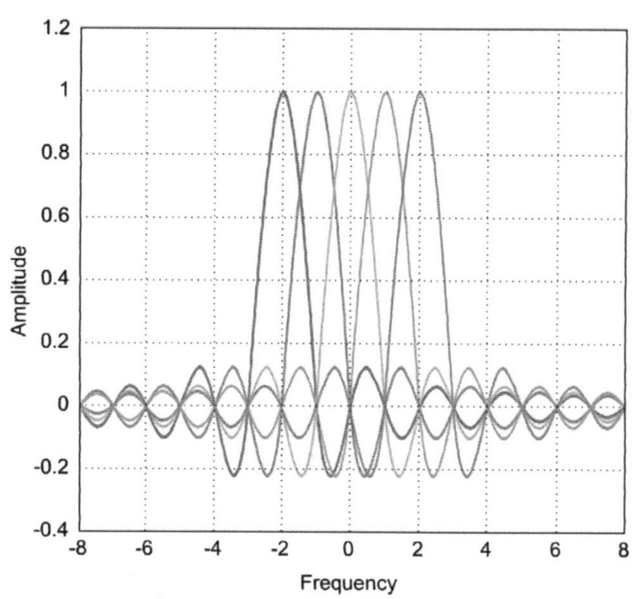

Figure 3 Spectrum diagram of OFDM

As shown in Figure 3, in the curves with different colors, the spectrum of adjacent subcarriers of the system overlaps at the place where the zero point and peak value pass. The carrier interval is t, and the tailing of subcarriers will not affect other subcarriers. Therefore, using OFDM Technology in the communication system has better high-frequency spectrum characteristics to a certain extent.

2.4 Bandwidth savings

FDM is frequency division multiplexing. A channel is divided into different users according to frequency bands. OFDM is orthogonal frequency division multiplexing, which divides the channel into several orthogonal sub-channels, converts high-speed data signals into parallel low-speed sub-data streams, and modulates them to transmit on each sub-channel.

At the receiving end, FDM requires a sufficiently large interval between carriers so that carriers do not overlap, so spectrum utilization is very low. The subcarriers of the OFDM system are orthogonal to each other, which can effectively save channel bandwidth.

Suppose that there are N subcarriers in an OFDM system. The subchannel symbol duration is T_B. Each subcarrier adopts M-base modulation, then the frequency bandwidth occupied by it is:

$$B_{OFDM} = \frac{N+1}{T_B} \tag{7}$$

Frequency band utilization is the bit rate transmitted per unit bandwidth:

$$\eta_{b/OFDM} = \frac{N log2^M}{T_B} * \frac{1}{B_{OFDM}} = \frac{N}{N+1} log2^M \approx log2^M \tag{8}$$

If the M base symbol of a single carrier is used for transmission, to obtain the same transmission rate, the symbol duration should be shortened to $\frac{T_B}{N}$, and the bandwidth occupied is equal to $\frac{2N}{T_B}$, so the frequency band utilization is:

$$\eta_{b/OFDM} = \frac{N log2^M}{T_B} * \frac{T_B}{2N} = \frac{1}{2} log2^M \tag{9}$$

By comparing the two formulas above, the frequency band utilization of the parallel OFDM system can be approximately doubled compared with the serial single-carrier system

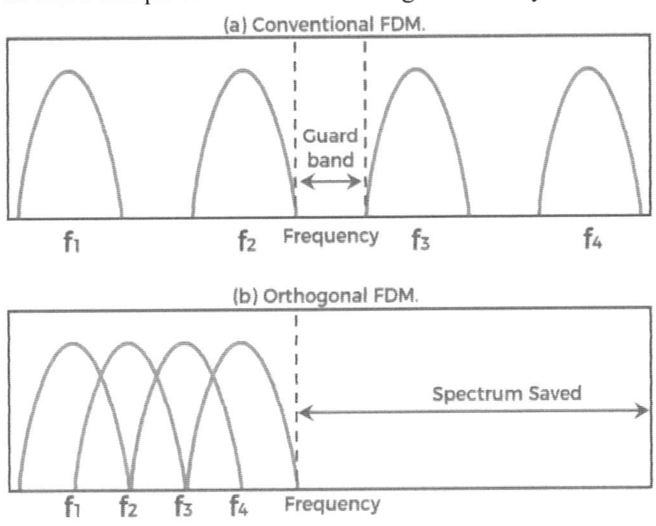

Figure 4 Bandwidth savings comparison

FDM is a combination of independent bands that do not coincide. At this time, to separate each channel of signals, only a band-pass filter of each band can be used, while the carrier segments of OFDM signals partially coincide. At this time, it is no longer possible to complete a few simple band-pass filters, but to extract each channel of signals with a product integration function as the filter.

Due to the orthogonality between their carriers, It can extract all signals well. Different from FDM, OFDM owns higher spectrum utilization. Its use in communication systems has greatly delayed the frequency selective fading rate and reduced the mutual interference between subcarriers.

3. Overview of spectrum sensing

3.1 Spectrum sensing

Spectrum sensing is used to locate free spectrum in the time domain, frequency domain, airspace, and spectrum hole. Unauthorized users can thus access the free spectrum resources in the authorized frequency band. The purpose of spectrum sensing technology is to discover spectrum opportunities accurately and rapidly, as well as to protect authorized users from harmful interference.

3.2 Energy detection

Energy detection is the most commonly used method for spectrum sensing and was proposed by H. Urkowitz in the 1960s[10]. This method assesses the presence of a signal based on the received signal's total energy or power level, characterizing the signal's field strength in the channel as occupied if it exceeds a particular threshold. Otherwise, the channel is assumed to be idle. The block diagram below depicts the various processes of the energy detection technique.

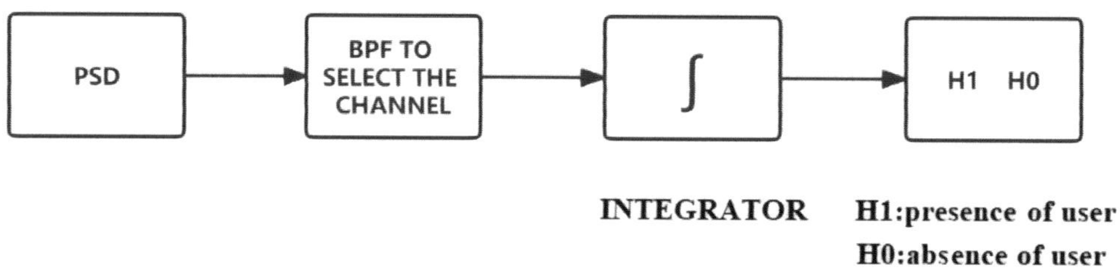

Figure 5. Energy detection block diagram.[11]

As shown in the figure, to eliminate the influence of interference outside the bandwidth, the received signal needs to pass the band-pass filter, and then the signal, after removing the interference, starts the sampling process. These sample points are integrated over a time period to obtain a specific energy value, and finally, a statistical comparison is made.

In the energy detection technique, the detection probability P_d is known as the probability that the primary user does exist even if the primary user is assumed to exist; the false alarm probability P_f is known as the probability that the primary user is incorrectly judged to exist when the primary user does not exist; and the missed detection probability $P_m = 1 - P_d$ is defined as the probability that the primary user is incorrectly judged not to exist when the primary user exists.

As one of the most common non-correlated detection methods, ED does not need any information about the detected signal and can effectively combat unknown dispersion channels and fading. For independent and co-distributed signals, ED is the best detection method [14]. However, detection performance is lowered for coupled signals. The confidence in the conclusion of energy detection drops towards the threshold value due to noise uncertainty. Yet, suppose the value gap between the energy detection and the threshold is large. In that case, the effect of noise uncertainty lessens, and the verdict can be formed efficiently utilizing the energy detection method. This method offers a straightforward process and high-speed calculation. But the noise immunity of this method is poor. In the case of a low signal-to-noise ratio, detection efficiency drops dramatically in communication environments, submerging the primary user signal in the noise and resulting in misjudgment [12].

3.3 Matched filter detection

The matched filter detection is to demodulate the target signal to maximize the signal-to-noise ratio, thus ensuring the accuracy of detection results in an additive Gaussian white noise environment, which requires knowledge of some prior information. The flow chart of its detection method is shown below.

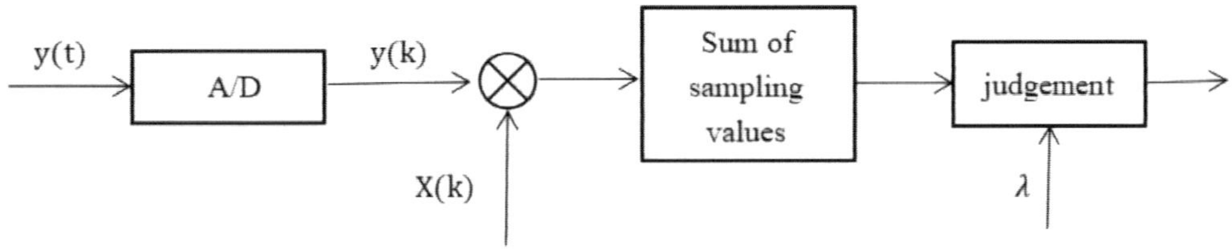

Figure 6. Matched filter block diagram.

As shown in the figure, the matched filter detection convolves the transmitted signal with the unknown signal after passing it through the analog-to-digital converter, and the detection statistics are then compared with the set threshold λ to determine whether the sensed signal can occupy the channel. The following expression gives the operation.

$$Y[n] = \sum_{k=-\infty}^{\infty} h[n-k]x[k] \tag{10}$$

where x is an unknown signal, the reference signal that matches the impulse response of the h convolution matching filter, maximizing the SNR.

The main advantage of MFD is that correlation detection is performed so that a high processing gain can be achieved relatively quickly. An accurate synchronization can detect any weak signal if the observation is long enough. However, a significant disadvantage of MFD is that CR will require a dedicated receiver for each primary user, making the system much more complex. It can produce a more idealized signal-to-noise ratio at the time of detection while not requiring a long sensing time, thus allowing the algorithm to achieve a more optimal result. The matched filter detection algorithm has excellent performance, but the prerequisites are demanding, and the overall computational effort is high [12].

3.4 Cyclostationary detection

Cyclic smoothness refers to the periodic and smooth variation of the statistical properties of a signal. In communication systems, the cyclic smoothness of the signal may come from the carrier, guide frequency, cyclic prefix, coding, etc. The smooth cyclic detection uses the spectral correlation function instead of the power spectral density, fully using the modulated signal's periodicity. The cyclic smooth characteristic is often an extra addition that one purposely gives to the communication signal for some purpose, which is generally not possessed by noise. Therefore, the cyclic smooth characteristic can be used as one of the features for signal extraction[13].

The statistics of a cyclic smooth signal vary with time in cycles or multiple cycles. Cyclic smoothing detection separates the signal from the noise based on the cyclic smoothing characteristics of the target signal. The algorithm does not need to know the modulation of the signal and has good noise immunity, but it is computationally overloaded. The noise signal is not correlated and is generated in a non-time-varying manner. However, the benefit of the periodic smooth feature detection method is that it can distinguish the noisy signal from other non-noisy signals. However, the method is computationally intensive, and the research computation period is very long, so the current research application of this method is not very wide.

3.5 Generalized Likelihood Ratio Detection

The purpose of the detection algorithm is to determine whether the primary user signal exists. While LRT is the best spectrum sensing algorithm, it requires knowledge of parameters such as noise variance, which is usually hard to acquire at the CR receiver side. When these parameters are unknown, a hypothesis test in which uncertain parameters exist is called a compound hypothesis test. In traditional detection theory, this problem can be solved by generalized likelihood ratio detection. To make the LRT more convenient, we can use the pre-estimated parameter values in the LRT [14].

In GLRT, the maximum likelihood estimates of the unknown parameters under H_0 and H_1 are first obtained, denoted as:

$$\begin{cases} \widehat{\Theta_0} = arg\max_{\Theta_0} p(x|H_0,\Theta_0) \\ \widehat{\Theta_1} = arg\max_{\Theta_1} p(x|H_1,\Theta_1) \end{cases} \quad (11)$$

where Θ_0 and Θ_1 denote the set of unknown parameters under H_0 and H_1 respectively, resulting in the test statistic of GLRT as:

$$T_{GLRT}(x) = \frac{p(x|\widehat{\Theta}_1,H_1)}{p(x|\widehat{\Theta}_0,H_0)} \quad (12)$$

where λ is the judgment threshold, and GLRT is judged at $T_{GLRT}(x) > \lambda$ as H_1 and vice versa as H_0. Since the unknown parameters in Θ_0 and Θ_1 are closely related to the statistical model of the noise and signal, their estimates are susceptible to model error.

3.6 Detection based on covariance matrix and eigenvalues

Covariance matrix-based detection is an emerging method for blind spectrum sensing. The basic principle is to make decisions by calculating the covariance matrix of the received signal and extracting the special statistics. This method does not require a priori knowledge of the noise variance, so it can effectively combat the uncertainty of the noise power.

In practice, based on a limited amount of signal samples, the covariance matrix is detected.

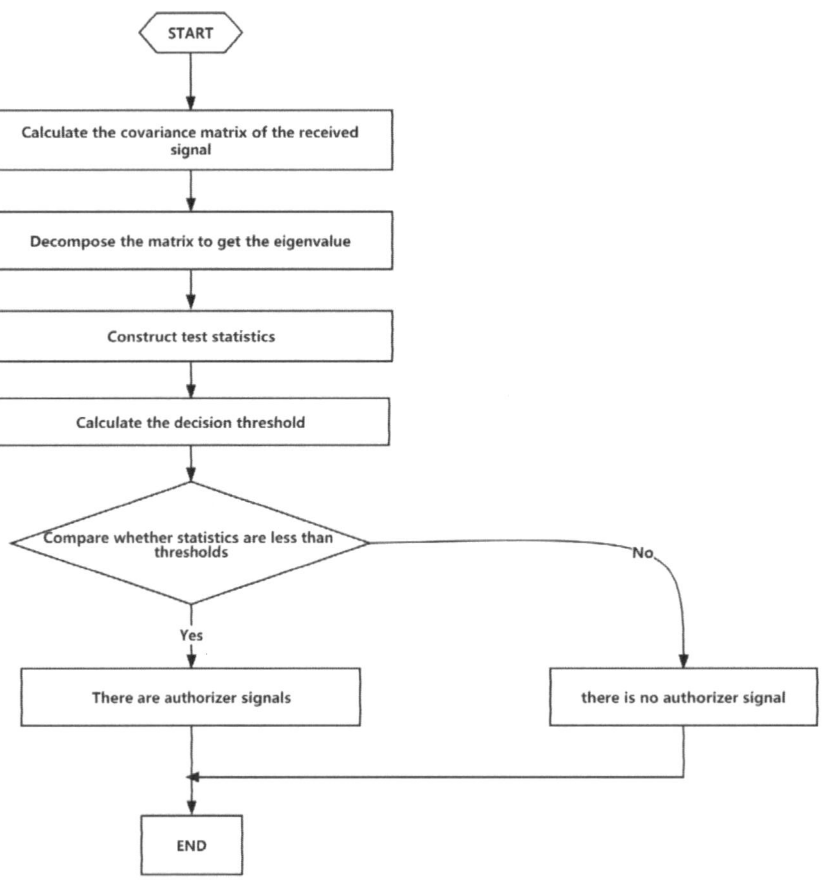

Figure 7. Flowchart based on covariance matrix and eigenvalues[15]

3.7 Comparison

This section summarizes the advantages and disadvantages of various detection algorithms in the above-spectrum sensing algorithms. Depending on their characteristics, they can be applied to various detecting environments: Energy detection is simple to implement, does not require prior knowledge, is generally applicable, and is typically employed when cognitive users do not know the primary user's prior knowledge. Matched filters have a rapid detection speed and strong performance, but they require knowledge of the primary user information, so they are only employed when the preceding information is available. Cyclostationary detection can effectively distinguish noise and signal, which is used when the signal has periodic autocorrelation characteristics. Because of its detection characteristics, generalized likelihood ratio detection is generally used to solve the hypothesis test with uncertain parameters [16]. The detection based on covariance matrix and eigenvalue does not need prior information and can estimate noise variance and signal power at the same time [17]. Therefore, it is often used to detect signals with uncertain noise power.

Table 1 The advantages and disadvantages of the detection methods.

Spectrum detection algorithm types	Advantages	Disadvantages
Energy detection	Does not need the prior information of the signal	Greatly affected by noise
Matched filter detection	Excellent algorithm performance and fast detection speed	High preconditions and complex overall calculation
Cyclostationary detection	Can distinguish different signal types	Large amount of calculation
Generalized likelihood ratio detection	Low complexity and good detection performance	Susceptible estimates to statistical model errors
Detection based on covariance matrix and eigenvalue	Does not need the prior information	High overall implementation complexity

4. Conclusion

In application, OFDM has significant advantages, such as anti-inter solid symbol interference, high anti-fading, and anti-narrowband interference ability. However, the existing OFDM systems also have drawbacks, such as requiring high-precision CSI estimation, being easily affected by frequency deviation, and having a high peak-to-average power ratio. On the other hand, the number of users in China is vast, and the regional distribution is highly uneven[18]. However, due to the lack of early research, the spectral division is not reasonable, there are some bandwidth congestion, some bandwidth idle, and the scheme based on signal power can not determine whether there are users in the target signal [12].

Compared with traditional FDM, the advantage of OFDM is that it can save some bandwidth and has higher spectrum efficiency. It is used to resist frequency selective fading in broadband mobile communication systems. The five methods listed above have their advantages and disadvantages. Energy detection does not need prior information about the signal. However, it is greatly affected by noise interference. Under a high signal-to-noise ratio, the performance of energy detection is reliable, and the energy detection method can be used to roughly judge the use of frequency bands. However, it is difficult to set the energy threshold. The detection performance can be improved by increasing the number of samples, but it also increases the complexity of signal processing. This part is still in urgent need of development[19]. Matched filter detection has both excellent algorithms and fast transmission speed, but it needs high premise requirements, and it is difficult to meet its requirements in hardware. Cyclic spectrum detection can distinguish different signal types, and its detection efficiency is superior, but it requires a large amount of calculation, and its real-time performance is poor.

To solve the problems mentioned above, this study focuses on the digital implementation of cyclostationary theory and optimizing the algorithm. The generalized likelihood ratio detection method has good detection performance and low complexity, but the error of the statistical model will cause many problems. The detection based on covariance matrix and eigenvalue does not need prior knowledge of noise variance and can effectively overcome the uncertainty of noise power, but it is difficult to realize. Although with society's increasing demand for communication digitalization, broadband, personalization, and mobility, the application prospect of OFDM Technology in the field of fixed wireless access and mobile access has been fully recognized by many universities, companies, and technicians. The development of OFDM technology has been carried out all over the world, which has successively solved the traditional problems of OFDM synchronization and high peak-to-average ratio, and carried out OFDM and space-time code, and joint transmission. The research work of combining joint detection and smart antenna[20], but judging from the shortcomings of the above five methods, OFDM spectrum sensing technology still needs to be improved. With its huge amount of computation, high hardware requirements, strong limitations, and other issues, this is a relatively hot research field at present, and many topics need further research.

References

[1] Li H. OFDM mobile communication technology principle and application analysis. *Popular Standardization* .2021.**10** p162-4

[2] Chen Q. Research on digital signal processing algorithm in high-frequency spectral efficiency optical OFDM communication system. Doctoral Dissertation. Hunan University.2018

[3] Chang R. W. Orthogonal Frequency Multiplex Data Transmission System.1970.

[4] Shan C., Wang Y., and Zhuang X. OFDM modulation technology. *The core technology of new generation mobile communication. Modern Electronic Technology*. 2007. **19** p34-7

[5] Cimini L. Analysis and Simulation of a Digital Mobile Channel Using Orthogonal Frequency Division Multiplexing. *IEEE Transactions on Communications.*1985. **33** p665-75

[6] Wu H., Meng J., Li Y. and Zhou L. Spectrum sensing method of OFDM signal based on deep learning(eds.) *Proceedings of 2019 national antenna Annual Conference (Volume II).*Xi'an University of Electronic Science and Technology Press. pp650-3.2019

[7] Wang D. Research on multi-mode of dm-im spectral efficiency optimization and detection algorithm.Master's thesis. Heilongjiang University. 2021

[8] Pan W. Spectrum sensing of OFDM signals in cognitive radio, Master's thesis, Hangzhou University of Electronic Science and Technology. 2012

[9] Zhang X., Yan J., Tian F., and Sun Z. Dual threshold cooperative spectrum sensing algorithm based on differential energy detection. *Journal of Instrumentation.*2014.**06** p1325-30

[10] Tan J. Performance analysis and verification of spectrum sensing method based on energy detection. Master's thesis. Hainan University. 2015

[11] Di Y. Spectrum sensing based on energy detection in OFDM system. Doctoral Observation. Wuhan University of Technology. 2015

[12] Song Y. Research on Key Technologies of non-cooperative spectrum utilization evaluation based on OFDM communication system. Master's thesis. Beijing University of Posts and Telecommunications. 2021

[13] Wang J., Yi J., Huang F., Chen Z., and Xu Y. Cyclostationary spectrum sensing method based on OFDM symbol width. *Journal of Electronics and Information.*2018.**40 (2)** p8

[14] Liu L. Research on spectrum sensing algorithm based on OFDM. Doctoral Dissertation. Xi'an University of Electronic Science and Technology.2013

[15] Gou C. and Jin M. A spectrum sensing method based on inverse eigenvalue of covariance matrix. CN108322277A. 2018

[16] Li Y. and Chen S. OFDM spectrum sensing algorithm in a fast-changing channel environment. *Modern Electronic Technology.* 2018. **41 (7)** p3

[17] Zhu F. Research on Key Technologies of fast spectrum detection in OFDM system. Doctoral Dissertation. University of Electronic Science and Technology. 2018
[18] Zhao M. Research on spectrum detection technology in cognitive radio. Doctoral Dissertation. Nanjing University of Posts and Telecommunications. 2012
[19] Deng W. Application of energy detection in cognitive radio spectrum sensing. *Communication Technology*. 2010. **8** p3
[20] Qin L. and Wang X. The application status and Prospect of OFDM technology. *China Science and Technology Information*. 2009

Journal of Physics: Conference Series **2384** (2022) 012010 doi:10.1088/1742-6596/2384/1/012010

Research on the Construction Method in Complex Electromagnetic Environment Based on Composite Signal

Zhijun Zhu[1], Jiwei Sun[1], Jian Yang[1*], Chenyong Li[1], Dongchu Su[1], and Lu Qiang[1]

[1]63870 Troops, Huayin, Shaanxi, 714200, China

[*]Corresponding author's E-mail: ssyn_yangjian@163.com

Abstract. Based on new signal generation technologies such as vector signal modulation and arbitrary waveform editing and playback, the transmitter resources are multiplexed by each signal component at the same time or sharing time to generate flexible and dense composite signals and solve the problem of complex electromagnetic environment construction under the limited hardware.

1. Introduction

In electronic equipment test and identification, it is necessary to construct an electromagnetic environment with certain complexity to test the working efficiency of the equipment in the complex electromagnetic environment. Due to the popularized utilization of electronic information equipment and various and widely distributed modern electromagnetic radiation sources and electromagnetic signals, a traditional construction scheme, which is based on the elements of a complex electromagnetic environment, requires many electromagnetic radiation sources. The construction cost is high, and it is difficult to deploy on the test task site.

Most of the existing high-power radiation sources have a single function and can only transmit a specific type of signal in a single frequency band, which seriously wastes transmitter resources and increases the difficulty to deploy field test tasks. Therefore, based on the vector signal modulation technology and arbitrary waveform editing and playback technology, the limited transmitter resources are multiplexed at the same or sharing time to generate a composite signal with flexible style and dense quantity to construct the complex electromagnetic environment under limited hardware. Moreover, equipment such as digital signal sources, power amplifiers, and antennae is used to build an electromagnetic environment construction system, and spectrum and signal analyzer are applied to monitor and analyze the constructed electromagnetic environment, verifying that the construction method is feasible.

2. Overview of complex electromagnetic environments

2.1. Characteristics of the complex electromagnetic environment

Complex electromagnetic environment mainly refers to the electromagnetic radiation signals with multiple types, full-spectrum, and high density in the environment, as well as the mutual influence and interference due to the extensive use of electronic equipment by oneself, resulting in a sudden change in the time domain, crisscross in the spatial domain, crowded overlap in the frequency domain, and uneven level in the energy domain, which seriously affects the performance of equipment [1].

A complex electromagnetic environment is formed under the joint action of lots of electronic information equipment. The main signal sources include frequency equipment used by oneself and the civil electromagnetic environment. From the perspective of influencing equipment efficiency, the main characteristics of a complex electromagnetic environment include:

(1) Signals with various types and large quantities. There are many types of signals, involving radar, communication, and navigation, and the signal style is more flexible. The number of signals is dense, and the electromagnetic radiation source works at the same time. Besides, the device is more likely to encounter a "best interference signal".

(2) Wide spectrum occupancy. Many signals coexist, and the spectrum is occupied from several MHz to tens of GHz. In particular, the involvement of electronic warfare equipment makes the spectrum almost occupied all the time, and the frequency security of the equipment is threatened.

(3) Interlaced overlapping in space. Radiation sources are scattered, and potential interference signals may come from all directions. Anti-interference measures of equipment may be difficult to work.

2.2. General construction method

There are three main ways to construct a complex electromagnetic environment:

(1) Real equipment construction. The environment built with electronic information devices has the highest fidelity and the most reliable effect, but it requires many devices, and the operation is complex. Therefore, the cost of workforce and material resources is quite huge.

(2) Simulator construction. The use of alternative equipment of various electromagnetic simulators for signal radiation has the advantages of convenient operation, but the simulation environment may fall in the risk of simple signal style and insufficient fidelity, and the construction cost due to many simulators cannot be ignored.

(3) Computer simulation. The way to digitally simulate a complex electromagnetic environment with the help of computer software and electromagnetic model algorithm consumes fewer resources, but the simulation system is difficult to realize and the effect is poor. It is generally not used as a test method in the equipment assessment stage.

To ensure the fidelity of the constructed electromagnetic environment to the electromagnetic environment, the construction mode combined with the actual installation and the simulator is generally adopted. The typical feature of the construction system is that it contains many high-power electromagnetic radiation sources[2]. For example, the US military electronic information test base has many electromagnetic radiation devices and thousands of independent radiation sources.

At present, the construction of a complex electromagnetic environment in China is mainly based on various types of simulators, including a small number of actual installations. The simulator is usually constructed according to the signal type and frequency band, such as a radar simulation system with a certain frequency band and a communication interference system with a certain frequency band. To construct various complex electromagnetic environments with a full band, many simulators are required[3]. The general construction scenario is shown in Figure 1.

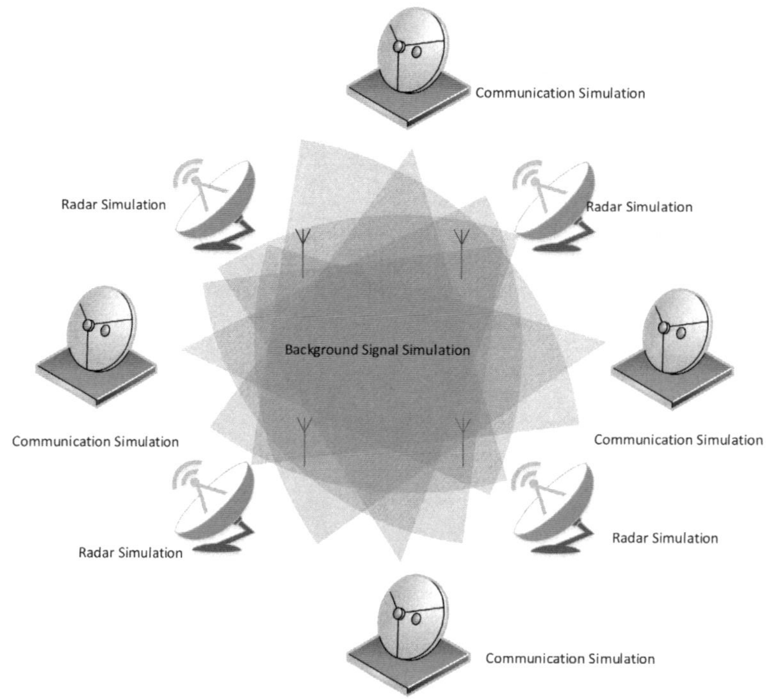

Figure 1. The general construction method of a complex electromagnetic environment

3. The construction method based on the composite signal

In the traditional construction method, considering complex electromagnetic environment factors such as the types and quantities of cover signals, frequency domain occupation, and spatial distribution, a certain number of simulation systems are usually required[4][5]. In addition, for the test and training tasks under complex terrain, deploying simulators in large numbers is costly and difficult to operate, which requires a relatively flexible alternative.

3.1. Generation of the composite signal

In this paper, a method replacing the single signal with the composite signal at the signal excitation end is proposed for the single simulator to transmit multiple signals, improving the utilization efficiency of simulator resources, and offsetting the deficiency in the number of simulators. The composite signal can be generated by multi-carrier modulation or formed by waveform sequences.

(1) Multi-carrier modulation

Multi-carrier signal refers to the signal containing multiple frequency components, and the parameters of each signal component (center frequency, bandwidth, modulation style, relative amplitude, etc.) can be set independently. Multi-carrier technology realizes the "one-to-many" of signal style and quantity and further realizes the "one-to-many" of the simulator. In a multi-carrier signal, each frequency component shares the transmitter resources at the same time, and the transmit power is distributed proportionally on each component.

At present, with the help of high sampling rate technology, digital signal sources can provide multi-carrier signals with frequency distribution up to several GHz to meet the requirements of broadband environment construction. Multi-carrier modulation effect is shown in Figure 2.

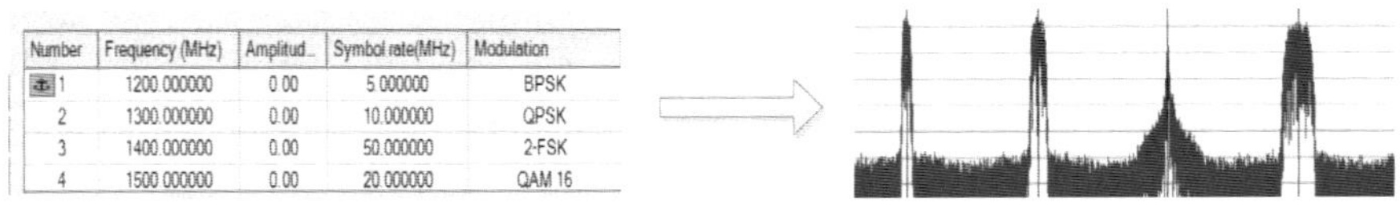

Multi-carrier setting Multi-carrier spectrum

Figure 2. The effect of multi-carrier modulation

(2) Waveform sequences

The editing tool of the digital signal can digitize the signal according to the modulation mode specified by the user, and the generated waveform file will be replaced by the digital signal source to output the specific signals. The waveform files of multiple signals are customized respectively, and the waveform files are arranged and replayed in a certain order to simulate the multi-signal environment. The waveform sequence can be cycled and jumped inside, and the frequency agility technology makes the waveform switch very fast in the digital signal source so that many signals can be sequentially transmitted, and further a single simulator can play multiple roles in a time-sharing manner and save the construction resources. The principle of waveform sequence playback is shown in Figure 3.

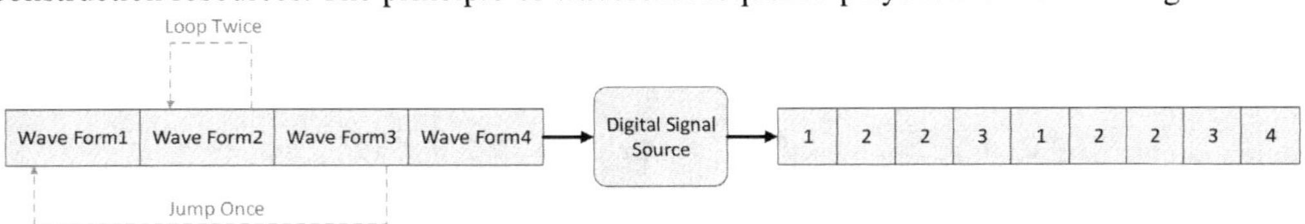

Figure 3. The principle of waveform sequence playback

In the composite signal composed of waveform sequence, the type and modulation mode of each component signal can be customized freely, such as pulse signal and continuous wave signal. Each component time-sharing monopolizes the transmitter resources in a time-sharing manner to achieve higher power radiation.

3.2. Construction method

To make full use of the limited construction resources, the environment is mainly based on the adversarial environment considering the correlation of equipment frequency. In other words, according to the frequency characteristics of the tested equipment, the interference environment is constructed in a targeted manner, and the interference frequency mainly covers the receiving radio frequency, intermediate frequency, and mirror frequency of the tested equipment[6][7].

The composite signal generator generates a variety of composite signals with a large number and wide frequency band, which are assigned to several transmitters deployed in a decentralized manner. Moreover, driven by different signal components, a single installed transmitter generates multiple virtual transmitters to complete the simulation tasks of multiple electromagnetic targets so that a complex electromagnetic environment can be constructed in a certain area. The schematic diagram of the construction method is shown in Figure 4.

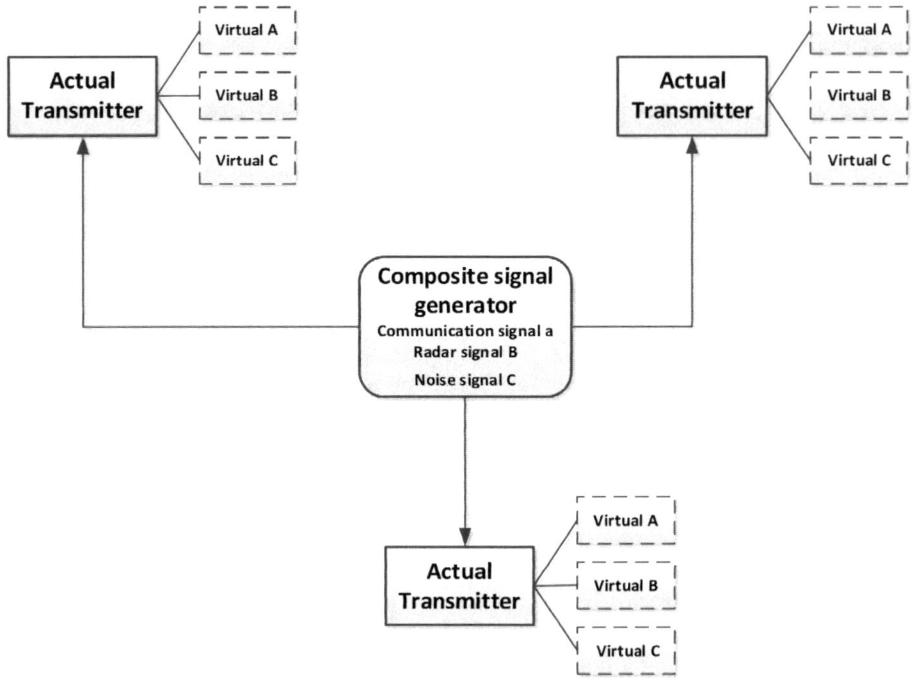

Figure 4. Schematic diagram of the construction method

The composite signal ensures that there are many types and quantities of signals with a wide frequency band, and through the distributed deployment of multiple transmitters, it is guaranteed that the transmitted signals overlap in space. The satisfaction that the construction method in this paper applies to the typical characteristics of a complex electromagnetic environment is shown in Table 1.

Table 1. Satisfaction of typical characteristics in the complex electromagnetic environment

Typical characteristics	The way to satisfaction
Signals with many types and quantities	Composite -signal technology
Wide spectrum occupancy width	Composite -signal technology
Interlaced overlapping in space	Decentralized deployment with multiple transmitters

4. Verification of construction method

Using the current mainstream instruments or equipment, a prototype system for constructing the complex electromagnetic environment is built, and considering the monitoring system, the construction method proposed in this paper is verified. The equipment used is shown in Table 2.

Table 2. Lists of equipment

Name of the equipment	Type
Digital signal generator	Tektronix AWG7122C, R&S SMW200A
Signal and spectrum analyzer	R&S FSW43

The equipment is connected as shown in Figure 5, and the digital signal source generates a composite signal radiated outward by the power amplifier and the antenna. In addition, the composite signal is composed of a digital modulation signal, analog modulation signal, pulse signal, white noise, etc. The monitoring system conducts spectrum scanning and further measures the parameters of each signal component. The measurement results are shown in Figure 6.

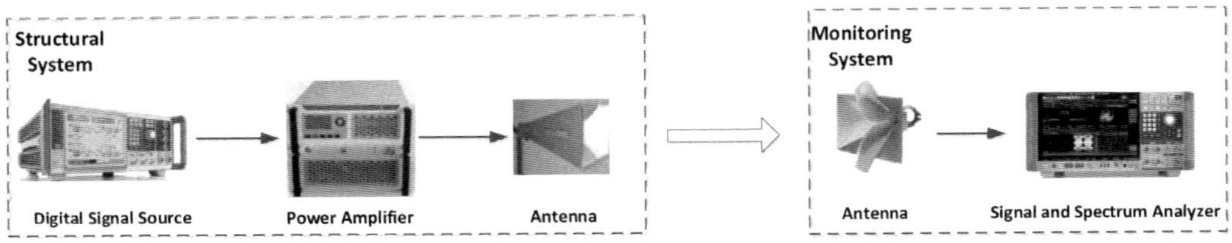

Figure 5. Diagram of the equipment connection

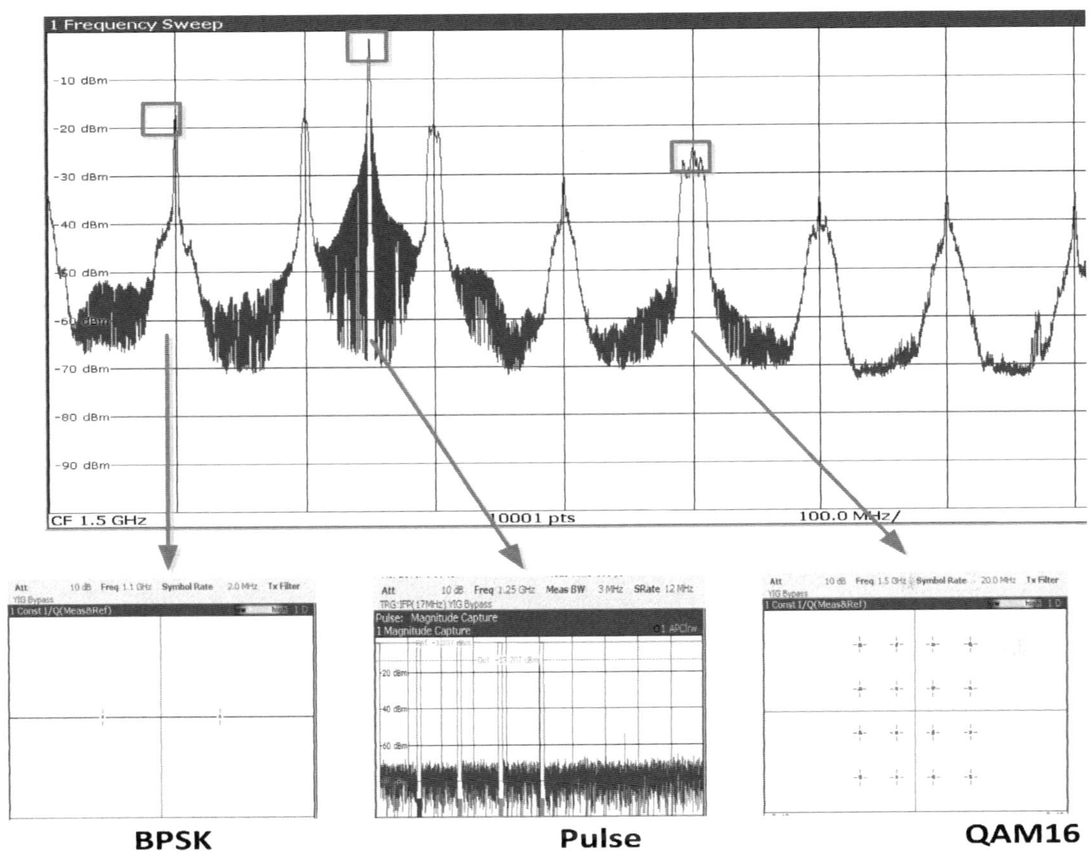

Figure 6. Results of spectrum monitoring and signal analysis

In summary, referring to the waveform sequence technology in the multi-carrier modulation domain of modern digital signal sources, a signal group containing hundreds of signals can be generated in a wider bandwidth (several GHz), and the signal parameters (frequency, bandwidth, modulation mode, amplitude, etc.) in the group can be independently set. In addition, the signal group radiates to space through the transmitter, and the electromagnetic environment with high complexity can be constructed in the region.

5. Conclusion

In terms of the difficulties such as the large demand for equipment and poor guarantee for complex terrain tasks when constructing complex electromagnetic environments, a construction method based on composite signals is proposed in this paper. In other words, multiple types and quantities of signal excitation are provided by composite signals to multiplex simulator resources and construct a complex electromagnetic environment. The new method proposed in this paper builds a simple construction system with the help of the existing instrument equipment. By monitoring and analyzing the constructed environment, it is found that the constructed environment meets some typical characteristics of the complex electromagnetic environment, which can effectively save valuable transmitter resources for the test.

References

[1] Zhu Y. G. (2012) *Construction of complex electromagnetic environment of shooting range. J.* Journal of Ordnance Equipment Engineering, 33(3).
[2] Zhang X., et al. (2013) *A construction method of complex electromagnetic environment. J.* Journal of Air Force Early Warning Academy, 27(5).
[3] Guo S. X., et al. (2013) *Research on simulation technology of complex electromagnetic environment. J.* Theory and Method, 32(7).
[4] Wu S. Y., et al. (2010) *Research on evaluation method of complex electromagnetic environment. J.* Informatization Research, 36(5).
[5] Wang J., et al. (2012) *Research on simulation technology of battlefield electromagnetic environment effect. J.* Electronic Measurement Technology,35(9).
[6] Wang R. Q. (2006) *Battlefield electromagnetic environment. M.* Beijing: The People' s Liberation Army Press, pp: 48-56.
[7] Wang L. D., et al. (2015) *Introduction to complex electromagnetic environment. M.* Beijing: National Defense Industry Press, pp: 94-101.

Construction and Simulation of Lanchester Battle Equations Based on Space-based Information Support

Yunzheng Liu[1a], Xiran Zhang[1b], Hongfeng Du[1], Gege Wang[1c] and Dexian Zeng[1*d]

[1] Space Engineering University, Beijing, 101400, China

[a] *Yunzheng Liu's e-mail: 2486829152@qq.com

[b*] Xiran Zhang's e-mail: 857093295@qq.com

[c*] Gege Wang's e-mail: 2506081376@qq.com

[d*] Dexian Zeng's e-mail: 2380152369@qq.com

Abstract: Modern warfare is a joint operation supported by space-based information, and space-based information is a "benefit multiplier" for joint combat effectiveness. In order to study the impact of space-based information support on joint operations, this paper improves the classical Lanchester equation [1] by introducing the space-based information support capability index, coordination index, decision-making index, etc. The Lanchester combat model was simulated and a scenario simulation was carried out. The simulation results show that the model can effectively describe the important support of space-based information for joint operations, and provides a theoretical reference for studying the winning mechanism of joint operations.

1. Introduction

With human's cognition and exploration of the space field, space-based information support has been widely used in joint operations, and the form of modern warfare has undergone qualitative changes. The main task of space-based information support is to use various types of spacecraft to provide battlefield situational information for joint operations, assist in combat mission decision-making, and coordinate troop actions. From the Gulf War, Kosovo War, and Afghanistan War to Russia-Ukraine special military operations, a large number of modern local wars have shown that space-based information support has been deeply integrated into all fields and levels of joint operations.

The Lanchester equation is a classical method to quantitatively study the principle of concentration of war and to find out the dominant factors in the confrontation process. Many military experts have improved and expanded on the basis of the classic Lanchester equation. Li H et al. [2] studied three applications of the Lanchester equation in war, mainly aimed at the key factors affecting the effectiveness of anti-terrorism operations, and initially established a mathematical model for predicting the outcome of anti-terrorist operations. Huang J C and Zhou D Y [3] carried out a forecast study on the process of UAV cooperative operation. On the basis of analyzing the characteristics of the UAV cooperative operation task, the Lanchester equation of UAV cooperative operation was designed. Gao X [4] proposes a discrete Lanchester model with support for the characteristics of modern naval warfare and deduces the mathematical expressions of different types of naval warfare models. Aiming at the problem that the land terrain has increased the constraints on combat in modern warfare, Feng H J et al, [5] constructed a land combat model reflecting terrain factors based on the classic Lanchester combat model, adding battlefield terrain factors.

Based on the analysis of the classic Lanchester linear law and square law, as well as many improved Lanchester equations, this paper believes that the current related equations are not enough to fully describe the combat style of modern battlefields. Therefore, this paper introduces space-based information by introducing space-based information. Support capability index, synergy factor, decision-making index, etc., the classical Lanchester equation has been appropriately improved, and a mathematical model that reasonably describes the winning mechanism of the joint combat battlefield under the condition of space-based information support is generated.

2. Lanchester Equation

2.1 The classical Lanchester equation

The classic Lanchester equation is mainly studied for two combat situations. The first situation is a linear law, which is mainly used in short-range line-of-sight combat. When the enemy positions carry out suppressive fire strikes, the changes in the loss of troops on both sides conform to a linear law [6]. Linear law is divided into first and second linear laws, wherein, the mathematical equation of the first linear law is:

$$\begin{cases} \dfrac{dy(t)}{dt} = -\alpha_l \\ \dfrac{dx(t)}{dt} = -\beta_l \end{cases} \tag{1}$$

In Equation (1), t represents the battle time, y(t) and x(t) represent the number of troops remaining in the battle at time t after the start of the battle, α_l、β_l respectively representing the proportion of the two sides killing each other's combat units in unit time (the damage coefficient).

The second linear law is the classic Lanchester guerrilla warfare model [7], and its applicable condition is that the enemy is exposed and our side is concealed. Without knowing the exact location of our combat units, the enemy launched an attack from the open area to our side in the bunker. At this time, the enemy is completely exposed to our field of vision, and we directly fire on the enemy and use the ignition power to fight. Its mathematical expression is:

$$\begin{cases} \dfrac{dm(t)}{dt} = -\beta_l n(t)\, m(t) \\ \dfrac{dn(t)}{dt} = -\alpha_l m(t) n(t) \end{cases} \tag{2}$$

In Equation (2), n (t) and m (t) are the configuration of the forces of the enemies β_l and α_l respectively; assuming that the number of initial combat units of the enemy and our side at the beginning of the battle is n(0)=N, m(o)=M, and the integral solution of Equation (2) can be obtained to obtain the state equation of both sides [8]:

$$\alpha_l \left[\mathrm{M} - m(t)\right] = \beta_l \left[N - n(t)\right] \tag{3}$$

From the state Equation (2), it can be inferred that when the relationship between the initial strength of the warring parties and the damage coefficient is satisfied $\alpha_l \boldsymbol{M} = \beta_l \boldsymbol{N}$, the two sides are tied, and there is no winner or loser; at that time, our side wins and the $\alpha_l \boldsymbol{M} < \beta_l \boldsymbol{N}$ $\alpha_l \boldsymbol{M} > \beta_l \boldsymbol{N}$ enemy is eliminated; upon victory, we are destroyed, as shown in Table 1:

Table 1 Combat Results (Linear Law)

The relationship between initial force and damage coefficient	battle result
$\alpha_l M = \beta_l N$	draw
$\alpha_l M < \beta_l N$	We win, the enemy loses
$\alpha_l M > \beta_l N$	The enemy wins, we lose

The second case is the square law, which is mainly used in long-distance over-the-horizon operations. The premise is that the enemy and us can see the target completely and can aim at the target, and the troops of both sides can fully utilize their numerical advantage to carry out long-range strikes against enemy firepower. The loss of troops on both sides follows the square law. Its mathematical equation is:

$$\begin{cases} \dfrac{dm(t)}{dt} = -\beta_s n(t) \\ \dfrac{dn(t)}{dt} = -\alpha_s m(t) \end{cases} \tag{4}$$

In Equation (4), n(t) and m(t) are the configuration of the forces of the enemy and us, respectively; β_s, α_s are the combat damage coefficients of the enemy and us, respectively. Assuming that the number of initial combat units on the enemy and our side at the beginning of the battle is n(0)=N, m(0)=M, and by integrating Equation (4), the state equations of the two warring parties can be obtained:

$$\alpha_s\left[M^2 - m(t)^2\right] = \beta_s\left[N^2 - n(t)^2\right] \tag{5}$$

From the state equation (4), it can be inferred that when the relationship between the initial force of the warring parties and the damage coefficient is satisfied $\alpha_s M^2 = \beta_s N^2$, the two sides are tied, and there is no winner; at that time, our side wins and the $\alpha_s M^2 < \beta_s N^2$ $\alpha_s M^2 > \beta_s N^2$ enemy is eliminated; upon victory, we are destroyed, as shown in table 2:

Table 2 Combat Results (Square Law)

The relationship between initial force and damage coefficient	Battle result
$\alpha_s M^2 = \beta_s N^2$	Draw
$\alpha_s M^2 < \beta_s N^2$	We win, the enemy loses
$\alpha_s M^2 > \beta_s N^2$	The enemy wins, we lose

3. Lanchester Joint Operations Equation with Space-Based Information Support

3.1 Lanchester Equation Command Information System Participation Model

Through in-depth analysis of the characteristics and laws of Lanchester's linear law and square law, combined with the characteristics and functions of information support, the command decision factor and information support factor are introduced into the classical Lanchester equation, and a command-and-information influence factor is established. Lanchester combat model [9]. Under this model, it can be considered that when both the enemy and our army have good information support capabilities, they can fully perceive the battlefield situation, and the two parties are in a situation of complete visibility, which satisfies the relevant conditions of Lanchester's square law; Neither of them has information support capabilities, and both the enemy and the enemy are in a completely concealed state on the battlefield, that is, they have no ability to perceive the battlefield situation, which satisfies the assumed conditions of Lanchester linear rate. The model can comprehensively reflect the significant impact of the command system's effectiveness and information acquisition capability on the three-dimensional battlefield. Its mathematical equation is:

$$\begin{cases} \dfrac{dm(t)}{dt} = -I_n \beta_s f(t) n(t) - (1 - I_n) \beta_l f(t) n(t) m(t) \\ \dfrac{dn(t)}{dt} = -I_m \alpha_s g(t) m(t) - (1 - I_m) \alpha_l g(t) m(t) n(t) \end{cases} \tag{6}$$

In Equation (6): I n and I m are the information support capability indices of the enemy and allies, respectively, $0 \le I\text{n} \le 1$, $0 \le I\text{m} \le 1$; n(t), m(t) are the military configuration of the enemy and allies, respectively situation; β_l、β_s respectively represent the damage coefficients of the enemy's exposed and concealed forces, α_l、α_s respectively represent our damage coefficients to the enemy's exposed and concealed forces, and $0 \le \beta_l \le 1$、$0 \le \beta_s \le 1$,$0 \le \alpha_l \le 1$,$0 \le \alpha_s \le 1$; f(t), g(t) are the command decision factors of the enemy and us, respectively.

3.2 The Lanchester equation for joint operations

In joint operations, both the enemy and our army have combat capabilities in different fields of sea, land, and air, and according to the degree of integration of the three armies of the enemy and us, there is a certain gap in the combat effectiveness of the two sides in the same dimension or in the same field. The Lanchester equation is expressed in It is established in the case of multi-dimensional space operations. In the case that both sides can attack each other's combat capabilities in different dimensions, only quantitative factors are considered, and non-quantitative factors, such as reinforcements and non-combat attrition on the battlefield, are not considered, which is in line with the basic assumptions of the Lanchester combat model.

The joint operation includes multiple static engagement planes and multiple linear confrontations, and each static engagement plane satisfies Lanchester's square law, and each linear confrontation satisfies Lanchester's linear law. Its mathematical equation is [10]-[12]:

$$\begin{cases} \dfrac{dm(t)}{dt} = -\beta n(t)^2 \\ \dfrac{dn(t)}{dt} = -\alpha m(t)^2 \end{cases} \tag{7}$$

In Equation (7): n(t) and m(t) are the configuration of the forces of the enemy and us, respectively; β and α are the combat damage coefficients of the enemy and us, respectively. It is assumed that the number of initial combat units of the enemy and allies at the beginning of the engagement is n(0)=N, m(0)=M.

3.3 Lanchester Equation Based on Space-Based Information Support

According to the analysis, space-based information support can improve the combat effectiveness of joint operations in the future informationized and intelligent battlefield, while the classical Lanchester linear law and square law cannot describe the impact of space-based information support on modern joint operations. In joint operations, the ability of space-based information support, the joint degree of the naval, land, and air forces, and the command and decision-making ability are the key factors that determine the future direction of the war [13]. The acquisition of space-based information is the basis for improving the effectiveness of the battlefield and the basis for commanding the actions of the troops; the degree of cooperation between the three armed forces directly depends on the decision-making and planning capabilities of the commander and the command organization and indirectly affects the trend of the battlefield situation [14]. Therefore, based on the above considerations, on the basis of ensuring that the model construction is relatively reasonable, the following assumptions are made:

1) For a certain period of time, the space-based information support capability has remained stable. Although the space-based information support capability may fluctuate due to the interference and confrontation of the enemy in the whole process of combat, it can be assumed to be stable within a certain period of time.

2) There is a certain influence between the various elements over time. The ability to obtain space-based information directly affects the commander's command and control ability, and indirectly affects the joint operations of the troops.

Based on the above assumptions and analysis, on the basis of the Lanchester equation for joint operations, the space-based information support capability index factor, coordination factor, and decision-making index factor are introduced to establish a Lanchester joint operations model based on space-based information support [15]. The pattern is shown in Figure 1:

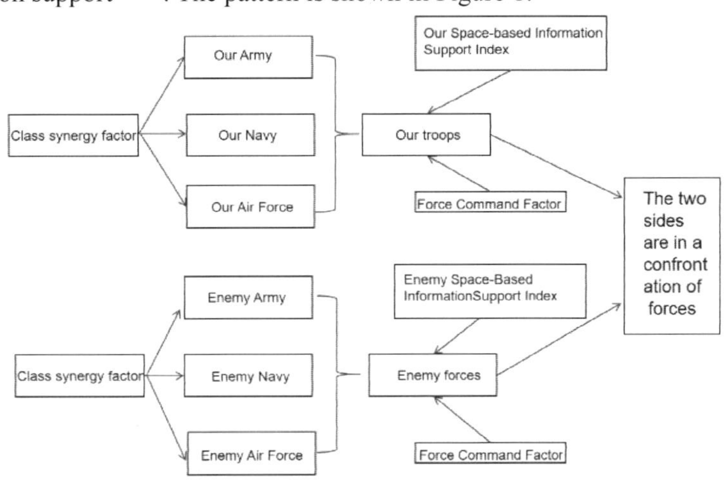

Figure 1 Joint operations model

The mathematical equation is obtained as follows (9):

$$\begin{cases} \frac{dm(t)}{dt} = -I_n \beta_s h(t) n(t)^2 - (1 - I_n) \beta_l h(t) m(t)^2 n(t)^2 \\ \frac{dn(t)}{dt} = -I_m \alpha_s k(t) m(t)^2 - (1 - I_m) \alpha_l k(t) n(t)^2 m(t)^2 \end{cases} \tag{8}$$

In the equation, I n and I m are the space-based information support capability indices of the enemy and allies, $0 \leq I\,\text{n} \leq 1$, $0 \leq I\,\text{m} \leq 1$; β_l、β_s and α_l、α_s are the damage indices of the enemy and our forces, respectively; n(t) and m(t) respectively represent the number of troops of both the enemy and allies that change with time; h(t) and k(t) are the command and control capability indices of the commanders of the enemy and allies, respectively, reflecting the influence of command decisions on operations. The mathematical expression is:

$$\text{h}(t) = \begin{cases} 0, 0 < t < t_1 \\ X_1, t_1 < t < T \end{cases} \tag{9}$$

$$\text{X}_1 = \begin{cases} \text{a}, t_1 < t < t_2 \\ \text{a+u}, t_2 < t < T \end{cases} \tag{10}$$

$$\mathrm{k}(t)=\begin{cases}0, 0<t<t_3\\ X_2, t_3<t<T\end{cases} \tag{11}$$

$$\mathrm{X}_2=\begin{cases}\mathrm{b}, t_3<t<t_4\\ \mathrm{b+v}, t_4<t<T\end{cases} \tag{12}$$

Among them, in Equations (9) and (11), t_1 and t_3 reflect the delay of the operation, and also reflects the delay of the combat actions of the enemy and the other side, and X_1, X_2 reflects the command ability of the commanders of the enemy and allies within a certain period of time. Combat effectiveness, T reflects the overall combat time. Due to the command and control ability of the commanders of both the enemy and the enemy, the change in the degree of the alliance of the three armed forces is caused. In Equations (10) and (12), t_2 and t_4 reflect the command and control delay of the enemy and allies, and a and b reflect the command ability of the enemy and allies. The constants, u and v are synergistic constants generated on the basis of the tri-army alliance.

In Equation (8), when I n = I m = 1, it means that both the enemy and the enemy have fully obtained space-based information support, and the information transparency is the highest, which means that both sides are completely exposed, which is in line with Lanchester's square law. When I n = I m = 0, it means that either side of the enemy or the enemy has not obtained space-based information support, and it can be considered that the two sides are completely concealed, which is in line with Lanchester's second linear law [16].

4. Model simulation and analysis

Under the premise of not considering the impact of the morale of the enemy and the reinforcement of troops, the simulation experiments are carried out in three hypothetical situations.

4.1 Simulation Assumption 1

Our space-based information support capability is superior to that of the enemy, and the enemy's military strength is superior to ours. The damage coefficient and command capability of the combat units on both sides are comparable. The impact of space-based information support capability and the number of troops on the combat process is analyzed.

Simulation conditions: our initial force M0=7 0, the enemy's initial force N0=10 0, the combat delay of the enemy and us t_1 = t_3 0, the command and control delay of the enemy and us t_2 = t_4 1, the command capability constant of the enemy and us a=b=1, the three-army coordination constant u=v=0.1 of the enemy and the enemy; the damage coefficient of the enemy and the enemy α_s = β_s =0.3, α_l = β_l=0.001, our space-based information support index Im is 0.8 and 0.5 respectively, and the corresponding enemy space-based information support index In is 0.1 respectively and 0.2, the simulation results are shown in Figure 2.1 and Figure 2.2.

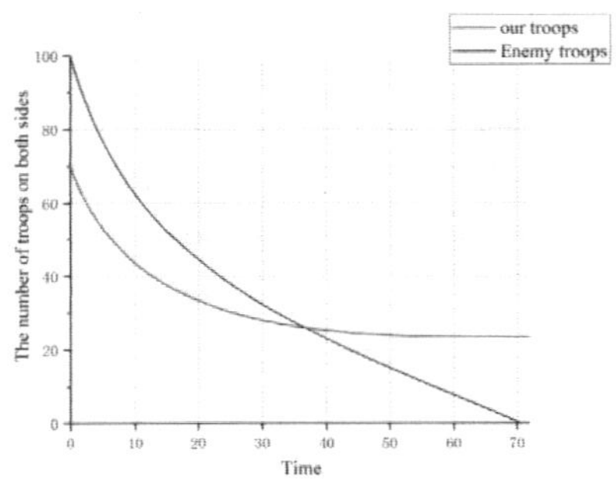

Figure 2 Im=0.8, In=0.1

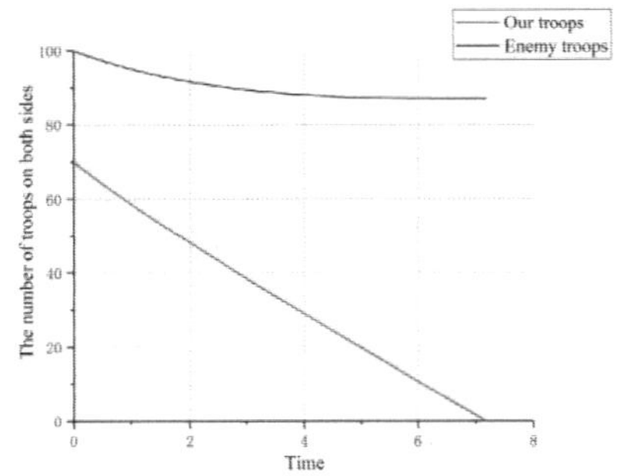

Figure 3 Im=0.5, In=0.2

As can be seen from Figure 2, when only our troops are at a disadvantage, and our command and coordination capabilities are basically the same as those of the enemy, but our space-based information support capabilities are slightly better than the enemy's, we lose 45 units of combat power, and the time used is about 80 Defeat the enemy in unit time. It can be seen that when the weaker party obtains the space-based information support capability, it achieves the goal of winning the weak with the weak, which further verifies the impact of space-based information support on the battlefield situation. However, as can be seen from Figure 3, when the information acquisition capabilities of the enemy and us are not much different, the future trend of the battlefield still depends on the initial number of troops of the enemy and us. Compared with Figure 2, it is found that when the enemy and our army are naturally When the basic information support capabilities are evenly matched, the enemy takes less than 8 units of time and consumes 15 units of troops to defeat us.

4.2 Simulation Assumptions 2

Simulation calculations are carried out according to the linear law of the Lanchester equation, the square law, and the Lanchester equation based on space-based information support, and the effects of different Lanchester equations on the combat process are compared.

Simulation conditions: our initial force M0=7 0, the enemy's initial force N0=10 0, the combat delay of the enemy and us $t_1 = t_3 0$, the command and control delay of the enemy and us $t_2 = t_4 1$, the damage coefficient of the enemy and us $\alpha = \beta 0.3$, the enemy command is always Quantity a = b = 0.7, the enemy's three-army coordination constant u = v = 0.07; our space-based information support index Im is 0.8 and 0.7 respectively. The enemy's space-based information support index In is 0.1, and the simulation results are shown in the figure.

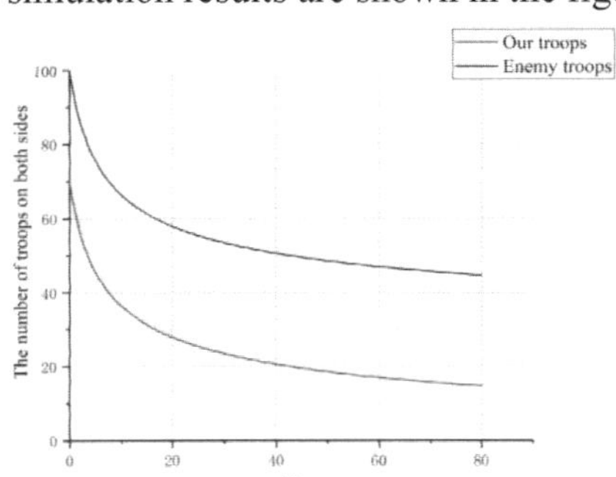

Figure 4 M0=70, N0=100, $t_1 = t_3 = 0$

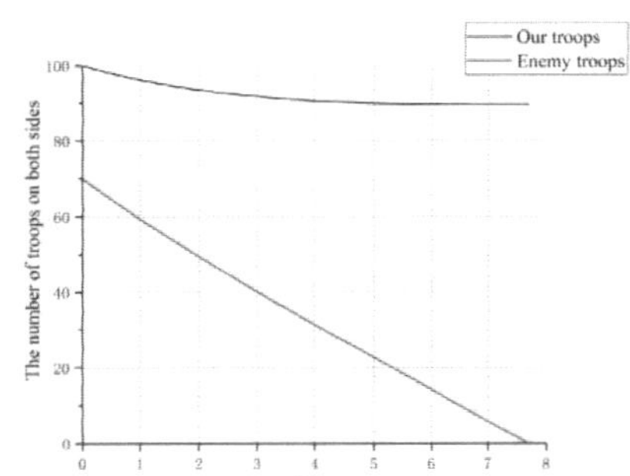

Figure 5 _ M0=70, N0=100, $t_2 = t_4 = 1$

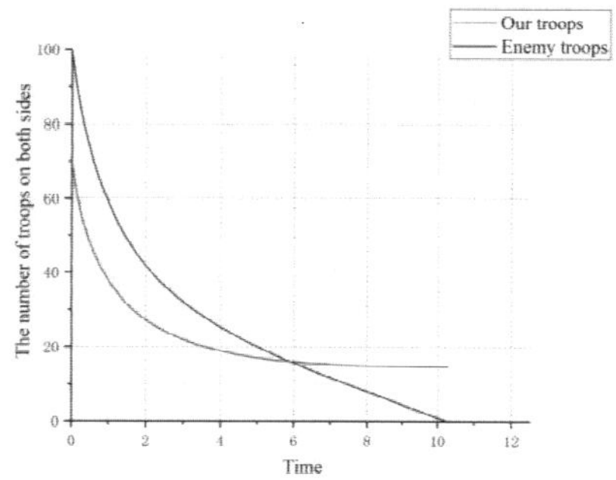

Figure 6 Im=0.8, In=0.1

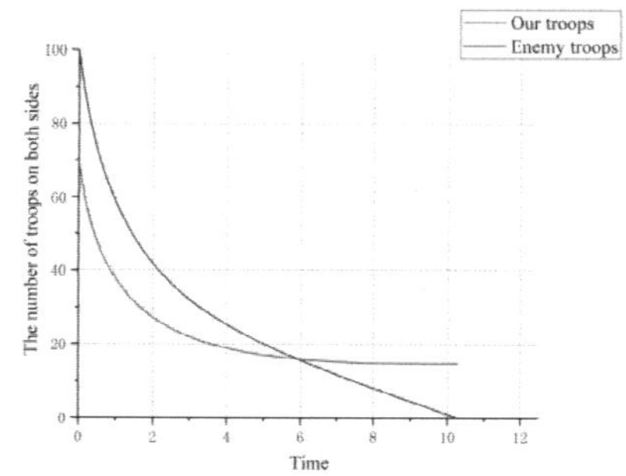

Figure 7 Im=0.7, In=0.1

From Figure 4 and Figure 5, we can see that our troops and command coordination capabilities are at a disadvantage, and we are defeated by the enemy without considering space-based information support capabilities. When considering space-based information support, our side has a strong space-based information support capability, which can make up for the shortage of troops and command decision-making to the greatest extent, and realize the reversal of the battle situation. It can be seen from Figure 6 and Figure 7 that when the enemy's space-based information support index remains unchanged, our space-based information support index decreases, and in the same period of time, the loss of our troops increases. The experimental results show that the space-based information support capability, as a support means on the modern battlefield, can provide the troops with strong combat power and help both sides of the war achieve related military goals.

5. Conclusion

The future battlefield puts forward higher requirements for combat simulation. To describe the war more accurately, more accurate mathematical equations are needed. Using mathematical models to analyze and predict the combat process has become an indispensable part of the future battlefield. In this paper, the Lanchester equation is improved, and the applicability of the model is further analyzed through three cases. It can be seen from the simulation results that the model has certain reference significance for the study of the new mechanism of joint operations in modern warfare, and can better reflect the relationship between the various elements of war. The organic combination of mathematical equations and actual combat to improve the training effect of troops has more important significance for joint operations under high-tech conditions.

References

[1] KAUP DJ. (2005) The Lanchester (n, 1) problem. Journal of the Operational Research Society, 56(12): 1399-1407
[2] Li H, Wu S, Wang G B. (2012) Research on Mathematical Model of Counterterrorism Based on Lanchester Mixture Law. Ship Electronic Engineering, 32(03):31-32+40.
[3] Huang J C, Zhou D Y. (2018) Lanchester equation design and combat process prediction for UAV cooperative combat. Electronics optics & control, 25(12):40-44.
[4] Gao X. (2018) Research on modern naval combat effectiveness based on Lanchester equation discretization. Command control & simulation, 40(05): 53-56.
[5] Feng H J, Zheng Y L, Yuan X M, Yang N. (2018) Combined with Lanchester equation to analyze the influence of battlefield terrain factors on war. National Defense Science and Technology, 2020,41(03):135-141.
[6] Ying L Z. (2020) A multi-domain system confrontation modeling method based on Lanchester equation. Chinese Institute of Command and Control. Proceedings of the 8th China Command and Control Conference. Chinese Institute of Command and Control:5.
[7] Yu B, Zhang Y H, Yu B C. (2018) Methods and applications of military operations research. Beijing: National University of Defense Technology Press.

[8] MACKAY N J. (2009) Lanchester models for mixed forces with semi-dynamical target allocation. Journal of the Operational Research Society, 60(10): 1421-1427
[9] DEITCHMAN S J. (1962) A Lanchester model of guerrilla warfare. Operations Research, 10(6): 818-827
[10] Bian L X, Luo X B, Liu G Q, Zhen J W. (2016) Lanchester Equation in SD-based Information Defense Operations. Firepower and Command and Control, 41(12): 105-107+112.
[11] Yang P, Zhao Z J, Ni X Q. (2016) Multi-dimensional combat space marine force combat simulation method. Command Control and Simulation, 38(02):76-79+114.
[12] Yang P, Ni X Q. (2016) Multidimensional Combat Power Index of Landing Operations—Lanchester Equation. Ship Electronics Engineering, 36(05):27-30.
[13] James G T. (2000) Lanchester-Type Models of Warfare. AD-A090842.
[14] Israel Mayk, Eng. Sc. D. (1998) The Command and Control Reference Model for Modeling, Simulations, and Technology Application. ADA278300.
[15] James M. (2006) Mathematical Modeling of Information Age Conflict. Journal of Applied Mathematics and Decision Sciences, 2006(3): 1-15.
[16] Cai L J, Dai G H, Xu X F, (2016) Ma Hualong. Lanchester combat model based on information-decision influencing factors. Command Control and Simulation,38(04):18-21.

Open-set HRRP recognition method based on joint sparse representation

Shuang Qu, Shengqi Liu*, Qiang Fu

National Key Laboratory of Science and Technology on Automatic Target Recognition, National University of Defense Technology

*Corresponding Author: Shengqi Liu, SQLiu@nudt.edu.cn

Abstract. Aiming at the problem of multi-view high-resolution range profile (HRRP) target recognition under open set conditions, we proposed an open set recognition method based on joint sparse representation (JSR), which solves the problem of low recognition rate of traditional methods under open set conditions. This method is applied to the background of radar single-station observation. JSR is used to solve the reconstruction error of multi-view HRRP by the over-complete dictionary, while extreme value theory (EVT) is used to model the reconstruction error tailing of matching and non-matching categories and transform the open set identification problem into a hypothesis testing problem. During recognition, we use the reconstruction error to determine the candidate class, the scores of the matching class and non-matching class are obtained according to the confidence of tail distribution, and the weighted sum of the two is used as the category criterion to finally determine the target or candidate class outside the library. This method can effectively use the relevant information between multi-view HRRPs to improve the performance of HRRP recognition under open set conditions. The algorithm is tested with HRRP data generated from MSTAR inversion, and the results show that the performance of the proposed method is better than the mainstream open set recognition method.

1. Introduction

High range resolution profile (HRRP) is the distribution of target echo obtained by broadband radar on different range units, reflects the scattering characteristics of radar targets under a certain angle of view, and contains important structural information of targets [1], which is very valuable for target recognition and classification.

Radar target recognition based on HRRP has produced rich research results [2-5], but the current results are mainly obtained under the assumption of closed sets, that is, assuming that the test samples are from the target categories contained in the training set, and then designing certain classification criteria to judge the test samples as the target categories with the best match. In practical application, the recognition system is often faced with non-cooperative targets. The samples to be recognized are likely to come from unknown categories outside the training set (open set). Traditional methods will judge the samples to be recognized as the target category with the highest built-in reliability in the training set, which will obviously get wrong results. Therefore, it is necessary to study the radar target recognition method under open sets.

Aiming at the problem of target recognition in an open environment, Scheirer and other scholars defined the problem of open set recognition (OSR) and established the theoretical framework of OSR [3]. Under this framework, scholars have proposed a series of algorithms, which can be roughly divided into the following categories:1) Based on support vector machine (SVM), such as W-SVM [4], PI-SVM

[5], etc.; 2) Based on sparse representation (SR), such as SR-OSR algorithm [6]; 3) Based on distance criteria, such as Nearest Non-Outlier[7], reverse k-nearest neighbor classifier[8], etc.; 4) Based on deep neural network (DNN), such as deep open classifier [9], category condition encoder [10], etc.; 5) Based on edge distribution, such as Extremum Value Machine algorithm[11]. Under the open set condition, the OSR method has achieved a good recognition effect, but the relevant achievements are mainly concentrated in the field of computer vision. The radar target recognition, especially the application of target recognition based on HRRP, is rarely reported.

In practical application, radar can obtain multiple HRRPs of targets within a certain angle range (such as multi-period observation obtained by radar continuously), and the comprehensive use of multi-observation information can improve the performance of target recognition. According to the characteristics of the radar target, the HRRP of the target in a small angle range is relevant. In order to make full use of the prior information related to multi-view HRRP in a small angle range, based on extreme value theory (EVT), we propose to apply the joint sparse representation (JSR) method to multi-view HRRP target recognition under open set conditions in the framework of OSR. EVT is used to fit the tail of reconstruction error obtained in the training stage, and the extreme value distribution of each training category is obtained. The placement reliability of the tail part of the test sample is taken as the score of the matching category and non-matched category, the final criterion value is obtained by weighted summation of the two scores, and then the target category is judged. The experimental results of the HRRP data set generated by MSTAR inversion show the superior performance of the algorithm.

2. Background

2.1. Joint Sparse Recognition

JSR is a sparse reconstruction algorithm for multiple observations. Its basic idea is to calculate the linear sparse approximation of each signal to ensure that the positions of non-zero elements in the sparse representation coefficients are the same.

The multi-observation signal is set to be $\mathbf{Y} = [\mathbf{y}_1, \ldots, \mathbf{y}_J] \in \mathbb{R}^{M \times J}$ (J is the number of observations), and JSR approximates $\mathbf{Y}$ by estimating the upper sparse representation coefficient $\boldsymbol{\alpha}$ of signals $\mathbf{Y}$ in dictionary $\mathbf{D}$, which represents the row sparse matrix. The above process can be expressed as:

$$\min \|\boldsymbol{\alpha}\|_{p,q} \quad s.t. \quad \|\mathbf{Y} - \mathbf{D}\boldsymbol{\alpha}\|_2^2 \le \varepsilon \tag{1}$$

Where $\|\boldsymbol{\alpha}\|_{p,q}$ is the mixed norm of $l_{p,q}$, which means that first take the l_p norm for the vector of row $\boldsymbol{\alpha}$, and then the l_q norm is taken for the result vector, that is:

$$\|\boldsymbol{\alpha}\|_{p,q} = \left(\sum_{i=1}^{N} \left(\|\boldsymbol{\alpha}(i,:)\|_q \right)^p \right)^{\frac{1}{p}} \tag{2}$$

To solute the sparse coefficient matrix, the multiple measurement vector (MMV) sparse reconstruction method can be used. Common methods include synchronous orthogonal matching pursuit (SOMP)[12] and compression sample match pursuit (Co- SaMP)[13].

2.2. Extreme Value Theory

Suppose the sequence of independent and identically distributed random variables $X_1, \ldots, X_n$ is set as the maximum value of $X_{n,n}$. Fisher-Tippett theorem[14]points out that if there is a constant column $a_n > 0$, b_n, the asymptotic distribution of $(X_{n,n} - b_n) / a_n$ is non-degenerate, that is:

$$P\left(\frac{X_{n,n}-b_n}{a_n}\le x\right)\to G(x), n\to\infty \tag{3}$$

Then, G must belong to the Frechet Pareto type, Gumbel type, or Weibull type distribution.

Fisher-Tippett theorem can be expressed by general extreme value distribution (GEV):

$$\mathrm{G}_\gamma\text{:}\ x\mapsto \exp\left(-(1+\gamma x)^{-\frac{1}{\gamma}}\right), \gamma\in\mathrm{R}, 1+\gamma x>0 \tag{4}$$

Where γ is the extreme value index, which depends on the distribution law of the original data. $\gamma>0$ corresponds to Frechet Pareto distribution, $\gamma<0$ corresponds to Weibull distribution, $\gamma=0$ corresponds to Gumbel distribution. γ can be estimated using the peaks over threshold (pot) method, which relies on the Pickands-Balkema-de Haan theorem[15], that is:

For a sufficiently large threshold value t, the distribution $\bar{F}_t(x)=P(X-t>x\mid X>t)$ of the excess value approximately obeys the general Pareto distribution (GPD):

$$\mathrm{G}_{\gamma,t,\sigma}(x)=\begin{cases}\left(1+\gamma\frac{x-t}{\sigma}\right)^{-\frac{1}{\gamma}} & \gamma>0, x>t\\ \exp\left(-\frac{x-t}{\sigma}\right) & \gamma=0, x>t\\ \left(1+\gamma\frac{x-t}{\sigma}\right)^{-\frac{1}{\gamma}} & \gamma<0, t<x<t-\frac{\sigma}{\gamma}\end{cases} \tag{5}$$

By selecting the threshold value t, the generalized Pareto distribution is estimated by using the observed value exceeding t. Record the overflow function $Y_j=X_j-t$, where $X_j>t$, $j=1,2,\ldots,N_t$ indicate that the jth overflow value is in the subscript corresponding to the X sequence, $N_t=\sum_{i=1}^{n} I(X_i>t)$, $I(\bullet)$ is an indicator function.

Its corresponding log maximum likelihood function is:

$$\log L(\sigma,\gamma)=-N_t\log\sigma-\left(\frac{1}{\gamma}+1\right)\sum_{i=1}^{N_t}\log\left(1+\gamma\frac{Y_j}{\sigma}\right) \tag{6}$$

Where $1+\gamma\frac{Y_i}{\sigma}>0, i=1,\ldots,N_t$. If $\gamma=0$, the likelihood function is:

$$\log L(\sigma,\gamma)=-N_t\log\sigma-\frac{1}{\sigma}\sum_{i=1}^{N_t}Y_j \tag{7}$$

According to the likelihood function, the estimated parameter $\hat{\sigma},\hat{\gamma}$ can be obtained.

3. Open-set HRRP recognition method based on joint sparse representation

For the identification under open set conditions, literature [6] describes the OSR problem as an open set risk function minimization problem:

$$\arg\min_f R_H(f)+\lambda R_\varsigma(f(\chi)) \tag{8}$$

Where f represents a measurable identification function, R_H and R_ς represent the corresponding open space risk and experience risk respectively, λ is a regularization constant, which is used to weigh R_H and R_ς. JSR uses the minimum reconstruction error to judge the category, the reconstruction error of the test sample for the corresponding category is much lower than that of other categories, so it can be used as the recognition function of open set multi-view HRRP classification.

For the input multi-view observation sample $\mathbf{Y}$, the joint sparse representation on the over-complete dictionary $\mathbf{D}$ can be modelled as an optimization problem of Formula (1). By solving the sparse coefficient $\boldsymbol{\alpha}$, the description error of the original signal is minimized. In this paper, the classical SOMP algorithm is used to solve Formula (1). After obtaining the sparse coefficient $\hat{\boldsymbol{\alpha}}$, the minimum reconstruction error criterion is used to determine the target category:

$$\hat{c} = \arg\min_{c\in\{1,\dots,C\}} \left\|\mathbf{Y} - \mathbf{D}\boldsymbol{\sigma}_c\left(\hat{\boldsymbol{\alpha}}\right)\right\|_2^2 \tag{9}$$

Where $\boldsymbol{\sigma}_\mathbf{c}\left(\hat{\boldsymbol{\alpha}}\right)$ means to retain the coefficient corresponding to class c in $\hat{\boldsymbol{\alpha}}$ and set other elements to zero.

Under the condition of an open set, it is necessary to ensure the minimization of the risk function of Formula (8) and the requirements of the identification function. This paper introduces the extreme value theory to describe the reconstruction error tailing of the output of the JSR algorithm. For the test samples, it is first to select the corresponding category of the minimum reconstruction error as the candidate category according to Formula (9), and then make the Kolmogorv-Smirnov statistical test for the reconstruction error of the samples to be identified according to the fitting results of the training stage, so as to obtain the confidence that it belongs to the corresponding tailing fitting distribution and judge whether it belongs to the candidate category or the target outside the library. Through the above steps, the open set identification problem is transformed into the following hypothesis testing problem:

$$\begin{aligned} &H_0 : G\left(r_c\right) \le \delta_g \\ &H_1 : G\left(r_c\right) > \delta_g \end{aligned} \tag{10}$$

Where the null hypothesis H_0 implies that the test data belongs to the matched class, the alternative hypothesis H_1 implies that the test data belongs to the non-matched class, $G\left(r_c\right)$ is the fitting of the tail of reconstruction error r_c, and δ_g is the rejection threshold.

3.1. Training

In the training phase, we preprocess the training sample $\mathbf{Y}_{tr} = [\mathbf{Y}_1, \dots, \mathbf{Y}_C]$, and first construct an over-complete dictionary. The dictionary construction method can adopt the manual construction method, the Meta faces learning method[16], etc. It is required to build various sub-dictionaries $\mathbf{D}_i \in \mathbb{R}^{N_i\times T}, i = 1,\dots,\mathrm{C}$ and integrate them into an over-complete dictionary $\mathbf{D}, \mathbf{D} = \left[\mathbf{D}_1, \mathbf{D}_2, \dots, \mathbf{D}_C\right]$. Considering that, the non-matched information can be used to strengthen the matched information, the matched reconstruction error $R_c^m = R\left(:,c\right)$ and the sum of non-matched reconstruction error $R_c^{nm} = \sum_{i:i\neq c} R\left(:,i\right)$ of each type of target are obtained respectively, and the tail size ρ is selected to obtain the corresponding threshold value $\mathrm{t}_c^m, \mathrm{t}_c^{nm}$. Since the data only overlap at the right tail of the matched reconstruction error and the left tail of the sum of the non-matched reconstruction error, the GPD distribution is fitted to the right tail of R_c^m and the left tail of R_c^{nm} respectively, and the corresponding extreme value parameters are obtained according to Formulas (6) and (7).

Algorithm 3.1: JSR-OSR training algorithm
1 Input: training sample $\mathbf{Y}_{tr} = [\mathbf{Y}_1, \dots, \mathbf{Y}_C]$, Label set $L_{tr} = [L_1, \dots, L_C]$, Dictionary matrix $\mathbf{D}$, Sparsity K, Number of multiple views n, tail size ρ

2 Divide the training sample into multi-view data $\mathbf{Y}_{11}=\left[\mathbf{y}_{11},\mathbf{y}_{12},\ldots,\mathbf{y}_{1(1+n)}\right]$, $L_{11}=\left[l_1,l_{12},\ldots,l_{1(1+n)}\right]$, …, $\mathbf{Y}_{ci}=\left[\mathbf{y}_{ci},\mathbf{y}_{c(i+1)},\ldots,\mathbf{y}_{c(i+n)}\right]$, $L_{ci}=\left[l_{ci},l_{c(i+1)},\ldots,l_{c(i+n)}\right]$

3 Calculate recognition error $R\leftarrow somp\left(\mathbf{Y},l,\mathbf{D},K\right)$

4 The matched recognition error: $R_c^m=R\left(:,c\right)$; the sum of non-matched recognition error: $R_c^{nm}=\sum_{i:i\neq c}R\left(:,i\right)$

5 Fitting tail extreme value parameters: $\mathrm{GPDfit}\left(R_c^m,\rho\right)\rightarrow\hat{\sigma}_c^m,\hat{\gamma}_c^m$

$$\mathrm{GPDfit}\left(-R_c^{nm},\rho\right)\rightarrow\hat{\sigma}_c^{nm},\hat{\gamma}_c^{nm}$$

6 Output the extremum parameter $\hat{\sigma}_c^m,\hat{\gamma}_c^m,\hat{\sigma}_c^{nm},\hat{\gamma}_c^{nm}$

3.2. Testing

In the testing phase, for a group of multi-view data $\mathbf{Y}_j=\left[\mathbf{y}_{j1},\mathbf{y}_{j2},\ldots,\mathbf{y}_{j(1+n)}\right]$ to be identified, the reconstruction error corresponding to the dictionary $\mathbf{D}$ is obtained by JSR. The traditional closed set method takes the class c corresponding to the minimum reconstruction error as the output class. In the open set environment, it also needs to consider the situation that the target comes from the class outside the library. We take class c as a candidate class to further determine whether it belongs to the category outside the library. According to the parameters $\hat{\sigma}_c^m,\hat{\gamma}_c^m,\hat{\sigma}_c^{nm},\hat{\gamma}_c^{nm}$ obtained in the training stage, the confidence of the corresponding tail distribution is obtained as the probability score $score_m$, $score_{nm}$ of matched class and non-matched class. The final discriminant score is obtained by using the weighted sum of two probability scores

$$score=score_m+w\cdot score_{nm} \tag{11}$$

Where $w=\mathrm{p}\bullet\left(1-\mathrm{openness}\right)$ is the weight, $0<\mathrm{p}\leq 1$, $\mathrm{openness}=1-\sqrt{\frac{2\times N_{TR}}{N_{TG}+N_{TE}}}$, N_{TR} represents the number of training categories, N_{TG} represents the number of target categories, and N_{TE} represents the number of table test categories [6].

By setting the decision threshold δ_g, the target category can be judged

$$\begin{cases} score>\delta_g, & newclass \\ score\leq\delta_g, & c \end{cases} \tag{12}$$

Algorithm 3.2: JSR-OSR classification method

1 Input: $\mathbf{Y}_j$, $\mathbf{D}$, K, $\hat{\sigma}^m,\hat{\gamma}^m,\hat{\sigma}^{nm},\hat{\gamma}^{nm}$, δ_g, w

2 Calculate recognition error $R\leftarrow\mathrm{somp}\left(\mathbf{Y}_j,\mathbf{D},K\right)$

3 Get candidate class $c^*=\arg\min_i R_i$

4 $r^m=R_{c^*}^m, r^{nm}=\sum_{i\neq c^*}R_i$

5 $score_m = G\left(r^m; \hat{\sigma}^m\left(c^*\right), \hat{\gamma}^m\left(c^*\right)\right),$
$score_{nm} = G\left(r^{nm}; \hat{\sigma}^{nm}\left(c^*\right), \hat{\gamma}^{nm}\left(c^*\right)\right)$

6 $score = score_m + w \cdot score_{nm}$

if $score > \delta_g$

$Y_j \in$ openset

else

$\text{Class}_{y_j} = c^*$

7 Output the class of Y_j

3.3. Performance evaluation

In this paper, the performance of the algorithm is comprehensively evaluated by means of accuracy and F value. Accuracy is defined as follows:

$$\text{accuracy} = \frac{\text{TP+TN}}{\text{TP+TN+FP+FN}} \tag{13}$$

Where TP, TN, FP, and FN are True Positive, True Negative, False Positive, and False Negative.

F-measure is expressed by accuracy and recall:

$$\text{F} = 2 \cdot \frac{\text{recall} \cdot \text{precision}}{\text{recall} + \text{precision}} \tag{14}$$

Where $\text{recall} = \frac{\text{TP}}{\text{TP}+\text{FN}}$, $\text{precision} = \frac{\text{TP}}{\text{TP}+\text{FP}}$.

4. Experimental Result

In this section, we use the HRRP data set retrieved from MSTAR to verify the effectiveness of the proposed algorithm. To reflect the effectiveness of the algorithm, three mainstream open set recognition methods are selected for comparison, including 1-vs-set [2], WSVM [3], and SR-OSR [5].

4.1. Data description

Due to the lack of HRRP measured data, we use MSTAR public data set to generate HRRP data from SAR data inversion for experiments. MSTAR includes 10 types of ground targets, namely BMP2, btr70, T72, BTR60, 2S1, BRDM2, D7, T62, ZIL, and ZSU. In the experiment, 17 ° data were used for training and 15 ° data were used for testing. The process of generating HRRP data by SAR inversion is shown in Fig. 1[17].

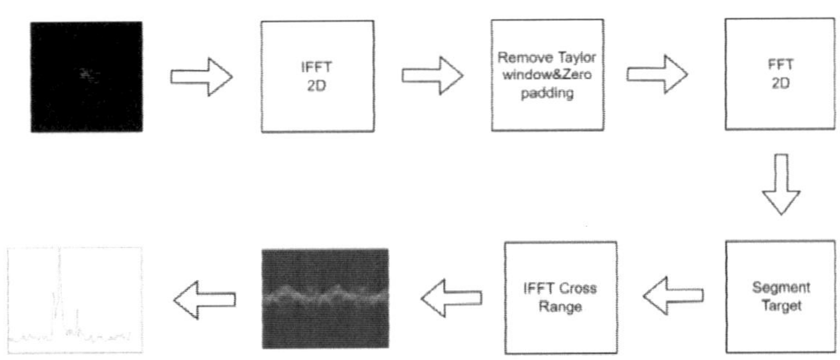

Figure 1. Flow chart of SAR inversion HRRP data

4.2. Experimental arrangement

After generating HRRP data through inversion, the incoherent average of HRRP within a certain angle range is taken to relax the attitude sensitivity of HRRP, the 2-norm amplitude normalization is taken to eliminate the intensity sensitivity, and the power transformation is performed on the data to enhance the data characteristics. In this paper, we use the target training samples to construct an over-complete dictionary. The HRRP data generated from each SAR image with an elevation angle of 17 ° is extracted at an azimuth interval of 3 ° and incoherently averaged to generate one HRRP data, which is sorted in azimuth order to build an over-complete dictionary.

In the training phase, the SAR image at a 17 ° elevation angle is used to retrieve the range profile data, and in the test phase, the SAR image at a 15 ° elevation angle is used to retrieve the range profile data. The experimental parameters are set as the number of multi-view observations $J=5$, the sparsity of SOMP $K=3$, the tail size $\rho=0.01$, and the decision threshold $\delta_g=0.9$.

4.3. Experimental results

In the experiment, BMP2, BTR70, and T72 were selected as the targets in the library for training, and the other 7 types of targets outside the library were added to the test stage (10 types in total). Table 1 shows the experimental results of JSR-OSR and the recognition accuracy of different algorithms. The first column represents the real category corresponding to the test sample, and the first row represents the output category of the classifier.

Table 1. JSR-OSR experimental results

	BMP2	BTR70	T72	Novel class
BMP2	0.890	0.026	0.031	0.053
BTR70	0.046	0.935	0.010	0.009
T72	0.005	0.005	0.985	0.005
BTR60	0.021	0.049	0.015	0.915
2S1	0.024	0.040	0.069	0.867
BRDM2	0.048	0.045	0.040	0.867
D7	-	0.010	0.048	0.942
T62	0.005	-	0.032	0.963
ZIL	-	-	-	1
ZSU	-	-	-	1
Average recognition rate:0.934				
Algorithms	JSR-OSR	SR-OSR	OneclassSVM	WSVM
Average recognition rate	0.934	0.881	0.784	0.786

According to the results shown in the table, JSR can improve the recognition accuracy by making use of the correlation between multiple observations, and the extreme value theory improves the classifier's judgment on extreme data by fitting tailing, so our method can obtain the best results among the four algorithms. According to the experimental results, the method in this paper makes use of the advantages of the JSR algorithm that can fully absorb the prior information related to multiple observations to improve the recognition performance. The experimental verification of HRRP recognition is carried out under the open set condition, and the highest F-measure and accuracy are achieved in all four algorithms, further illustrating the effectiveness of the proposed method.

5. Conclusion

In order to make effective use of the relevant prior information among HRRP multiple observations, this paper uses JSR to reconstruct the signal and uses extreme value theory to describe the error tail distribution. JSR is applied to multi-view recognition under open set conditions. The experimental results on the MSTAR dataset show that the proposed method is superior to many open-set recognition methods.

The method in this paper is to improve the target recognition performance under the open set condition by using the prior information with the correlation between adjacent multiple observations. In the case of large angle differences obtained by multi-view observation (such as multi-station radar observing the same target), there may be a great difference between different observations. At this time, the JSR-OSR algorithm is no longer applicable. A possible solution to this problem is to extend the JSR method so that the algorithm has certain robustness.

Acknowledgments
This work was partially supported by the National Science Foundation of China (No. 62001486).

References

[1] Li, Z. X., Xi'an University of Electronic Science and technology, (2021). Research on online learning method of radar high resolution range profile target recognition. Doi: 10.27389/d.cnki.gxadu.2021.001762
[2] Liu, J., Fang, N., Wang, B. F., & Xie, Y. J. (2016). Scale-space theory-based multi-scale features for aircraft classification using hrrp. Electronics Letters, 52(6), 475-477.
[3] Scheirer, W. J., Anderson, D. R. R., Sapkota, A., & Boult, T. E. (2013). Toward open set recognition. Pattern Analysis & Machine Intelligence IEEE Transactions on, 35(7), 1757-1772.
[4] Scheirer, W. J., Jain, L. P., & Boult, T. E. (2014). Probability models for open set recognition. IEEE Transactions on Pattern Analysis and Machine Intelligence, 36(11), 2317-2324.
[5] Jain L P., Scheirer W J., Boult T E. (2017). Multi-class open set recognition using probability of inclusion. In European Conference on Computer Vision, 393–409.
[6] Zhang, H., & Patel, V. M. (2017). Sparse representation-based open set recognition. IEEE Transactions on Pattern Analysis and Machine Intelligence, PP (8), 1-1.
[7] Bendale, A., & Boult, T. (2015). Towards Open World Recognition. Computer Vision & Pattern Recognition (pp.1893-1902). IEEE.
[8] Sadhukhan, P. (2020). Can reverse nearest neighbors perceive unknowns?. IEEE Access, PP(99), 1-1.
[9] Shu, L., Xu, H., & Liu, B. (2017). Doc: deep open classification of text documents. Conference on Empirical Methods in Natural Language Processing, 2905-2910.
[10] Oza, P., & Patel, V. M. (2019). C2AE: Class Conditioned Auto-Encoder for Open-Set Recognition. 2019 IEEE/CVF Conference on Computer Vision and Pattern Recognition (CVPR). IEEE.
[11] Rudd, E. M., Jain, L. P., Scheirer, W. J., & Boult, T. E. (2015). The extreme value machine. IEEE Transactions on Pattern Analysis and Machine Intelligence, PP (99), 1-1.
[12] Zhu, H., Chen, W., & Wu, Y. (2020). Efficient implementations for orthogonal matching pursuit. Electronics, 9(9), 1507.
[13] Huang, F., Tao, J., Xiang, Y., Peng, L., & Wang, L. (2017). Parallel compressive sampling matching pursuit algorithm for compressed sensing signal reconstruction with opencl. Journal of Systems Architecture, 72(C), 51-60.
[14] Pappada, R., Perrone, E., Durante, F., & Salvadori, G. (2016). Spin-off extreme value and archimedean copulas for estimating the bivariate structural risk. Stochastic Environmental Research and Risk Assessment, 30(1), 327-342.
[15] K Underlíková, & R Bartková. (2018). The Pickands–Balkema–de Haan theorem for intuitionistic fuzzy events.
[16] Liu, Z. Jiangnan University, (2021). Research on some new methods of image classification based on sparse representation. Doi: 10.27169/d.cnki.gwqgu.2021.001957
[17] Miao, W. X. Harbin Institute of technology, (2019). Research on SAR image echo inversion and typical target feature extraction. Doi: 10.27061/d.cnki.ghgdu.2019.001165

Design of a Miniaturized Cavity Bandpass Filter

Shanshan Du, Qinghui Yang, Ming Wang, Xinan Fan

State Key Laboratory of Electronic Thin Films and Integrated Devices, University of Electronic Science and Technology of China, Chengdu 610054, China

yangqinghui@uestc.edu.cn

Abstract. A novel design of a miniaturized cavity filter structure is proposed in this paper. The filter adopts a combination of interdigital and combines coupling structures instead of the traditional filter structure to reduce the filter size while ensuring excellent performance. The size is further reduced by adding metal rectangular blocks on the filter covers to increase the capacitance between the open end of the resonant rod and the ground and the coupling between the resonant rods. The critical parameters were designed and optimized by CST Studio Suite. Consequently, a nine-stage bandpass filter was fabricated and realized at 4.75 GHz with a fractional bandwidth of 10.5% and insertion loss of less than 0.7 dB.

1. Introduction

Filters effectively block out unnecessary frequency signals in circuits, such as spurious and harmonic signals, which are present in almost all microwave communication systems of receivers and transmitters. In order to meet the system requirements for miniaturized, it is necessary to overcome the contradiction between size and performance to maximize the structural optimization of the components [1].

Many methods for reducing filter size have been published. Chen Yi-Ming et al. [2] presented an improved step-impedance combining bandpass filter that uses an array of step-impedance resonators with tapped transformer coupling at both the input and output. Maharjan et al. [3] proposed a novel microstrip cross-coupled interdigital half-wavelength stepped impedance resonator bandpass filter. However, microstrip line filters are not easy to tune and have low power capacity and high insertion loss. Substrate integrated waveguide is a popular research direction for filter miniaturization [4]. This type of filter has lower insertion loss and higher quality factor compared with microstrip line filters, but it is less used in practical engineering applications. The cavity filter has the advantages of high Q factor, high suppression of stopband, small insertion loss, and high power capacity. However, cavity filters are generally large in size, and miniaturization has been the focus of the research. In [5], coaxial cavity filters based on SIR structure are designed to reduce the size. In addition, the miniaturization of filters with waveguide or cavity structures using the capacitive loading technique is one of the effective methods [6]. All these methods are relatively effective in reducing the size, but they are more complicated in design and fabrication.

A new compact filter structure is proposed in this paper. The first two quarter-wavelength resonators are interdigital structures, and the remaining resonators are combined structures. The metal rectangular blocks are arranged on the filter covers corresponding to the open end of each resonant rod to increase the capacitance between the resonant cavity and the resonant rod. This structure achieves high suppression and effectively reduces the size.

2. Filter Design

Figure 1 shows the proposed cavity bandpass filter structure, which includes nine rectangular resonant rods in a metallic cavity. The tuning screws corresponding to each resonant rod are employed to adjust the overall performance of the filter. The structure is centrosymmetric, with an interdigital coupling structure for the first two and the last two resonant rods and a combined coupling structure for the rest. Each resonant rod has corresponding metal rectangular blocks on both sides of the open-circuit end.

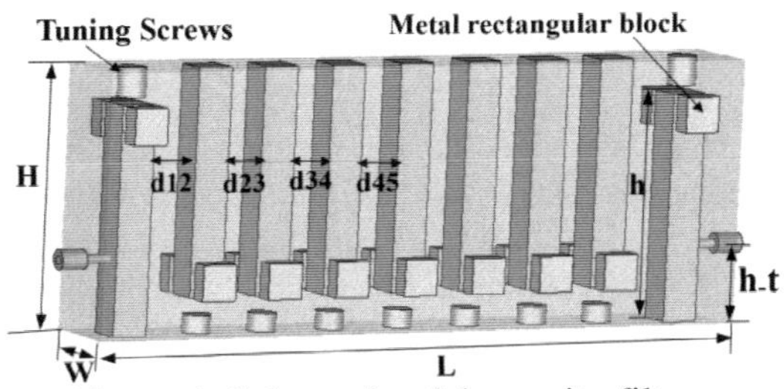

Figure 1. Schematic of the cavity filter.

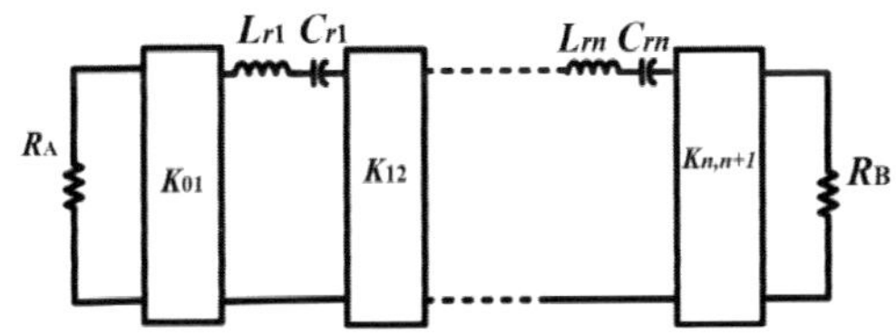

Figure 2. The equivalent circuit of the filter

The bandpass filter is designed to have a center frequency of 4.75GHz, a fractional bandwidth of 10.5%, an insertion loss of 1dB, and a return loss of less than -15dB. The Chebyshev-type response is employed. The coupling coefficients($k_{i,i+1}$) and loaded quality factors(Q_l) are given as follows [7].

The equivalent circuit of the bandpass filter is shown in Figure 2. L_{rj} and C_{rj} are the series inductance and capacitance of the resonator. ω_0 is the center frequency of the filter. The reactance slope parameter of the circuit when resonance occurs is

$$\chi_j=\omega_0 L_{rj}=\omega_0(C_{rj})^{-1} \tag{1}$$

The impedance of the impedance converter is

$$K_{01}=K_{n,n+1}=(R_A\chi_1\chi_1)^{1/2}(g_{0(n)}g_{1(n+1)})^{-1/2} \tag{2}$$

$$K_{j,j+1}=w(\chi_j\chi_{j+1})^{1/2}(g_jg_{j+1})^{-1/2} \tag{3}$$

where g_0 to g_n are the normalized lowpass component values (i=1,2,3.... n), and w is the fractional bandwidth. Figure 2 shows that the inductive reactance of the first resonator is χ_1 and the resistance coupled to this resonant circuit is $K_{01}{}^2/R_A$, so the loaded Q factors of the first and last resonators are

$$Q_l=\chi_1 R_A K_{01}{}^{-2}=\chi_n R_B K_{n,n+1}{}^{-2}=g_{0(n)}g_{1(n+1)}w^{-1} \tag{4}$$

The coupling between the two resonators is carried out by $K_{j,j+1}$. The coupling coefficient is

$$k_{i,i+1}=K_{j,j+1}(\chi_j\chi_{j+1})^{-1/2}=w\,(g_ig_{i+1})^{-1/2}=(f_2^2-f_1^2)(f_2^2+f_1^2)^{-1} \tag{5}$$

where f_1 and f_2 are frequency values obtained using the coupled resonator model simulation.

The center frequency of the filter depends on the height of the resonant rods. CST is used to build a single resonant cavity with an air cavity length and width of 6mm and a resonant rod length and width of 2 mm. The desired resonant frequency is obtained by adjusting the height of the resonant rod. As shown in Figure 3, the capacitance value between the open end of the resonant rod and the ground is increased by adding metal rectangular blocks at the cavity positions corresponding to both sides of the open end of the resonant rod. The resonant frequency is lower for the same resonant rod height. The size of the metal rectangular block can be adjusted according to different design frequencies.

The loaded Q factor depends on the height of the input taps, as shown in Figure 4. The resonant frequency of the two resonant rods near the input and output will become smaller after the microwave excitation signal is added. The resonant frequency can be adjusted without changing the size of the cavity and resonant rods by the size of the metal rectangular blocks. Finally, the size of the corresponding metal rectangular blocks at the first and the last resonant rod is set to 2mm×2mm×1.1mm, and the dimensions of the remaining metal rectangular blocks are 2mm×2mm×1mm.

The filter coupling bandwidth depends on the distance between the resonant rods. The interdigital and combined coupling models with metal rectangular blocks are shown in Figures 5(a) and 6(b), which were used to simulate the coupling bandwidth. Figure 6 shows the coupling bandwidth of the interdigital and combined coupling structures versus the distance for the same resonant rod size. The coupling bandwidth of the combined structure is smaller than that of the interdigital structure for the same distance. In addition, the coupling bandwidth of the combined coupling structure with metal rectangular blocks is larger than that without metal rectangular blocks at the same distance due to the coupling between adjacent metal rectangular blocks. According to the filter bandwidth, although the distance between the resonant rods using the combined structure is smaller compared to the interdigital structure, the coupling bandwidth between the first and the second resonant rods is large, and the distance is close, which makes manufacturing difficult. Therefore, the first two resonant rods adopt the interdigital structure, and the rest adopt the combined structure, which can ensure the filter is reduced in size and easy to fabricate.

The filter model was built based on the initial dimensions, as shown in figure 1. The filter was optimized by adjusting the depth of the tuning screws reaching into the cavity, the distance between the resonant rods, and the height of the taps. The final optimized parameters are shown in Table 1.

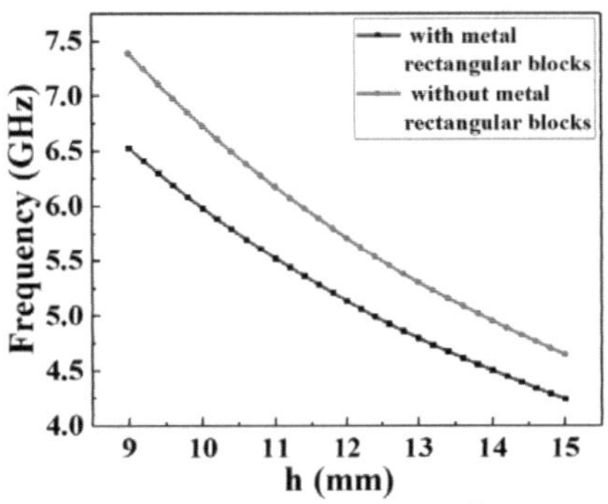

Figure 3. Simulated resonant frequency depends on different h.

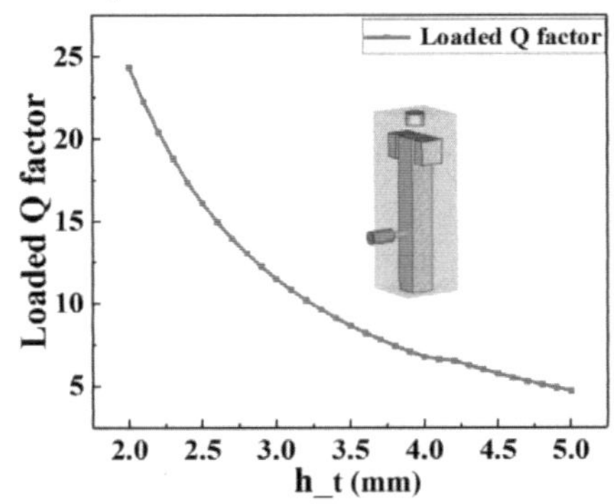

Figure 4. Simulated loaded quality factor depends on different h_t.

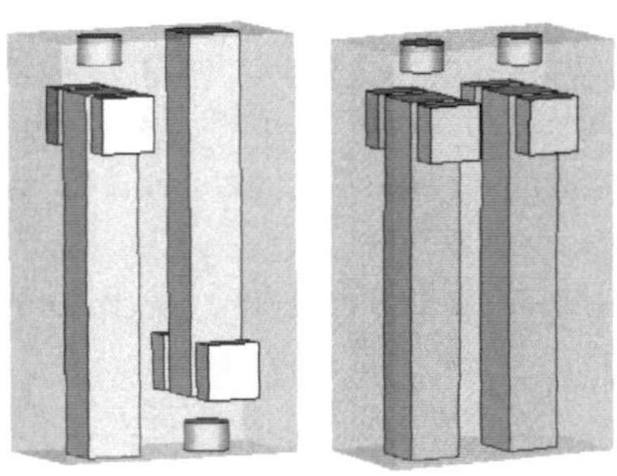

(a) (b)

Figure 5. (a) Interdigital structure model, (b) Combline structure model.

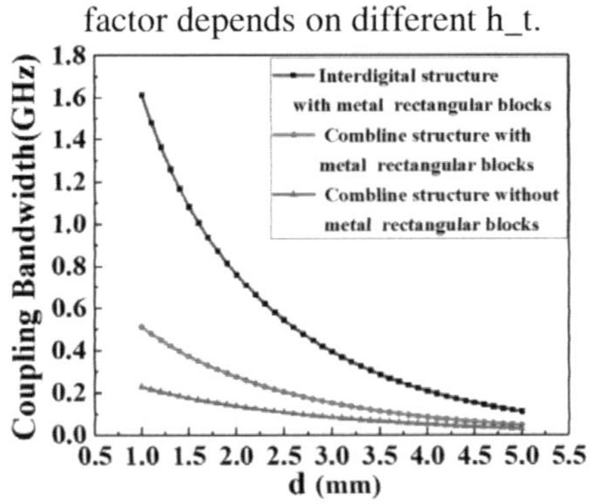

Figure 6. Simulated coupling bandwidth depends on different d.

Table 1. Filter Physical Parameters.

Parameters	W	L	H	h	h_t	d12	d23	d34	d45
Dimensions(mm)	5	39.04	14.65	12.65	4.15	2.63	2	2.16	2.23

3. Filter Fabrication and Experimental Results

The fabricated filter is shown in Figure 7. The figure shows the internal cavity structure with the filter cover removed. Metal rectangular blocks are on both sides of the filter covers. The whole cavity surface is silver-plated to improve the insertion loss performance of the filter.

The simulated and measured S-parameters of the bandpass filter are shown in figure 8. All the measurements are taken using a vector network analyzer (ZNB40). The frequency of the filter is 4.5-

5GHz, the return loss of the filter is greater than 17dB, the insertion loss is about 0.7dB, the out-of-band suppression is about 116dBc near 2.4GHz, and the out-of-band suppression is about 109dBc near 7.2GHz. Compared to the interdigital filter with the same performance specifications, the filter size is 45 mm*8 mm*22.5 mm, reducing the size by 30%.

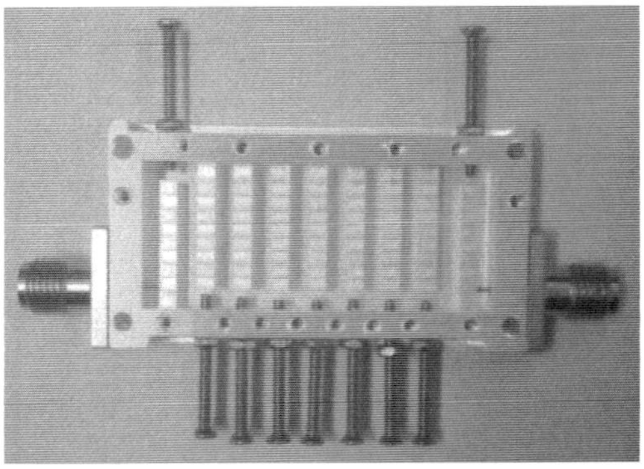

Figure 7. The fabricated cavity filter

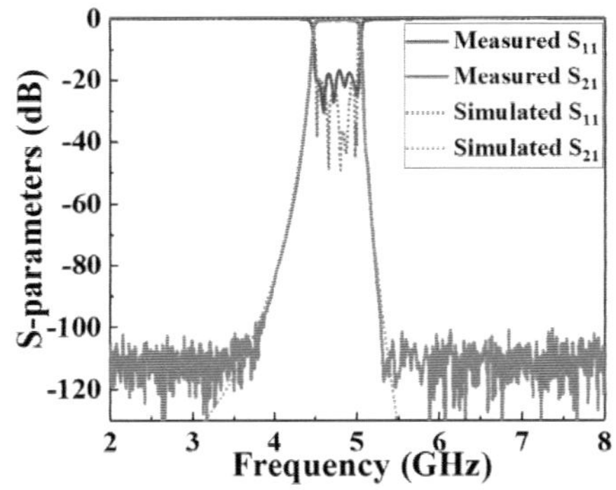

Figure 8. Simulated and measured S parameters of the bandpass filter

4. Conclusion

A cavity filter with small size, low loss, and high out-of-band rejection is presented. The additional metal rectangular blocks on the filter covers reduce the height of the resonant rod due to the loading capacitance with the resonant rod. It is easier to fabricate than conventional T-type or disc-type capacitor loading structures. Filters with a combined coupling structure can obtain a wider bandwidth by the coupling between metal rectangular blocks. Moreover, the combination of interdigital and combined coupling structures can effectively reduce the size of filters with medium bandwidth. The filter structure has good application prospects in miniaturized wireless communication systems.

Acknowledgments

This work was supported by the International Cooperation Project under Grant No. SQ2018YFE02056 00, National Natural Science Foundation of China under Grant Nos. 52072062, 51472046.

References

[1] Hunter, I.C, Billonet, L, Jarry, B, Guillon, P. (2002). Microwave filters-applications and technology. *IEEE Transactions on Microwave Theory and Techniques,* 50(3), 794-805.

[2] Chen, Yi-Ming, Chang, Sheng-Fuh, Chang, Chia-Chan, & Hung, Tin-Jae. (2007). Design of Stepped-Impedance Combline Bandpass Filters With Symmetric Insertion-Loss Response and Wide Stopband Range. *IEEE Transactions on Microwave Theory and Techniques, 55*(10), 2191-2199.

[3] Maharjan, Ram Krishna, Shrestha, Bhanu, Kim, Nam-Young. (2012). Microstrip Cross-coupled Interdigital SIR Based Bandpass Filter. *Radio engineering,* 21(3), 881-885.

[4] Hao, Zhang-Cheng, Ding, Wen-qi, Hong, Wei. (2016). Developing Low-Cost W -Band SIW Bandpass Filters Using the Commercially Available Printed-Circuit-Board Technology. *IEEE Transactions on Microwave Theory and Techniques, 64*(6), 1775-1786.

[5] Dongxu Feng, Zhelei Xia, Fanghua Ling, Shuang Liu. (2011). Miniaturization coaxial cavity filter design used in 3G communication. *2011 Second International Conference on Mechanic Automation and Control Engineering,* 1397-1400.

[6] Zhang, Xianrong, Wang, Qingyuan, Cao, Jinhua, Peng, AnJ. (2012). Design of a Compact Bandpass Filter with Capacity Loaded Ridged Waveguide. *Microwave Journal (International Ed.),* 55(10), 124-134.

[7] Puglia, KV. (2001). A general design procedure for bandpass filters derived from low pass prototype elements: Part II. *Microwave Journal,* 44(1), 114-136.

Cascade detector for oriented airplane detection in remote sensing images

Yangte Gao[1], Jianfeng Gao[2], Xiaoyu Nie[3], Jianhong Han[1*], Fukun Bi[4]

[1] School of Information and Electronics, Beijing Institute of Technology, Beijing, 100081, China

[2] UAV Research Institute, Changfeng Electronic Technology Co., Ltd, Gansu, 730070, China

[3] Party school of Liaoning Provincial party committee, Liaoning Academy of Governance, Liaoning, 110031, China

[4] School of Information Science and Technology, North China University of Technology, Beijing, 100144, China

*Corresponding Email: hanjianhong1996@163.com

Abstract. Although the current target detection field has made remarkable progress, airplane detection in large field optical remote sensing images is still challenging. To this end, we innovatively propose an oriented airplane detector to solve the two difficulties. In response to the slow detection speed of optical remote sensing images, we use an anchor-free structure to avoid complex calculations related to anchor boxes and significantly reduce the use of memory and time in computing. Since the airplane target are distributed with arbitrary orientation, the accuracy of the general detector is reduced. We alleviate this problem by replacing the standard convolution with asymmetric convolution. In addition, we use a cascade structure in prediction, which can further improve accuracy with almost no increase in the calculation. Abundant experiments and detection results show our methods effectiveness. In airplane detection task, our proposed model achieves 88.14 mAP on DOTA-v1.0 and 90.46 on UCAS- AOD.

1. Introduction

Whether military or civilian, airplane detection has gained much attention in computer vision. Unlike natural images, airplane targets in optical remote sensing images usually account for a small proportion of pixels and arbitrary orientation because the images perspective. The above problems make it difficult for a general detector to achieve satisfactory performance both in accuracy and speed.

At present, the traditional method and CNNs method mainly constitute the aircraft target detection methods system. The traditional methods mainly use the features designed manually by prior knowledge and establish complex detection rules to localize airplane targets. However, with the rise of stochastic neural network techniques, the performance of such methods has been far surpassed by deep learning-based methods.

Recently, several detection methods have been explicitly proposed for airplane targets. However, these methods all rely heavily on anchors. Such methods usually require densely placed anchor boxes on the input to ensure high recall, which yields enormous computation and memory footprints [1].

Moreover, airplane targets in remote sensing images usually have angular attributes from 0° to 360°.

Since the regular CNNs are not equivariant to the rotation, the extracted target features are usually not robust for the the orientation.

Therefore, we propose an oriented airplane detection method in response to the above two problems. Our proposed detection framework enjoys the following advantages.

(1) Asymmetric convolution blocks are adopted [2] in the networks that can boost the model's robustness to the airplane which has arbitrary direction.

(2) Our detection framework is an anchor-free-based method, which reduces the complex computation associated with IOU, especially in the training phase. Furthermore, to further improve the accuracy, we design a cascaded structure using deformable convolution that only increases a small amount of computation.

2. Method

In this section, we describe the entire network architecture in detail. Firstly, we boost the network's ability to extract rotated target features by utilizing the asymmetric convolution blocks in the backbone. As discussed, we adopt an anchor-free-based architecture to process remote sensing images more rationally with large fields of view. Finally, our prediction head adopts a cascaded architecture to increase accuracy further. The entire network architecture is shown in Figure 1:

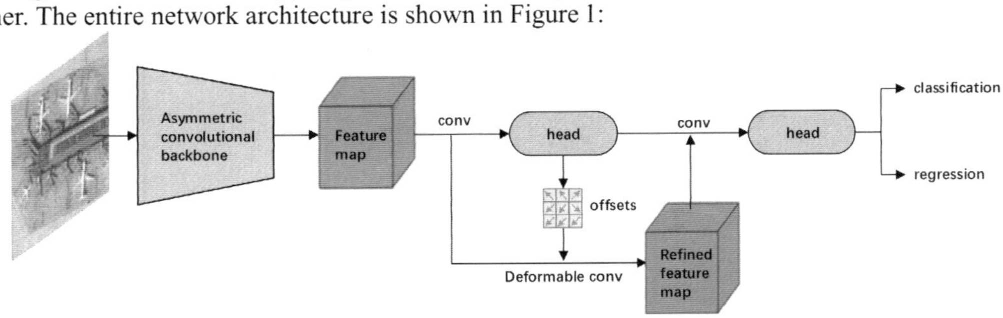

Figure 1. Overview of the proposed detection framework.

2.1. Asymmetric convolutional backbone

Modern detectors use convolutional neural networks to extract images' deep semantic features by learning, e.g., the widely used ResNet [3]. We use it as the baseline and replace the standard convolution block with asymmetric convolution. Its specific implementation is shown in Figure 2:

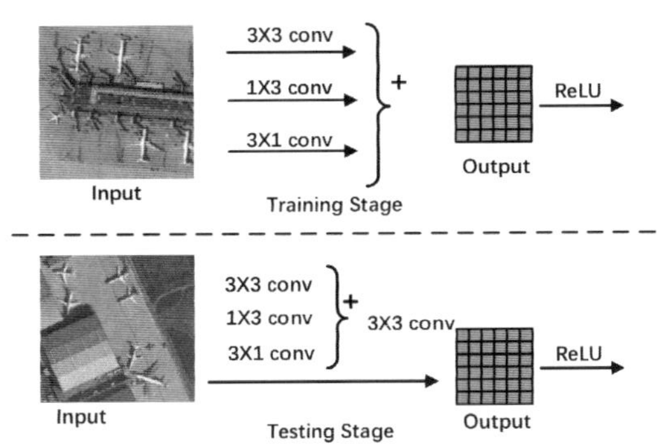

Figure 2. Implementation of asymmetric convolution.

Specifically, in the training stage, the final output is obtained by fusing these calculation results by replacing the standard 3x3 convolution with three convolutions of 3x1, 1x3, and 3x3. In the testing stage, the kernel parameters of the above three convolutions are fused to achieve the complete consistency of the network structure with the original network. Therefore, this structure can enhance the robustness of the network to rotational distortion without adding any parameters in the testing stage.

2.2. Cascade Anchor-Free prediction head

FCOS is currently a classical and advanced detector of Anchor-Free architecture. We take it as the baseline and add a refinement module based on deformable convolution to the prediction head.

We further revised the predicted value obtained by the first-level prediction to get the final result. That cascade method can revise the targets search region and improve the accuracy of the result. An obvious problem that arises in such structures is the feature misalignment problem. It is caused by the mismatch between the revised prediction and the used feature map. Therefore, we use the deformable convolution [4] in this module to modify the used feature map to solve this problem. For a specific sampling point P in the feature map, the transform equations are as follows:

$$Y(F_0) = \sum_{F_n \in R} w(F_n) \bullet x(F_0 + F_n + \Delta F_n)$$
$$X(F) = \sum_h O(w,h) \bullet x(h) \tag{1}$$

Where is w an enumeration over all integer positions, $O(w,h)$ is a two-dimensional bilinear interpolation kernel, and ΔF is the predicted offset value.

For each prediction head, both in classification and regression, its structure is consistent with the FCOS. For the regression branch, a 4-d regression vector $d^* = (l^*, t^*, r^*, b^*)$ is predicted to the offsets of generated anchors. For the classification branch, binary classifiers are trained to discriminate the airplane.

Overall, the loss of network architecture can be computed as ：

$$L = \alpha_{ref} L'_{ref} + \alpha_p L'_p \tag{2}$$

Where α_{ref} , α_p are two ratio in order to adjust the weight of each loss. We set them to 0.5 and 0.5, respectively. L'_{ref} , L'_p the calculation function is the same, defined as follows:

$$L'(\{p_{x,y}\},\{t_{x,y}\}) = \frac{1}{N}(L_c(p_{x,y}, c_{x,y}) + L_r(t_{x,y}, t^*_{x,y})) \tag{3}$$

Where L_c is the focal loss [5] for classification, $p_{x,y}$, $c_{x,y}$ are predicted class scores and class labels, respectively. L_r is the loss of the regression box, using the IOU loss [6], $t_{x,y}$ is bounding box coordinates, calculated only for positive samples. N is the number of positives.

3. Experimental results and analysis

We conducted amount experiments to evaluate our method on two publicly authoritative remote sensing datasets, UCAS-AOD [7] and DOTA [8]. This chapter first introduces the dataset-related information and experiment settings. Then, we provide a detailed ablation study and show the overall capability of our proposed model.

3.1. Data Set

3.1.1. UCAS-AOD. Data source is Google Earth, and the size of all images is approximately 1280 × 659. The dataset consists of 1000 aerial images and 7482 airplane samples, all of which have been carefully selected to give a uniform distribution of orientations. We randomly divided the entire dataset into three parts: training、 validation and testing with a ratio of 6:2:2.

3.1.2. DOTA. This dataset contains 2,806 aerial images with image sizes ranging from 800 to 4000 from different platforms. The dataset contains objects of various scales, orientations, and shapes with 188,282 instances. We select all the airplane targets for the experiment, 197 images with 8055 airplane samples and 70 images with 2531 airplane samples.

3.2. Ablation Study

We provide ablation experiments on the DOTA test set to prove the effectiveness of each module. The mAP metric is used to evaluate the result.

We use the FCOS with resnet-50 backbone as the baseline and evaluate the impact of each improvement module. The results as shown in table 1, a +1.48 mAP improvement using an asymmetric convolution block replaces the standard convolution(ResNet-50-AC), and the change of prediction head from once to cascade can bring a +2.34 mAP improvement.

Table 1. Performance verification of improved modules

Backbone	Cascade	Map
ResNet-50 (baseline)	×	84.64
ResNet-50	√	86.98
ResNet-50-AC	×	86.12
ResNet-50-AC	√	88.14

Zhang et al. [9] have demonstrated that due to the different distribution of positive and negative samples, anchor-based and non-anchor methods have different accuracy. So we only show the speed difference between the two results. Specifically, we use Retinanet [5] as the baseline of the anchor-based method. In order to make the comparison as fair as possible, we only use one square anchor box per location to predict the result. The result as shown in table 2, the Anchor-free-based method greatly improves the detector's speed by about 25% due to its simple decoding.Moreover, our cascaded modules can increase the accuracy with little impact on the detection speed.

Table 2. Comparison of detector's speed in inference.

Method	Backbone	Cascade	FPS
Retinanet (A=1)	Resnet-50	×	16.9
FCOS	Resnet-50	×	21.3
Our method	Resnet-50	√	20.4

3.3. State-of-the-art Comparison

We also compare it with some classic and high performance detection systems to comprehensively evaluate our method, such as SSD [10], RetinaNet and FCOS. As shown in table 3, our proposed model is competitive on both the aforementioned public datasets, and achieves 88.14 mAP and 90.46 mAP, respectively.

Table 3. Performance of the proposed Detector on DOTA-V1.0 and UCAS-AOD.

Method	Backbone	Cascade	mAP on different Datasets	
			DOTA-V1.0	UCAS-AOD
SSD	VGG16	×	77.76	79.94
Retinanet	Resnet-50	×	84.05	84.67
FCOS	Resnet-50	×	84.64	85.59
Our method	ResNet-50-AC	√	88.14	90.46

4. Conclusion

We propose an end-to-end detector for oriented airplane target in remote sensing images with a large field of view, which consists of two parts: the asymmetric convolutional backbone and the cascade Anchor-Free prediction head. The former can can boost the model's robustness to the airplane which has arbitrary direction. The latter can predict in a multi-level stage and Significantly reduces the amount of calculations without use anchors. Meanwhile, considering the shortcoming of misalignment, we design a feature refinement module to revise the feature maps which used by cascade regression.The elaborated experiments on multiple public detection data sets DOTA-v1.0 and UCAS-AOD, demonstrating that our method has competitive accuracy with high efficiency.

Acknowledgments
This work was supported in part by the National Natural Science Foundation of China (Grant No. 61971006).

References

[1] Tian Z, Shen C, Chen H and He T. 2019. Fcos: fully convolutional one-stage object detection. **In:** International Conference on Computer Vision. Seoul, South Korea. 9627-36.
[2] Ding X, Guo Y, Ding G and Han J. 2019. Acnet: strengthening the kernel skeletons for powerful CNN via asymmetric convolution blocks. **In:** International Conference on Computer Vision. Seoul, South Korea.1911-20.
[3] He K, Zhang X, Ren S, and Sun J. 2016. Deep residual learning for image recognition. **In:** IEEE Conference on Computer Vision and Pattern Recognition. Las Vegas, USA. 770-8.
[4] Dai J, Qi H, Xiong Y, Li Y, Zhang G, Hu H and Wei, Y. 2017. Deformable convolutional networks. **In:** International Conference on Computer Vision. Venice, Italy. 764-73.
[5] Lin T Y, Goyal P, Girshick R, He K, and Dollár P. 2017. Focal loss for dense object detection. **In:** IEEE Conference on Computer Vision and Pattern Recognition. Hawaii, USA. 2980-8.
[6] Zheng Z, Wang P, Liu W, Li J, Ye R and Ren D. 2020. Distance-iou loss: faster and better learning for bounding box regression. **In:** AAAI Conference on Artificial Intelligence. New York, USA. 12993–13000.
[7] Zhu H, Chen X, Dai W, Fu K, Ye Q and Jiao J. 2015. Orientation of robust object detection in aerial images using a deep convolutional neural network. **In:**IEEE International Conference on Image Processing. Quebec City, Canada. 3735-9.
[8] Xia G S, Bai X, Ding J, Zhu Z, Belongie S, Luo J B, Datcu M H, Pelillo M and Zhang L P. 2018. Dota: a large-scale dataset for object detection in aerial images. **In:** IEEE Conference on Computer Vision and Pattern Recognition. Salt Lake City, USA. 3974-83.
[9] Zhang S F, Chi C, Yao Y Q, Lei Z and Li S Z. 2020. Bridging the gap between anchor-based and anchor-free detection via adaptive training sample selection. **In:** IEEE Conference on Computer Vision and Pattern Recognition. Seattle, USA. 9759-68
[10] Liu W, Anguelov D, Erhan D, Szegedy C, Reed S, Fu C Y and Berg A C. 2016. Ssd: single shot multibox detector. **In:** European conference on computer vision. Amsterdam, Netherlands. 21–37.

Journal of Physics: Conference Series **2384** (2022) 012015 doi:10.1088/1742-6596/2384/1/012015

A Video Description Model with Improved Attention Mechanism

Feiyan Huang[1,a], Shangyou Zeng[1,*b], Jie Ke[1], Songtong Lei[1], JinJin Wang[1]

[1] School of electronic engineering, Guangxi Normal University, Guilin, Guangxi, 541001, China

[a] Email: 18593938623@163.com

[*b] Corresponding author's e-mail: zsy@mailbox.gxnu.cn

Abstract. Video description generation refers to the automatic generation of text descriptions of videos by computers, which belongs to the intersection of computer vision and natural language processing. Aiming at the problem that the traditional attention mechanism has insufficient ability to extract video features, the model is complex and the description quality is not high, this paper proposes a video description model with an improved attention mechanism. The model is based on the encoder-decoder structure, uses inception-v4 as the encoder to extract features, and introduces a lightweight coordinate attention module (CA) into the attention mechanism, which improves the feature extraction effect and reduces the model complexity, and sends the extracted important feature information into the decoder long short-term memory network (LSTM) to generate the description sentence corresponding to the video. The model is validated on the MSVD dataset using various evaluation metrics (BLEU, ROUGEL, CIDEr, METEOR). The experimental results show that the improved attention mechanism of the video description model proposed in this paper has better accuracy in different performance metrics and can further improve the performance of video description.

1. Introduction

Video understanding technology has great application in life. In the Internet field, video understanding technology has applications such as video search, question answering systems, and video special effects. Understanding technology has applications such as navigating blind people and describing movies or short videos to blind people. At present, video understanding has developed into a very broad academic research and industrial application direction.

Current video description algorithms are mainly based on an encoder-decoder architecture. The encoder uses a 2D/3D convolutional neural network to get features from the video, encodes the video information, and uses a recurrent neural network to decode the resulting features into descriptive sentences. In 2014, Subhashini et al. sent the result of pooling the convolutional features of the video frame to the decoder as coding features, while the subsequent S2VT model [1] argued that this approach ignored the temporal relationship before the video frame, so LSTM was used on the decoding end. In order to better relate the video features to the language model, literature[2] embedded the attention mechanism into the next generation of the video for the first time, which can enable the next generation model of video to determine the weights of video features when generating text sequences. The literature [3] proposes the SENet module, which focuses on channel relationships and automatically acquires the interconnections between each analogue channel by means of learning for the purpose of re-calibrating the channel characteristic response. However, he only noticed the impact of re-weighting each channel

Journal of Physics: Conference Series **2384** (2022) 012015 doi:10.1088/1742-6596/2384/1/012015

by modelling channel relationships, forgetting that positional information actually has an impact too, which cannot well connect the features and word features in the video.

In response to the above dilemma, this paper introduces the CA (coordinate attention) module [5] based on the literature [4], which embeds the position information into channel attention. Decomposing it into two 1D feature encoding processes, along the Aggregating features in two directions can be effective in obtaining the more important video frames. The model adopts a variety of evaluation indicators to verify the model on the MSVD [6] public dataset. The results show that the proposed method has achieved good results.

2. Video description related work

In recent years, with the development of deep learning, the literature on video description is also increasing. The video description task itself is a multi-modal problem, because the problem of two different modalities of video and text must be dealt with at the same time, and the conversion between them must be realized[7]. This task needs to grasp the video Semantic information to generate human-readable natural description language[8].

Most of the existing video description generation algorithms use the encoder-decoder structure framework based on CNN-RNN, which can achieve end-to-end training through the CNN structure. It is used to get and encode the features of the video data in order to fully capture the temporal information in the video, and then the RNN or its variant LSTM is used as the decoder to generate the final description statement[9].

The attention model (AM) is derived from human visual intuition. Used to automatically learn and calculate the size of the contribution of the input data to the output data, because the target area has more attention resources. The ability to focus precisely on the important parts while suppressing other useless information. In order for the video description model to selectively focus on the more salient features in the video, AM is applied to the video description[10].

3. Improved video description methods

The method proposed in this paper is based on an Encoder-Decoder framework. For the encoding part, the input video is first pre-processed, and the CNN inception-v4 as the encoder to encode the video. In the decoding part, a decoder network composed of LSTM is used to decode the visual information obtained in the pre-processing stage to generate video description sentences[11]. In this paper, by embedding the CA module into the encoder-decoder framework, the model can dynamically focus on more relevant video frame subsets when generating words at the current moment, thereby generating more accurate video description sentences. Figure 1 is the overall framework of the video description model.

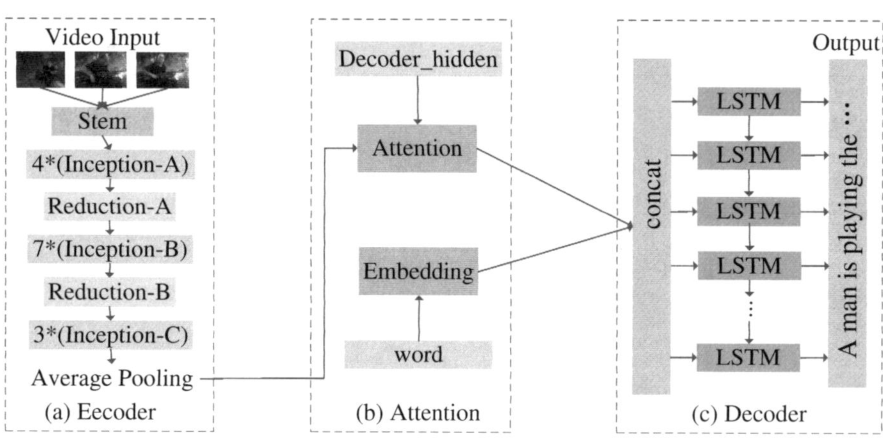

Figure 1. The overall framework of the proposed model

3.1 Encoder-Decoder Module

As the current dataset size for the video description task is still relatively small, it is difficult to design a complex neural network model to accomplish end-to-end learning as in other domains. Therefore, this paper then takes Inception-v4 [12], pre-trained on the ImageNet classification task dataset, to get visual information. Inception architecture has excellent performance and low computational cost, and compared with inception-v3, it has a more ordered and simple framework and rich inception modules, increasing the network depth and width while increasing the adaptability to image size.

The decoder used in this paper is LSTM. The LSTM network can fully learn and utilize the temporal features and can capture the temporal structure of the target video at the global level while retaining the temporal features. LSTM controls whether information needs to be retained through the "gate", so that important information can be selectively retained, which solves the problem that RNN cannot handle long-distance dependencies.

3.2 Coordinate attention module

So as to improve the correlation between video features and word features of the video description model, this paper further improves on the basis of the literature [5] and introduces the CA attention module. The improved CA module is shown in Figure 2(a), and the coordinate attention (CA) structure diagram is shown in Figure (b).

The CA module is contains two parts: Coordinate Information Embedding (CI) and Coordinate Attention Generation (CA). First, in the CI part, the global pooling is split into a pair of 1D feature encodings so that the AM module has the ability to obtain spatial long-range dependencies with deterministic location information. it means that for input P, each channel is first encoded along the horizontal and vertical coordinate directions using pooling kernels of dimensions *(M, 1)* and *(1, N)*. Thus, the output of the c-th channel with height m and width n are respectively represented as follows.

$$V_c^m(m) = \frac{1}{N}\sum_{0\le i<N} P_c(m,i) \tag{1}$$

$$V_c^n(n) = \frac{1}{M}\sum_{0\le i<M} P_c(m,i) \tag{2}$$

Secondly, in the CA module, the two feature maps generated in the previous module are transformed with a shared 1x1 convolution, where the generated intermediate feature map is represented by the generated $l \in K^{c/k\times(m+n)}$, as follows.

$$\mathrm{l} = \delta(\mathrm{F}_1([\mathrm{z}^{\mathrm{m}}, \mathrm{z}^{\mathrm{n}}])) \tag{3}$$

Finally, Finally, dividing f into two independent tensors, $l \in K^{c/k\times M}$ and $l \in K^{c/k\times N}$, along the spatial dimension gives

$$q^m = \delta(L_m(L^m)) \tag{4}$$

$$q^n = \delta(L_n(L^n)) \tag{5}$$

Then q^{m} and q^{n} are expanded into attention weights and the final output formula for the CA module is shown below.

$$R_c(i,j) = P_i(i,j) \times q_c^m(i,j) \times q_c^n(i,j) \tag{6}$$

The AM of the CA module embeds positional information into channel notes, aggregating features along different directions, one spatial direction capturing long-distance dependencies and the other retaining precise positional information, through the application of input feature maps to enhance the presentation of important objects.

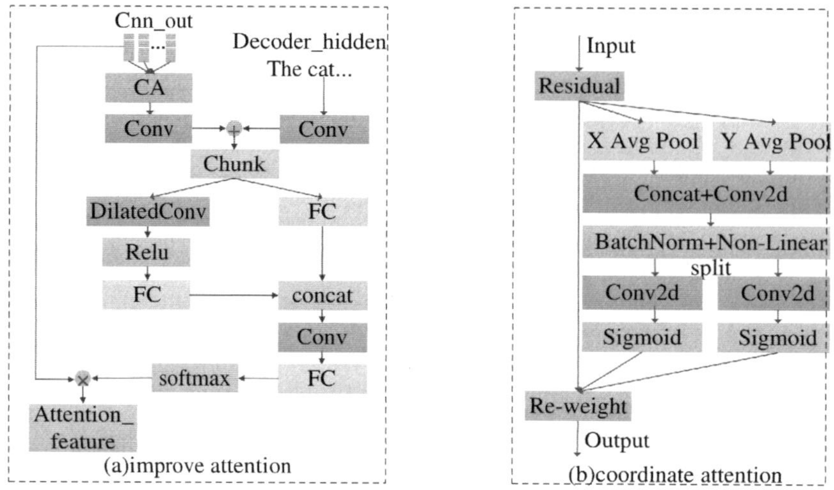

Figure 2. Improved attention mechanism(a) and coordinated attention (b)

4. Experimental Details

The experimental environment is 64-bit Window10, using PyTorch as the underlying framework for deep learning, CUDA8.0 as the GPU acceleration library, the computer memory is 32GB RAM, Intel i7-6700K quad-core CPU, and NVIDIA-GTX1080Ti GPU.

In the network training, the input dimension of the word vector and decoder in this experiment is 468, the hidden layer unit is 512, and the batch size is 32. The parameters of the model are optimized using the AMSGrad optimizer, and the epoch is set to 50.

4.1 Experimental Dataset *And Evaluation Metrics*

The model uses the public dataset MSVD to validate the model, which contains 1970 short videos, each video clip contains 40 sentences, and each sentence has an average of 7 words. The dataset is divided into the training set, validation set, and test set, of which the training has 1200 short videos, the validation has 100 short videos, and the test has 670 short videos.

The evaluation metrics used in this paper are the text quality evaluation toolkit Microsoft COCO Caption Evaluation provided by Microsoft, which includes evaluation metrics such as BLEU(n), ROUGE-L, METEOR, and CIDEr.

4.3 Experimental Results and Analysis

To further demonstrate the effectiveness of the proposed model, the scores of the video description model proposed in this paper compared with six models in recent years on various evaluation metrics on the MSVD test dataset are shown in Table 1.

Table 1. Comparison of evaluation indicators of this model and other models on MSVD %

Model	BLUE_4	ROUGE_L	CIDEr	METEOR
S2VT(inception-v4)	39.6	67.5	66.7	31.2
SA-LSTM (inception-v4)[13]	45.3	64.2	76.2	31.9
PickNet[4]	46.1	69.2	76.0	33.1
TDDF[14]	45.8	**69.7**	73.0	33.3
LSTM-E[4]	45.5	—	—	31.0
MMI[4]	46.7	65.0	76.8	33.6
Our method	**48.9**	69.2	**77.1**	**34.6**

As shown in the figure, compared with other video description models, the model in this paper has good performance in all index scores, proving that by introducing the CA attention module, which pays attention to the channel information along with the location information. It can better focus on the more important part of the video frames, allowing the model to better get the video frame features in the encoding part.

5. Conclusion

This paper proposes a video description model with an improved attention mechanism. The model uses the convolutional neural network Inception-v4 to get visual features, and can better focus on more important video frames by introducing the CA module. LSTM is used to decode the visual information obtained in the preprocessing stage to generate video description sentences. Experiments were conducted on the MSVD video dataset and the results showed a significant improvement compared to the previous method.

Acknowledgments

This work was supported by the National Natural Science Foundation of China (61976063).

References

[1] Qi, B. (2021) Research on the key issues of video retrieval technology [J]. Information Recording Materials, 22(08):66-68.
[2] Li, W H. Zeng, S Y. Wang, J J. (2021) An image description generation algorithm based on improved attention mechanism[J]. Computer Applications, 41(05):1262-1267.
[3] Su, Q S. (2022) Design of video retrieval algorithm based on attention mechanism[J]. Digital Technology and Applications, 40(04):143-145.
[4] Wang, J J. Zeng, S Y. Li, W H. et al. (2021) Attention Mechanism Video Description Model Based on Dilated Convolution. Electronic Measurement Technology, 44(23):99-104.
[5] Hou, Q., Zhou, D., Feng, J., 2021. Coordinate attention to efficient mobile network design. In: Proceedings of the IEEE/CVF conference on computer vision and pattern recognition, Nashville. 13713-13722.
[6] Chang C, Zhao, D X. (2020) A review of research on video description methods based on deep learning[J]. Journal of Tianjin University of Technology, 36(06):17-23.
[7] Zhang, Y D. Jiang, X Y. Shi, X Y. (2022) Perceptual lexical information for action detection in video description[J]. Fujian Computer, 38(06):95-98.
[8] Cao, L. Wan, W G. (2021) A multilingual video description algorithm based on dynamic visual attention[J]. Industrial Control Computer, 34(07):62-64.
[9] Li, X l. Zhao, B. (2021) Video extraction[J]. China Science:Information Science,51(05):695-734.
[10] Zhang, L H. Cao, L B. (2018) Research on video description method based on deep migration learning[J]. Journal of Testing Technology, 32(05):443-449.
[11] Xiao, H H. Shi, J L. (2018) Video description based on C3D and visual elements[J]. Journal of South China University of Technology (Natural Science Edition), 46(08):88-95.
[12] Zhang, J Y. Liang, C. Wu, P. Chen, L. Liu, Y T. (2020) Research on facial expression recognition algorithm based on improved Inception V4[J]. Optical Technology Applications, 35(01):56-63.
[13] Fu, Y. Ma, Y. Ye, O. (2021) A video description method integrating deep learning and visual text[J]. Science Technology and Engineering, 21(14):5855-5861.
[14] Ding, E J. Liu, Z Y. Liu, Y F. Yu, W L. (2020) A video description method based on multidimensional and multimodal information[J]. Journal of Communication, 41(02):36-43.

Shift Gait: A Skeleton-Based Gait Recognition Method

Yao Tan, Lei Yu, Datao Xie and Baoxiang Tang

College of Computer and Information Science, Chongqing Normal University, Chongqing 401331, China

ylcqnu@163.com

Abstract—Gait recognition is a promising biometrics identification technology that is widely used in criminal investigations, surveillance systems, social security, and other fields. Most existing methods of gait recognition are appearance-based methods that use silhouettes as the input of networks to acquire gait features. However, silhouettes will be lost fine-grained spatial information. In addition, the silhouette contains other information that is not related to gait, such as dress, backpack, etc., therefore, it cannot be treated as a pure "gait" recognition. At the same time, most methods do not take into account the complexity of background extraction in the context of dynamic changes. To solve the above mentioned problems, ShiftGait, a method based on skeletal features, is proposed, which extracts body posture from RGB images using 2D pose estimator and feeds the extracted skeletal sequences into graph convolutional network to acquire the features of skeletal sequences. Finally, the extracted features are used for classification. Experiments on the CASIA-B dataset indicates that our network is superior and more robust to dressing changes.

1. Introduction

The purpose of gait recognition is to identify pedestrians using gait sequences captured by cameras. Gait recognition has advantages over other biometric technologies: (1) remote acquisition without subject cooperation; (2) difficult to disguise or imitate; (3) low resolution recognition. There are two types of gait recognition in existence: appearance-based methods and model-based methods. The former often use silhouette as the network's input. The silhouette is effective for describing the gait in a single frame with low computational cost when the subject and the environment do not change much. But the appearance of gait is often influenced by other factors, for example, carrying, clothing, and camera view, etc., thus increasing the difficulty of gait recognition. The latter, on the other hand, recognize the human's identity based on its structure and movements, and thus have greater stability for changes in the appearance of the subject or changes in viewpoint. In recent years, the emergence of human pose estimators has also driven the development of the model-based methods. Human pose estimators are more robust to complex background changes, occlusions, and viewpoint changes, etc. The features in the skeleton sequence contain only gait information, which is more "pure" and beneficial to cope with changes in covariates. In order to efficiently perform gait recognition, a new model-based approach is proposed in this paper, ShiftGait. This method makes skeleton-based gait recognition a true "gait" recognition.

2. The proposed method

The silhouettes contain information irrelevant to gait, and to enhance the robustness of gait recognition methods under occlusion, viewpoint changes, background changes, etc., so, we propose a model-based method ShiftGait. This method has three main steps: (1) extracting skeleton sequences; (2) requiring features of skeleton sequences; (3) classification

2.1. The network structure

The 2D human pose information is easily accessible and contains only pose information. Therefore, we use High-Resolution Net (HRNet) [1] to extract 2D skeleton sequences. Meanwhile, shift graph convolutional network (ShiftGCN) [2] is chosen in this paper to acquire the spatiotemporal features of gait sequences. The network structure of ShiftGait is shown in Figure 1.

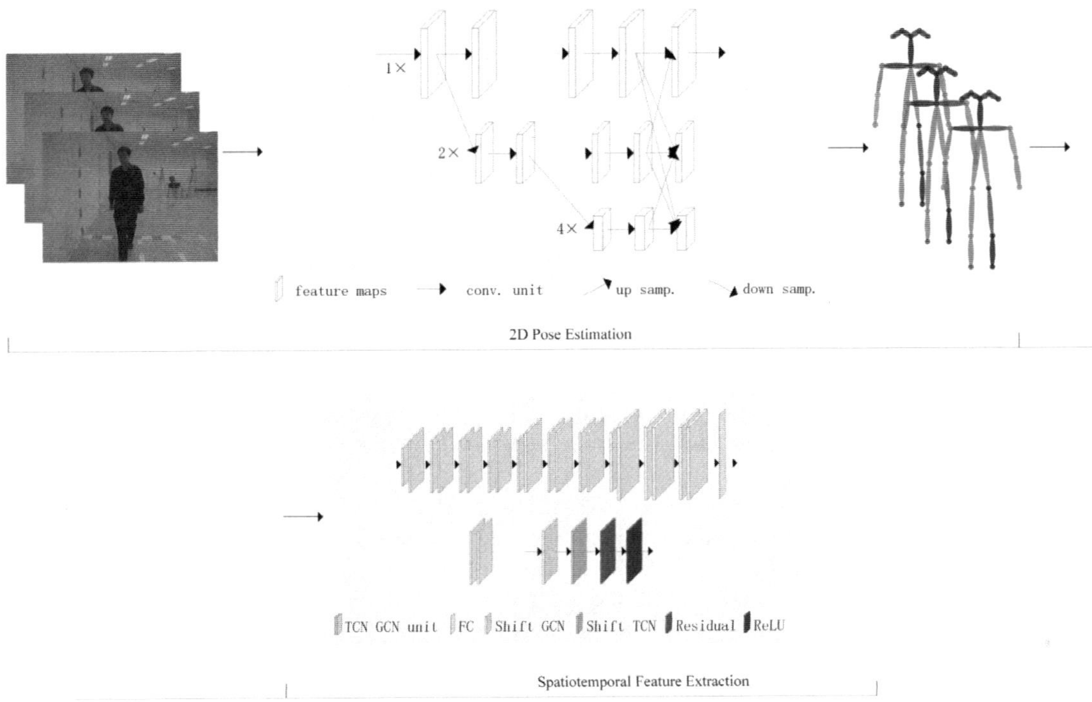

Figure 1. ShiftGait Network architecture.

2.2. 2D pose estimation

For purpose of decreasing the impact of various covariates, this paper introduces human pose estimator to extract the skeleton sequence of gait. Compared with single-person gait estimation, the multi-person gait estimator has better performance. Therefore, a 2D human pose estimation method, the high-resolution network (HRNet) [1], is used in proposed gait recognition network, which takes a high resolution subnet for the first phase and gradually adds high-resolution to low-resolution subnets to generate more phases. The multi-resolution subnets happen at the same time and are connected. Throughout this process, the network is capable to keep the high-resolution features. Meanwhile, it exchanges information in multi-resolution subnetworks in parallel to generate iterative multiscale fusion. Ultimately, the network outputs a high-resolution feature as the key points estimation's way and obtain more accurate key point localization. HRNet was pre-trained on the COCO dataset [3], which has pose annotation consisting of 17 key points, and the skeleton connectivity uses the most commonly used one, as shown in Figure 2.

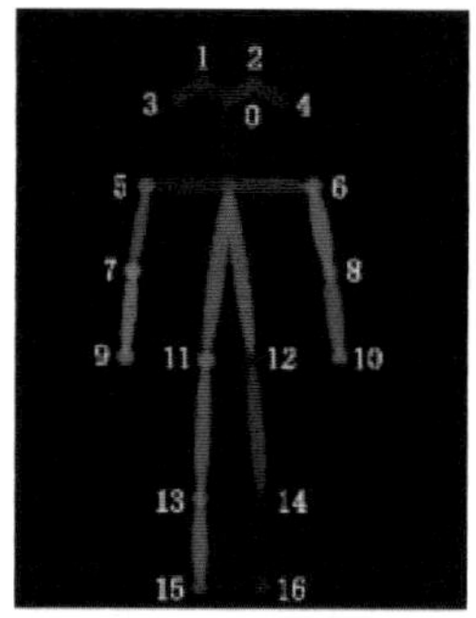

Figure 2. Link of COCO bones.

2.3. Spatiotemporal feature extraction

Skeleton sequences are natural spatiotemporal graph data, they can be modeled by spatiotemporal graph convolution network (GCN), such as ST-GCN proposed by Yan et al. [4]. ST-GCNs are modeled with regular GCN in the space domain and one-dimensional convolution in the time domain. Subsequently, many variants of ST-GCNs have been proposed [5][6]. However, these networks have two main drawbacks: (1) excessive computational complexity; (2) needing artificially pre-set the spatiotemporal perceptual fields. In order to avoid the above-mentioned shortcomings, we use ShiftGCN as a spatiotemporal feature extraction network. ShiftGCN replaces the regular spatiotemporal convolution with spatiotemporal shift convolution and point-by-point convolution, which can provide flexible perceptual fields with lower time complexity. In the spatial shift convolution, part of the information of the current node is obtained from the neighboring nodes, and the temporal shift convolution constructs the graph by connecting adjacent frames in the temporal dimension. The shift convolution is combined with point-by-point convolution in two ways, as shown in Figure 3. After the second connection method, i.e., another shift operation after the point convolution, the output features have a larger perception field and usually results in improved performance. Besides, there is also a multi-stream fusion structure in the network for improving the correct rate of the recognition.

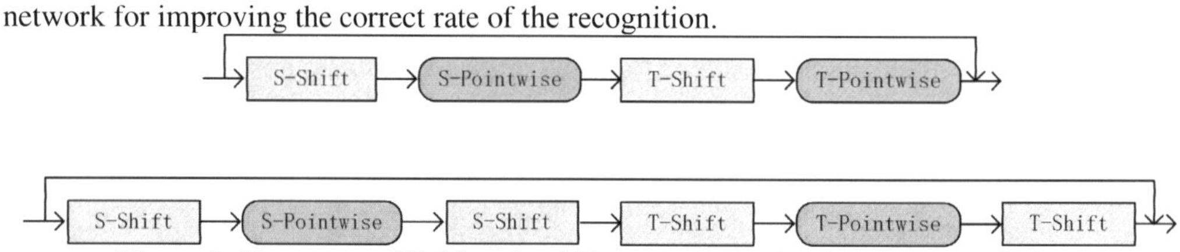

Figure 3. Two ways of linking the shift operation and pointwise convolution.

3. Experimental results and analysis

3.1. Dataset

For effective training of deep neural networks, huge datasets are generally chosen. Meanwhile, in order to extract the human pose, the dataset must provide RGB images. Therefore, we choose to conduct experiments on a commonly used public dataset, CASIA-B [7]. xunlianji [7] dataset is used frequently in gait recognition. This d ` atabase involves 124 participants, which are presented as RGB videos and silhouettes. It is a huge multi-view gait dataset capturing from 11 views. In addition, the dataset contains three various walking states, that is normal (NM), clothing (CL), and carrying (BG) conditions, each with 6, 2, and 2 sequences from 11 view. We use the most commonly used division method for the training set and test set, i.e., subjects 1-74 is training set, 75-124 is test set, and NM 1-4 of test set is probe set, NM 5-6, CL 1-2, BG 1-2 is gallery set. The final identification effect is measured and displayed using Top-1 accuracy rates.

3.2. Training

In the training phase, the frames need to be extracted from the videos and then the RGB gait images are fed into the network. The network is trained with an SGD optimizer and Nesterov momentum (0.9) for 2000 epochs, and calculate the loss using the cross-entropy loss function. The learning rate is initially set to 0.1. At the 800th, 1200th, and 1600th epochs decrease to 0.01, 0.001, and 0.0001, respectively. The batch size is appointed to 64 for the CASIA-B dataset.

3.3. Experimental results on the CASIA-B dataset

Table 1 shows Top-1 accuracy of the proposed method ShiftGait for 11 viewpoints with 0° to 180° increments of 18 under normal, carrying, and clothing condition, respectively. In the experiments, the former 4 sequences of NM are the probe set; the last 2 sequences of normal walking, the 2 sequences of carrying bag, and the 2 sequences of taking jacket are gallery set for three sets of experiments, respectively. From Table 1, we can see that the accuracy changes with the angle; meanwhile, the accuracy of the gallery set as normal walking is significantly improved than the gallery sets as carrying bag and taking jacket. This indicates that changes in covariates such as view and clothing can influence the accuracy extremely, and our method can reduce the impact of changes in these covariates.

Table 1. Top-1 accuracy in ShiftGait based on CASIA-B dataset.

	0	**18**	**36**	**54**	**72**	**90**	**108**	**126**	**144**	**162**	**180**	**Mean**
NM#5-6	69.3	70.8	77.0	76.7	71.4	68.4	71.6	69.0	75.0	71.6	68.5	**71.8**
BG#1-2	49.7	49.6	52.4	52.9	44.8	45.0	50.7	50.2	50.8	52.3	46.6	**49.6**
CL#1-2	45.0	49.7	52.9	50.8	47.5	44.1	50.3	50.9	53.7	57.2	46.8	**49.9**

3.4. Comparative experiments

Table 2 represents the ablation experiments, and higher values in the table are shown in bold. As is indicated in the table 2, our method, namely ShiftGait, has greater Top-1 accuracy than that of PoseGait [8] for all viewpoints and variation conditions. Both methods use an accurate pose estimator for acquiring skeletal sequences, which proves advantage of the GCN network applied in this paper in feature extraction and improves performance of whole gait recognition network.

The features used in our method ShiftGait are model-based features. Table 3 display the comparison experiment between model-based and appearance-based methods so as to show the advantages of model-based methods. As is indicated in the Table 3, ShiftGait has higher accuracy than the appearance-based methods - SPAE [9] and GaitGANv2 [10] - and also has higher accuracy than the model-based methods - PoseGait. It is particularly noteworthy that ShiftGait is much more accurate than the other methods when the gallery set is CL (wearing a coat). Compared with other methods, although the GaitSet [11] method has higher accuracy, this is mainly due to two aspects: (1) GaitSet uses appearance-based high-dimensional features; (2) it trains CNNs in a validation manner and uses a large number of training combinations.

Table 2. Top-1 accuracy comparison of ShiftGait and PoseGait based on CASIA-B dataset.

		0	**18**	**36**	**54**	**72**	**90**	**108**	**126**	**144**	**162**	**180**	**Mean**
NM#5-6	PoseGait	53.9	64.8	69.7	69.4	64.0	62.5	65.7	64.4	69.9	65.3	52.1	63.8
	ShiftGait	**69.3**	**70.8**	**77.0**	**76.7**	**71.4**	**68.4**	**71.6**	**69.0**	**75.0**	**71.6**	**68.5**	**71.8**
BG#1-2	PoseGait	36.4	45.2	48.2	47.5	44.4	44.1	42.5	43.6	45.4	39.8	30.6	42.5
	ShiftGait	**49.7**	**49.6**	**52.4**	**52.9**	**44.8**	**45.0**	**50.7**	**50.2**	**50.8**	**52.3**	**46.6**	**49.6**
CL#1-2	PoseGait	23.0	26.3	35.7	36.0	35.0	32.8	38.1	37.8	35.8	30.4	20.9	32.0
	ShiftGait	**45.0**	**49.7**	**52.9**	**50.8**	**47.5**	**44.1**	**50.3**	**50.9**	**53.7**	**57.2**	**46.8**	**49.9**

Table 3. Comparison of top-1 accuracy between model-based method and appearance-based method on the CASIA-B dataset.

Types	Methods	Gallery		
		NM 5-6	BG 1-2	CL 1-2
Appearance-based Methods	SPAE	62.8	40.4	26.0
	GaitGANv2	66.3	46.2	25.9
	GaitSet	**95.1**	**90.9**	**79.0**
Model-based Methods	PoseGait	63.8	42.5	32.0
	ShiftGait	**71.8**	**49.6**	**49.9**

3.5. Ablation experiments

Compared with the regular spatiotemporal convolution in ST-GCN, the shift spatiotemporal convolution in the ShiftGCN network are more efficient. In this paper, ablation implementation of the spatiotemporal shift graph convolution network is performed on the CASIA-B [7] dataset, and the experimental results can be seen in Table 4, with higher values in bold. In this experiments, the spatial and temporal shift convolution in the ShiftGCN network are replaced with regular spatial and temporal convolution, respectively, and four sets of comparison experiments are performed. Table 4 displays the experiments result, the accuracy is the lowest when using the regular spatial and temporal convolution, and accuracy improves after replacing one of the regular convolutions with shift convolution, while Top-1 accuracy is the highest when using the shift spatiotemporal convolution.

The ShiftGCN network uses a multi-stream fusion strategy. In order to prove that the multi-stream fusion structure in the ShiftGCN can improve the network effect network, some experiments are conducted in this paper. In those experiments, all other settings were left unchanged, and only the Top-1 accuracy using joint data, bone data, joint motion data, bone motion data, and multi-stream fusion data on the CASIA-B dataset was compared. From the Table 5 indicates that the Top-1 accuracy using joint data is the highest among the four single-stream data, but the accuracy is significantly and substantially improved using multi-stream fusion, which can prove the effectiveness of the multi-stream fusion structure in the ShiftGCN network.

Table 4. Ablation study of spatiotemporal shift graph convolution on CASIA-B dataset.

Space Module	Temporal Module	Gallery		
		NM	BG	CL
Regular spatial GCN	Regular temporal GCN	61.5	43.8	39.0
Shift spatial GCN	Regular temporal GCN	71.0.	49.1	49.1
Regular spatial GCN	Shift temporal GCN	69.9	49.3	49.4
Shift spatial GCN	Shift temporal GCN	**71.8**	**49.6**	**49.9**

Table 5. Ablation study of multi-stream strategy on CASIA-B dataset.

Features	Gallery		
	NM	BG	CL
joint	61.3	38.6	36.0
bone	59.0	41.5	35.3
joint motion	54.9	37.1	39.3
bone motion	55.4	35.1	37.3
ensemble	**71.8**	**49.6**	**49.9**

4. Conclusion

We propose ShiftGait, an efficient gait recognition network based on skeleton features, which applies skeleton features to gait recognition so that the input of gait recognition contains only gait information of human posture without considering the complexity of contour extraction under dynamic background changes. Also, it is more robust to changes in covariates such as clothing. In addition, the graph convolution network ShiftGCN used in this paper replaces the regular convolution with shift convolution and point-by-point convolution, providing flexible perceptual fields in space and time with lower time complexity. Experimental results on the classical gait recognition dataset CASIA-B indicate that our method works more efficient than other model-based gait recognition methods while being more accurate than some appearance-based methods.

References

[1] Sun K, Xiao B, Liu D and Wang J 2019 Deep high-resolution representation learning for human pose estimation *Proc. of the IEEE/CVF conf. on computer vision and pattern recognition* pp 5693-703

[2] Cheng K, Zhang Y, He X, Chen W, Cheng J and Lu H 2020 Skeleton-based action recognition with shift graph convolutional network *Proc. of the IEEE/CVF Conf. on Computer Vision and Pattern Recognition* pp 183-92

[3] Lin T Y, Maire M, Belongie S, Hays J, Perona P, Ramanan D and Zitnick C L 2014 Microsoft coco: Common objects in context *European conf. on computer vision* (Switzerland: Cham) pp 740-55

[4] Yan S, Xiong Y and Lin D 2018 Spatial-temporal graph convolutional networks for skeleton-based action recognition *Thirty-second AAAI conf. on artificial intelligence*

[5] Si C, Chen W, Wang W, Wang L and Tan T 2019 An attention enhanced graph convolutional lstm network for skeleton-based action recognition *Proc. of the IEEE/CVF conf. on computer vision and pattern recognition* pp 1227-36

[6] Wen Y H, Gao L, Fu H, Zhang F L and Xia S 2019 Graph CNNs with motif and variable temporal block for skeleton-based action recognition *Proc. of the AAAI Conf. on Artificial Intelligence* vol 33 pp 8989-96

[7] Yu S, Tan D and Tan T 2006 A framework for evaluating the effect of view angle, clothing and carrying condition on gait recognition *18th Int. Conf. on Pattern Recognition* (ICPR'06) vol 4 pp 441-44

[8] Liao R, Yu S, An W and Huang Y 2020 A model-based gait recognition method with body pose and human prior knowledge *Pattern Recognition* 98 107069

[9] Yu S, Chen H, Wang Q, Shen L and Huang Y 2017 Invariant feature extraction for gait recognition using only one uniform model *Neurocomputing* 239 pp 81-93

[10] Shiqi C and Haifeng 2019 GaitGANv2: Invariant gait feature extraction using generative adversarial networks *Pattern Recognition* 87 pp 179-89

[11] Chao H, He Y, Zhang J, and Feng J 2019 Gaitset: Regarding gait as a set for cross-view gait recognition *Proc. of the AAAI conf. on artificial intelligence* vol 33 pp 8126-33

Transformer-based radio modulation mode recognition

Lingyun Li[1,2], Chanchan Qin[1,2], Guoqing Li[3,4], Shengbo Hu[1,2], Yike Xie[1,2] and Zhenwei Lei[1,2]

[1] School of Big Data and Computer Science, Guizhou Normal University, The University Town, Guian New Area, Guiyang, Guizhou 550025, China

[2]Center for RFID and WSN Engineering, Department of Education, Guizhou Normal University, The University Town, Guian New Area, Guiyang, Guizhou 550025, China

[3]College of Physical Science and Technology, Central China Normal University, NO.152 Luoyu Road, Wuhan, Hubei 430079, China

[4]Key Laboratory of Quark and Lepton Physics (MOE) and College of Physical Science and Technology, Central China Normal University, NO.152 Luoyu Road, Wuhan, Hubei 430079, China

Correspondence should be addressed to Chanchan Qin; 201407141@gznu.edu.cn

Abstract. The wireless communication technology develop rapidly, signal modulation mode technology becomes increasingly important. Therefore, in order to enhance the ability of radio signal modulation mode recognition in the complex electromagnetic environment, a novel radio signal modulation mode recognition method is proposed on account of Transformer network. Firstly, raw data were split with a fixed size window. Then, the segmented data was projected into a vector sequence and fed into the Transformer module to model and mine the relationship between signal waveform and modulation method to achieve identification. Extensive experiments were conducted on the RML2016.10a data set generated by GNU radio in the real system. The experimental results show that the proposed Transformer-B/8×2 achieves optimal results for modulation mode recognition with signal-to-noise from -20dB to 18dB, and recognition accuracy reaches 86.75%, verifying the practicability and validity of the forenamed algorithm.

1. Introduction

Recently, with the gradual promotion of information technology strategy and the continuous growth of radio communication technology, the wireless communication system has penetrated into all areas of life. Meanwhile, the growth of radio terminals and the Internet of things devices make the communication environment increasingly complex and changeable, and radio waves have become an important carrier to connect everything.

Communication signals often use complex and diverse modulation modes, and its main task is to determine the modulation mode of the received and other parameters in an environment with many signals, so as to ensure the accuracy and authenticity of the radio signal for deeper analysis and processing of the signal by technicians. It is now crucial to monitor and identify these communication signals effectively and quickly.

In the 1990s, domestic and foreign experts and scholars began to research modulation mode recognition technology. The early methods have split into Likelihood-Based [1] and Feature-Based [2] approaches. These two methods are easily limited by the subjective factors of the technicians in the implementation process, so its identification technology requires not only a high level of operation and professional knowledge of the technicians, but also a certain amount of modulation experience.

The computer technology has ballooned in recent years, Deep Neural Networks (DNN) [3] have been applied to image processing [4], speech recognition [5], and natural language processing with unprecedented success. The burtsting of deep learning techniques has brought new opportunities for radio signal modulation mode recognition technology. O'Shea [6-9] and his team have published several articles that have opened a new era in the recognition of radio signal modulation modes from empirically driven to data-driven. Eman et al. [10] proposed radio signal modulation mode recognition using Convolution Neural Networks (CNN) and Long-Short Term Memory (LSTM). Although this method slightly improves the recognition accuracy, the unit considers the global information of the signal and has high computational complexity.

Lately, Vaswani et al. [11] has been put forward a global attention-based Transformer, which overcomes the problem that recurrent networks cannot be parallelized and has been extensively used in computer vision [12-13].

Therefore, in order to ameliorate the recognition of radio signal modulation mode in a complex electromagnetic environment, this article raises a way to build a Transformer network to solve the problem of modulation mode recognition. The main contributions contain: 1) An idea to fuse the time series relationship of the signals is proposed. 2) The signals are identified by modulation mode using the Transformer encoder, and the best window settings are obtained on the RML2016.10a dataset used in this paper while achieving better performance than other existing methods.

2. Related Work

Transformer: The Transformer originated in 2017 for solving sequence-to-sequence problems. The Transformer standard model composes an encoder module and a decoder module [14], which contains multiple encoder/decoders of the same architecture. However, instead of using network architectures such as convolution and pooling, the encoder and decoder rely entirely on a parallelized architecture with a self-attentive to capture sequence dependencies.

Recently, a Transformer based on a global self-attention mechanism has achieved good results in many sequential pattern recognition tasks. For example, machine translation [15], speech recognition [16]. The Vision Transformer (ViT) model proposed by Dosovistskit [13] was the first time to use the encoder part of a Transformer directly for image classification, and after a long period of practice, the Transformer also proved to have stronger performance than traditional CNN models in computer vision, which provided a foundation for the subsequent Transformer vision research has laid the foundation for subsequent research on Transformer vision. Although Transformer is developing rapidly in computer vision field, there is still much room for development in the recognition of radio signal modulation modes.

3. Method

In this article, we put forward a way of radio signal modulation mode recognition on account of a Transformer network, which mainly uses a multi-head attention mechanism to identify types of ground signal modulation mode existing in the environment under a complex electromagnetic environment. Figure 1 shows the architecture diagram based on Transformer, and the architecture primarily contains of two sections: pre-processing module, feature extraction, and recognition module.

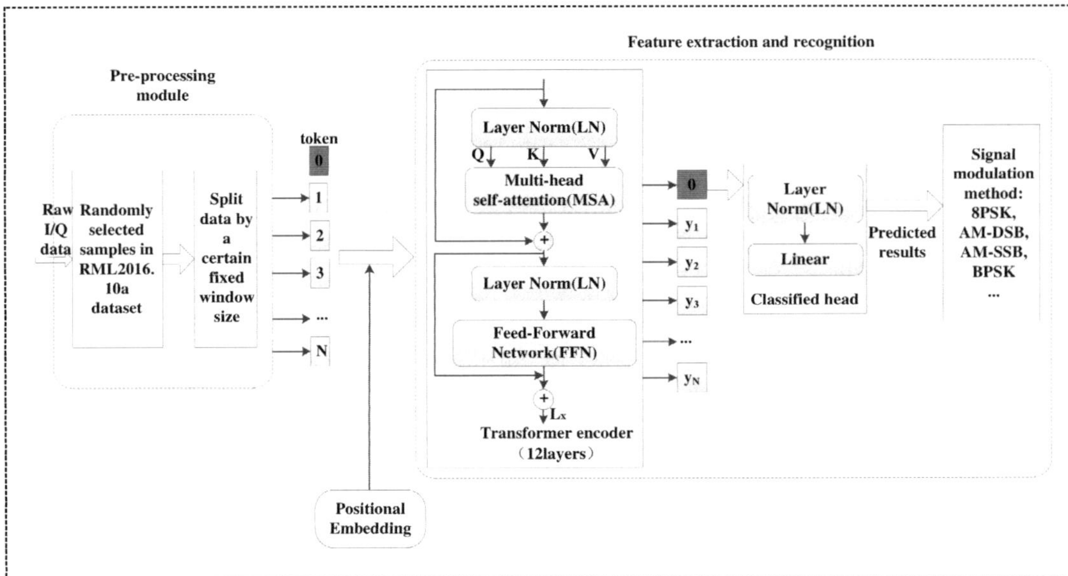

Figure 1. The flowchart of Transformer-based.

3.1. Pre-processing Module

Running the Transformer method starts with data pre-processing, which is aimed at chunking the original I/Q sampled data. In order to take full advantage of the Transformer network in feature extraction, the researcher wants the pre-processing process to be lossless to the original data. Based on the characteristic that the original sampled samples are composed of two paths with 128 points each, this paper adopts a two-dimensional signal processing approach. Specifically, the two-dimensional signal 128×2 is divided into a series of signal block sequences by a fixed window size*of* $P\times2$. The number of signal blocks is thus obtained as $N=128/P$.

3.2. Feature Extraction and Recognition Module

After pre-processing, the radio-modulated signal is divided into blocks of signals, and the next step is to perform feature extraction on the blocks to identify the modulation mode of the signal. The essence of this work is exactly the same as the pattern recognition task in computer vision, with the difference that this input is sequential data rather than pictures in nature. In this paper, we use the currently popular ViT [13] as a basis for constructing a network model for radio signal modulation mode recognition by modifying several hyper-parameters.

The Transformer encoder used in this paper is composed in the order of Layer-Norm (LN) [17], Multi-head Self-Attention (MSA) module, Layer-normalization, and Feed-Forward Network (FFN) module. Among them, the important part of Transformer encoder is MSA, which is a combination of 12 scaled point multiplication attention modules. Firstly, the input matrices are input into 12 groups of transform matrices respectively, and each group gets the query matrix Q, key matrix K, and value matrix V by three different weights of transform matrices, and calculates the self-attention feature map to get 12 weighted feature matrices ($head_1$~$head_{12}$). Then the 12 feature matrices are spelled into a large feature matrix by column. Finally, the feature matrices are fully connected by one layer to gain the ultimate self-attentive outcome, and the operation process is as equation (1) and equation (2). FFN mainly contains three layers, the first layer is fully connected, the second layer is linearly activation ReLU [18], and third layer is fully connected, and the calculation formula is as in equation (3). LN can prevent the over-fitting of the model and speed up the convergence of the model to some extent.

$$head = Attention\left(X_iW_i^Q, X_iW_i^K, X_iW_i^V\right) \tag{1}$$

$$MultiHead\left(Q,K,V\right)=Concat\left(head_1,head_2,...,head_h\right)W^0 \quad (2)$$

$$FFN\left(X\right)=\max\left(1,XW_1+b_1\right)W_2+b_2 \quad (3)$$

where X_i is the input matrix of the ith attention module. W_i^Q, W_i^K, W_i^V are the linear transformation matrix of the ith group, respectively. W^0 is matrix of a linear transformation. X is the input vector, W_1 and b_1 are the weight and bias of the first fully connected layer, W_2 and b_2 are the weight and bias of the second fully connected layer.

The Transformer encoder consists of 12 coding layers and uses residual connections behind the MSA module and the FFN module in each coding layer, where the residual structure is a superposition of the initial input and the result of the operation performed on the initial input. The input of the first encoding layer consists of an additional embedding token, the processed signal sequence segment, and the positional embedding, denoted by the operator formula in equation (4). The output of the multi-headed attention residual connection of each layer and the input of the second layer normalization of each layer are denoted by the formula in equation (5). The output of each layer is denoted by the formula in equation (6).

$$Z=\left[x_{class};x_p^1E;x_p^2E;...;x_p^NE\right]+E_{pos} \quad (4)$$

$$Z_l^{'}=MSA\left(LN\left(Z_{l-1}\right)\right)+Z_{l-1} \quad (5)$$

$$Z_l=FFN\left(LN\left(Z_l^{'}\right)\right)+Z_l^{'} \quad (6)$$

where *Epos* denotes the location embedding. *E* is a linear projection that maps the signal segment to D dimensions. In addition, a Transformer layer is employed to learn the feature representation. Since all Transformer layers have a global receptive domain, the problem of finite receptive domain present in CNN-based methods is thus solved.

The first three layers of the Transformer encoder are used for feature selection, and the last nine layers are applied to recognition. The output of the last layer of the encoder is passed through a classification head containing only layer normalization and full connectivity to obtain the recognition result y, which is attached to it. The operational formula is shown in equation (7).

$$y=LN\left(Z_L^0\right) \quad (7)$$

4. Experiments Result and Discussion

4.1. Datasets

The dataset, generated by GNU [8-9] Radio and called RML2016.10a, contains 8 digital and 3 analog modulations. These modulations are all widely used in the wireless communication systems around us, specifically BPSK, QPSK, 8PSK, QAM16, QAM64, GFSK, CPFSK, PAM4, WBFM, AM-SSB and AM-DSB.

In this dataset, each modulated signal contains 20 signal-to-noise ratios, each with 1000 samples, and each sample has two signals, I and Q. Each signal contains 128 points. The RML2016.10a dataset has a total of 220,000 data, which 154,000 data are selected as the training set, and the remaining 66,000 data are used to testing.

4.2. Experimental Setup

In all experiments of this paper, the Transformer encoder is chosen as the main part of the network. In this module, the hidden size is set to 768, the dimension of the FFN is 3072, the number of heads is 12, and the number of layers of the network is 12. In addition, the hyper-parameters settings of the training process are as follows: epoch is 100, the batch size is 100, and the hyper-parameters for testing are as follows: batch size is 1000. The parameters are updated every mini-batch during the experiment. At the same time, the loss mode is cross entropy, and the activation function uses Adam's [19] optimizer. The relevant paramenters of the Adam optimizer are as follows: the learning rate is 0.001 and the weight decay is 5e-5.

In the model evaluation part, Accuracy and multi-classification confusion matrices under each SNR dataset are used as evaluation metrics.

4.3. Window selection experiment

In this section, we discuss the choice of window size to illustrate the effect of the signal block size or sequence length on the classification accuracy of the model.

According to the original input sequence size, we mainly use the following window sizes for training and testing. The window sizes are 32×2, 16×2, 8×2, 4×2, and 1×2. The classification accuracy with increasing SNR for different windows is obtained, as shown in Figure 2.

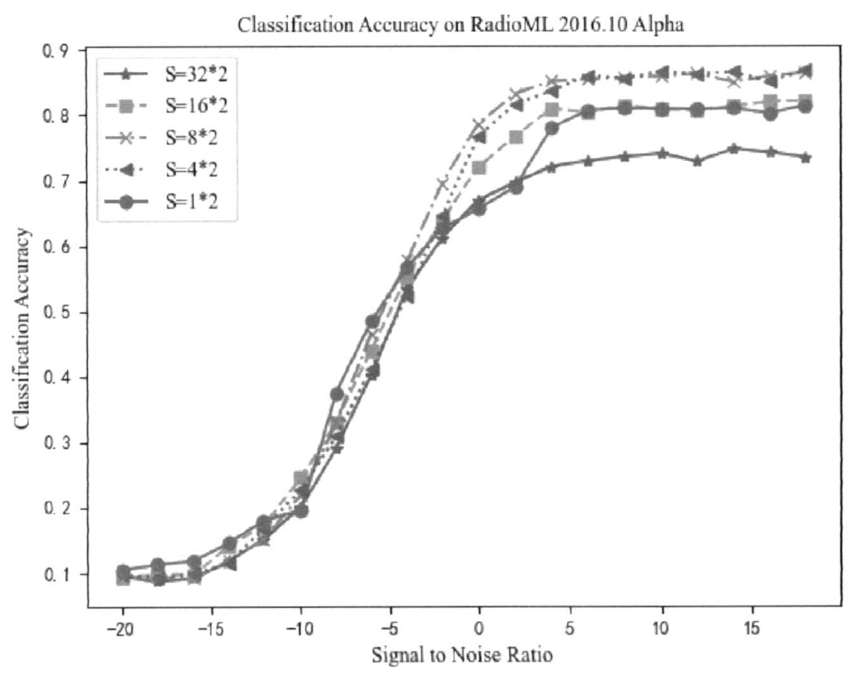

Figure 2. Accuracy vs. Signal sequence size(S).

Figure 2 illustrates how model function varies with the size of the time sample period adopted for each instance in a single classification used. Here, accuracy improved by about 4-7% for every doubling of the signal segment size, and when we reach 4×2, the returns do not change significantly. However, the returns decrease significantly as we reach 1×2. Therefore, we find that the window size, i.e., the signal block size, will have an impact on the classification accuracy of the model, and on balance, we choose a window of size of 8 × 2.

4.4. Comparison with other methods

To verify the effectiveness of the Transformer-B/8×2 method, a quantitative comparison is done on the dataset RML2016.10a with other methods, which are CNN2, LSTM, and CLDNN. In this paper, the training set, test set, and hyper-parameters of all methods are the same as the model. The classification accuracy on the dataset RML2016.10a is shown in Figures 3.

Figure 3 displays connection of four methods on different signal-to-noise radio data sets. When SNR is more than 10dB, the Transformer-B/8×2 is always higher than the other methods.When SNR is +2dB, the Transformer-B/8×2 is 12.53% higher than the CLDNN.When SNR is +8dB, the Transformer-B/8×2 is 12.94% higher than the CLDNN.When SNR is +18dB, the Transformer-B/8×2 is 13.33% higher than the CLDNN.

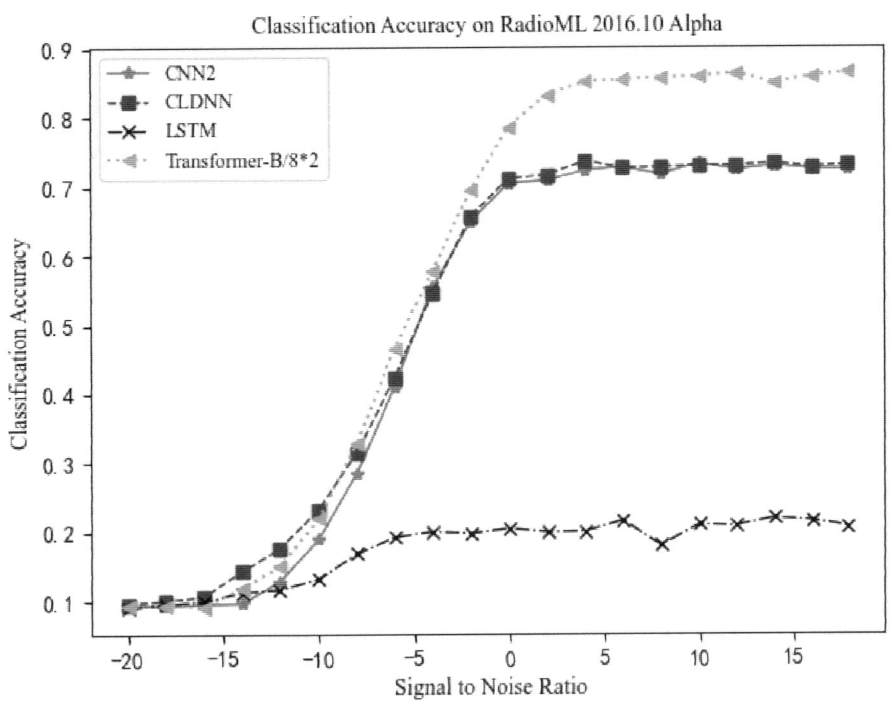

Figure 3. Accuracy vs. SNR in all models.

Figure 4, 5, 6 show the multi-classification confusion matrix of four methods at different SNRs which are -6dB, 8dB, and 18dB.Where, the vertical axis represents the recognized modulation type, the horizontal axis represents the real modulation type, and the coclor density of the grid represents the recognition effect of this type.

As shown in Figure 4, it can be seen that the modulation methods AM-DSB and AM-SSB are better identified among the four methods. Figure 4(a) shows that the WBFM modulation method is obviously misidentified as the AM-DSB modulation method. From Figures 4(b), (c), and (d), it is observed that CNN2, CLDNN, and Transformer-B/8×2 also identify PAM4 better and the latter two methods also identify QAM64 better. In addition to this, it is also observed that confusion exists mainly between QAM16 and QAM64, between CPFSK and GFSK, and between AM-DSB and WBFM.

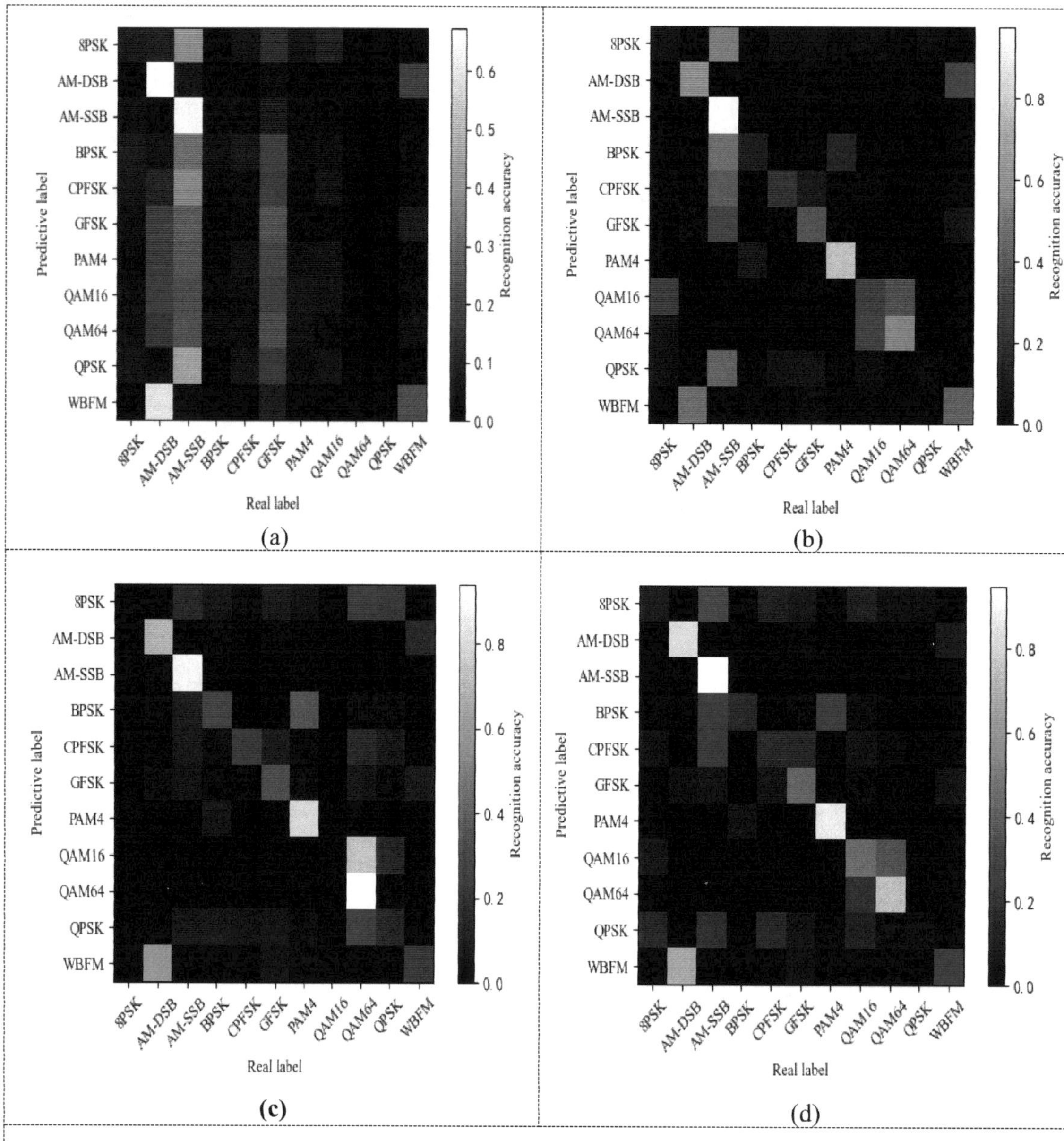

Figure 4.Confusion Matrix for SNR=-6dB of four networks. (a) LSTM(b)CNN2(c) CLDNN (d) Transformer-B/8×2.

Observing Figure 5(a), we find that the LSTM method can only identify AM-DSB and AM-SSB better; while observing Figure 5(b), (c), and (d), we find that the other three methods can identify not only AM-DSB and AM-SSB better, but also BPSK, CPFSK, GFSK, PAM14, etc. well. In addition, the Transformer-B/8×2 method can better identify QPSK and 8PSK compared with CNN2, better identify 8PSK compared with CLDNN, and distinguish QAM16 and QAM64.

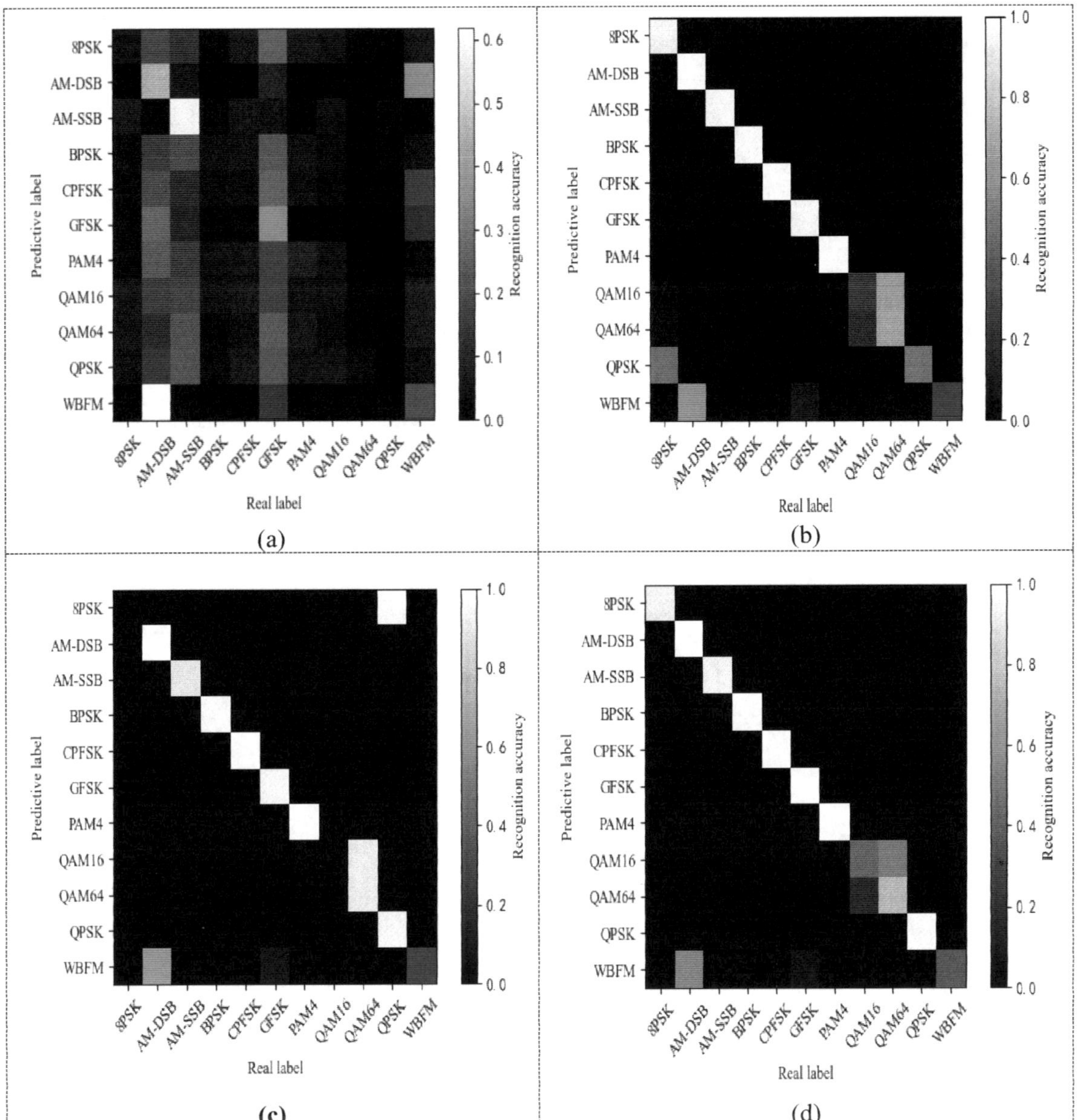

Figure 5. Confusion Matrix for SNR=+8dB of four networks. (a) LSTM(b)CNN2(c) CLDNN (d) Transformer-B/8×2.

Figure 6 shows that all four methods have confusion between WBFM and AM-DSB, while CCN2 and CLNDD methods also have confusion between 8PSK and QPSK, QAM16, and QAM64.The Transformer-B/8×2 method also has confusion between QAM16 and QAM64. Observing Figures 6(b), (c), and (d), it is found that CNN2, CLDNN, and Transformer-B/8×2 methods can better identify AM-DSB, AM-SSB, BPSK, CPFSK, GFSK, PAM14, QAM16, and QPSK while Transformer-B/8×2 method can also better identify 8PSK.

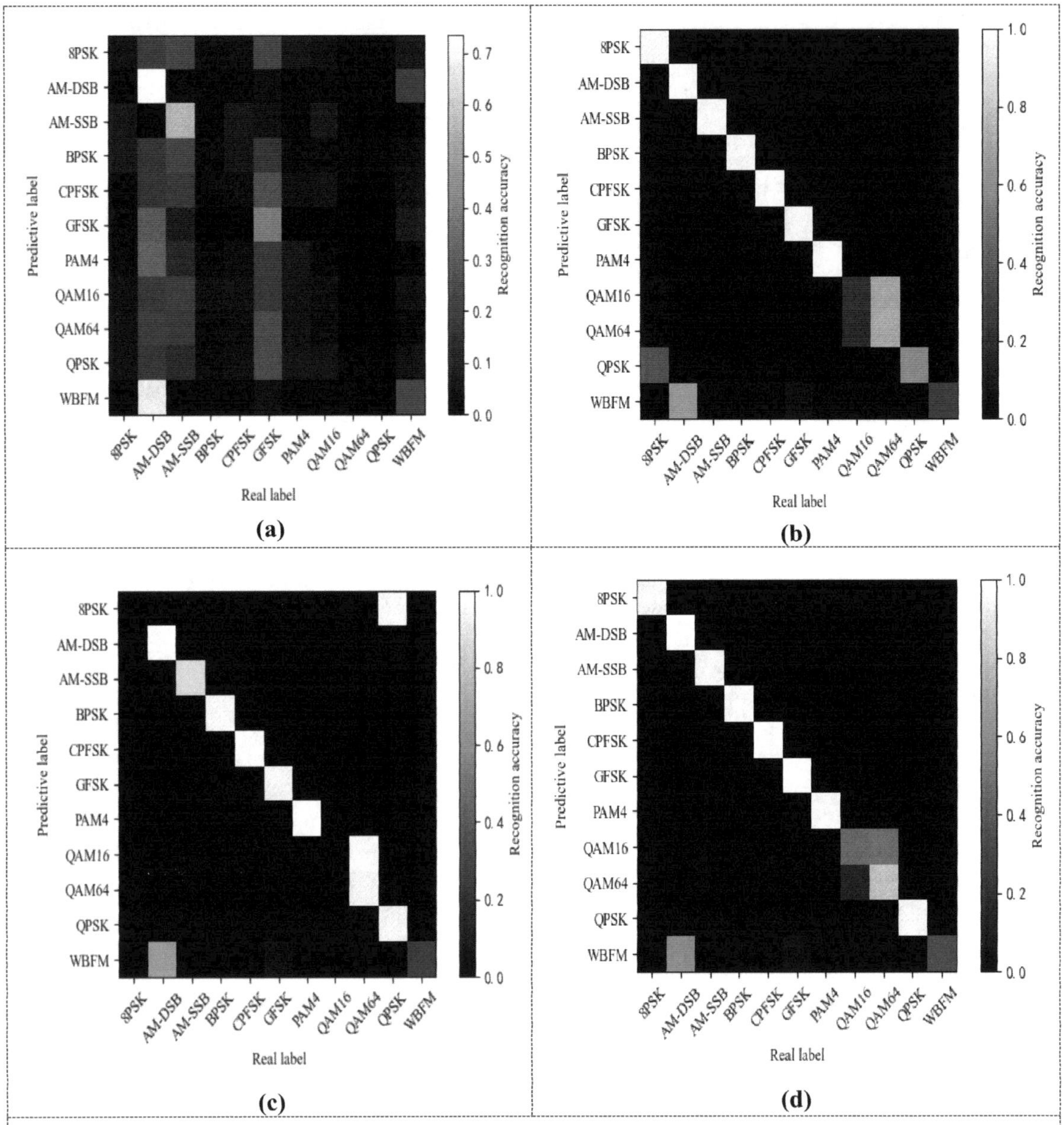

Figure 6.Confusion Matrix for SNR=+18dB of four networks. (a)LSTM(b)CNN2(c) CLDNN (d) Transformer-B/8×2.

5. Conclusions

In this paper, we employ Transformer-based architecture for automatic modulation mode recognition. Compared with CNN guidelines, pure LSTM, and CLDNN, the advantage of the proposed architecture lies in the self-attention. At the high SNR, the recognition accuracy improved from 73% to 86.4%.

Furthermore, based on the study of structural parameters of ViT, we recommend the ViT-B/8×2 architecture for automatic modulation mode recognition.

In this paper, we explore the direct application of Transformer to modulation recognition. We interpret the signal as a series of signal segments and process it through the standard Transformer encoder used in the NLP.

While these initial results are encouraging, many challenges remain. For example, without robust competitive datasets in the radio field, it is hardly to compare function directly with current methods. Besises, the performance of the influence of signal segment sequence size and deeper Transformer encoder layer, in turn, opens up a new idea for obtaining better performance at low SNR values for future scientific research.

Acknowledgments

The authors gratefully acknowledge funding from the Doctoral Scientific Foundation of Guizhou Normal University and the Natural Science Research Foundation of the Education Department of Guizhou Province. Finally, the writer be gradeful the editor and paper reviewer for their considerate comments and constructive feedback.

References

[1] Dobre OAand Hameed F2006Likelihood-based algorithms for linear digital modulationclassification in fading channels*Proc. Cana. Conf. Elect.& Comput. Engineer.* (Piacataway)pp 1347-50
[2] Dobre O A, Abdi A, Bar-Ness Y and Su W 2007 Survey of automatic modulation classification techniques: classical approaches and new trends*IETCommun.*1(2):137-56
[3] Sainath T N, Vinyals O, Senior A, and Sak H2015 Long short-term memory, fully connected deep neural networks*Pro. IEEEInt. Conf. on Acoustics, Speech, and Signal Process.* (Piscataway)pp 4580-84
[4] Krizhevsky A, Sutskever I and Hinton G E2017 Imagenet classification with deep convolutional neural networks*Commun. of the acm* 60(6):84-90
[5] Sainath T N, Weiss R J, Senior A W, Wilson K W and Vinyals O2015 Learning the speech front-end with raw waveform cldnns*Pro. Conf. Int. Speech Commun. Association*(ISCA)pp 1-5
[6] O'SheaT J, Corgan J and Clancy T C2016 Unsupervised representation learning of structured radio communication signals *Pro. Int. Workshop on Sensing and Learning for Intelligent Machines*(Piscataway)pp 1-5
[7] O'Shea T, West N, Vondal M, and 2017 Semi-supervised radio signal identification *Pro. Int. Conf. Advanced Commun. Tech.*(Piscataway)pp 33-38
[8] O'Shea T and West N 2016 Radio machine learning dataset generation with gnu radiohttps://pubs.gunradio.org/index.php/grcon/article/view/11
[9] O'shea T J, Corgan J, and Clancy T C 2016 Convolutional radio modulation recognition network *Pro. Int. Conf. Engineer. Appli. Neural Networks* (Cham)pp 213-26
[10] Emam A, Shalaby M, Aboelazm M A, Bakr H E A, and Mansour H A A2020 A Comparative Study between CNN, lstm, and cldnnmodels in the context of radio modulation classification*Pro. Int. Conf. Electrical Engineering*(Piscataway)pp 190-95
[11] Vaswani A, Shazeer N, Parmar N, Uszkoreit J, Jones L, Gomez A N, Kaiser L, and Polosukhin I2017 Attention is all you need. *Neural Information Processing Systems*30:5998-6008
[12] Zhu X, Su W, Lu L, Li B, and Dai J 2020 Deformable detr: deformable Transformers for end-to-end object detection https://arvix.org/pdf/2010.04159.pdf
[13] Dosovitskiy A, Beyer L, Kolesnikov A, Weissenborn D and Houlsby N 2020 An image is worth 16x16 words: Transformers for image recognition at scalehttps://arvix.org/pdf/2010.11929v2.pdf
[14] Bello I, Zoph B, Le Q, Vaswani A, and Shlens J 2019 Attention augmented convolutional networks *IEEE Int. Conf. Computer Vision* (Piscataway)pp 3285-94
[15] Devlin J, Change M W, Lee K and Tputanova K 2018 Bert: pre-training of deep bidirectional Transformers for language understanding https://arvix.org/pdf/1810.04805v1.pdf
[16] Dong L, Xu S and Xu B 2018 Speech-Transformer: a no-recurrence sequence-to-sequence model for speech recognition *Pro. IEEE Int. Conf. Acoustics,Speech, and Signal Process.*(Piscataway)pp 5584-888

[17] Ba J, Jr K, and Ge H 2016 Layer normalization https://arvix.org/pdf/1607.06450.pdf
[18] Nair V and Hinton G E2010 Rectified linear units improve restricted Boltzmann machines*Int. Conf. on Machine Learning*(Omnipress)pp 807-14
[19] Kingma D and Ba J 2014 Adam: amethod for stochastic optimization http://arxiv.org/pdf/1412.6980.pdf

Journal of Physics: Conference Series **2384** (2022) 012018 doi:10.1088/1742-6596/2384/1/012018

A Kind of Antenna Array Based on Dielectric Antenna with Small Size

Chang Zhou[1], Jing Xia[1], Yaohua Li[1] and Nan Wang[1]

[1] The Fifth Research Lab, Wuhan Second Ship Design and Research Institute, Wuhan, Hubei, 430074, China

Corresponding author's e-mail: rabbite6@126.com

Abstract. Based on the antenna array design method, a new antenna array is designed in this paper. The optimization design results show that the existence of dielectric materials can effectively compress the electromagnetic field, which could greatly reduce the antenna size. The design results show that the size of this type of antenna array is 126mm * 126mm, which is only 70% of the conventional antenna array, the working frequency is 2.4GHz, the return loss is not more than -10dB, the gain is not less than 12dBi, the half power beam width is not more than 40 degrees and the design results show that the dielectric antenna array has excellent performance and good application prospect in the modern communication system.

1. Introduction

The antenna is the key link of the modern communication system and the communication efficiency and operation mode of the antenna directly determine the working efficiency of the communication system [1-5]. Therefore, all communication system antennas need to be strengthened according to their own needs and meet all relevant indicators. At present, the focus of antenna design is to improve the gain and directivity of the antenna. High gain could increase the working distance and efficiency of the antenna and directivity can reduce the interference between antennas which is very important in the complex communication system. Among kinds of antennas, reflector antenna, lens antenna, and array antenna have certain advantages in the above data [6-9]. Among the above antennas, the reflector antenna has stable electrical performance, low price, and convenient installation. It is the most widely used antenna with high gain and high directivity at present. However, the reflector antenna itself is large and needs to occupy a larger area when turning in work and it is difficult to be widely used in complex systems. The lens antenna is also large in size and high in price. The array antenna can be made into a plane conformal state, which has considerable advantages in space. Therefore, the array antenna has been studied extensively.

Recently, the array antenna based on the dielectric antenna elements has received more attention and research due to its excellent performance. Its characteristic is that the dielectric antenna can effectively compress the electromagnetic field which makes the array antenna based on the dielectric antenna elements have a smaller area [10]. Based on the above theory, this paper designs an antenna array based on the circular dielectric antenna elements and the performance of the antenna array is also optimized and analyzed. The design results show that, compared with the traditional array antenna, the electrical performance of the array antenna based on the dielectric antenna elements is basically the same, but it has obvious advantages in volume.

2. Models of Antenna

The cylindrical antenna array is composed of nine antenna elements. The antenna elements are cylindrical dielectric antennas. The schematic diagram of the antenna array and dielectric antenna elements is shown in Figure 1.

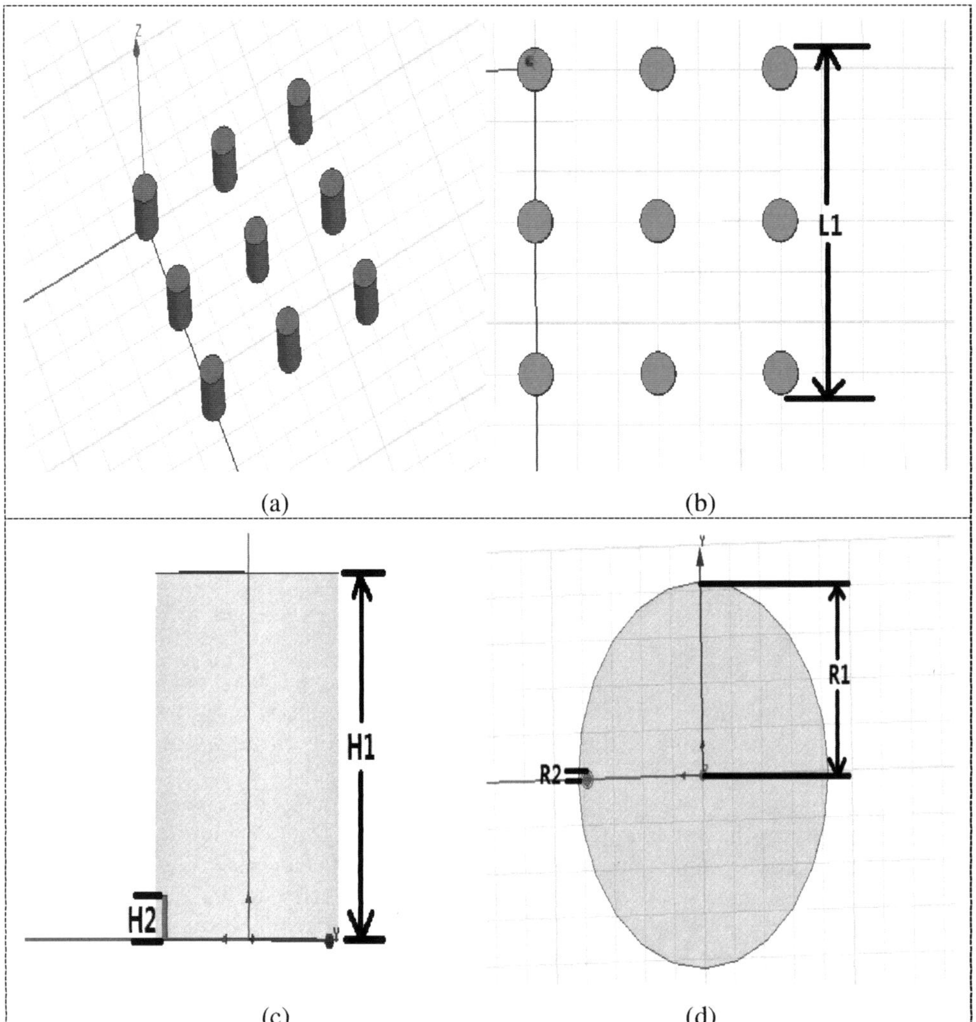

Figure 1. The models of the antenna array and antenna element.

3. Simulation and Discussion

According to the design requirements, the working frequency of the antenna array is 2.4GHz, the gain is not less than 12 dBi, and the half-power beam width is not more than 40 degrees.

In order to ensure the realization of the above requirements, the design indexes of the cylindrical dielectric antenna element are preferably as follows: the working frequency is 2.4GHz, the gain is not less than 5dBi and the half power beam width is not more than 80 degrees.

Based on the above design requirements, the cylindrical dielectric antenna element is firstly designed. The initial parameters are as follows: the height of the cylindrical antenna is 50mm, the radius is 21mm, and the height of the energy probe is 6mm.

The preliminary simulation results are shown in Figure 2.

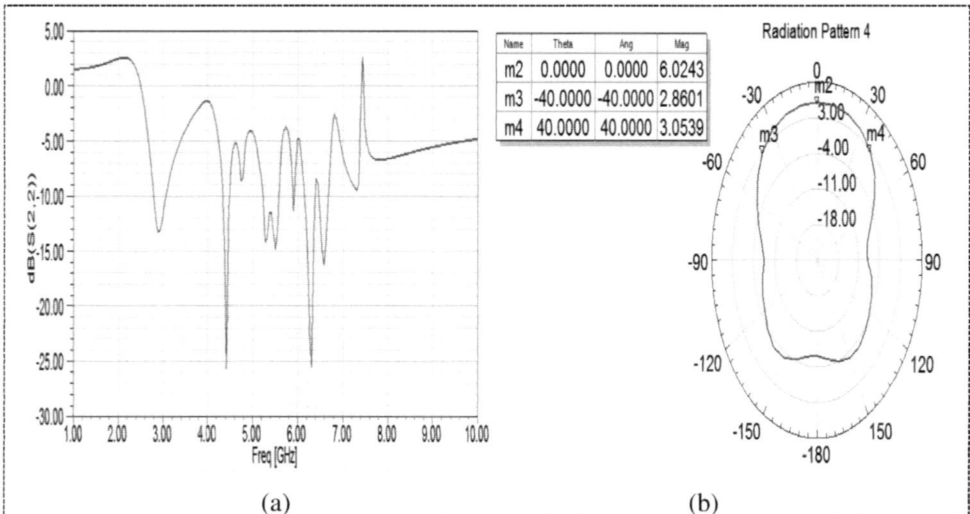

Figure 2. The simulation results of the dielectric antenna element.

According to the analysis of the calculation results, the performance of the cylindrical dielectric antenna element is: the gain of the antenna could reach 6dbi, the half power beam width is not more than 80 degrees and its far-field performance basically meets the requirements. On the other hand, the return loss greatly exceeds the standard. Therefore, it can be judged that the dielectric material and the overall configuration of the cylindrical dielectric antenna meet the design requirements, but the position and size of the energy probe need to be further optimized. Therefore, the dielectric material and external configuration parameters of the cylindrical antenna element remain constant, the position and size of the energy probe are adjusted, and the optimal simulation results are also shown in Figure 3.

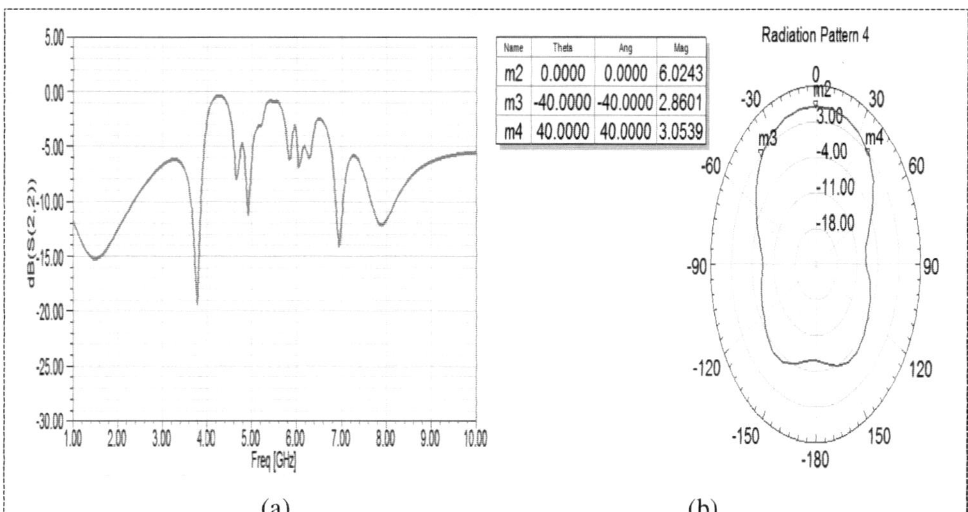

Figure 3. The optimal simulation results of the dielectric antenna element.

According to the calculation results, it could be seen that the return loss parameters can be effectively optimized by changing the position and size of the energy probe and the initial far-field performance of the antenna element is basically not affected.

On the basis of the above results, the cylindrical dielectric antenna element is formed into an antenna array according to Figure 1. Based on the antenna array design method, the amplitude and phase distribution of each antenna element are as follows: element 1 phase is 0 and amplitude is 0.4373 ; element 2 phase is 0 and amplitude is 0.6613 ; element 3 phase is 0 and amplitude is 0.4373 ; element 4 phase is 0 and amplitude is 0.6613 ; element 5 phase is 0 and amplitude is 1 ; element 6 phase is 0 and amplitude is 0.6613 ; element 7 phase is 0 and amplitude is 0.4373; element 8 phase is 0and amplitude is 0.6613 ; element 9 phase is 0 and amplitude is 0.4373;

The simulation results of the antenna array are shown in Figure 4.

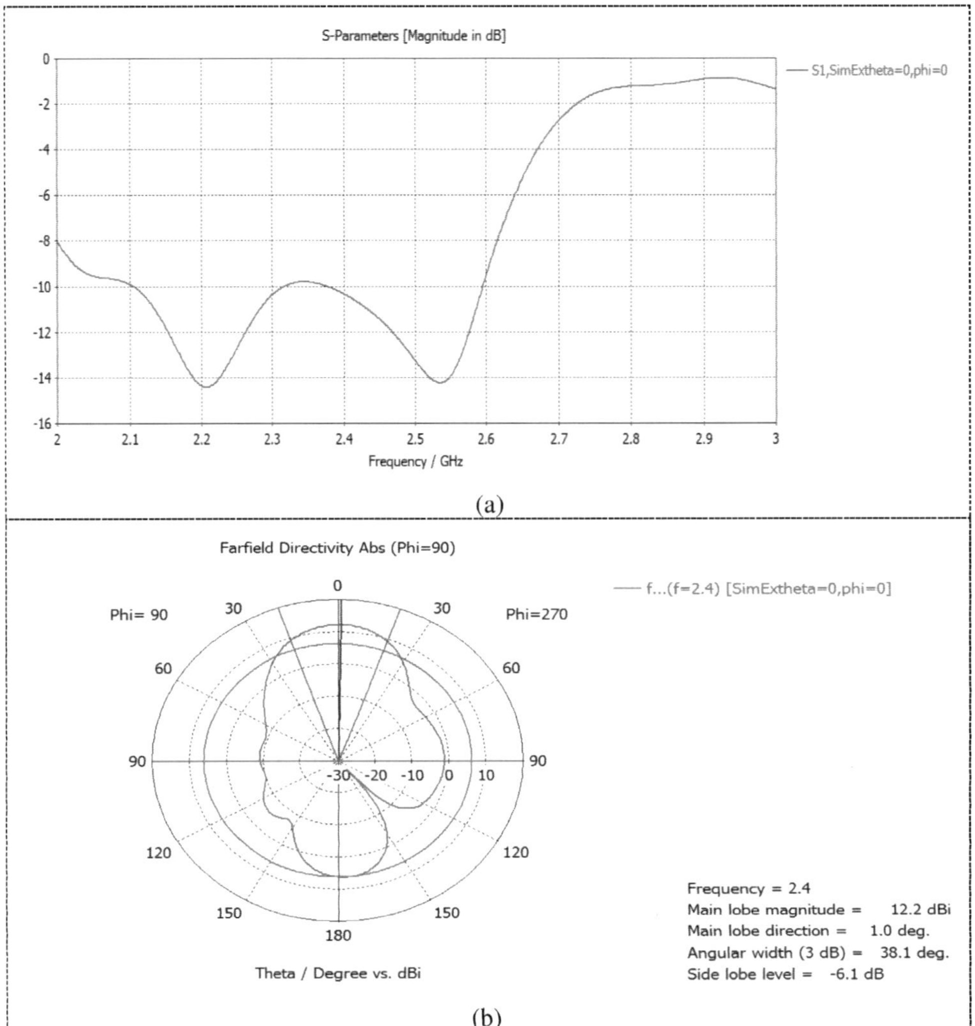

Figure 4. The optimal simulation results of the antenna array.

According to the results, the working frequency of the antenna array is 2.4GHz, the return loss is not more than -10dB, the gain is not less than 12dBi, the half power beam width is not more than 40 degrees and the design results all meet the requirements.

4. Conclusions

In conclusion, through reasonable material selection and design, the size of the cylindrical dielectric antenna is much lower than that of the conventional antenna and the performance can also meet the design requirements. The size of the antenna array designed in this paper is 126mm * 126mm, which is only 70% of the conventional antenna array, the working frequency is 2.4GHz, the return loss is not more than -10dB, and the gain is not less than 12dBi, the half power beam width is not more than 40 degrees and the design results show that the antenna array based on the dielectric antenna has small size and excellent performance, which plays an important role in improving the performance of communication system.

References

[1] Liu Y, Jiao Y, Weng Z, Zhang C and Zhang H 2018 A high-gain dielectric resonators antenna array fed by back-cavity 2018 Int. Workshop on Antenna Technology (iWAT) (Nanjing) pp 1-3

[2] Kumar V S and Kurup D G 2017 Design of dielectric rod antenna for ground station applications. IEEE Int. Conf. on antenna Innovations & Modern technologies for Ground, Aircraft and Satellite Applications (iAIM) (Bangalore) pp 1-4

[3] Zou M and Pan J 2015 Investigation of resonant modes in wideband hybrid omnidirectional rectangular dielectric resonators antenna IEEE Trans. Antennas Propag. 63 3272-75

[4] Kiyani Nasimuddin and Esselle K P 2018 A wideband circularly polarized dielectric resonators antenna over a metasurface IEEE Int. Symp. on Antennas and Propagation & USNC/URSI National Radio Science Meeting (Boston) pp 2085-6

[5] Li R Y, Jiao Y C, Zhang Y X, Wang H Y and Cui C Y 2020 A wideband MIMO triangular dielectric resonators antenna with a simple decoupling structure 9th Asia-Pacific Conf. on Antennas and Propagation (APCAP) (Xiamen) pp 1-2

[6] Sun W, Yu Y, Hu Y, Yang W and Chen J 2022 A novel kind of wideband low-profile dielectric resonators antenna suitable for beam-scanning applications China Commun. 19 98-107

[7] Zhou C, Fan Y W, Yang H R, Xu R Y, Liu G and Qu D 2015 GPS antenna based on low loss hexaferrite Int. Conf. on Energy Equipment Science and Engineering (Guangzhou) pp 1549-52

[8] Zhou C, Ji C and Wu G P 2017 An antenna with good electrical properties and artistic shape for lampshade Proc. of The 2017 3rd Int. Conf. on Architectural, Civil and Hydraulics Engineering (ICACHE 2017) (Yunnan) pp 162-7

[9] Gan L, Jiang W, Chen Q, Li X, Zhou Z and Gong S 2021 Method to estimate antenna mode radar cross section of large-scale array antennas IEEE Trans. Antennas Propag. 69 7029-34

[10] Yang Y, Ban Y L, Yang Q, Lian J W, Sun Q and Wu G 2022 Millimeter wave wide-angle scanning circularly polarized antenna array with a novel polarizer IEEE Trans. Antennas Propag. 70 1077-86

Design of branch line directional coupler for 5G millimeter wave communication

Aiguo Shen [1,a], Yukai Lin [1,b], Chendong Yang [2,c], Guangsong Yang [1,d], Gaiyan Hong [1,e*], Dezhi Wei [1,f*]

[1] Chengyi University College, Jimei University, Xiamen, China

[2] School of Ocean Information Engineering, Jimei University, Xiamen, China

[a]email: jakefunny@163.com, [b]email: 2451878635@qq.com, [c]email: 1434492017 @qq.com, [d]email: gsyang@jmu.edu.cn, [e]email: wtzhgy@163.com, [f]email: weidezhi@163.com

*Corresponding author: [e]email: wtzhgy@163.com, [f]email: weidezhi@163.com

Abstract: A directional coupler based on a microstrip branch line for 5G millimeter wave communication is designed. The branch line structure is simple and easy to implement. The comprehensive decomposition design method of the microwave system is proposed to improve the design efficiency. Based on the theoretical analysis of microwaves, the design method of combining automatic tuning and manual tuning optimization is used to optimize the schematic simulation. The layout simulation adopts iterative optimization to approximate the design index and reduce the influence of branch line connection loss. Finally, a millimeter wave branch line directional coupler is designed with a bandwidth of 24.25ghz to 27.5ghz, a coupling degree and insertion loss of about 5dB, and an isolation degree of more than 20dB. In the coming 6G, this design method has a certain forward-looking significance.

1. Introduction

As an important passive device in a 5G microwave circuit, the directional coupler can realize power distribution between input and output ports and maintain a certain phase difference. It is widely used in the mixer, balanced amplifier, and array antenna feeding network [1]. At present, some new progress has been made in the study of couplers as follows.

In [2], a kind of stripline-line forward wave directional coupler with a short circuit branch line is proposed. The frequency band range is 6.1-6.7 GH. A short circuit branch line is added to the coupler to enhance the working bandwidth of the unit cell design. A new method of realizing a coupled-line co-directional (CL-COD) coupler is proposed for the first time [3]. The experimental results show that compared with the traditional single section $\lambda/4$ branch line coupler, the size of the designed single section CL-COD coupler is reduced by 64.1%, the bandwidth is increased by 13%, and the second harmonic suppression exceeds 20 dB.

An artificial neural network (ANN) is used for the first time to extract the transfer function of the proposed coupler resonator so that the value of the transmission zero can be located within the desired frequency range [4]. A microstrip coupler with a simple structure, compact structure, T-shaped resonators, and open-ended stubs at 1.8 GHz is designed and fabricated. The coupler produces two transmission zeros at 3.4ghz and 3.87ghz with an attenuation level of more than 50 dB and has a good suppression effect on the second harmonic. In [5], a 28 GHz Butler matrix switching beam antenna suitable for 5G wireless applications is proposed. A broadband two-segment branch line coupler is used to reduce the phase and amplitude errors and the Butler matrix variation. The aluminum shell stabilizes electrical properties, reduces side lobes, and improves structural stability. The measurement results show that the four-way switched beam antenna successfully demonstrates the advantages of high integration, showing high-performance 5G wireless applications at 28 GHz. In [6], a novel microstrip high harmonic suppression branch line coupler (BLC) with high performance is designed, which uses T-shaped resonators, open stubs cross-shaped resonators, and radial resonators. The BLC is designed to operate at a 1GHz frequency and can suppress up to 5 spurious harmonics with an attenuation level of 20 dB. A 180-degree branch-line coupler based on the K band is proposed, which consists of two types of iris-loaded waveguides [7]. The coupler is based on the optimal electrical length, and the 180-degree output phase difference can be achieved without the need for an additional phase shifter.

In [8], a kind of miniaturized lumped component branch coupler with a PI equivalent circuit is designed. TSV technology is used to realize the lumped component in the branch coupler. The industrial simulation software HFSS is used to simulate and verify the coupler in S-band. The simulation results show that the return loss of the circuit is greater than 19 dB, the isolation performance is better than 20 dB, and the insertion loss is less than 1.53 dB at 2.1-2.4 GHz. A novel dual-frequency/dual-mode compact hybrid coupler is proposed, which acts as a dual-frequency branch line coupler at low frequency and a rat-type coupler at high frequency [9]. The design structure is implemented using an artificial transmission line (ATLs) based on an open split ring resonator (OSRR), while the output is on the same side in each operating mode.

The above studies mainly involve the low-frequency band, but the millimeter wave above 20GHz is rarely involved. The rest of this paper mainly explains the advantages of the branch line and related theoretical analysis. On this basis, how to design a millimeter wave coupler is introduced in detail.

2. Design of 5G millimeter wave directional coupler

2.1. Branch line directional coupler

The main forms of the directional coupler are branch line directional coupler, coupling line directional coupler, Bethe hole directional coupler, and Lange coupler. It is not convenient to make and apply the geometric shape of the Bethe hole coupler. For the Lange coupler, the distance between the coupling lines is very narrow. Before the production of the Lange coupler, it is necessary to deal with the processing problem of connecting lines between the coupling lines. In the process of realizing the integration and miniaturization of the 5G millimeter-wave circuit, they are not as simple and convenient as the branch line directional coupler. Therefore, we use a branch line directional coupler to design a 5G millimeter-wave circuit, which has significant advantages in terms of process and implementation.

2.2. Design Process

The design of the coupler uses the comprehensive decomposition method of the microwave system to streamline the whole design process, which greatly improves the design efficiency. The overall design process is as follows.

1)First of all, the technical index of the branch line directional coupler should be determined. Only after the technical index is determined, the parameter optimization of the branch line coupler can be carried out with purpose. After all, the characteristics of the branch line coupler will change with the frequency. The main coupling coefficients are substrate parameters, center frequency F, characteristic impedance Z of each port, and coupling degree C.

2)The normalized admittance and characteristic impedance of the branch line and main line are calculated by theoretical analysis. Then the ADS software calculates the required microstrip line size after setting the substrate parameters and characteristic impedance.

3)The ADS software is used to design and simulate the schematic diagram, and the S parameter waveform is obtained to confirm whether it meets the expected index. If not, fine-tuning and optimization of the microstrip line size should be carried out to make it conform to the set index. Then the corresponding layout should be generated after the schematic diagram optimization.

4)The layout simulation adopts the electromagnetic simulation of the moment method, which is iteratively optimized according to the set index until it meets the requirements.

2.3. Design indicators

ITU (International Telecommunication Union) decided to use 24.25GHz-27.5 GHz, 37GHz-43.5 GHz, and 66GHz-71GHz millimeter wave bandwidth of 14.75 GHz for 5G commercial application. Therefore, the expected operating band of the millimeter wave directional coupler is set as 24.25 GHz-27.5 GHz. Its center frequency is set to 26GHz.

Substrate selection is an important factor for directional couplers. When designing a directional coupler, it is necessary to understand the influence of the PCB layout on the final performance of the directional coupler. It is known that the dielectric constant and dielectric loss of the Substrate will directly affect the insertion loss of the directional coupler, and the dielectric constant of some Substrates will change with temperature. Therefore, select the Substrate with high dielectric constant stability to reduce the insertion loss of the final directional coupler. After several adjustment tests, RO4350B is selected as the substrate in this paper, because its dielectric constant stability is relatively high. The standard value is 3.48 at 10GHz, and the dielectric constant will decrease with the increase of frequency. The dielectric constant at 24GHz frequency is 3.47, which is 0.01 lower than that at 10GHz frequency. Since the frequency of this design is higher than 24GHz, and the theoretical simulation has a certain deviation from the actual simulation, the dielectric constant is set as 3.66.

In summary, the expected design indicators are as follows:

1) The operating bandwidth of the branch line directional coupler is between 24.25GHz and 27.5GHz, and the center frequency is 26GHz.

2) The coupling degree of the branch line directional coupler is 5dB.

3) The insertion loss of the branch line directional coupler is 5dB.

4) Isolation degree of the branch line directional coupler is greater than 20dB.

5) A load of each port of the branch line directional coupler is 50 ohms.

6) The substrate parameters of the branch-line directional coupler are the dielectric constant of 3.66, the dielectric layer thickness of 0.5mm, and the loss Angle tangent (TanD) value of 0.02.

2.4. Normalized admittance

When the coupling degree is given, the normalized admittance value of each branch line of the directional coupler can be calculated by Equations (1), (2), and (3).

$$|U_{r3}| = \sqrt{\frac{1}{\lg^{-1}\frac{C(dB)}{10}}} \tag{1}$$

$$b = \frac{1}{\sqrt{1-|U_{r3}|^2}} \tag{2}$$

$$a_1 = b|U_{r3}|, a_2 = \frac{a_1}{R} \tag{3}$$

Where C is the coupling degree, $|U_{r3}|$ is the reflected voltage, R is the variable resistance ratio, and B, a_1 and a_2 are the normalized admittance values of the main line and the branch line respectively. Through calculation, we can get the following results.

$$a_1 = a_2 = 1, b = \sqrt{2} \tag{4}$$

The characteristic impedance formula is

$$Z_{a_1} = \frac{1}{Y_{a_1}} = \frac{1}{a_1 Y_0}, Z_{a_2} = \frac{1}{Y_{a_2}} = \frac{1}{a_2 Y_0} \tag{5}$$

$$Z_b = \frac{1}{Y_b} = \frac{1}{bY_0} \tag{6}$$

We can get

$$Z_{a_1} = Z_{a_2} = 50\Omega, Z_b = 35.3\Omega \tag{7}$$

Finally, the ADS calculation software "LineCalc" is used to calculate the width and length of the branch line as 1.140mm and 1.656mm, and the size width and length of the main line as 1.924mm and 1.613mm.

3. Simulation and optimization

This section introduces the simulation and optimization of the branch line directional coupler. Based on the directional coupler parameter data obtained in the previous section, the schematic of the branch line directional coupler and the size of the substrate can be drawn in the software ADS, and then the simulation can be adjusted to achieve the set index according to the S parameter waveform.

3.1. Optimized Schematic simulation

After drawing the schematic diagram, the frequency interval is set to 0.01GHz. S parameter waveforms such as S11, S21, S31, and S41 can be obtained by running simulations, as shown in Figure 1.

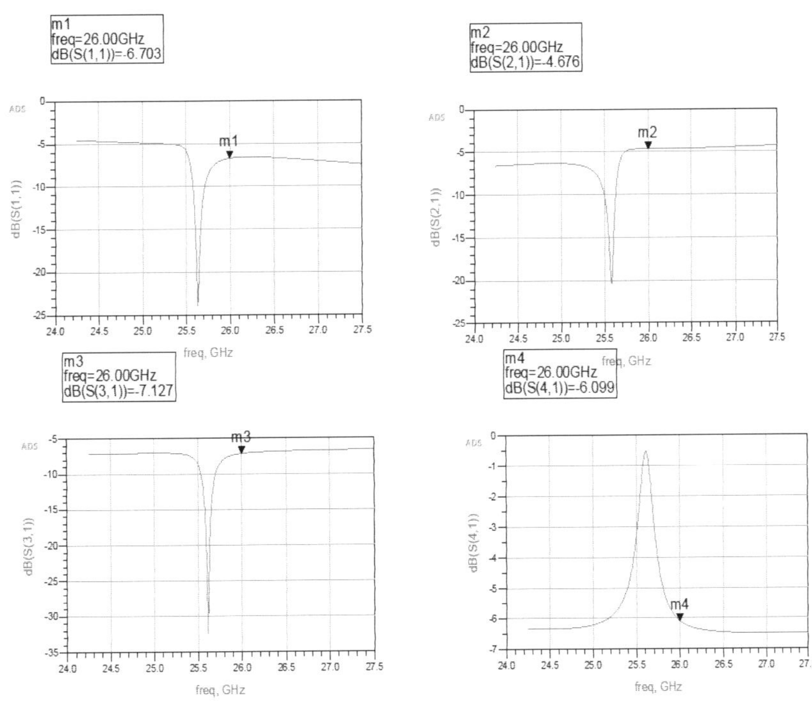

Figure 1. S parameter waveform of the initial schematic simulation

At the point close to 26GHz, the insertion loss is greater than 4dB in the S21 waveform, and the coupling degree is greater than 7dB in the S31 waveform, which obviously does not conform to the set indicators. So tuning optimization is needed. The advantage of manual tuning is that you can debug a rough waveform. But the disadvantage is that it is difficult to make the waveform more accurate, and it may not be possible to know where to start when there are too many unknown parameters to adjust. The automatic tuning is based on computer calculation of automatic fine-tuning, and self-calculated optimization to meet the expected requirements. But the automatic tuning can not meet the requirements of the waveform at one time, need to be tuned many times. If the expected waveform has not been obtained, manual fine-tuning is carried out to achieve the desired effect under the existing parameters of the optimized waveform. Manual tuning and automatic tuning do not conflict, the combination of the two is a more accurate tuning method. The optimization effect of manual tuning and automatic tuning is shown in Figure 2.

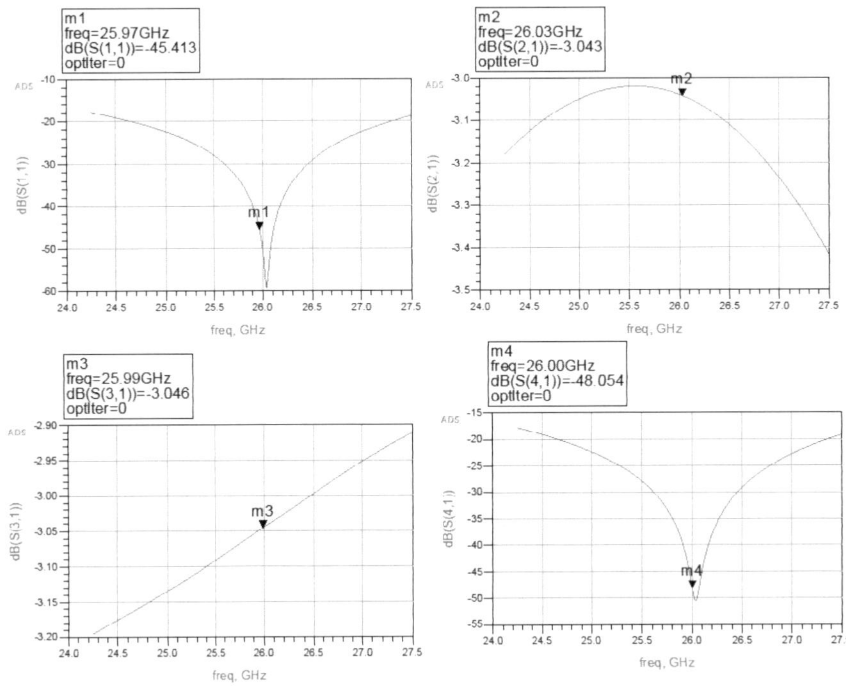

Figure 2. S parameter waveform of the optimized schematic simulation

We can see that at 26GHz, the insertion loss is 3.043dB, the coupling degree is 3.046dB, and the isolation degree is -48dB, which exceeds the target.

3.2. Optimized Layout simulation

After schematic optimization, the layout is generated. Simulation results of the S parameter are shown in Figure 3.

The simulation of the schematic is carried out ideally without the slightest disturbance. The simulation of layout is closer to reality, which fully takes into account some interference and material loss in the transmission of millimeter waves, as well as connection loss. Compared with schematic simulation, the S parameter waveform of layout simulation has different degrees of deviation.

It can be seen from Figure 3 that at 26GHz, the insertion loss in S21 is -8db, the coupling degree in S31 is -5db, and the isolation degree in S41 is -10db. The simulation effect of the branch line directional coupler deteriorates sharply, which is far from reaching the expected index in this paper. Therefore, the layout needs to be optimized.

In this paper, the layout is iteratively optimized according to the expected index until the expected index is reached. The optimized results are shown in Figure 4. It can be seen that after layout optimization, the insertion loss and coupling degree are about -5dB, and the isolation degree is about 21dB, which is within the controllable range of the design index. At this point, the layout is regenerated.

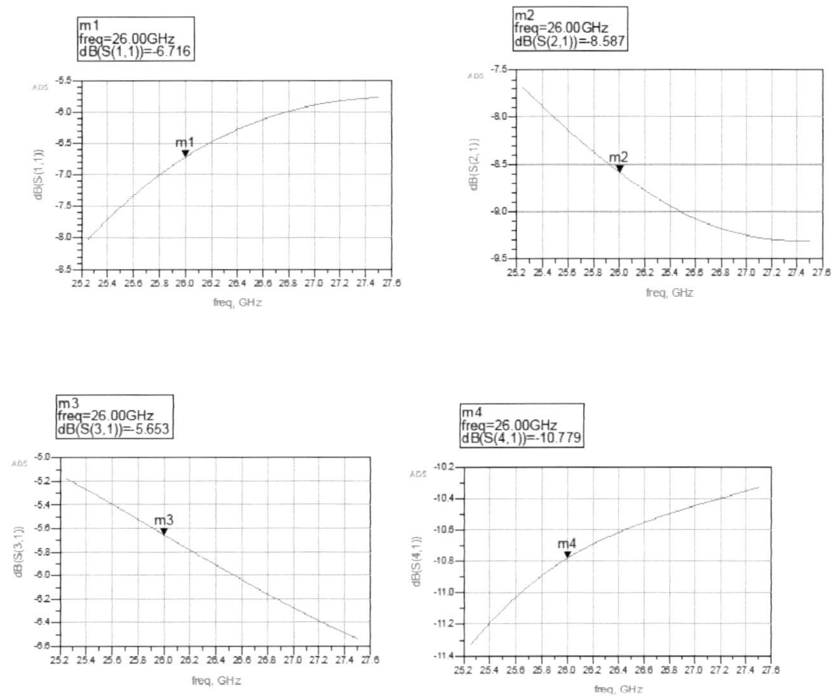

Figure 3. S parameter waveform of the initial layout simulation

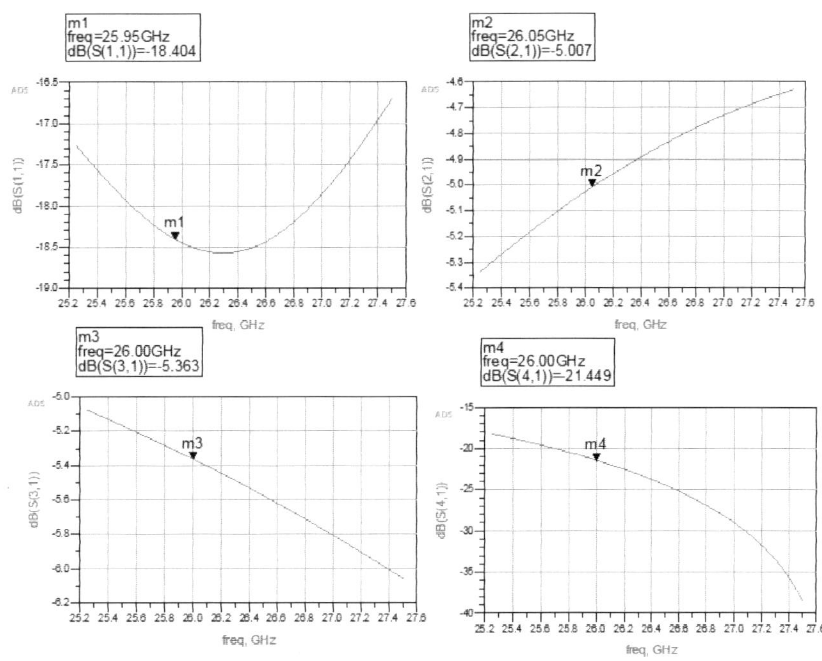

Figure 4. S parameter waveform of the optimized layout simulation

4. Conclusion

With the advent of the 5G and 6G era, the frequency band requirement of microwave devices has been raised to millimeter waves. Therefore, it is very meaningful to design a millimeter wave branch directional coupler.

In real life, it is impossible to make a completely lossless directional coupler, there are a lot of interference factors. The deviation of the S-parameter waveform in the acceptable range appeared in the subsequent improved schematic diagram and layout. Finally, the simulation results show that the designed branch line directional coupler basically meets the expected index. The deviation of the S-parameter waveform is caused by the loss at the junction of branch lines, which is within the acceptable range.

Acknowledgments

This work is supported by the Education and Scientific Research project for young and middle-aged teachers in Fujian Province (JAT210669), the Innovation and Entrepreneurship training program for college students in Fujian Province (S202213471033, S202113471028), and the Natural Science Foundation of Fujian Province (No. 2021J01865).

References

[1] Seko M H, Correra F S. (2019) Dual-band branch-line coupler with shorted stepped-impedance stubs arranged in a π-shaped topology. Microwave and Optical Technology Letters, 61(5): 1154-1160.
[2] Yu Z, Shi Y, Shen R, et al. (2021) Stripline forward-wave directional coupler based on double multi-via mushroom and short-circuited branch line. International Journal of RF and Microwave Computer-Aided Engineering, 31(11): e22850.
[3] Liu H, Guan M, Wang X, et al. (2021) Coupled-line co-directional coupler with inherent harmonic suppression and its wideband implementation. AEU-International Journal of Electronics and Communications, 138: 153854.
[4] Roshani S, Azizian J, Roshani S, et al. (2022) Design of a miniaturized branch line microstrip coupler with a simple structure using artificial neural network. Frequenz, 76(5-6): 255-263.
[5] Lee S, Lee Y, Shin H. (2021) A 28-GHz switched-beam antenna with integrated butler matrix and switch for 5g applications. Sensors, 21(15): 5128.
[6] Roshani S, Yahya S I, Roshani S, et al. (2022) Design of a Modified Compact Coupler with Unwanted Harmonics Suppression for L-Band Applications. Electronics, 11(11): 1747.
[7] Yukawa H, Ushijima Y, Yoneda N, et al. (2020) 180-Degree Branch Line Coupler Composed of Two Types of Iris-Loaded Waveguides. IEICE Transactions on Electronics.
[8] Wang F, Xiao S, Yin X, et al. (2021) A miniature TSV-based branch-line coupler using π equivalent circuit model for the transmission line. IEICE Electronics Express, 18.20210515.
[9] Ghaffarian M S, Moradi G, Khajehpour S, et al. (2021) Dual-Band/Dual-Mode Rat-Race/Branch-Line Coupler Using Split Ring Resonators[J]. Electronics, 10(15): 1812.
[10] Shen A, Wei D, Yang G. (2021) Design of bandpass filter for 5G millimeter wave[C]//Proceedings of the 2021 5th International Conference on Electronic Information Technology and Computer Engineering. pp.97-102.

A Traceable Self-Generated Pseudonym Scheme in VANETs

Bintao He, Shangping Wang

School of Automation and Information Engineering, Xi'an University of Technology Xi'an, China

hebintao@snut.edu.cn

Abstract. The generation of the pseudonym in vehicular ad hoc networks (VANETs) usually requires the participation of a trusted center (TA), and it needs multiple interactions with the TA. To address these problems, we propose a novel traceable self-generated pseudonym based on the non-interactive zero-knowledge proof (NIZK) and consortium blockchain in VANETs. In our scheme, a self-generated pseudonyms method is introduced to eliminate excessive communication burden, and a consortium blockchain is composed of multiple roadside units (RSUs) to replace the TA. RSUs in the consortium blockchain verify the effectiveness of OBU's self-generated pseudonyms through the consensus mechanism. Multiple trace authorities (TRAs) are designed to jointly tracking of the real identity of the OBU, which effectively prevents the abuse of supervision by the only supervisor.

1. Introduction

Vehicular ad hoc networks (VANETs) have become one of the research hotspots in recent years, which can provide safe and efficient traffic information systems. The onboard unit (OBU) interacts with the roadside unit (RSU) through dedicated short-range communication (DSRC) in VANETs. To protect privacy, the OBU usually uses pseudonyms to interact with other entities in VANETs. Zhang et al. proposed an anonymous authentication protocol with conditional privacy protection [1]. There were frequent interactive operations in the pseudonym authentication, which is inconsistent with the real-time goal in VANETs. Literature [2-4] proposes privacy protection schemes that support batch verification in VANETs, but these schemes belong to traditional privacy protection schemes that require trusted centers, prone to a single point of failure. Blockchain is decentralized, data-consistent, and tamper-proof, which can be an excellent solution to the distributed authentication of pseudonyms. However, the existing pseudonyms authentication schemes based on the blockchain usually need authorization from the authority, resulting in a centralized pseudonyms authentication scheme in essence. Literature [5-6] used blockchain to design two distributed pseudonyms publishing schemes, but these two schemes do not consider the regulation of pseudonyms. Literature [7] studied the self-generated pseudonyms schemes. The pseudonym updating is based on the updated database of the TA. It still was a centralized pseudonym management scheme. Literature [8] used the non-interactive zero-knowledge proof (NIZK) to generate pseudonyms to reduce the communication overhead caused by multiple interactions. In literature [9], the centralized TA was replaced by a distributed RSUs association. With the threshold mechanism, the RSUs association issued the certificate for OBU, without interaction with the online TA. This paper uses the NIZK to design a traceable self-generated pseudonym scheme in VANETs.

Journal of Physics: Conference Series **2384** (2022) 012020 doi:10.1088/1742-6596/2384/1/012020

2. Preliminaries

In this section, we will introduce some basic theories in cryptography.

2.1. *Zero-knowledge proof*

In a zero-knowledge protocol, the prover P proves to the verifier V the statement that he knows the secret w without revealing anything about w. We convert to non-interactive zero-knowledge proof (NIZK) by applying the Fiat-Shamir heuristic. It enables the prover P to complete the proof by sending a proof message to the verifier V, without interaction and without returning a challenge. Simulate a random oracle approximated by a secure hash function, P keeps the private key $sk = w$, discloses public $pk = w \cdot P$, randomly selects r, generates a commitment $com = r \cdot P$, calculates $c = hash(com \| pk)$, and the proof $\pi = r + c \cdot w$, P sends c, π to V, V calculates $com_1 = \pi \cdot P - c \cdot pk$, $c_1 = hash(com_1 \| pk)$, and judges whether $c_1 = c$ is established. If so, receive the statement that P knows the private key w. It can be seen from figure 1 that in the interactive zero-knowledge proof process, P and V need to interact three times. The NIZK makes the proof process more efficient and secure. The prover only needs to provide evidence once to achieve the purpose. It can effectively prevent communication content from being tampered with and intercepted and can effectively prevent security risks.

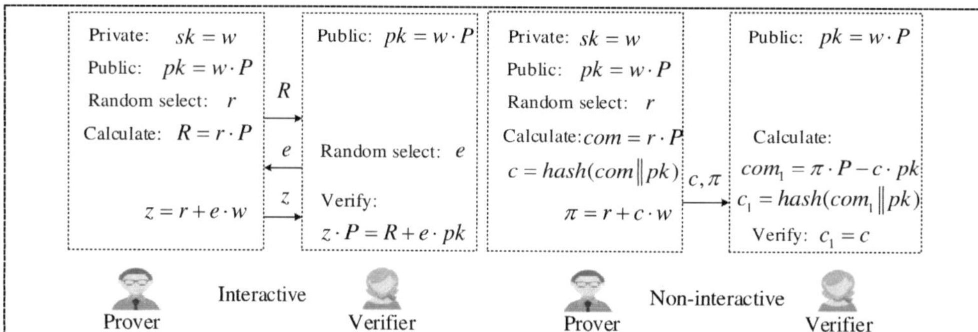

Figure 1. Interactive zero-knowledge proof and non-interactive zero-knowledge proof.

2.2. *PBFT consensus mechanism*

The PBFT (Practical Byzantine Fault Tolerance) algorithm is a distributed consensus algorithm that can tolerate byzantine faults. The consensus process is shown in Figure 2. It is divided into five stages: Request, Pre-Prepare, Prepare, Commit, and Reply. We assume that F is the number of byzantine nodes. If a node receives $2F+1$ commit message, it can commit the new block.

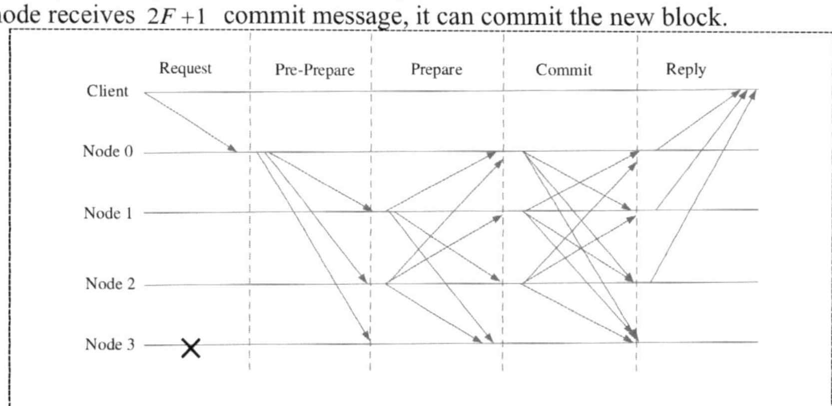

Figure 2. Stage chart of PBFT consensus mechanism.

3. The proposed scheme

In this section, we describe the symbols used and the proposed scheme in detail.

3.1. Notation illustration

We list the notations mainly used in this paper in Table 1.

Table 1. Notation illustration.

Notation	Description
pk_{OBU}, sk_{OBU}	Public key and private key of the OBU
PID	The pseudonym of the OBU
RSU_j	The j th roadside unit
TRA_i	The i th trace authority
pk_{TRA_i}, sk_{TRA_i}	Public key and private key of the TRA_i
pk_{TRA}, sk_{TRA}	Public key and private key of the $TRAs$ association
$PID \| Enc_{pk_{TRA}}(pk_{OBU})$	Concatenate operation of PID and $Enc_{pk_{TRA}}(pk_{OBU})$
$l_j = \prod_{j=1, j\neq i}^{t_1} \frac{j-0}{j-i}$	Lagrange coefficient

3.2. Introduction to entities in the system

This scheme includes four entities: consortium blockchain (CB), onboard unit (OBU), roadside unit (RSU), and trace authority (TRA). The system model is shown in Figure 3.

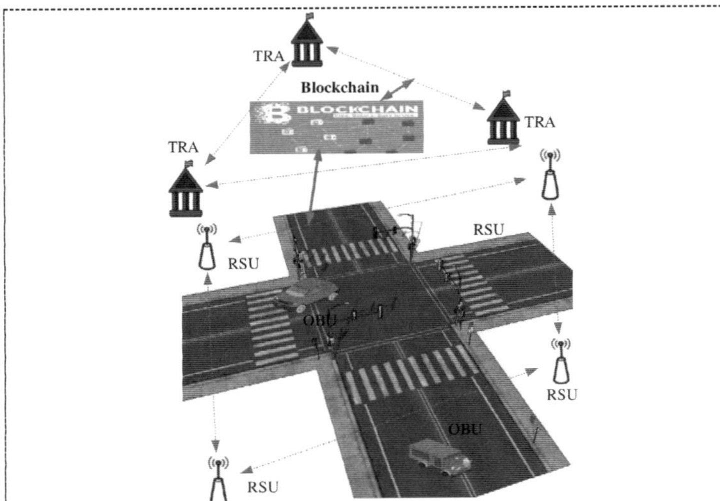

Figure 3. Stage chart of PBFT consensus mechanism.

- **Consortium blockchain (CB).** CB comprises k_1 roadside units（RSUs）and k_2 trace authorities (TRAs). The RSU association will jointly agree on the public parameters of the system and initialize the consortium blockchain. The k_2 trace authorities (TRAs) publish their partial public keys and jointly generate the public key of the TRAs association. The public key of the TRAs association as the system's tracking public key will be published on the blockchain.
- **Roadside unit (RSU).** The association composed of k_1 RSUs verifies the legitimacy of

OBU pseudonym generation process and the correctness of OBU's identity public key encryption through the PBFT consensus mechanism. In addition, when the RSU finds that there is illegal behavior under a pseudonym, it can publish the suspected illegal pseudonym to the blockchain network and request the TRAs to track the real identity of the OBU.

- **Onboard unit（OBU）.** The OBU can self-generate verifiable and traceable pseudonyms. OBU uses the NIZK to prove the validity of pseudonym generation and the correctness of identity public key encryption to RSUs.
- **Trace authority (TRA).** Each TRA generates its own part of the tracking public key and the tracking private key when the system is initialized. Using the Pedersen (t_2, k_2) secret sharing scheme, more than t_2 partial tracking public keys can generate the tracking public key of the TRAs association, and publish the tracking public key to the blockchain. When the tracking function needs to be realized, more than t_2 TRAs in the system can jointly generate the tracking private key of the TRAs association to trace the real identity of the OBU.

3.3. Scheme description
The scheme is divided into four phases: the system initialization phase, the OBU pseudonym generation and identity public key encryption phase, the pseudonym and identity public key ciphertext verification phase, and the OBU's real identity trace phase.

3.3.1. System initialization
The k_1 RSUs jointly agree on the public parameters of the system and initialize the consortium blockchain. System parameters involved in our construction are $\{E, G, P, p, q, a, b, H\}$, where p, q are two large prime numbers, E is a non-singular elliptic curve defined by $y = x^3 + ax + b \bmod p(a, b \in F_p)$, G is a cyclic group which consists of all points on E, P is the generator of G with order q, $Q \in G$. The k_2 TRAs jointly generate and publish the public key on the blockchain. The OBU's private key is sk_{OBU}, and the public key is $pk_{OBU} = sk_{OBU} \cdot P$. We assume that the public key pk_{OBU} represents the real identity of the OBU.

3.3.2. The OBU pseudonym generation and identity public key encryption
The OBU self-generates pseudonym, encrypts its public identity key with the public key of TRAs association, and uses the NIZK to provide corresponding proofs without revealing its real identity. The whole process is divided into the following nine steps.

Step1: The OBU secret selects a pseudonym private key $r \in Z_q^*$, and generates a pseudonym $PID = sk_{OBU} \cdot P + r \cdot Q$.

Step2: The OBU selects two random numbers $\rho_1, \rho_2 \in Z_q^*$ and calculates the commitment $Com_1 = \rho_1 \cdot P + \rho_2 \cdot Q$.

Step 3: The OBU calculates the hash value $hash_1 = H_1(Com_1 \| PID)$.

Step 4: The OBU generates the proof of the master private key $\pi_{sk_{OBU}} = \rho_1 + hash_1 \cdot sk_{OBU}$.

Step 5: The OBU generates the proof of the pseudonymous private key $\pi_r = \rho_2 + hash_1 \cdot r$.

Step 6: The OBU randomly selects $\tau \in Z_q^*$, adopts the El-Gamal encryption algorithm to encrypt the identity public key pk_{OBU_a}, pk_{TRA} is the public key of the TRAs association.

$$Enc_{pk_{TRA}}(pk_{OBU_a}, \tau) = (CT_1, CT_2) = (\tau P, pk_{OBU} + \tau \cdot pk_{TRA})$$

Step 7: The OBU randomly selects $\rho_3, \rho_4, \rho_5 \in Z_q^*$, calculates commitments

$Com_2 = \rho_3 \cdot P + \rho_4 \cdot pk_{TRA}$, $Com_3 = \rho_4 \cdot P$, $Com_4 = \rho_3 \cdot P + \rho_5 \cdot Q$,

$hash_2 = H_2(Com_2 \| Com_3 \| Com_4 \| Enc_{pk_{TRA}}(pk_{OBU}, \tau) \| PID)$.

Step 8: The OBU generates a new proof of the master private key $\pi'_{sk_{OBU}}$, the proof of the random number π_τ, and the proof of the pseudonym private key π'_r. $\pi'_{sk_{OBU}} = \rho_3 + hash_2 \cdot sk_{OBU}$, $\pi_\tau = \rho_4 + hash_2 \cdot \tau$, $\pi'_r = \rho_5 + hash_2 \cdot r$.

Step 9: The OBU submits the pseudonym request $PID\_Qst = (PID, hash_1, \pi_{sk_{OBU}}, \pi_r)$ and the proof of identity public key encryption $Proof\_Enc = (PID, Enc_{pk_{TRA}}(pk_{OBU}, \tau), \pi'_{sk_{OBU}}, \pi_\tau, \pi'_r, hash_2)$ to the consensus network consisting of k_1 RSUs for confirmation.

3.3.3. The pseudonym and identity public key ciphertext verification

The verification algorithm includes two aspects: the legality verification of the OBU pseudonym generation process and the correctness verification of the OBU identity public key encryption, which is completed by the k_1 RSUs association through the PBFT consensus protocol.

Step 1: The RSU_j (leader node) in the consortium blockchain broadcasts PID_Qst and $Proof\_Enc$ to other consensus nodes $RSU_i(i = 1, \cdots, j-1, j+1, \cdots, k_1)$ after receiving the PID_Qst and $Proof\_Enc$ from the OBU.

Step2: Each consensus node first parses $(PID, hash_1, \pi_{sk_{OBU}}, \pi_r)$ from, determines whether the pseudonym has appeared in the blockchain network, if so, stops the consensus protocol and requires, if the pseudonym PID has not appeared in the blockchain, Each RSU calculates $Com'_1 = \pi_{sk_{OBU}} \cdot P + \pi_r \cdot Q + hash_1 \cdot PID$ and $hash'_1 = H(Com'_1 \| PID)$, determines whether $hash_1$ in PID_Qst is equal to the recalculated $hash'_1$, if not, stops the consensus protocol. If $hash_1 = hash'_1$, It indicates that the pseudonym generation process is legal. Continue to step 3 to determine whether the OBU's identity public key encryption is correct.

Step3: Each consensus node parses $(PID, Enc_{pk_{TRA}}(pk_{OBU}, \tau), \pi'_{sk_{OBU}}, \pi_\tau, \pi'_r, hash_2)$ from $Proof\_Enc$, and first confirms that the pseudonym PID supplied in $Proof\_Enc$ is the same as that in the PID_Qst, then calculates the commitment value and hash value

$$Com'_2 = \pi'_{sk_{OBU}} \cdot P + \pi_\tau \cdot pk_{TRA} + (-hash_2) \cdot CT_2,\ Com'_3 = \pi_\tau \cdot P + (-hash_2) \cdot CT_1,$$

$$Com'_4 = \pi'_{sk_{OBU}} \cdot P + \pi'_r \cdot Q + (-hash_2) \cdot PID,\ hash'_2 = H_2(Com'_2 \| Com'_3 \| Com'_4 \| Enc_{pk_{TRA}}(pk_{OBU}, \tau) \| PID).$$

Each consensus node determines whether $hash_2$ in $Proof\_Enc$ is equal to the recalculated $hash'_2$. If not, stop the consensus protocol. If they are equal, submit their verification conclusion to other consensus nodes.

Step4: According to the PBFT protocol, when $\frac{2}{3}k_1$ RSUs have passed the legitimacy verification of pseudonym and the correctness verification of identity public key encryption, the concatenation of pseudonym and identity public key encryption $PID \| Enc_{pk_{TRA}}(pk_{OBU})$ is added to the leaf node of the Merck tree on the blockchain. The pseudonym PID is considered a valid pseudonym verified by the consensus network only if $PID \| Enc_{pk_{TRA}}(pk_{OBU})$ appears in the leaf node of the Merck tree in Figure 4. The RSUs association returns the root hash value of the Merck tree where the pseudonym is located and the path of the pseudonym $(PID, root, path)$ to the OBU. OBU can generate multiple pseudonyms and corresponding validity proofs at the same time. $PID_1 \| Enc_{pk_{TRA}}(pk_{OBU})_1, \ldots, PID_2 \| Enc_{pk_{TRA}}(pk_{OBU})_w$ will appear in the leaf nodes of the Merkle tree. When interacting with different RSUs, different pseudonyms can be used to realize the unlinkable of pseudonyms.

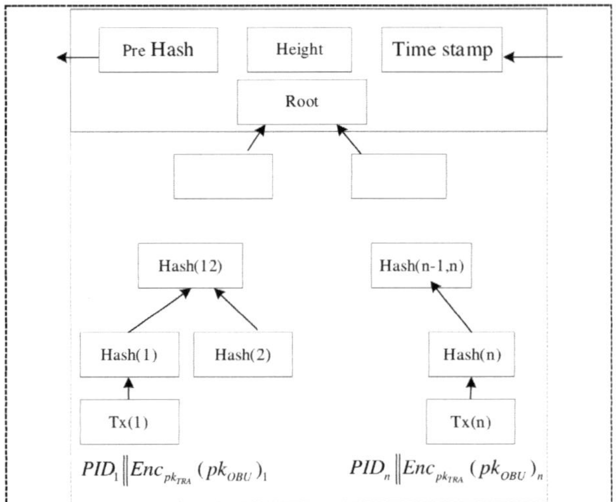

Figure 4. Valid pseudonyms in the Merck tree.

3.3.4. The OBU's real identity trace

The trace algorithm $Trace(PID^*, sk_{TRA}) \rightarrow (pk_{OBU}^*)$ is jointly implemented by the TRAs association. When RSU finds that a pseudonym PID^* publishes false or illegal information, the RUS publishes the pseudonym PID^* of the OBU suspected of illegal behavior to the blockchain network, t_2 TRAs can jointly construct the tracking private key sk_{TRA} through the Pedersen secret sharing protocol, $sk_{TRA} = \sum_{i=1}^{t_2}(\prod_{j=1, j\neq i}^{t_2} \frac{j-0}{j-i})sk_{TRA_i}$, and download the identity public key ciphertext $Enc_{pk_{TRA}}(pk_{OBU})^*$ bound by pseudonym PID^* from the blockchain, Where $Enc_{pk_{TRA}}(pk_{OBU})^* = (\tau P, pk_{OBU}^* + \tau pk_{TRA})$. Furthermore, the tracking private key sk_{TRA} is used to decrypt the public identity key pk_{OBU}^* and obtain the real identity of the tracked OBU.

$$\begin{aligned} pk_{OBU}^* + \tau \cdot pk_{TRA} - sk_{TRA} \cdot \tau \cdot P &= pk_{OBU}^* + \tau \cdot pk_{TRA} - \tau \cdot sk_{TRA} \cdot P \\ &= pk_{OBU}^* + \tau \cdot pk_{TRA} - \tau \cdot pk_{TRA} \\ &= pk_{OBU}^* \end{aligned} \tag{1}$$

4. Security analysis

4.1. Proof of correctness

The correctness includes the correctness of pseudonym generation and the correctness of identity public key encryption:

(1) The correctness of identity public key encryption:

$$\begin{aligned} hash_1' &= H_1(Com_1' \| PID) \\ &= H_1(\pi_{sk_{OBU}} \cdot P + \pi_r \cdot Q + (-hash_1) \cdot PID \| PID) \\ &= H_1((\rho_1 + hash_1 \cdot sk_{OBU}) \cdot P + (\rho_2 + hash_1 \cdot r) \cdot Q + (-hash_1) \cdot (sk_{OBU} \cdot P + r \cdot Q) \| PID) \\ &= H_1(\rho_1 \cdot P + \rho_2 \cdot Q \| PID) \\ &= H_1(Com_1 \| PID) \\ &= hash_1 \end{aligned} \tag{2}$$

(2) The correctness of identity public key encryption:

$$\begin{aligned}
hash_2' &= H_2(Com_2' \| Com_3' \| Com_4' \| Enc_{pk_{TRA}}(pk_{OBU_a},\tau) \| PID) \\
&= H_2(\pi'_{sk_{OBU}} \cdot P + \pi_\tau \cdot pk_{TRA} + (-hash_1)\cdot CT_2 \\
&\quad \| \pi_\tau \cdot P + (-hash_2)\cdot CT_1 \\
&\quad \| \pi'_{sk_{OBU}} \cdot P + \pi'_r \cdot Q + (-hash_2)\cdot PID \\
&\quad \| Enc_{pk_{TRA}}(pk_{OBU_a},\tau) \| PID) \\
&= H_2((\rho_3 + hash_2 \cdot sk_{OBU})\cdot P + (\rho_4 + hash_2\cdot\tau)\cdot pk_{TRA} + (-hash_1)\cdot(pk_{OBU} + \tau\cdot pk_{TRA}) \\
&\quad \| (\rho_4 + hash_2\cdot\tau)\cdot P + (-hash_2)\cdot\tau\cdot P_1 \\
&\quad \| (\rho_3 + hash_2\cdot sk_{OBU})\cdot P + (\rho_5 + hash_2\cdot r)\cdot Q + (-hash_2)\cdot(sk_{OBU}\cdot P + r\cdot Q) \\
&\quad \| Enc_{pk_{TRA}}(pk_{OBU_a},\tau) \| PID) \\
&= H_2((\rho_3\cdot P + \rho_4\cdot pk_{TRA}) \| \rho_4\cdot P \| (\rho_3\cdot P + \rho_5\cdot Q) \| Enc_{pk_{TRA}}(pk_{OBU},\tau) \| PID) \\
&= H_2(Com_2 \| Com_3 \| Com_4 \| Enc_{pk_{TRA}}(pk_{OBU},\tau) \| PID) \\
&= hash_2
\end{aligned} \tag{3}$$

4.2. Proof of security characteristics

Our scheme meets the requirements of unforgeability, unlinkability, and distributed traceability.

- **Unforgeable:** Under the assumption of the discrete logarithm, we construct the NIZK of OBU's master private key sk_{OBU}. Our zero-knowledge proof adopts the random oracle model, attackers can only forge zero-knowledge proofs by identifying random numbers in commitments, but the probability of such forgery occurring is negligible under the discrete logarithm assumption.
- **Unlinkable:** The OBU can self-generate multiple pseudonyms $PID_1, PID_2, \ldots, PID_w$ at the same time. The same identity public key pk_{OBU} will generate a different ciphertext. As long as the ciphertext is honestly encrypted under the tracking public key pk_{TRA}, the correctness of the encryption process can pass the verification of the system. EL-Gamal encryption satisfies (IND-CPA) security. It is impossible to infer any useful information about the real identity from the public key ciphertext published on the blockchain.
- **Distributed traceability:** If the OBU publishes false or illegal information, then more than t_2 TRAS can jointly calculate the tracking private key of the system, and the OBU's real identity can be traced. If more than t_2 TRAs are corrupted, the system will be broken. We simulate the security of the (t_2, k_2) threshold mechanism under various threshold values t_2 in the case of the total number of TRAs $k_2 = 9,15$, respectively. It can be seen from Figure 5 and Figure 6, the abscissa p represents the enemy's attack ability and the security of distributed trace scheme is $\sum_{i=0}^{t_2-1} C_{k_2}^i p^i (1-p)^{k_2-i}$. The security of the tracking system increases with the increase of the value of t_2. When the value of t_2 is greater than one-third of k_2, even if the enemy's attack ability p reaches 0.3, the distributed tracking mechanism is also secure.

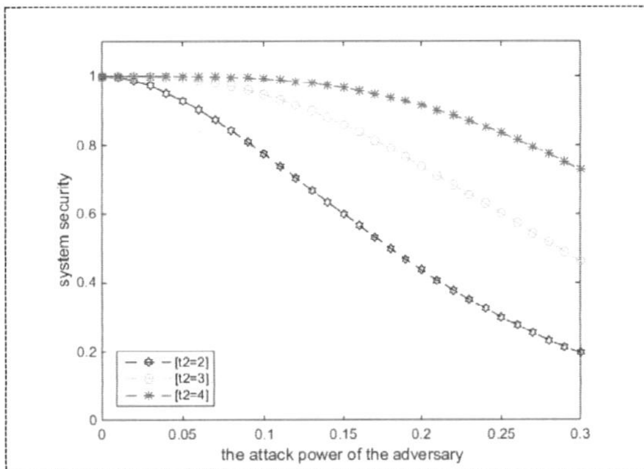

Figure 5. The security of (t_2, k_2) threshold trace mechanism when $k_2 = 9, t_2 = 2,3,4$.

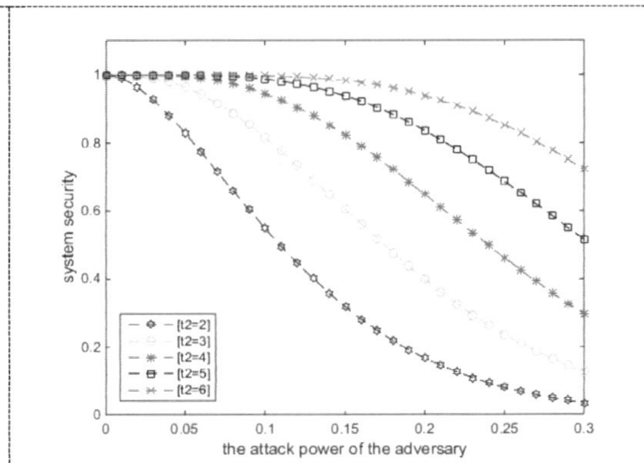

Figure 6. The security of (t_2, k_2) threshold trace mechanism when $k_2 = 15, t_2 = 2,3,4,5,6$.

5. Performance Analysis

We use T_{pm} to express the processing time of a point multiplication in G. We test the computational cost of T_{pm} by the cryptography library PBC. According to our experimental results, $T_{pm} \approx 0.96$ ms. Our scheme does not involve the bilinear pair operation, which requires a high computational cost. The computation cost in each stage of this scheme is relatively small, and the specific calculation cost is shown in table 2. The calculation time of each stage is less than 10 ms.

Table 2. The computational cost of each stage.

Notation	Communication overhead	Computational cost
PID_Qst	$G + 3Z_q^* = 100$ Bytes	$4T_{pm} \approx 3.84$ ms
$Proof\_Enc$	$3G + 4Z_q^* = 200$ Bytes	$7T_{pm} \approx 6.72$ ms
PID_Ver		$3T_{pm} \approx 2.88$ ms
Enc_Ver		$8T_{pm} \approx 7.68$ ms
$Trace$		$T_{pm} \approx 0.96$ ms

It is assumed that the element size in the elliptic curve group G is 40 Bytes, the element size in the bilinear transformed group G_1 is 128 Bytes, and the size for general hash function: $\{0,1\}^* \rightarrow Z_q^*$ and timestamp T are considered to be 20 Bytes and 4 Bytes. In this paper, the signature of a message m with a pseudonym PID is $\sigma(PID, m, root, path, T, sig_{sk_{OBU}}(m))$, PID and m are belongs to the element in the group G, which is 40 Bytes, the root is 20 Bytes, the path is 200 Bytes, $sig_{sk_{OBU}}(m)$ belongs to Z_q^*, which is 20 Bytes, Timestamp T is 4 Bytes. According to the calculation in Table 3, the pseudonym signature length of this scheme is 324 Bytes. It is slightly higher than the scheme [7], but this scheme adds the distributed tracing function and realizes the tracing of the real identity of the pseudonym without obviously increasing the communication cost. Therefore, this scheme has certain practicality.

Table 3. Comparison of the communication cost of each scheme.

Scheme	Message signature	Traceable
Scheme [2]	$3G_1+3Z_q^*=404$ Bytes	NO
Scheme [3]	$3G_1+Z_q^*+4=408$ Bytes	NO
Scheme [7]	$3G_1+3Z_q^*+4=320$ Bytes	NO
Ours	$2G+Z_q^*+200+20+4=324$ Bytes	Yes

6. Conclusion

This paper designs a consortium blockchain composed of multiple RSUs to replace the TA in the traditional centralized scheme. The OBU uses the NIZK method to generate verifiable pseudonyms. Each RSU in the consortium blockchain verifies the effectiveness of OBU's self-generated pseudonyms through the consensus mechanism. The distributed tracing function is realized through the threshold mechanism. The new scheme meets the requirements of anonymity, unforgeability, unlinkability, and distributed traceability.

Acknowledgments

This research was funded by the National Natural Science Foundation of China under Grant 61572019, the Science and Technology Project of Shaanxi Province 2022GY-040.

References

[1] Zhang M, Zhu B, Li Y and Wang Y 2022 TPM-Based Conditional Privacy-Preserving Authentication Protocol in VANETs. *J. Symmetry*. **14** 1123

[2] Liu J, Yu Y, Zhao Y, Jia J and Wang S 2018 An efficient privacy-preserving batch authentication scheme with detectable function for VANETs. *C. International Conference on Network and System Security*. Springer, Cham. 288-303

[3] Liu, F, and Wang Q 2019 IBRS: an efficient identity-based batch verification scheme for VANETs based on ring signature. *C. IEEE Vehicular Networking Conference* (VNC).1-8

[4] Deng X, Gao T, Guo N, Zhao C, and Qi J 2022 PCP: A Pseudonym Change Scheme for Location Privacy Preserving in VANETs. *J. Entropy*. **24** 648

[5] Johar S, Ahmad N, Durrani A and Ali G 2021 Proof of pseudonym: Blockchain-based privacy-preserving protocol for the intelligent transport system. *J. IEEE Access,* **9** 163625-39

[6] Chulerttiyawong D and Jamalipour A 2021 A Blockchain Assisted Vehicular Pseudonym Issuance and Management System for Conditional Privacy Enhancement. *J. IEEE Access*. **9** 127305-19

[7] Qi J, and Gao T 2022 An Anonymous Authentication Scheme Based on Self-generated Pseudonym for VANETs. *C. International Conference on Innovative Mobile and Internet Services in Ubiquitous Computing*. Springer, Cham. 75-84

[8] Bogatov D, Caro A D, Elkhiyaoui K, and Tackmann B 2021 Anonymous transactions with revocation and auditing in hyperledger fabric. *C. International Conference on Cryptology and Network Security*. Springer, Cham. 435-59

[9] Shen H, Zhou J, Cao Z, Dong X and Choo K K R 2020 Blockchain-based lightweight certificate authority for efficient privacy-preserving location-based service in vehicular social networks. *J. IEEE Internet of Things J.* **7** 6610-22

Journal of Physics: Conference Series **2384** (2022) 012021 doi:10.1088/1742-6596/2384/1/012021

A Miniaturized Broadband Omnidirectional Monopole Antenna Operating at UHF Band in Wireless Communication

Xuan Zhang[a]*, Chunqi Zhao[b], Yang Yang[c], Ren Wang[d], Zhaojing Zhang[e], Jialei Li[d]

Xi'an Branch of China Unicom, Xi'an, China

[a*]616256640@qq.com, [b]zhaocj94@chinaunicom.cn, [c]327179534@qq.com, [d]358982709@qq.com, [e]1301889287@qq.com, [d]947556727@qq.com

Abstract—In the paper, a miniaturized broadband omnidirectional monopole antenna is designed. The two bending E-shaped structures are introduced into the common two-strip structure monopole antenna. The radiation patches can be effectively distributed on two sides of the dialectic body by four shorting pins. Thus, the proposed antenna has the function of miniaturization and integration. According to the introduced structures, the whole antenna has a smaller size of 318mm*12.2mm, which can improve a 34.2% integration degree. Finally, the present monopole antenna is manufactured, and the measured data shows that it has stable gains and good omnidirectional patterns from 463MHz to 513MHz.

1. Introduction

Owing to the high-performance requirement of wireless ultra high frequency (UHF) communication systems, many studies have been conducted on smaller monopole antenna designs for better system integration. The monopole antenna has been regarded as a good choice for the UHF band because of its advantages, such as its simple manufactured structure, omnidirectional radiation pattern, and higher gain. Some methods have been applied to miniaturize the dimension of the UHF band frequency antenna. A meander-line monopole and metal shielding box was applied to miniaturize the volume, but the bandwidth covered 5MHz [1]. Though the antenna working at UHF band frequency was miniaturized by the present high-order fractal structure in [2], a slightly bigger volume somehow limits the system integration degree. In literature [3], the asymmetric planar dipole antenna was proposed for miniaturizing the size. However, the present new structure's dimension is slightly larger in working bandwidth. Some other ways [4-5] were also introduced to lessen the antenna's size, but the bandwidth of the monopole antenna still needs to be improved. More monopole antennas are designed in [6] [7] [8], and compact size and wide bandwidth are obtained, but the manufacturing difficulty is slightly higher.

In this research, a miniaturized broadband omnidirectional printed monopole antenna is proposed. In the printed monopole antenna, two E-shaped radiation patches are blended into two sides of the dialectic body, and they are connected by four shorting pins. Thus, the proposed antenna has the function of miniaturization and integration.

2. Antenna Configuration and Mechanism

2.1. Antenna Configuration

Fig.1 depicts the configuration of the miniaturized monopole antenna. The printed antenna contains three layers: the top layer is a thin conductor patch, shown in Fig. 1(a); the second layer is the dielectric substrate (=2.55), plotted in Fig. 1 (b); the third layer is a thin conductor patch, shown in Fig. 1 (c). The top thin conductor comprises two E-shaped conductors, which are back to back. The length of the E-shaped conductor is one-quarter of the wavelength responding to the resonant frequency. The dielectric substrate (Arlon AD255A) is with 1mm thickness, whose relative constant is 2.55 and loss tangent is 0.0015. Actually, the whole bending E-shaped patches are distributed on the two-side of the dielectric substrate, and they are connected by four short pins. According to the above integrated bent E-shaped patches, the antenna has the function of miniaturization. The compact printed structure is also easy to manufacture. The detailed parameters' values of the presented antenna are depicted in Table 1.

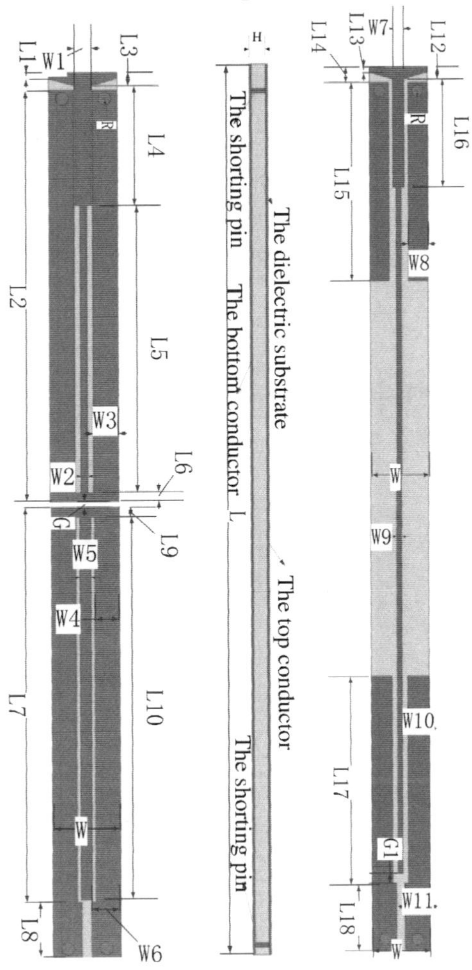

Fig. 1. The presented antenna: (a) The top view of the presented antenna; (b) The side view of the presented antenna; (c) The bottom view of the presented antenna.

TABLE 1 DIMENSIONS OF THE PROPOSED ANTENNA

Parameter	L	W	G	L1	L2	L3
Value(mm)	318	12.2	0.5	2.3	311.4	5
Parameter	L7	L8	L9	L10	L11	L12
Value(mm)	135.2	20.5	3.2	131.2	145.9	5.2
Parameter	L16	L17	L18	W1	W2	W3
Value(mm)	45	62.5	20.5	2.9	1.2	4.45
Parameter	W7	W8	W9	W10	W11	R
Value(mm)	2.3	4	1	4.45	5.2	1

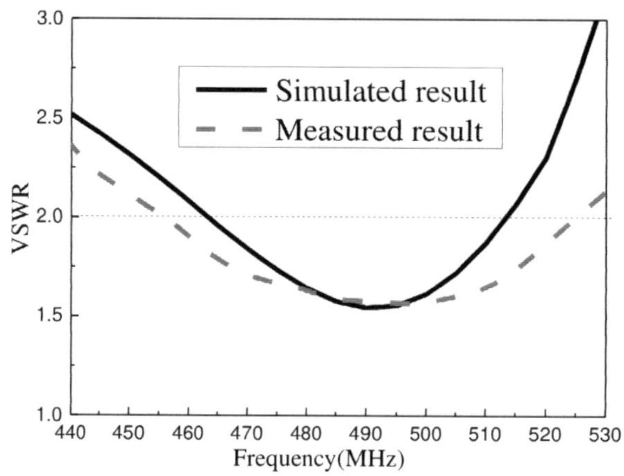

Fig. 2. The simulated and measured VSWR of the presented antenna

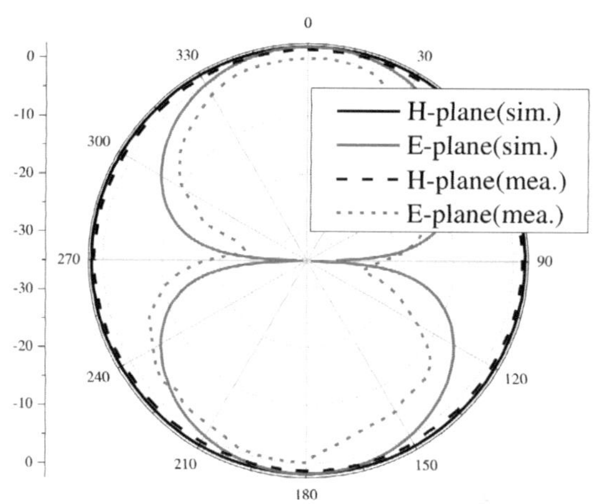

(a)

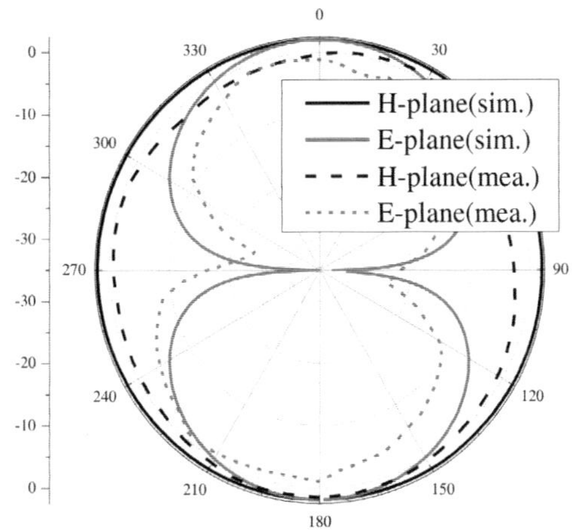

(b)

Fig. 3. The simulated and measured radiation pattern of the presented UHF antenna. (a) The radiation pattern at 0.47GHz. (b) The radiation pattern at 0.50GHz.

2.2. *Antenna Mechanism*

The presented antenna is simulated by adopting the EM simulation software HFSS v13. The E-shaped length is selected to ensure the antenna resonates around the UHF band frequency from 463MHz to 513MHz. The parameters of the bending E-shaped strips are optimized to obtain better EM performances.

(a) (b)

Fig. 4. The current distribution of the antenna at 0.49GHz: (a) The top view; (b) The bottom view.

The antenna is simulated and measured, and its voltage standing wave ratio (VSWR) is shown in Fig.2. From the picture, we can see that the antenna can operate well in 463MHz-513MHz at the UHF band. The simulated and measured radiation patterns at 470MHz and 500MHz are shown in Fig. 3, from which the good agreements and the good omnidirectional radiation patterns are achieved. From the current distributions in Fig. 4, the introduced bending E-shaped patches effectively realized the effect of bending current, which explains the miniaturization mechanism well.

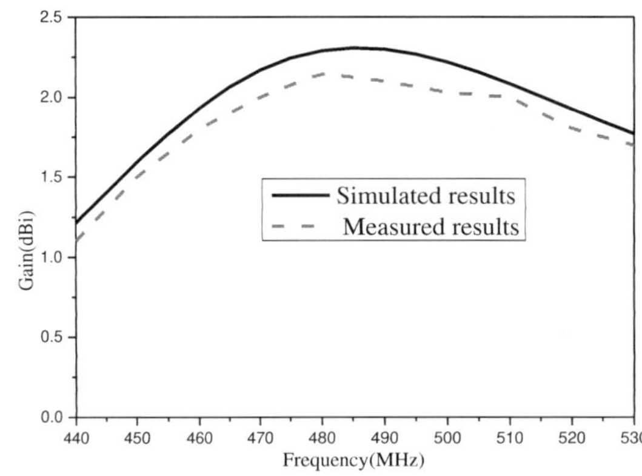

Fig. 5. The simulated and measured gain of the presented UHF antenna.

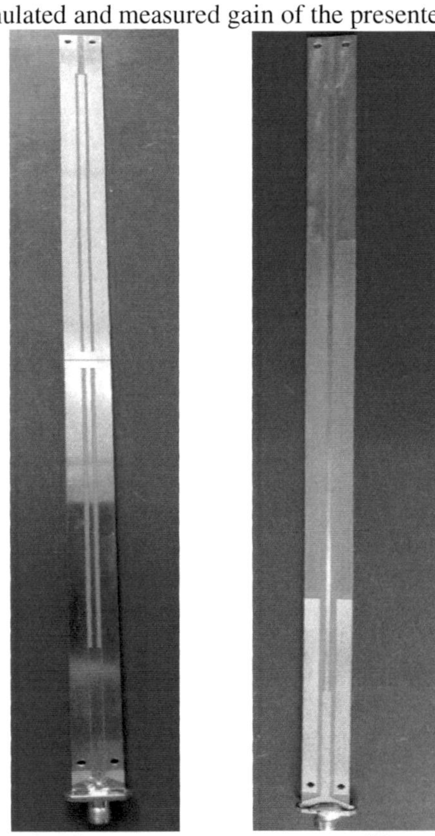

(a) (b)

Fig. 6. The presented UHF antenna: (a) Top view; (b) Bottom view.

3. Experimental Verification

The equipment Agilent 8719ES manufactured by Agilent company is adopted to test the VSWR of miniaturized monopole antenna. From Fig. 2, we can see that the proposed antenna can operate from 456MHz to 524MHz, which has a relative bandwidth of 13.87%. The measured radiation pattern in Fig. 3(a) and Fig. 3(b) show that E-plane has an "8" shape radiation pattern and H-plane has an "O" shape radiation pattern. The gain of the presented antenna is simulated and measured, and the results are shown in Fig. 5, from which the present antenna has stable gain in operating bands. From the above research, good consistencies between simulated and measured results have been obtained. At the same time, the picture of the manufactured antenna is shown in Fig. 6.

4. Conclusion

A miniaturized broadband omnidirectional monopole antenna operating at UHF for wireless communication is proposed in this paper. The two bending E-shaped structures are introduced into the common two-strip structure monopole antenna. Four shorting pins are used to connect two bending E-shaped patches, which are distributed on two sides of the dielectric body. The measured data show the antenna works in 456MHz-524MHz with good omnidirectional radiation and stable gain. The entire antenna size is reduced by 34.2%.

References

[1] W. Liu, Z. Zhang, and Z. Feng, "ISM 433-MHz Miniaturized Antenna Using the Shielding Box of Mobile Terminals," IEEE Antennas Wireless Propagation Letters., vol. 11, 2012, pp. 330-333.

[2] Y. Yu, and W. Wu, "Miniaturized Peano Fractal Loaded Monopole Antenna," 2009 Asia Pacific Microwave Conference. In Singapore, 2009, pp. 2668-2700.

[3] H.-R. Lee, and J.-M. Woo, "Asymmetric planar dipole antenna on the surface of conducting plane for RFID tag," 2009 Asia Pacific Microwave Conference., In Singapore, 2009, pp. 633-636.

[4] N. Ripin, A. A. Sulaiman, N.E.A. Rashid, and M. F. Hussin, "Miniaturization of Printed Monopole Antenna through Fractal Geometry and Partial Cutting Methods for UHF Application," 2015 International Conference on Computer, Communications, and Control Technology, in Malaysia,2015, pp. 26-30.

[5] W. B. Hong, and S. Kamal, "Low-Profile, Multi-Element, Miniaturized Monopole Antenna," Applied Physics., vol. 57, 2009, pp. 72-80.

[6] A. D.Johnson, J. A.Caripidis, S. B. Venkatakrishnan, M. Ali, and J. L.Volakis, "Deployable Inverted-Hat Monopole With 3:1 Constant Gain Bandwidth," IEEE Antennas Wireless Propagation Letters., vol. 19(6), 2020, pp. 935-938.

[7] X. Zhang, F. Sun, G. Zhang, and L. Hou, "Compact UHF/VHF Monopole Antennas for CubeSats Applications," IEEE Access., vol. 8, 2020, pp. 133360-133366.

[8] X. H. Zhang, X. W. Dai, W. Sun, H. Jin, and G. Q. Luo, "Sleeve Monopole Antenna With Choke Structure for Broadband Application," IEEE Antennas Wireless Propagation Letters., vol. 20(7), 2021, pp. 1230-1233.

Journal of Physics: Conference Series **2384** (2022) 012022 doi:10.1088/1742-6596/2384/1/012022

Performance analysis of RSMA-based cognitive networks with random distribution of eavesdroppers

Qinying Li, Yue Tian, Xianling Wang and Chen Zhu

Fujian Key Laboratory of Communication Network and Information Processing, Xiamen University of Technology, Xiamen 361024, China

Corresponding author: Xianling Wang (e-mail:xianling.wang@ieee.org)

Abstract. Due to the pace of various emerging wireless communication technologies and new application services innovation, network information security, and capacity requirements have become a hot issue of concern. In order to improve network security performance and transmission capacity, researchers have proposed cognitive wireless network and physical layer security as potential solution. Most of the current research work focuses on the assumption that the location of the eavesdropping node is fixed. Starting from the physical layer security and based on the random geometry theory, this paper studies the security performance of cognitive wireless networks under the scenario of random distribution of eavesdropping nodes based on rate splitting multiple access, proposes a system model, and deduces the closed form solutions of connection interruption probability and security interruption probability of secondary users in the cognitive wireless network. The results show that the security interruption probability is related to the density and security of eavesdropping nodes. Finally, the correctness of the formulas is verified by simulation analysis.

1. Introduction

A cognitive radio network (CRN) is a wireless communication network with cognitive characteristics. In future communications networks, the demand for wireless communications continues to increase. Spectrum resources are becoming increasingly scarce. CRN can observe the surrounding wireless network environment, using environmental awareness to obtain spectrum usage information.

In [1], the current status of spectrum sensing technology in CRN was not only introduced but also compared. Energy efficiency and spectral efficiency are becoming more and more important. Therefore, many researchers have begun to study a cognitive wireless network based on non-orthogonal multiple access (NOMA), which can simultaneously transmit information and capabilities [2]. In [3], the spectrum allocation algorithm of cognitive radio networks using reinforcement learning is introduced in detail.

Underlay cognitive radio network (UCRN) is a more novel approach in cognitive networks. In a UCRN, the cognitive networks improve spectrum utilization by allowing the primary user access to the authorized spectrum to transmit information when the interference from the transmit power to the primary user is manageable. Secrecy analysis of a single input multiple outputs (SIMO) based URCN was discussed in [4-5]. In addition, the performance in terms of primary networks (PNs) and secondary networks (SNs) in URCN. The effect of incremental amplification and forwarding (IAF) relay in imperfect channel state information (CSI) was studied in [6].

Traditional measures to secure wireless communications have been through encryption of the upper layers of the network and the establishment of an encryption regime and a system of security protocols.

As science and the popularity of smart terminals improves by leaps and bounds, this method of using computing power in exchange for information security is at risk of being breached. Since wireless networks have an open broadcast nature, the security of wireless communication networks is easily threatened, and therefore, wireless communication networks are faced with various security issues. In recent years, physical layer security (PLS) technologies based on information theory have generated a lively debate. In [7-8], a NOMA-based coordinated multipoint system is studied to improve performance. Furthermore, mutual-aid NOMA-based cooperative networks perform significantly in computational complexity and quality of service [9-10]. The study of PLS in cognitive networks has been a popular discussion recently. The authors in [11] calculated the secrecy performance of CRN-based Internet of Things (IoT) in the presence of an eavesdropper.

Rate splitting multiple access (RSMA), a new multiple access technique combining the advantages of existing multiple access techniques, has attracted extensive attention [12]. [13] proposed an RSMA-based secure beamforming design, where user messages are split and encoded into common and private streams. Another study [14] examined the advantages of applying RSMA in multi-cell dense networks.

The broadcast nature of the wireless channel allows for various attacks, such as identity impersonation, wireless eavesdropping, and service denial, to seriously threaten the secure operation of wireless networks. Private information transmitted between nodes can easily be eavesdropped on by some illegal nodes (eavesdropping users), thus posing a threat to the secure transmission of information. The physical layer uses the characteristics of the wireless channel itself, i.e., the difference between the eavesdropping channel and the normal communication channel of a normal user, to transmit information securely. However, in many studies on physical layer security, the eavesdropping user usually remains silent and does not send any signals, i.e., the location distribution of the eavesdropping user is difficult to obtain. This paper investigated the security performance under the condition of random distribution of eavesdroppers' locations in RSMA-based cognitive networks.

2. System model

The proposed system for the UCRN is shown in Figure 1, where the system consists of a primary network (PN) and a secondary network (SN). In Figure 1, PT and PR represent the primary transmitter and primary receiver in the primary network. Consider one layer rate-splitting (RS) secondary network where a secondary transmitter, ST, equipped with N_t antennas servers K single-antenna secondary receivers. All secondary receivers are denoted by $SR_k\left(k=1,2,\cdots,\ K\right)$. Moreover, E_j represents eavesdroppers in a secondary network, equipped with a single antenna. The primary network allows the secondary network to communicate in Underlay mode, i.e., PR in the primary network receives a signal from PT and allows the secondary users to occupy the primary user's frequency band when the power at PR is below the threshold, denoted by I_t. As a result, the signal from the ST does not interfere with the PN. The transmitter power at ST is denoted as P.

On the side of secondary receivers, the distinct messages of all users are denoted by $M_k\left(k=1,2,\cdots,\ K\right)$. The message of each receiver is divided into two parts, common messages as M_k^c, and private messages as M_k^p. Then, all common messages are encoded together into a common stream s_0 for the secondary receivers. All secondary receivers will eventually decode the stream s_0. The private information of every secondary receiver is encoded into a separate private stream $s_k\left(k=1,2,\cdots,\ K\right)$. The overall date streams to be transmitted based on RS is $s=\left[s_0,s_1,\cdots s_K\right]^T$. The date streams are linearly precoded in the precoder and $\mathrm{P}=\left[p_0,p_1,\cdots p_K\right]$. Finally, the signals are superimposed as $\mathrm{x}=\mathrm{P}s$. In order to satisfy the constraint of instantaneous interference, I_t. Let the transmit power of ST be:

$$P=\frac{I_t}{\left|h_{SP}\right|^2 d_{SP}{}^{-\alpha}} \tag{1}$$

where d_{SP} represents the distance between ST and PR, $h_{sp} \sim CN(0,1)$.

This work assumes that the information transmission channel is composed of path-loss and quasi-static Rayleigh fading channel superposition.

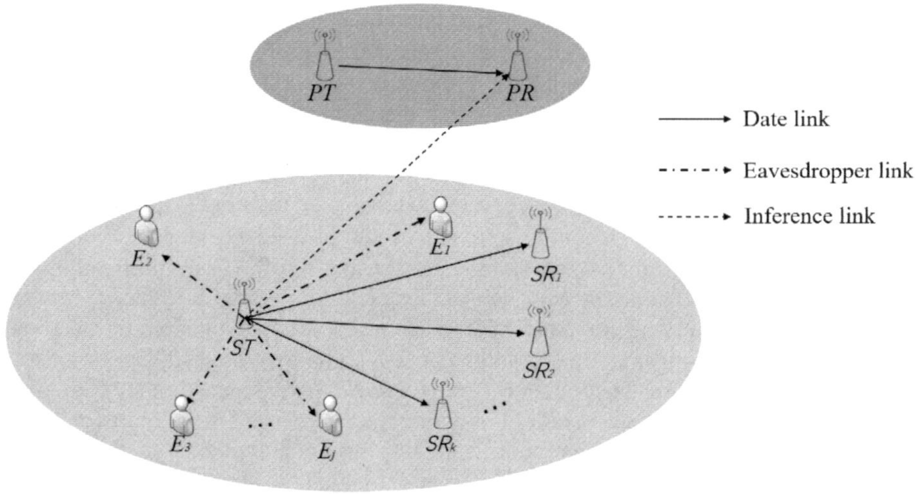

Figure 1. RSMA-Based underlay cognitive radio network with eavesdroppers.

2.1 Full Activity Strategy

For the full activity strategy, the secondary transmitter ST could not detect the presence of eavesdroppers, therefore, ST cannot determine the specific location of the eavesdroppers. At the same time, the channel state information (CSI) between any secondary receiver and ST is not clear. As a result, under the premise of satisfying the power limit, the ST always transmits confidential information to the $SR_k\ (k=1,2,\cdots,\ K)$. Therefore, the transmitted power of ST conforms to (1). In reality, eavesdroppers usually remain silent and do not send any signals. We define Φ_E as the location set of the eavesdroppers, which is modeled as an HPPP with density λ_E. Then, let us assume that all eavesdropping nodes can communicate with each other, then the combination of all eavesdroppers can be seen as one eavesdropper equipped with multiple distributed antennas, denoted by E_{joint}. The SNRs at the SR_k is given by

$$\gamma_{R_k} = \frac{I_t \left|h_{SR_k} p_K\right|^2 d_{SR_k}{}^{-\alpha}}{\sigma_{R_k}{}^2 \left|h_{SP}\right|^2 d_{SP}{}^{-\alpha}} \tag{2}$$

The SNRs at the eavesdropper E_{joint} is given by

$$\gamma_E = \frac{I_t \sum_{E_j \in \Phi_E} \left|h_{SE_j}\right|^2 d_{SE_j}{}^{-\alpha}}{\sigma_{E_j}{}^2 \left|h_{SP}\right|^2 d_{SP}{}^{-\alpha}} \tag{3}$$

where α denotes the path loss exponent, d_{SR_k} is denoted as the distance between ST and SR_k, d_{SE_j} denotes the distance between ST and E_j. h_{SR_k} and h_{SE_j} are denoted as the channel coefficients for the channel from ST to SR_k and the channel from ST to E_j. In addition, $\sigma_{R_k}{}^2$ and $\sigma_{E_j}{}^2$ denote additive white Gaussian noise (AWGN) variances E_j and SR_k, respectively. In addition, $\sigma_{R_k}{}^2 = \sigma_{E_j}{}^2 = \sigma^2$.

2.2 Secrecy Guard Zone Strategy

Such a scenario is assumed that where the ST is aware of the presence or absence of eavesdropping users within a certain range. Then, we model it as a secrecy protection circle $\Re$ of radius r in a finite range around ST . The ST sends confidential information only when the eavesdropper is not detected within the scope of the secrecy guard.

The SNRs at SR_k , eavesdropper E_{joint} are given by

$$\gamma_{R_k} = \frac{I_t \left|h_{SR_k} p_K\right|^2 d_{SR_k}{}^{-\alpha}}{\sigma_{R_k}{}^2 \left|h_{SP}\right|^2 d_{SP}{}^{-\alpha}} \mathbf{1}_{(\text{condition})} \tag{4}$$

and

$$\gamma_E = \frac{I_t \sum_{E_j \in \Phi_E} \left|h_{SE_j}\right|^2 d_{SE_j}{}^{-\alpha}}{\sigma_{E_j}{}^2 \left|h_{SP}\right|^2 d_{SP}{}^{-\alpha}} \mathbf{1}_{(\text{condition})} \tag{5}$$

The $\mathbf{1}_{(\text{condition})}$ in (4) and (5) denote the event that no eavesdropper is in the secure domain, which is given by

$$\mathbf{1}_{(\text{condition})} = \begin{cases} 1, \text{if the condition is valid} \\ 0, \text{otherwise} \end{cases} \tag{6}$$

3. Performance analysis

3.1 Full Activity Strategy

The secondary transmitter ST *keeps transmitting messages to* SR_k. The connection outage probability (COP) refers to the possibility that the legal channel capacity is no more than the given actual coding rate, which is given by $P_{co} = \mathrm{P}\{C_B < R_B\}$. Where C_B represents the legal channel capacity and R_B represents the given actual coding rate. *The derivation process of COP is*

$$\begin{aligned} P_{CO} &= \mathrm{P}(\log(1+\gamma_{R_k}) < R_B) \\ &= \mathrm{P}\left(\left|h_{SR_k} p_K\right|^2 < \frac{(2^{R_B}-1)\ \sigma^2 d_{SR_k}{}^{\alpha}}{I_t d_{SP}{}^{\alpha}} \left|h_{SP}\right|^2 \right) \\ &= \frac{\left(2^{R_B}-1\right)\sigma^2 d_{SR_k}{}^{\alpha}}{\left(2^{R_B}-1\right)\sigma^2 d_{SR_k}{}^{\alpha} + I_t d_{SP}{}^{\alpha}} \end{aligned} \tag{7}$$

The secrecy outage probability (SOP) indicates the possibility that the legal channel capacity of the system is less than the set security rate threshold, which is given by $P_{so} = \mathrm{P}\{C_E < (R_B - R_s)\}$. Where the eavesdrop channel capacity is denoted by C_E , secure encoding rate is denoted by R_S .

The secrecy outage probability (SOP) is derived as

$$\begin{aligned}
P_{SO} &= \mathrm{P}(R_B - R_S < \log(1+\gamma_E)) \\
&= \mathrm{P}\left(R_B - R_S < \log_2\left(1 + \frac{I_t}{\sigma^2 |h_{SP}|^2 d_{SP}^{-\alpha}} \sum_{E_j \in \Phi_E} |h_{SE_j}|^2 d_{SE_j}^{-\alpha} \right)\right) \\
&= \mathrm{E}_{\Phi_E}\left\{ 1 - \exp\left(-\frac{I_t \sum_{E_j \in \Phi_E} |h_{SE_j}|^2 d_{SE_j}^{-\alpha} d_{SP}^{\alpha}}{(2^{R_B - R_S} - 1)\sigma^2} \right) \right\} \\
&= 1 - L_{\sum_{E_j \in \Phi_E} |h_{SE_j}|^2 d_{SE_j}^{-\alpha}}\left(\frac{I_t d_{SP}^{\alpha}}{(2^{R_B - R_S} - 1)\sigma^2} \right)
\end{aligned} \tag{8}$$

3.2 Secrecy Guard Zone Strategy

As we know, ST only transmits confidential messages when no eavesdropper is detected by the secrecy guard. The number of eavesdroppers is denoted as N, which is a Poisson random variable with a mean $\pi r^2 \lambda_E$. Based on (5), the location set Φ_E follows a Poisson distribution with density λ_E outside the security guard zone $\Re(0,r)$. ST is at the origin O of the coordinate axis, then we denote the secrecy guard zone around the ST with $\Re(0,r)$. The COP is the same as (7). The SOP is driven by

$$P_{so} = 1 - L_{\sum_{E_j \in \Phi_E} |h_{SE_j}|^2 d_{SE_j}^{-\alpha}}\left(\frac{I_t d_{SP}^{\alpha}}{(2^{R_B - R_S} - 1)\sigma^2} \right) \tag{9}$$

The Laplace transform of $\sum_{E_j \in \Phi_E} |h_{SE_j}|^2 d_{SE_j}^{-\alpha}$ is derived as

$$L_{\sum_{E_j \in \Phi_E} |h_{SE_j}|^2 d_{SE_j}^{-\alpha}}(S) = \exp\left[-\lambda_E \int_{R^2 \setminus \Re(o,r)} \left(1 - \frac{1}{1 + sx^{-\alpha}} \right) dx \right] \tag{10}$$

4. Numerical results

In this section, the performance of the above system is verified by simulation. The following network parameter values were used in all figures: $\alpha = 4$, $I_0 = 10dB$, $R_S = 0.5$, $R_b = 1.5$.

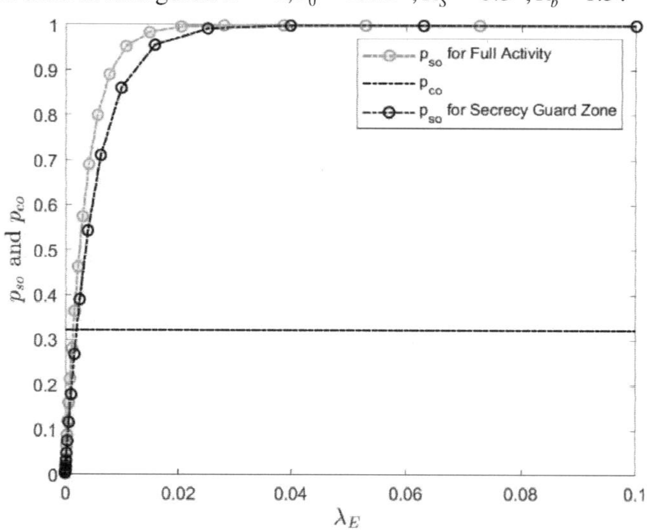

Figure 2. COP and SOP versus the eavesdropper density for two strategies.

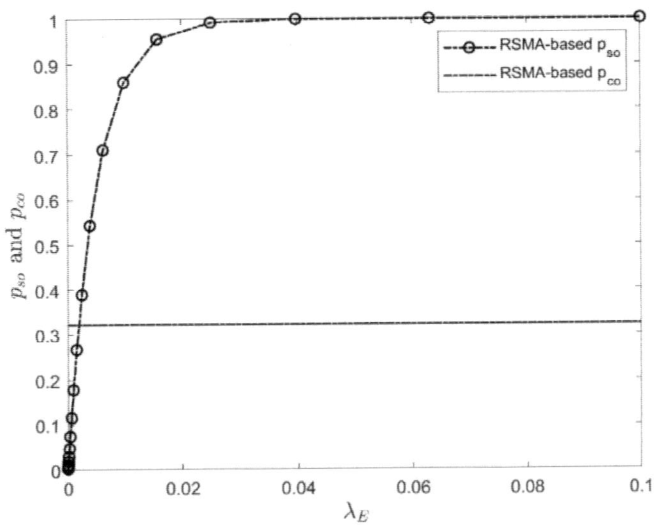

Figure 3. COP and SOP versus the eavesdropper density with $D = 0$.

The impact of reliability and confidentiality should be our primary research. Figure 2 shows the various parameters required λ_E for the two strategies. In terms of the full activity strategy, the secrecy guard zone strategy decreases SOP. The result shown in Figure 3 is the network with $D = 0$. The COP remains constant as the eavesdropper density increase. However, the SOP increased as eavesdropper density increased. As the number of eavesdropping users increases, legitimate users are more likely to be eavesdropped by eavesdropping users. Then the security of the system is degraded.

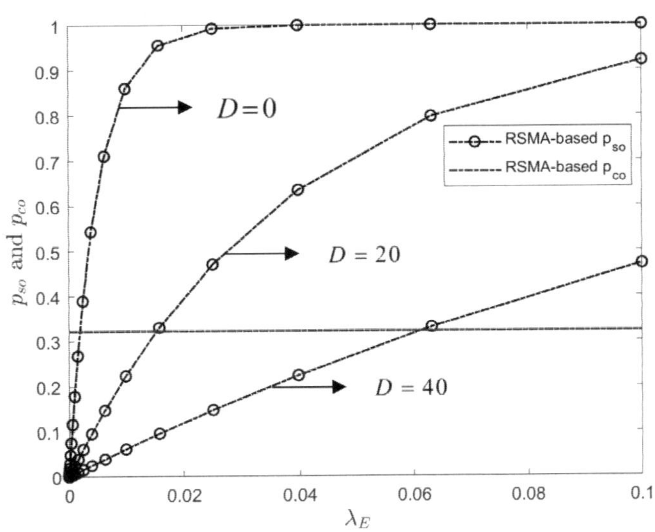

Figure 4. Comparison of eavesdropper density in different security zones about the COP and SOP.

Figure 4 depicts the COP and SOP versus the eavesdropper density λ_E, with different secrecy guard zone D. With λ_E increases, we note that the SOP decreases in all different secrecy guard zone. When the interference tolerance I_t is fixed, the COP remains unchanged, and the SOP decreases when the radius of the protection domain increases. In this case, the system is more secure. In addition, communication security is enhanced.

5. Conclusion

The study investigated the physical layer transmission performance and security performance of cognitive wireless networks with the random distribution scenario of eavesdroppers. Simulations are performed based on the system model, and the results verify the correctness of the closed-form solutions for the connection outage possibility and the security outage possibility, leading to the conclusion. Provided that the communication can be connected properly, p_{so} increases when the density of eavesdroppers increases, and the security performance of the system decreases. Based on this scenario, the secrecy guard zone is changed, and the results show that p_{so} decreases when the secrecy guard zone increases, which leads to the improvement of the security performance of the system.

Acknowledgment

This work is supported by the Youth Innovation Foundation of Xiamen (3502Z20206067) and the Natural Science Foundation of Fujian Province, China (2022J011276, 2021J011219). This work is supported in part by the Natural Science Foundation of China under Grant 62201482.

References

[1] Fatima Salahdine, "Spectrum sensing techniques for cognitive radio networks." *arXiv preprint arXiv:1710.02668* (2017).

[2] Yuhao Wang, Yuhang Wu, Fuhui Zhou, Zheng Chu, Yongpeng Wu, and Fei Yuan, "Multi-Objective Resource Allocation in a NOMA Cognitive Radio Network With a Practical Non-Linear Energy Harvesting Model," in *IEEE Access,* vol. 6, pp. 12973-12982, 2018.

[3] Yonghua Wang, Zifeng Ye, Pin Wan, and Jiajun Zhao, A survey of dynamic spectrum allocation based on reinforcement learning algorithms in cognitive radio networks. Artif Intell Rev 51, 493–506 (2019).

[4] Suneel Yadav and Devendra Singh Gurjar, "Secrecy Performance of SIMO Underlay Cognitive Radio Networks Over $\alpha-\mu$ Fading Channels," in *IEEE Access,* vol. 9, pp. 62616-62629, 2021.

[5] Deemah H. Tashman and Walaa Hamouda, "Physical-Layer Security on Maximal Ratio Combining for SIMO Cognitive Radio Networks Over Cascaded κ-μ Fading Channels," in *IEEE Transactions on Cognitive Communications and Networking,* vol. 7, no. 4, pp. 1244- 1252, Dec. 2021.

[6] Lam Thanh TU, Phan Lam Tung, Trinh Van Chien, Tran Trung Duy, and Nguyen Tien Hoa, "Performance Evaluation of Incremental Relaying in Underlay Cognitive Radio Networks with Imperfect CSI," *2020 IEEE Eighth International Conference on Communications and Electronics (ICCE)*, 2021, pp.472-477.

[7] Yue Tian, Andrew R Nix, and Mark Beach, "On the Performance of Opportunistic NOMA in Downlink CoMP Networks," in *IEEE Communications Letters,* vol. 20, no. 5, pp. 998-1001, May 2016.

[8] Yue Tian, Andrew R Nix, and Mark Beach, "On the Performance of a Multi-Tier NOMA Strategy in Coordinated Multi-Point Networks," in *IEEE Communications Letters,* vol. 21, no. 11, pp. 2448-2451, Nov. 2017.

[9] Yue Tian, Xianling Wang, Zhanwei Wang, and Yau Hee Kho, "On the Performance of Mutual-Aid NOMA Strategy in Cooperative Networks," in *IEEE Communications Letters,* vol. 24, no. 2, pp. 282-286, Feb. 2020.

[10] Yue Tian, Baiyun Xiao, Xianling Wang, Yau Hee Kho, and ChenTian, "Performance analysis of opportunistic NOMA strategy in uplink coordinated multi-points systems, "*Computer Communications,* Volume 177, pp207-212, 2021.

[11] Van Nhan Vo, Nguyen Quoc Long, Viet-Hung Dang, Chakchai So-In, Anh-Nhat Nguyen and Hung Tran, "Physical Layer Security in Cognitive Radio Networks for IoT Using UAV with Reconfigurable Intelligent Surfaces," *2021 18th International Joint Conference on Computer Science and Software Engineering (JCSSE),* 2021, pp. 1-5.
[12] Yijie Mao, Bruno Clerckx and Victor O K Li, Rate-splitting multiple access for downlink communication systems: bridging, generalizing, and outperforming SDMA and NOMA. *J Wireless Com Network* 2018**,** 133 (2018).
[13] Huiyun Xia, Yijie Mao, Bruno Clerckx, Xiaokang Zhou, Shuai Han, and Cheng Li, "Weighted Sum-Rate Maximization for Rate-Splitting Multiple Access Based Secure Communication," *2022 IEEE Wireless Communications and Networking Conference (WCNC),* 2022, pp. 19-24.
[14] Qiao Zhu, Zhihong Qian, Bruno Clerckx and Xue Wang, "Rate-Splitting Multiple Access in Multi-cell Dense Networks: A Stochastic Geometry Approach." *arXiv preprint arXiv:2207.11430* (2022).

BPSK/BOC Modulation Signal System for GPS Satellite Navigation Signals

Haochen Liu[†]

Beijing Dublin International College, Beijing University of Technology, Beijing, China

Liuhc@emails.bjut.edu.cn

Abstract. With the continuous research and development of the global positioning system in various countries, the requirements for the accuracy and efficiency of GPS signals are getting higher and higher. To improve the performance of GPS signal modulation, this research focuses on the BPSK modulation signal and the BOC modulation signal and introduces the basic principles and practical uses of the two modulation signals. Based on the disadvantage of weak anti-noise of BPSK modulated signal and high ambiguity of BOC modulated signal, two algorithms of signal-to-noise ratio estimation and parallel phase code acquisition are introduced in this paper. By introducing the principles and functions of the two algorithms, the improvement of the optimization algorithm for the two modulated signals is expounded. This paper introduces the working principle of the GPS signal, shows the advantages and future development prospects of the BPSK modulation signal and the BOC signal based on the GPS modulation signal, and compares the two modulation signals with the optimization algorithm. The BOC modulated signal shows superior performance. Combined with the optimization algorithm, high-precision transmission signals can be achieved, but the requirements for the receiver are high. The BPSK modulation signal combined with the optimization algorithm can meet most of the needs in life. Still, it is slightly worse than the BOC modulation signal in the face of high-precision and high-demand signal transmission.

1. Introduction

In the 1990s, the Global Positioning System (GPS) was introduced. It is high-precision radio navigation and positioning system based on artificial earth satellites, interacting with users through signal transmission and modulation [1]. Traditional GPS uses binary phase shift keying (BPSK) signals. BPSK is the basic PSK modulation method. PSK's advantages are that the receiver pressure is smaller and the signal processing is easy. However, the anti-noise ability is poor, making mistakes in the signal modulation process easy. GPS technology is too mature compared to other navigation systems and gradually becomes a monopoly. Therefore, countries continue to build their satellite navigation systems.

With the advent of different navigation systems, GPS was progressively transformed into the Global Navigation Satellite System (GNSS), whose emergence indicates the rising global demand for signal positioning. The BOC modulation technology is used in GPS signal modulation to ensure that the limited spectrum can support the transmission of a large amount of information. Broadcasts modulated with Binary Offset Carrier (BOC) are commonly used to distinguish between military and civilian signals [2]. The BOC modulation signal offers superior noise immunity and higher positioning accuracy than the BPSK signal; however, the signal processing is biased due to the ambiguity induced by the multimodality of the BOC modulation signal's autocorrelation function [3]. BOC modulation also places additional strain on the receiver [4].

Reviewing the previous research, this paper combines the SNR estimation algorithm [5] and the improved code phase acquisition algorithm [3] to improve the bit error rate of BPSK modulation, improve the capture efficiency of BOC modulated signals, and reduce ambiguity. Based on previous research, it is believed that BOC modulation has higher artificial noise immunity and transmission accuracy than BPSK modulation, and BOC modulation will occupy the mainstream position in the development of GNSS [2]. This paper will first introduce the basic principles of BPSK modulation [5] and BOC modulation [3] and the principles and functions of the two optimization algorithms. This paper introduces the composition and principle of GPS signal [2] and the components of BPSK modulation and BOC modulation for GPS signal [6]. In future applications, BPSK modulated, and BOC modulated signals will be mainly used for the transmission of military and civil signals [7], combined with optimization algorithms to achieve precise signal transmission, modulation, and identification. The optimized modulated signal can meet the requirements of the satellite communication system for a high signal rate and large bandwidth [8], which will provide a technical foundation for the future development of satellite communication systems and satellite signal receivers.

2. BPSK Modulation

BPSK can convert analog signals into data values and uses a combination of out-of-phase complex waves to represent the information key phase-shifting method. As navigation systems evolve, BPSK modulation technology is employed in satellite navigation systems. The BPSK modulation method improves the spectral properties and bandwidth utilization, boosting the performance and accuracy of the navigation signal [9]. This chapter will demonstrate the BPSK modulation concept and introduce Improved SNR Estimation Algorithm to improve the performance of BPSK modulation.

2.1. The principle of BPSK Modulation

BPSK is a binary digital modulation. During the modulation process, the phase of the sinusoidal carrier will change discretely with the binary digital baseband signal to generate a BPSK signal, which is a 0-1 signal. The signal time domain of BPSK in the function of $S_{BPSK}(t)$ can be expressed as [10]

$$S_{BPSK}(t) = (\sum_n a_n g(t - nT_s)) * cos\omega_0 t \tag{1}$$

where a_n is in accordance with the characteristic statistical variable, when the binary number is 0 $a_n = 1$, and the binary number is 1, $a_n = -1$ [11], T_s is the pulse width.

Table 1. BPSK signal mapping relationship.

Binary Signal	a_n	Phrase	Modulated waveform
0	1	0	$s(t) = \cos(\omega_0 t + \varphi)$
1	-1	π	$s(t) = -\cos(\omega_0 t + \varphi)$

For the change of the binary digital signal, the mapping signal of BPSK can be obtained when the phase is 0 and π [12], Table 1 shows the corresponding values when the binary input signal is 0 and 1, and the calculation method to convert the binary signal into a BPSK mapped signal. When the binary signal is 0, it corresponds to the mapping signal in the form of cos, and when the binary signal is 1, it corresponds to the mapping signal in the form of $-cos$.

In the initial stage of BPSK modulation, the transmitter will input a series of random binary sequence signals, which can be converted into a waveform function between -1 and 1 by using a sine function as a carrier. During the actual signal transmission process, the input signal in the channel will be superimposed with the Gaussian white noise in the signal to form the received signal, which can be expressed as

$$r(t) = s(t) + n(t) \tag{2}$$

The $r(t)$ is the receiving signal, $s(t)$ is the sending signal, and $n(t)$ is the Gaussian white noise.

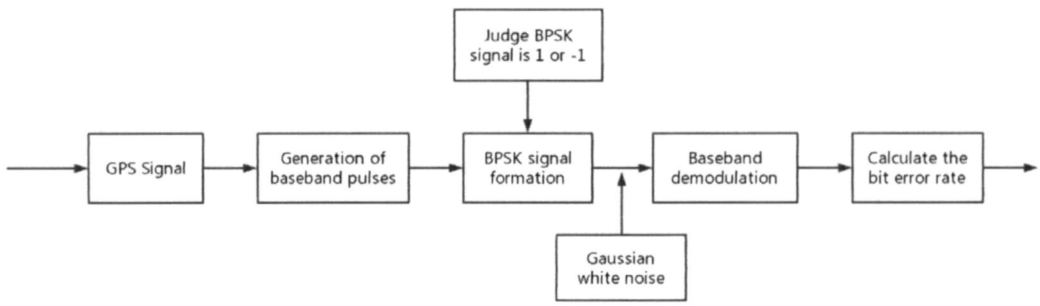

Figure 1. Flow chart of BPSK modulation.

Figure 1 succinctly shows the whole process of BPSK modulation. The system filters and demodulates the superimposed white noise signal and coherently demodulates it with the carrier wave to obtain a demodulated signal. In the process of BPSK modulation, there is no obvious carrier component in the power spectrum, and the carrier extraction can be realized through the Costas loop tracking loop. The sampling decision will be determined according to the phase during modulation, and the signal will be processed 0-1. When the signal is less than 0, the digital signal that the system will pay is set to 0. Otherwise, it is set to 1. For the demodulated signal, the probability of 0 and 1 are equal. Error-recovered digital signals are called bit errors, also known as the "inverted phenomenon."

To study the difference between the theoretical SNR and the actual SNR of the BPSK modulated signal, the theoretical model was drawn, and the real situation was simulated for the BPSK modulated signal. When the signal-to-noise ratio is small, it is considered a high-noise situation, and the bit error rate of BPSK transmission is high. When the signal-to-noise ratio is improved, and the transmission is low noise, BPSK shows a lower bit error rate and better transmission efficiency [13]. According to the simulation, the bit error rate of the modulation effect in practical application slightly deviates from the theoretical value.

2.2. Raised cosine roll-off filter

As mentioned above, a modulation signal must be formed through a filter when the transmitted signal is superimposed with Gaussian white noise. This section mainly introduces the working principle of the filter. The initially transmitted signal is difficult to modulate and identify when superimposed with the Gaussian white noise. A raised cosine roll-off filter usually filters the baseband signal to reduce noise interference and improve the power utilization rate and information accuracy. The effect of noise on the initial signal increases the authenticity of sampling points and information [14]. The expression of raised cosine filter in the time-domain is

$$\begin{cases} T,\ |f| \leq \frac{1-a}{2T} \\ T\cos^2\frac{\pi T}{2a}\left(|f|-\frac{1-a}{2T}\right), \frac{1-a}{2T} \leq |f| \leq \frac{1+a}{2T} \\ 0, |f| \geq \frac{1+a}{2T} \end{cases} \tag{3}$$

Among them, a is the roll-off factor, ranging from $0{\sim}1$, and T is the symbol period. The transmitter increases the signal power several times and sends it out through the radio frequency antenna. At the same time, it is ensured that the signal is not affected by the spectrum of the sideband signal. Then the modulated signal is processed by a band-pass filter, and a square root raised cosine roll-off filter is added at the receiving end to reduce inter-symbol interference [14]. The main parameter of the raised cosine filter is its roll-off coefficient, which determines the bandwidth. The larger the roll-off factor a, the wider the signal bandwidth; otherwise, the narrower the signal bandwidth.

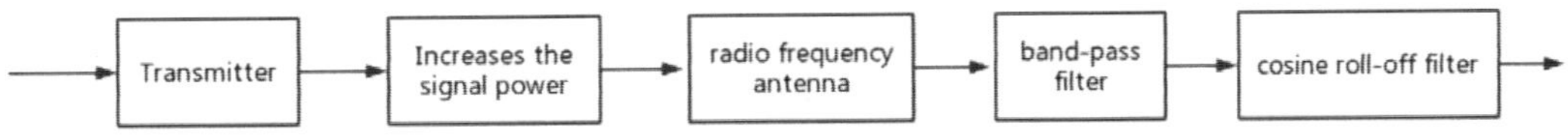

Figure 2 Raised cosine roll-off filter flow chart.

For the raised cosine filter used in practice, it is impossible to realize the sharp drop of the edge in the ideal model, and the practice shows that the roll-off has produced a good effect. The advantage of using the raised cosine roll-off filter is that the tail attenuation is faster, which is beneficial to reduce inter-symbol interference and bit timing errors, and it has high-frequency band utilization.

2.3. Improved SNR Estimation Algorithm

BPSK modulation technology is not ideal for high noise resistance and cannot show good results in complex signal transmission. In order to increase the accuracy and efficiency of BPSK modulation, this section will introduce an improved SNR estimation algorithm to improve the performance of BPSK modulation.

The maximum likelihood estimation is suitable for Gaussian noise. The maximum likelihood criterion is used to obtain the amplitude, and then the power is obtained. After the signal power is obtained, the noise power can be obtained by the difference between the total power and the signal power. The traditional maximum likelihood criterion estimation will produce errors as the phase bias increases. To eliminate the error caused by phase offset, this section adopts the pre-sampling signal-to-noise ratio estimation algorithm proposed by LIU Yuanyuan et al. to improve the BPSK modulation performance [5].

In the traditional algorithm, the symbol-spaced sampled signal can be expressed as:

$$r_k = Aa_k + n_k \tag{4}$$

The new algorithm introduces phase offset, so in this algorithm, the signal received by the receiver is expressed as

$$r_k = A_k^i a_k^i e^{j\theta} + n_k \;\; i \in (1 \ldots N_{ss} - 1) \tag{5}$$

For the signal r_k before sampling, due to the root raised cosine shaping derivative, because the amplitude of each point is different, the direct use of the maximum likelihood criterion to derive the SNR estimator cannot be calculated accurately. In this regard, r_k is squared, and the average power of the signal is obtained by an indirect method. It is assumed that the noise after the secondary processing approximately conforms to the characteristics of Gaussian white noise.

$$r_k' = r_k^2 = (A_k^i)^2 a_k^2 e^{2j\theta} + n_k^2 + 2A_k^i a_k e^{j\theta} n_k \;\; i \in (1 \ldots N_{ss} - 1) \tag{6}$$

A_k^i is the amplitude after root raised cosine shaping, and N_{ss} is the sampling multiple of the sending end, the same as the signal amplitude change period after processing. Due to the introduction of phase offset, the actual signal power is divided into two parts $A_k' = A^2 a_k^2 e^{2j\theta}$, $N_k' = n_k^2 + 2Aa_k e^{j\theta} n_k$.

$$A_k' + N_k' = (A_{Ik}' + N_{Ik}') + j(A_{Qk}' + N_{Qk}') \tag{7}$$

Calculate the probability density function for the two parts I and Q, take the partial derivative of the log-likelihood function and make it 0. The formula can be expressed as

$$A_I' = \frac{1}{L} \sum_{k=0}^{L-1} Re\{y_k'\} \tag{8}$$

$$A_Q' = \frac{1}{L} \sum_{k=0}^{L-1} Im\{y_k'\} \tag{9}$$

Therefore, the signal power formula can be obtained as

$$P_A = [(A_I')^2 + (A_Q')^2]^{1/2} \tag{10}$$

Combining the above calculation process, the algorithm finds that the signal power of each group of signals is

$$P_i = (A^i)^2 = |\frac{1}{N_{sym}}\sum_{i=0}^{N_{sym}-1} r'_{i+lN_{ss}}| \tag{11}$$

N_{sym} is the number of raw symbols and $L = N_{sym}N_{ss}$, the average signal power is

$$P_i = \frac{1}{N_{ss}}\sum_{i=0}^{N_{ss}-1}\{\left|\frac{1}{N_{sym}}\sum_{i=0}^{N_{sym}-1} r'_{i+lN_{ss}}\right|\} \tag{12}$$

so that the signal-to-noise ratio estimator of the signal before sampling is:

$$r_B = P(\frac{1}{L}\sum_{k=0}^{L-1}|r_k|^2 - P)^{-1} \tag{13}$$

The algorithm is not affected in the case of phase bias and shows good performance in the condition of a low signal-to-noise ratio. This algorithm does not need symbol synchronization and has the characteristics of a wide estimation range, high precision, and low complexity, and is highly practical [5].

For the above optimization algorithm, the Field Programmable Gate Array (FPGA) was used to verify the improved algorithm. In the FPGA module, the input terminal obtains the imaginary and real two-part signals of the received signal, and the two parts of the signal are squared and accumulated. When the accumulated value reaches the set value, the corresponding low-order data is truncated to complete the mean value processing. The signal power is obtained through summation and square root, and then the noise power is obtained. Simulation experiments show that the algorithm in this paper has low complexity, is easy to implement on FPGA, and has been stably applied in a certain type of project [5]. The optimization algorithm can be applied to the signal receiver to identify and compile the signal.

3. BOC Modulation

With the continuous establishment of different GPS technologies in various countries, GPS is gradually transformed into GNSS. Facing the ever-increasing demand for signal transmission, BOC modulation reduces the pressure of signal transmission in a limited spectrum, so BOC modulation is the mainstream technology in the future. This section will focus on the principle of BOC modulation and the capture technology of BOC signals.

3.1. BOC Modulation basic model

Compared with BPSK modulation, BOC modulation adds a pseudo-code step and performs a secondary spread spectrum on the subcarrier of the spread spectrum signal before carrier hopping. The BOC received signal can be expressed as:

$$r(t) = Ad(t-\tau)c(t-\tau)sc(t-\tau)cos\,((W_{IF} + 2\pi f_d)t + \emptyset) + N(t) \tag{14}$$

A represents signal amplitude; $d(t)$ represents data; $c(t)$ represents spread spectrum code; $sc(t)$ represents BOC subcarrier; W_{IF} represents intermediate frequency signal angular frequency; $\emptyset$ represents the phase. represents carrier Doppler; τ represents signal Delay.

Let $sc(t)$ denote the subcarrier signal. The function expression can be expressed as

$$\begin{cases} sc(t) = sign(sin(2\pi f_s t)) \\ sc(t) = sign(cos(2\pi f_s t)) \end{cases} \tag{15}$$

The corresponding expression of the BOC modulated baseband signal $s(t)$ is

$$s(t) = \sum_{k=-\infty}^{\infty} c_k \mu_{T_c}(t - kT_c)sc(t) \tag{16}$$

c_k is a pseudo-code sequence of modulated data values, T_c is the chip width, μ_{T_c} is a rectangular pulse signal with a pulse width of T_c, and $sc(t)$ is the offset binary subcarrier.

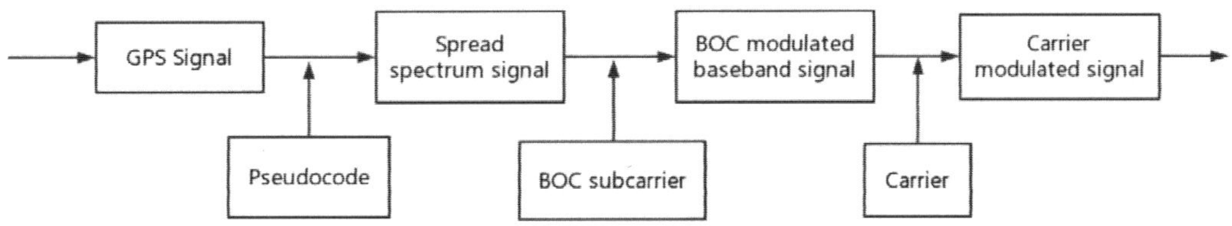

Figure 3 Flow chart of BOC modulation.

Figure 3 shows the process of BOC modulation. Adding pseudo code and subcarrier is a very important link in BOC modulation. The pseudo-code is a spread spectrum modulated by the square wave subcarrier. The spectrum is moved to realize the symmetrical characteristic of both ends of the BOC modulated signal.

To understand the distribution characteristics of the BOC signal, the spectral distributions of the two BOC modulated signals, BOC(1,1) and BOC(2,1), are measured here. Unlike the BPSK modulated signal, the BOC modulated signal presents left-right symmetry, and the occupation is concentrated on both sides. The spectral occupancy in the central region is significantly lower than that of the BPSK modulated signal. This distribution state can improve the utilization of the spectrum.

3.2. BOC modulation signal acquisition algorithm

The acquisition algorithm can improve the accuracy of signal modulation and help the receiver to transmit and modulate the signal. This section will introduce the parallel code phase search algorithm. Referring to the algorithm improvement scheme proposed by Zhu Yanbing et al., we propose a way to improve the efficiency and accuracy of BOC modulation [15].

3.2.1. Parallel Code Phase Search Algorithm

There are two kinds of FFT parallel acquisition algorithms: frequency domain parallel search and parallel code phase search. In general, parallel code phase search shows higher efficiency. The principle of the parallel code phase search algorithm [15] is mainly to mix the received signal with the local carrier signal and remove the carrier, then perform FFT (fast Fourier transform) transformation between the signal and the pseudo-code sequence and perform the conjugate operation on the bit code sequence. Then we multiply the processed signal by the calculated pseudo-code sequence, perform IFFT (Inverse fast Fourier transform) change, and determine whether the signal is captured by judging the peak value. Finally, we set two sequences x(n) and y(n), with period N. The formula of the related function is expressed as

$$r_{xy}(n) = \frac{1}{N}\sum_{m=0}^{N-1} x(m)y(m-n) \tag{17}$$

The DFT (Discrete-time Fourier Transform) change of the correlation function can be obtained

$$R_{xy}(k) = \frac{1}{N}\sum_{n=0}^{N-1}\sum_{m=0}^{N-1} x(m)y(m-n)e^{-j2\pi nk/N} \tag{18}$$

Further calculation of Equation 13 yields the following relation

$$R_{xy}(k) = \frac{1}{N}\sum_{m=0}^{N-1} x(m)e^{-2\pi mk/N}\sum_{n=0}^{N-1} y(m+ n)e^{j2\pi(m-n)k/N} = \frac{1}{N}X(k)Y^*(k) \tag{19}$$

$X(k)$ and $Y(k)$ are the DFT transform of x, y, and Y^* is the conjugate of Y.

To test the capture effect of the phase parallel code, the following four cases are set: BOC (1,1) signal under the condition of SNR= -28dB capture for simulation, the capture of BOC(6,1), BOC(14,2) and BOC(7,1) signals under the condition of SNR = -20dB. The results show that the effective acquisition of the signal cannot be accomplished by using the parallel code phase search algorithm. To realize the BOC signal acquisition under the low signal-to-noise ratio, the algorithm should be improved by combining the BOC and the pseudo-code cross-correlation function. The following section will introduce algorithmic improvements to the above problems.

3.2.2. Algorithm improvement and optimization

For a BOC signal with a modulation factor of k, the cross-correlation result with its pseudo-code includes 2k peaks, with k positive and negative peaks each. The improved algorithm is used for IFFT transformation on the above cross-correlation results. We derive the transformed signal and then compare the BOC signal with the pseudo-code. The cross-correlation sequence operates on correlation, peak detection results, and subsequent processing [15]. The improved algorithm can effectively acquire the BOC signal, and the captured code phase is consistent with the simulation's code phase, which improves signal acquisition's sensitivity and accuracy. At the same time, by optimizing the acquisition program, the pressure on the receiver can be reduced.

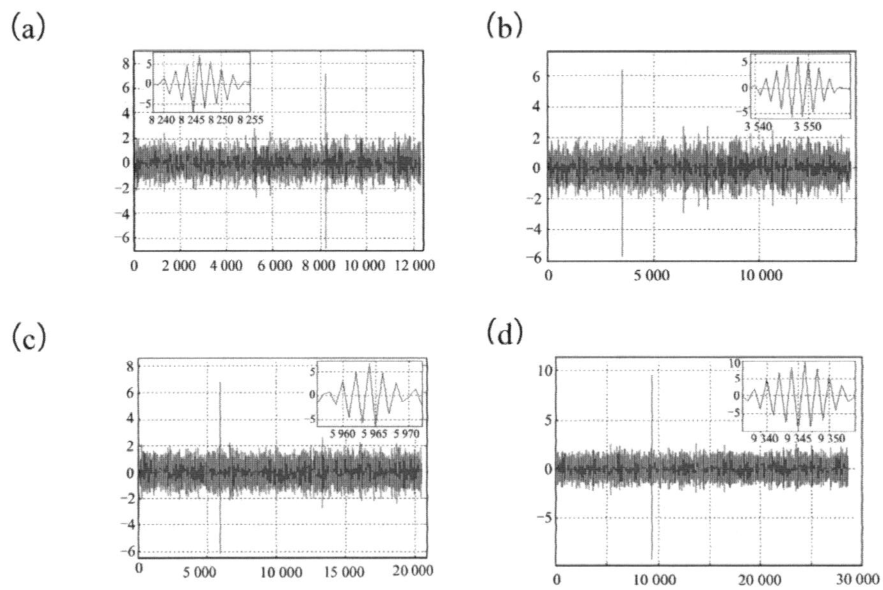

Figure 4 Optimization algorithm result graphs. The resulting graphs shown Figure (a) is BOC(6,1), Figure (b) is BOC(7,1), Figure (c) is BOC(10,1) and Figure (d) is BOC(14,1) [15].

Figure 4 shows different BOC modulation signals. Each BOC modulation signal diagram shows the simulation setting code and the captured code. The improved algorithm can effectively capture the BOC signal and the captured code phase and simulation. The set codes are consistent, proving the optimization algorithm's feasibility.

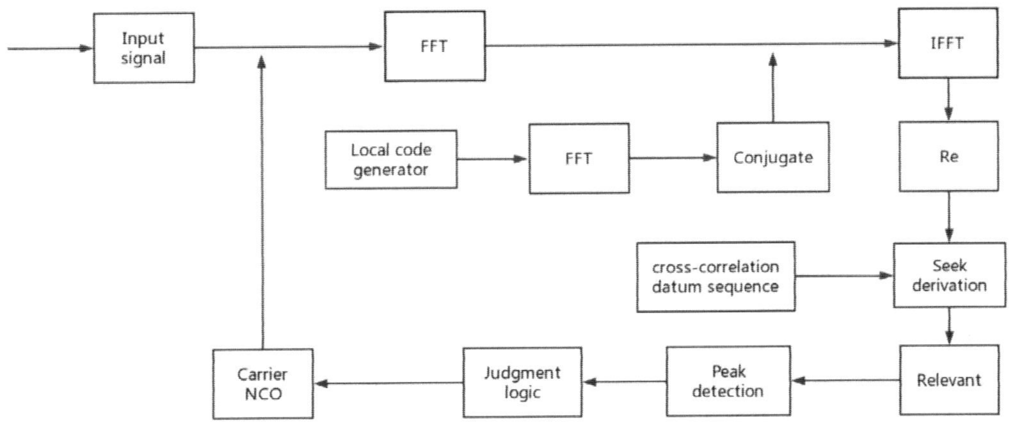

Figure 5. The whole process of the optimization algorithm.

To explore the applicable scope of the algorithm, the BOC (5, 2) signal is simulated and captured using the classical and improved algorithms. The results show that simply transplanting the parallel code phase search algorithm is entirely inapplicable. For the BOC ($\boldsymbol{p_n}$, $\boldsymbol{n}$) signal ($\boldsymbol{p}$ is not an integer), and simple transplantation cannot complete the signal acquisition by using the parallel code phase search algorithm [15].

The optimized algorithm can capture the BOC ($\boldsymbol{p_n}$, $\boldsymbol{n}$) signal ($\boldsymbol{p}$ is not an integer). Compared with the classical algorithm and the traditional BOC signal capture that is usually used, when the capture signal-to-noise ratio is above 10dB, the algorithm shows higher efficiency and lower requirements for equipment. In practical application, the algorithm can replace the classical capture algorithm to realize the capture of BOC signal. The hardware requirements of the BOC signal capturer are also reduced, reducing the manufacturing cost and technical requirements.

4. GPS signal transmission system

The previous section introduced the principle of BPSK modulation and the principle of BOC modulation. Both BPSK and BOC modulation are components of GPS signals. Two optimization algorithms can improve the effect of GPS transmission and modulation. The following will introduce GPS signals and positioning and explain how BPSK and BOC are used in GPS signals.

The working principle of GPS is that the satellite sends a signal to the ground, and the ground control department receives, modulates, measures, determines the position, and returns the signal to the satellite. The satellite relays the orbit information of the satellite for the signal. Finally, the user determines his position by receiving signals from the visible satellites [1]. Modulation and identification are important parts of satellite positioning to transmit satellite signals. Low bit error rate and high-efficiency signal acquisition technology can allow users to obtain more precise positioning. Currently, most research use BPSK and BOC signals to simulate GPS signals. In this section, we hope to conduct a comparative study of BPSK modulation and BOC modulation and discuss the GPS signal modulation method that integrates BPSK and BOC modulation.

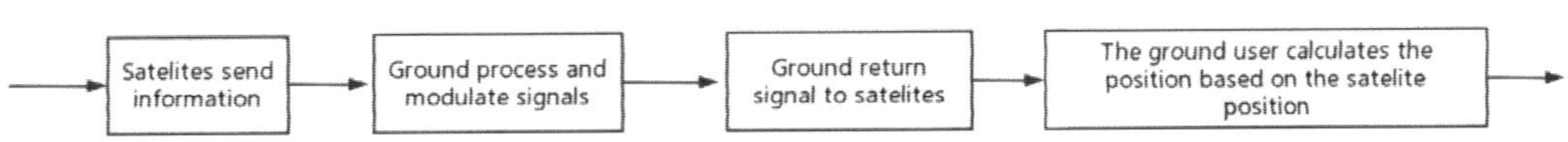

Figure 6. GPS positioning process.

Figure 6 shows the whole process of GPS positioning. BPSK modulation and BOC modulation are mainly used in signal transmission and the receiver of satellite and ground receiving information.

4.1. Modulation method performance analysis

The principles of BPSK modulation and BOC modulation are mainly introduced in the second and third subsections. BPSK modulation is a relatively basic modulation method that can transmit binary signals. Regarding GPS signals, BPSK modulation can guarantee the basic characteristics and functions [1]. In the case of low noise, it shows a good bit error rate and transmission efficiency. The receiver has low requirements on the receiving end, and the pressure on the receiver is small, which is more suitable for the cost and technical requirements. However, it does not show good performance in terms of noise immunity. Compared with other PSK modulation methods, BPSK has poor noise immunity in the case of strong noise. In addition to its performance, the energy of BPSK is mainly concentrated in the main lobe bandwidth, which causes great pressure on the spectrum when multiple modulation signals coexist. Compared with BPSK modulation, BOC modulation adds a binary subcarrier based on BPSK and readjusts the spectrum of the baseband signal after spread spectrum modulation. Compared with the BPSK signal occupying the center of the spectrum, the original spectrum of BOC modulation is split to both sides of the center frequency, and there are two main peaks to form a symmetrical form [4]. At the same time, because of its split spectrum characteristics, the BOC modulation signal will not cause serious interference to the original GPS signal. However, the peak value of the BOC modulated signal is relatively dense, which is easy to cause interference to the catcher and cause capture errors. To improve this phenomenon, the technical requirements for the receiver are relatively high, and the cost and technical requirements are relatively high. BPSK modulation is generally suitable for businesses with low precision requirements and low consumption costs, while BOC modulation is suitable for businesses with high precision requirements and sufficient expenses.

To explore the mixing effect of the BPSK modulation signal and BOC modulation signal, the experiment modulates the BPSK modulated signal and the BOC (10,10) modulated signal [16]. These two signals can combine the advantages of the two modulated signals [17]. The results show the difference in spectrum distribution between BOC modulated signals and BPSK modulated signals: BOC modulated signals are mainly distributed on both sides. BPSK modulated signals are concentrated in the central area. BPSK modulated signal peaks appear at frequency offsets of 0MHz, $\pm$15MHz, $\pm$25MHz, and $\pm$35MHz relative to the carrier. The density is decreasing. The peak value of the BOC modulation signal appears at $\pm$8MHz and $\pm$35MHz and has no overlap with the BPSK modulation signal.

5. Applications of BPSK and BOC modulation signal

In practical applications, BPSK and BOC modulation signals are used for GPS positioning or various national industrial projects. As more and more industries use Beidou positioning, people have higher and higher requirements for signal transmission of Beidou positioning. Beidou positioning adopts a BPSK modulation signal [18], and the receiver adopts the optimization algorithm and compression technology to filter out the artificial interference signal and natural interference signal in the signal and extract the BPSK signal that transmits information. In the face of industrial use, the B1I signal and suppression information used by Beidou satellites are transmitted using BPSK-modulated signals. The optimization algorithm's job is to ensure the accurate identification of BPSK. At the same time, the BPSK modulated signal is also used for the acoustic wave transmission in the downhole drill string [19]. The adaptive BPSK and optimization algorithms are used to ensure the efficient transmission of the acoustic wave in the drill string channel. As an essential part of navigation system design, GNSS involves almost every field of the national economy and social development. Its signal transmission generally adopts a BOC modulation signal to solve the signal transmission under the condition of a limited spectrum. The optimized phase parallel codes improve the capture performance of BOC modulated signals and enable them to be more accurately identified [20]. BOC modulated signals are mainly used in global positioning systems to achieve high-precision information transmission between the ground and satellites.

Table 2. Application of BPSK and BOC modulation signal.

	BPSK	BOC
Performance	High transmission efficiency and easy compilation.	Strong anti-noise, high transmission accuracy, difficult to compile.
The occupied area of the spectrum	Mainly occupy the central part, the middle is high, and the two sides are low.	Mainly occupy the two sides of the spectrum, and can make full use of the limited spectrum.
Scope of application	Wide range, such as GPS and drilling, radar.	Mainly used for GNSS.
Applicable fields	High transmission efficiency and fast speed are required for information transmission.	High-precision transmission, high technical level, and configuration requirements.
Optimized algorithm	Improved SNR Estimation Algorithm.	Optimized Phase Parallel Code.
Prospects	Technology has matured.	May become the mainstream technology of GPS in the future.

6. Conclusion

This paper mainly introduces the basic principles of BPSK modulation and BOC modulation. It introduces the improved SNR estimation algorithm of BPSK and the improved code phase acquisition algorithm of BOC to improve the accuracy and efficiency of information modulation. The research shows that the improved SNR estimation algorithm of BPSK reduces the bit error rate of BPSK modulation and improves the accuracy of signal transmission. Two algorithms achieve accurate information transmission and identification. The comparison results of previous studies of BOC modulation and BPSK modulation show that BOC modulation is better than BPSK modulation in terms of performance. The future development trend will be dominated by BOC modulation. However, BOC modulation will bring greater pressure and technology to the receiver. Require. In future research, the authors believe it is possible to study and analyze GPS signals combining BPSK modulation and BOC modulation and set up the receiver.

References

[1] M. P. Ananda, H. Bernstein, K. E. Cunningham, W. A. Feess, E. G. Stroud 1990 "Global Positioning System (GPS) autonomous navigation," *IEEE Symposium on Position Location and Navigation. A Decade of Excellence in the Navigation Sciences,* pp 497-508, doi: 10.1109/PLANS.1990.66220.

[2] Fantino, M., Dovis, F., Presti, L. L., & Wang, J. 2004 Acquisition performance analysis for BOC modulated signals. *The 2004 International Symposium on GNSS/GPS* **(p. 22832695)**. Sydney, Australia: GNSS.

[3] Zhang H L, Ba X H, Chen J & Zhou H 2017 Unambiguous acquisition technology for BOC(m,n) signal. *Chinese Journal of Aeronautics and Astronautics* **(04)**, pp 222-231.

[4] Sun X Y, Song S J, Ji Y F, Liang W B, Li Y M 2022 A fuzzy-free tracking algorithm for BOC and its derived signals[J].*Journal of Xidian University*, **49(02)** pp 58-66.

[5] Liu Y Y, Jiang X, Xie X Z, Lei W J 2014 An improved BPSK signal-to-noise ratio estimation method and its FPGA implementation *J. Journal of Chongqing University of Posts and Telecommunications* (Natural Science Edition), **05** pp 630-635.

[6] Yu Y, Li X Y, Zhang X C. BPSK, QPSK, UQPSK, 64QAM 2015 Signal Automatic Modulation Identification *J. Electronic Science and Technology,* **(01)** pp 49-52.

[7] Delmas, J. P., & Abeida, H. 2005 Cramer-Rao bounds of DOA estimates for BPSK and QPSK modulated signals. *IEEE Transactions on Signal Processing,* **54(1)** pp 117-126.
[8] Tao G T 2015 Research and implementation of BPSK/QPSK demodulation system for satellite communication receiver D. Chongqing University.
[9] Meng Q F, Jiao G T, Wang L X, Geng S Q, & Ding J 2013 AltBOC Navigation Signal Quality Assessment and Analysis. *Journal of North Central University (Natural Science Edition),* **34(4)** pp 474-480.
[10] Ding K 2021 Simulation analysis of BPSK bit error rate in AWGN channel *J. Microprocessor,* **(03)** pp 23-26.
[11] Al Safi, A., & Bazuin, B. 2016 FPGA based implementation of BPSK and QPSK modulators using address reverse accumulators. *In 2016 IEEE 7th Annual Ubiquitous Computing, Electronics & Mobile Communication Conference (UEMCON)* pp 1-6. IEEE.
[12] Zhao L J 2020 Modeling of BPSK digital communication system based on SIMULINK *J. Modern Industrial Economy and Informatization,* **(05)** pp 24-25+29.
[13] Yang L, Hu G B 2017 An Improved BPSK/QPSK Signal Modulation Identification Algorithm *J. Telecommunications Technology,* **57(08)** pp 896-902.
[14] M. A. Rihani, J. -C. Prévotet, F. Nouvel, M. Mroue and Y. Mohanna 2016 "ARM-FPGA based platform for automated adaptive wireless communication systems using partial reconfiguration technique," *2016 Conference on Design and Architectures for Signal and Image Processing (DASIP),* pp. 113-120, doi: 10.1109/DASIP.2016.7853806.
[15] Zhu Y B, Li B, Liu R 2015 Research on BOC Signal Improved Code Phase Acquisition Algorithm *J. Journal of Wuhan University(Information Science Edition),* **03** pp 406-411.
[16] Liu C, Liu K Q, Hua Y 2021 Research and analysis of the anti-multipath performance of pseudolites based on BOC modulation *J. Electronic Design Engineering,* **(19)** pp 168-173.
[17] Kawalec, A., & Owczarek, R. 2004 Radar emitter recognition using intrapulse data. *In 15th International Conference on Microwaves, Radar and Wireless Communications (IEEE Cat. No. 04EX824)* **(2)** pp 435-438. IEEE.
[18] Yuan C X, Su D C, Yang J, Wang R G, Xie S J 2021 Research on the Influence of BPSK Modulation Suppression Interference on the B1I Signal of Beidou Receiver *J. Electronic World,* **(11)** pp 79-82. DOI: 10.19353/j.cnki.dzsj .2021.11.037.
[19] Zhou J, Tian H, Gao J B, Qiu B 2016 Application of Adaptive BPSK in Downhole Drill String Acoustic Transmission *J.Journal of Southwest Petroleum University(Natural Science Edition),* **38(02)** pp162-168.
[20] Yang Z X, Yang J W, Zheng X D, Liu M Y, Zhang L 2021 Theory and application of modern GNSS signal acquisition performance evaluation *J.Science in China: Physics Mechanics Astronomy,* **51(01)** pp187-200.

Joint Interference Cancellation and Symbol Detection for Multistatic Backscatter System

Wanxian Liu, Xiuli He, Ze Wang, Li Zhao, Hongbo Xu

The Department of Electronics and Information Engineering, Central China Normal University, Wuhan 430079, China

e-mail: liuwx@mails.ccnu.edu.cn, 2021112301h@mails.ccnu.edu.cn, wangze0205@mails.ccnu.edu.cn, zhli@mails.ccnu.edu.cn, xuhb@ccnu.edu.cn

Abstract. Backscatter communication is a promising passive radio technique for the green Internet of Things (IoT) due to its low cost and low power consumption. However, the limited communication range is a challenging issue because of the double fading of the backscatter link. This paper proposes a multi-static backscatter system in which multiple readers illuminate a tag simultaneously to strengthen the received backscatter signals. In this system, direct links greatly impact the backscatter links, making it difficult to recover the tag information. The interference cancellation (IC) is exploited to cancel out the interference of the direct link and improve the detection probability. Then the expectation maximization (EM) algorithm is used for symbol detection, in which latent variable is introduced to form simpler distribution, and parameters are updated based on their probability. The numerical results verify the better performance of our proposed method.

1. Introduction

As a promising technique for connecting massive devices, IoT can exchange information among different devices to realize intelligent identification, positioning, tracking, supervision, and other applications [1,2]. However, most devices consume high power. Therefore, backscattering communication [3], a low-cost and low-power scheme, is emerging, which reflects radio waves and transmits information bits to readers by changing the impedance of the antenna.

However, the limited range is the key challenge for backscatter communication due to the double fading of the backscatter link. Wang et al. [4] considered a monostatic architecture. The reader consists of the transmitter that emits the signal to the tag and the reader that decodes the modulated signal. In addition, the tag that places close to the reader benefits, while the tag that places far is difficult to communicate with the reader. This results in a limited distance between the tag and the reader. Alevizos et al. [5] proposed a multistatic system for extended coverage reliability, which consists of multiple reader-tag pairs to extend the communication range. However, the system does not eliminate the strong interference of the direct link.

In this paper, we consider a multistatic backscatter system for extending range, in which several dislocated readers illuminate a tag to enhance the received backscatter signals. Moreover, the system further expands the communication range by using cooperative readers. The reader comprises a transmitter and a receiver, generating a radio frequency (RF) wave to the tag. The tag receives the signal from the reader, then modulates and reflects the signal to the reader. Therefore, the reader receives two signals: the direct link signal from the RF source and the backscatter signal from the tag.

Journal of Physics: Conference Series **2384** (2022) 012024 doi:10.1088/1742-6596/2384/1/012024

Owing to the double fading of the backscatter link, the direct link is much stronger than the backscatter link[6]. In this system, assuming that the channel statistical knowledge is unknown, the least squares (LS) estimation method is chosen to eliminate the interference of the direct link[7]. In addition, we consider the channel estimation method based on pilot symbols. However, the backscatter link is very weak compared with the direct link. Using long pilot sequences may further lead to efficiency loss. Comparatively, the short pilot sequences not only can make use of the pilot symbol and data symbol to estimate the channel, but also decrease efficiency loss [8]. Therefore, we adopt the LS estimation based on short pilot sequences. After eliminating the interference of the direct link, it becomes easier to detect tag symbols from the residual signal. This paper considers the expectation maximization (EM) algorithm[9], which detects tag symbols by the maximum likelihood function.

The main contributions are summarized as follows:

- Because of the double fading of the backscatter link, the direct link is much stronger than the backscatter link, which makes it difficult to detect the tag symbol. Therefore, we use the LS method to eliminate the interference of the direct link and obtain the residual received signal before detection.
- Then, a tag symbol detection method based on the EM algorithm for a multistatic backscatter system is proposed. The latent variable is introduced to form simpler distribution, and parameters are updated based on their probability to improve detection probability.

The rest of the paper is organized as follows. Section Π presents the multistatic backscatter system. Section III proposes joint channel estimation and symbol detection, which uses the IC method to eliminate the influence of the direct link and the EM algorithm to detect tag symbols. Section IV analyses the effectiveness of the proposed schemes. Section V concludes this paper.

2. System model

As depicted in Figure 1, we consider a multistatic backscatter system, which consists of K single-antenna readers and one single-antenna tag. The K readers illuminate a tag simultaneously to strengthen the received backscatter signals. The reader works in full-duplex mode, which transmits and receives simultaneously. In this article, we consider $K=3$. Three readers transmit RF source signals simultaneously, and the tag modulates its information bits by switching its antenna impedance to reflect RF source signals. Each reader receives two types of signals: the direct link signal from the other two readers and the backscatter signal from the tag.

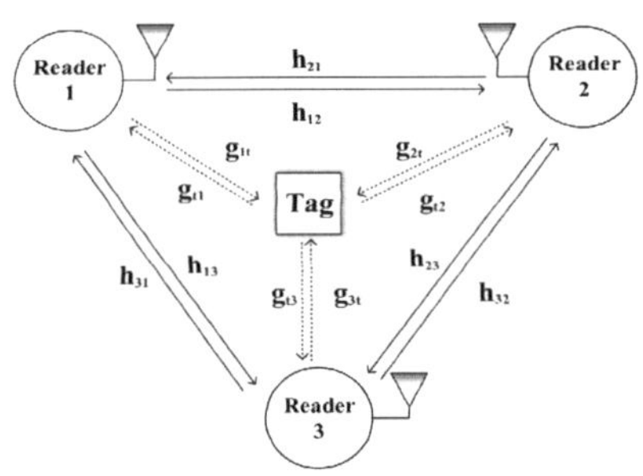

Figure 1. System model of multistatic backscatter communication

Since the direct link signal is significantly stronger than the backscatter link, the tag reflects the signal at a rate lower than the sampling rate of the RF source signal. Assume that one frame consists of I tag symbols, and one tag symbol period covers N RF source symbol periods, as illustrated in Figure 2. Specifically, we consider a block-fading channel model, where the channel remains unchanged during one frame and varies independently among different frames.

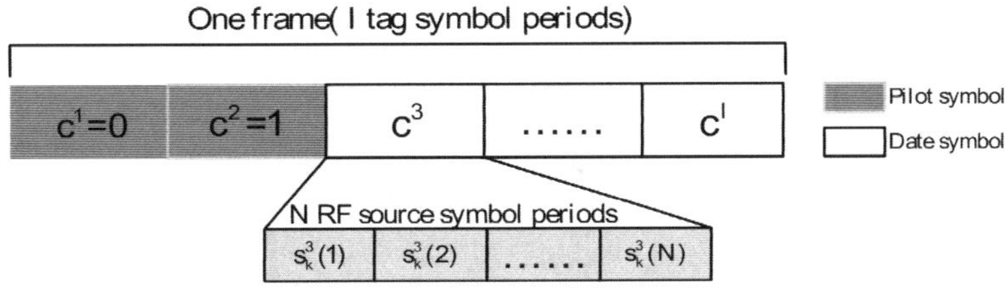

Figure 2. The frame structure of the backscatter system.

Let c^i be the tag signal at the ith tag symbol. The tag adopts Binary Phase Shift Keying (BPSK) modulation, so we have $c^i \in \{0,1\}$. When bit '0' is transmitted, there is no reflection. When bit '1' is transmitted, the RF source signals are reflected by the tag. Assume that one frame consists of I backscatter symbols $\mathbf{c}=[c^1,\cdots,c^i,\cdots c^I]$. The first two pilot symbols $c^1=0$ and $c^2=1$ are used for channel estimation and symbol detection and $c^3,\cdots,c^I$ are data symbols.

Let $s_k^i(n), n=1,\cdots,N$ be the RF source signal at the time instant n within the ith tag symbol for the reader $k, k=1,2,3$. Assume that $s_k^i(n)$ is independently and identically distributed at different time instants and $s_k^i(n)$ follows the CSCG distribution with the mean zero and the variance one, i.e., $s_k^i(n)\sim\mathcal{CN}(0,1)$. We denote $\mathbf{s}^i(n)=[s_1^i(n),s_2^i(n),s_3^i(n)]\in\mathbb{C}^{3\times1}$ as the signal collected by the three readers at the time slot n during the ith tag symbol. $\mathbf{S}^i=[\mathbf{s}^i(1),\ldots,\mathbf{s}^i(n),\ldots,\mathbf{s}^i(N)]\in\mathbb{C}^{3\times N}$ denotes the RF source signal collected by three readers during the ith tag symbol. Then, we define the channel from reader j to reader k as h_{jk}, $j,k=1,2,3$. The channel from reader k to tag and from tag to the reader k are denoted as g_{kt} and g_{tk} respectively. Due to the channel reciprocity, $h_{jk}=h_{kj}, g_{kt}=g_{tk}$.

For reader 1, during i^{th} tag symbol, the received signal at the time constant n can be expressed as

$$y_1^i(n)=\sqrt{P_s}h_{11}s_1^i(n)+\sqrt{P_s}(h_{21}s_2^i(n)+h_{31}s_3^i(n))+\alpha\sqrt{P_s}g_{t1}\mathbf{g}\mathbf{s}^i(n)c^i(n)+w_1^i(n), \tag{1}$$

where P_s is the transmit power from the reader, the transmit power of each reader is assumed to be the same. $\sqrt{P_s}h_{11}s_1^i(n)$ stands for the self-interference of reader 1. Owing to h_{11} is known to reader 1, $\sqrt{P_s}h_{11}s_1^i(n)$ can be canceled. $\sqrt{P_s}(h_{21}s_2^i(n)+h_{31}s_3^i(n))$ stands for the interference from reader 2 and from reader 3, respectively. $\alpha\sqrt{P_s}g_{t1}\mathbf{g}\mathbf{s}^i(n)c^i(n)$ stands for the received signal from the backscatter link, where $\mathbf{g}=[g_{1t},g_{2t},g_{3t}]\in\mathbb{C}^{1\times3}$ is the backscatter link vector and α is the reflection coefficient. $w_1^i(n)$ stands for the complex additive white Gaussian noise (AWGN) of reader 1 at n^{th} RF source period during i^{th} tag symbol, which follows the CSCG distribution with the mean zero and the variance σ^2, i.e., $w_k^i(n)\sim\mathcal{CN}(0,\sigma^2)$.

Then, the received signal collected by three readers at a time instant n during i^{th} the tag symbol can be written as $\mathbf{y}^i(n)=[y_1^i(n),y_2^i(n),y_3^i(n)]^T\in\mathbb{C}^{3\times1}$. And the received signal during i^{th} the tag symbol can be represented as $\mathbf{Y}^i=[\mathbf{y}^i(1),\ldots,\mathbf{y}^i(n),\ldots,\mathbf{y}^i(N)]\in\mathbb{C}^{3\times N}$. Similarly, we can express $\mathbf{w}^i(n)=[w_1^i(n),w_2^i(n),w_3^i(n)]^T\in\mathbb{C}^{3\times1}$ and $\mathbf{W}^i=[\mathbf{w}^i(1),\ldots,\mathbf{w}^i(n),\ldots,\mathbf{w}^i(N)]\in\mathbb{C}^{3\times N}$. Suppose that channel prior knowledge is unknown and readers cooperate. The received signal (1) can be expressed as a matrix form

$$\mathbf{Y}^i=\sqrt{P_S}\mathbf{H}^i\mathbf{S}^i+\sqrt{P_S}\mathbf{G}^i\mathbf{S}^ic^i+\mathbf{W}^i, \tag{2}$$

where the backscatter link matrix is defined $\mathbf{G}^i=\alpha\mathbf{g}^T\mathbf{g}\in\mathbb{C}^{3\times3}$, the direct link channel matrix is defined as $\mathbf{H}^i=\begin{bmatrix}0 & h_{21} & h_{31}\\ h_{12} & 0 & h_{32}\\ h_{13} & h_{23} & 0\end{bmatrix}\in\mathbb{C}^{3\times3}$.

Owing to the fact that the direct link is much stronger than the backscatter link, which makes it difficult for symbol detection. In this paper, our goal is to eliminate the strong interference of the direct link and detect the tag symbol from the residual signal.

3. Joint channel estimation and symbol detection

In this section, we propose to cancel out the interference of the direct link based on the IC method and to detect tag symbols based on the EM algorithm.

3.1. Interference Cancellation(IC)

Due to the strong interference of the direct link, it is difficult to detect the tag symbol from the received signal. To overcome this problem, we propose to use the LS method to estimate the channel state information (CSI) and cancel out the interference of the direct link. In this paper, assuming that prior knowledge is unknown, we estimate the CSI according to the pilot symbols.

When $c^i = 0$, (2) can be rewritten as

$$\mathbf{Y}^i = \sqrt{P_s}\mathbf{H}^i\mathbf{S}^i + \mathbf{W}^i, \tag{3}$$

which constructs a linear observation equation

$$\mathbf{y}^i = \mathbf{X}^i\mathbf{h}^i + \mathbf{w}^i, \tag{4}$$

where $\mathbf{y}^i = vec(\mathbf{Y}^i) \in \mathbb{C}^{KN\times 1}$, $\mathbf{X}^i = \sqrt{P_s}\mathbf{S}^{iH} \otimes \mathbf{I}_K \in \mathbb{C}^{KN\times KK}$, $\mathbf{h}^i = vec(\mathbf{H}^i) \in \mathbb{C}^{KK\times 1}$ $\mathbf{w}^i = vec(\mathbf{W}^i) \in \mathbb{C}^{KN\times 1}$. The estimated channel vector $\hat{\mathbf{h}}^i$ can be obtained using the LS estimation by the received signal vector $\mathbf{y}^i$ and the transmitted signal matrix $\mathbf{X}^i$. Then, $\hat{\mathbf{h}}^i$ is obtained through minimizing the $J(\hat{\mathbf{h}}^i)$, where $J(\hat{\mathbf{h}}^i)$ is mean square error function,

$$J(\hat{\mathbf{h}}^i) = \left\|\mathbf{y}^i - \mathbf{X}^i\hat{\mathbf{h}}^i\right\|_2^2. \tag{5}$$

According to the law of derivation of vectors, letting the first-order derivative of Function (5) with regard to $\hat{\mathbf{h}}^i$ equal to zero and estimate channel parameter $\hat{\mathbf{h}}^i$,

$$\hat{\mathbf{h}}^i = [\mathbf{X}^{iH}\mathbf{X}^i]^{-1}\mathbf{X}^{iH}\mathbf{y}^i. \tag{6}$$

Then, we can cancel out the interference of the direct link to improving the detection probability. By subtracting $\mathbf{X}^i\hat{\mathbf{h}}^i$ from $\mathbf{y}^i$, we can get the residual signal $\tilde{\mathbf{y}}^i$,

$$\tilde{\mathbf{y}}^i = \mathbf{y}^i - \mathbf{X}^i\hat{\mathbf{h}}^i. \tag{7}$$

We reshape $\tilde{\mathbf{y}}^i$ vector into the matrix $\tilde{\mathbf{Y}}^i \in \mathbb{C}^{3\times N}$ for further tag symbol detection.

3.2. Tag Symbol Detection based on EM Algorithm

After eliminating the interference of the direct link, we consider the EM algorithm to detect the tag symbol. To facilitate signal detection, we use pilot symbols to initial covariance matrix. In addition, due to the backscatter link being much weaker than the direct link, using long pilot sequences may suffer serious efficiency loss. So, we adopt the first two tag symbols of each frame as pilot symbols to detect tag symbols.

Since $\mathbf{S}^i$ and $\mathbf{W}^i$ follow CSCG distribution, we find $\tilde{\mathbf{Y}}^i$ given that $c^i = m$, $m = 0,1$ also follows CSCG distribution with mean zero and the covariance matrix $\mathbf{R}_m$. The probability density function (PDF) of $\tilde{\mathbf{Y}}^i$ can be represented as

$$p(\tilde{\mathbf{Y}}^i) = \sum_{m=0}^{1}\pi_m\mathcal{N}(\tilde{\mathbf{Y}}^i \mid 0, \mathbf{R}_m), \tag{8}$$

where $\mathcal{N}(\tilde{\mathbf{Y}}^i \mid 0, \mathbf{R}_m)$ is the PDF of Gaussian distribution. $\mathbf{R}_m (m = 0,1)$ is the channel statistical covariance matrix.

Assume that the tag transmits bit '0' and bit '1' with equal probability, i.e., $\pi_m = \frac{1}{2}$. Then the log-likelihood function is given by

$$\ln p(\tilde{\mathbf{Y}}^i) = \ln \sum_{m=0}^{1} \frac{1}{2} \mathcal{N}(\tilde{\mathbf{Y}}^i \mid \mathbf{R}_m). \tag{9}$$

Since (9) is a non-convex function, it is difficult to calculate parameters directly. Therefore, we use the EM algorithm, this method introduces a 2-dimensional latent variable z in which a particular element z_m is equal to one, and other elements are equal to zero [9,10]. The introduction of latent variables allows a form simpler distribution. The values of z_m satisfies $z_m \in \{0,1\}$ and $\sum_m z_m = 1$. The marginal distribution over z is represented as $p(z_m = 1) = \pi_m$. The steps of the tag symbol detection method are given as follows:

1) Initialization: In every iteration, the tag transmits two pilot symbols $c^1 = 0$, $c^2 = 1$. And according to the two pilot symbols, calculate the covariance matrixes,

$$\begin{cases} \mathbf{R}^1 = \frac{1}{N} \sum_{n=1}^{N} \tilde{\mathbf{y}}^1(n) \tilde{\mathbf{y}}^1(n)^H, c^1 = 0 \\ \mathbf{R}^2 = \frac{1}{N} \sum_{n=1}^{N} \tilde{\mathbf{y}}^2(n) \tilde{\mathbf{y}}^2(n)^H, c^2 = 1, \end{cases} \tag{10}$$

where $\mathbf{R}^1$, $\mathbf{R}^2$ is the covariance matrix of $\tilde{\mathbf{y}}^1(n)$, $\tilde{\mathbf{y}}^2(n)$ respectively, $\tilde{\mathbf{y}}^i(n) = [\tilde{\mathbf{y}}_1{}^i(n), \tilde{\mathbf{y}}_2{}^i(n), \tilde{\mathbf{y}}_3{}^i(n)]^T \in \mathbb{C}^{3\times1}, i = 1,2$. Then initialize $\mathbf{R}_0 = \mathbf{R}^1, \mathbf{R}_1 = \mathbf{R}^2$.

2) E step: Calculate the probability of $\tilde{\mathbf{Y}}^i$ belonging to z_m, where $m = 0,1$.

$$\gamma(z_{im}) \triangleq p(z_m = 1 \mid \tilde{\mathbf{Y}}^i) = \frac{p(\tilde{\mathbf{Y}}^i \mid c^i = m)}{\sum_{j=0}^{1} p(\tilde{\mathbf{Y}}^i \mid c^i = j)}, \tag{11}$$

where $\tilde{\mathbf{Y}}^i$ follows CSCG distribution with conditional PDF given c^i,

$$\begin{aligned} p(\tilde{\mathbf{Y}}^i \mid c^i = 0) &= \prod_{n=1}^{N} \frac{1}{\pi \mid \mathbf{R}_0 \mid} e^{-\tilde{\mathbf{y}}^i(n)^H \mathbf{R}_0^{-1} \tilde{\mathbf{y}}^i(n)} \\ p(\tilde{\mathbf{Y}}^i \mid c^i = 1) &= \prod_{n=1}^{N} \frac{1}{\pi \mid \mathbf{R}_1 \mid} e^{-\tilde{\mathbf{y}}^i(n)^H \mathbf{R}_1^{-1} \tilde{\mathbf{y}}^i(n)}. \end{aligned} \tag{12}$$

3) Clarification: For $i = 3,4,\ldots,I$, we classify tag symbols according to the following formula.

$$\begin{cases} \gamma(z_{i0}) > \gamma(z_{i1}), c^i = 0 \\ \gamma(z_{i0}) < \gamma(z_{i1}), c^i = 1. \end{cases} \tag{13}$$

4) Update the CSI: Estimate the channel parameter again and cancel out the interference of the direct link. For all $c^i = 0$ cases, we estimate $\hat{\mathbf{h}}^i$ again according to (6), and average them by $\hat{\mathbf{h}}^i = \frac{1}{N_h} \sum_{n=1}^{N_h} \hat{\mathbf{h}}^i$, where N_h is the number of tag symbols when $c^i = 0$. Then for $i = 1,2,\ldots,I$, we use (7) to regenerate the vector $\tilde{\mathbf{y}}^i$, and reshape $\tilde{\mathbf{y}}^i$ vector into the matrix $\tilde{\mathbf{Y}}^i$ for the next iterations.

5) M step: Let the first-order derivative of the log-likelihood function (9) regarding $\mathbf{R}_m$ equal to zero. Then channel statistical covariance matrixes are given as

$$\begin{cases} \mathbf{R}_0 = \frac{1}{N_m} \sum_{i=1}^{I} \gamma(z_{i0}) \mathbf{R}^i \\ \mathbf{R}_1 = \frac{1}{N_m} \sum_{i=1}^{I} \gamma(z_{i1}) \mathbf{R}^i, \end{cases} \tag{14}$$

where $N_m = \sum_{i=1}^{I} \gamma(z_{im}), m = 0,1$ is the sum of $\gamma(z_{im}), m = 0,1$. $\mathbf{R}^i = \frac{1}{N}\sum_{n=1}^{N} \tilde{\mathbf{y}}^i(n)\tilde{\mathbf{y}}^i(n)^H$ is the covariance matrix of $\tilde{\mathbf{y}}^i(n)$.

The detail of the tag symbol detection method is described in Algorithm 1.

Algorithm 1 Algorithm for channel estimation and tag symbol detection

Input: all received signals $\{\mathbf{Y}^i\}_{i=1}^{I}$.

1: Estimate channel $\hat{\mathbf{h}}^i$ according to (6), regenerate $\tilde{\mathbf{y}}^i$ by (7), and reshape $\tilde{\mathbf{y}}^i$ into $\tilde{\mathbf{Y}}^i$.

2: Calculate $\mathbf{R}^i$ based on (10) for $i = 1,2$.

3: Initialize channel statistical covariance matrixes $\mathbf{R}_0 = \mathbf{R}^1$, $\mathbf{R}_1 = \mathbf{R}^2$.

5: E step: Calculate the probability of $\tilde{\mathbf{Y}}^i$ belonging to $c^i = m$ according to (11) and (12).

6: Classification: According to (13), we classify $c^i, i = 3,4,\ldots,I$.

7: Update the CSI: For all $c^i = 0$ cases, estimate channel parameters $\hat{\mathbf{h}}^i$ again and average them. Then cancel out the interference of direct link by (7).

8: M step: Update $\mathbf{R}_0$ and $\mathbf{R}_1$ according to (14).

9: **Until** the algorithm meets convergence.

Output: c^i, $i = 1,2,\ldots,I$.

4. Numerical results

In this section, the simulation results verify the validity of the proposed transceiver design. In the simulations, we consider a multistatic backscatter system with three readers and one tag. We place three readers on a circle with a radius of d centered on the tag, and the distance between any two readers is the same. The distance between readers is 50 m. We set the backscatter frame consisting of 100 backscatter symbols and a backscatter symbol containing 100 RF source symbol periods. In addition, we consider the Rayleigh fading channels. The large-scale fading for each link is modeled as $L(d) = C_0(d / D_0^{-\alpha})$, where C_0 = -30 dB, D_0 = 1 m, α = 2.5, which represent path loss, reference distance, and path loss factor, respectively. We assume that the reflection coefficient $\alpha = 0.5$, the noise variance $\sigma^2 = -120$ dBm, and the transmit power P_s = 20 dBm. All the numerical results are averaged over 10000 Monte Carlo realizations.

Figure 3 presents the results for the BER performance versus the transmit power P_s. To illustrate the performance of the algorithm, we compare our proposed multi-reader joint optimal detector (MJOD), single-reader optimal detector (SROD), and detector without inter-reader interference cancellation (DWIC). The SROD is the single-reader optimal detector with $K = 1$. And the DWIC is a detector that doesn't consider strong interference of the direct links. That would lead to the following. One reader views the signal with the strong direct-link interference from two other readers as noise. Since the interference of the direct link is usually much stronger than the backscatter signal, this results in a very high bit error rate (BER). We observe that for MJOD and SROD, the BER decreases as the transmit power P_s increases. Moreover, the performance of MJOD is better than the other two. When P_s=20dBm, MJOD has a BER of 0.02, and SROD has a BER of 0.08, which demonstrates the effectiveness of our proposed detector.

Figure 4 illustrates the BER performance of the detector under different RF source symbol periods N. Firstly, the performance of MJOD and SROD improves with the increase of N. Secondly, the performance of MJOD is better than SROD and DWIC. For $N = 250$, the BER of MJOD is about 0.01, while the BER of SROD and DWIC is higher than that of MJOD. Thirdly, the BER of DWIC decreases very slightly when N increases. In addition, when N is greater than 80, the BER of MJOD increases slowly with the increase of N.

Figure 5 plots the BER versus the distance between readers and tag. We change the distance between the readers and the tag by changing the circle's radius d. We observe that BER increases with distance for MJOD. As the distance increases, the signal will have a more significant attenuation, so the signal strength is much lower, which leads to an increase in BER. The conclusion is that compared with DWIC, the BER performance and operating range of MJOD are improved.

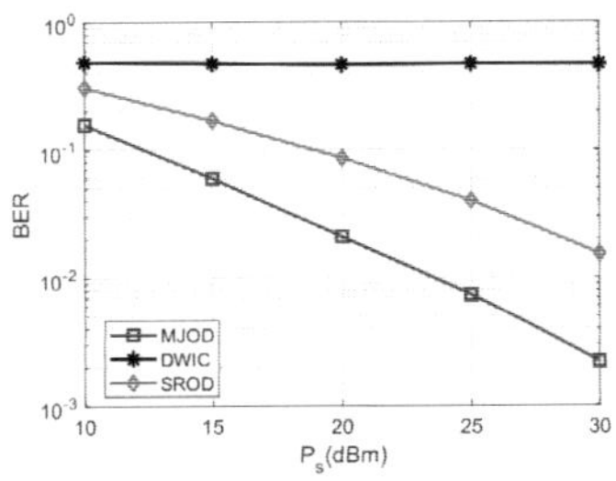

Figure 3. The BER versus P_s

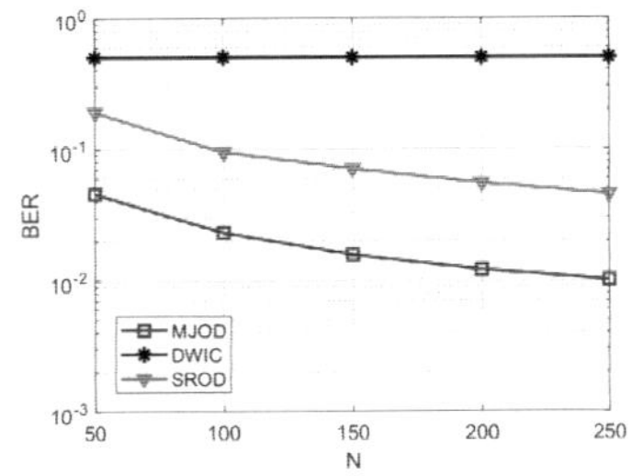

Figure 4. The BER versus N

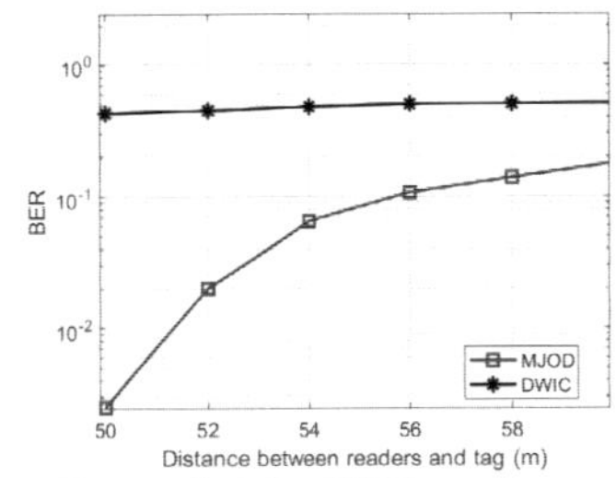

Figure 5. The BER versus d

5. Conclusion

This paper considers a multistatic backscatter system, in which multiple readers illuminate a tag simultaneously to extend the communication range. However, due to the strong interference of the direct link, it is difficult to detect the symbol information. Therefore, we consider using the IC method to eliminate the interference of the direct link and the EM algorithm to detect the tag symbol. Specifically, we update the covariance matrix at every iteration based on their statistical probability to further improve the detection accuracy. The numerical results verify the better performance of our proposed method.

References

[1] T. Islam, S. C. Mukhopadhyay, and N. K. Suryadevara, "Smart sensors and internet of things: A postgraduate paper," IEEE Sensors Journal, vol. 17, no. 3, pp. 577–584, 2016.

[2] R. Mehta, J. Sahni, and K. Khanna, "Internet of things: Vision, applications and challenges," Procedia computer science, vol.132, pp. 1263–1269, 2018.

[3] J.-P. Niu and G. Y. Li, "An overview on backscatter communications," Journal of Communications and Information Networks, vol. 4, no. 2, pp. 1–14, 2019.

[4] J. Wang, J. Zhang, R. Saha, H. Jin, and S. Kumar, "Pushing the range limits of commercial passive RFIDs," in 16th USENIX Symposium on Networked Systems Design and Implementation (NSDI 19), 2019, pp. 301–316.

[5] P. N. Alevizos, K. Tountas, and A. Bletsas, "Multistatic scatter radio sensor networks for extended coverage," IEEE Transactions on Wireless Communications, vol. 17, no. 7, pp. 4522–4535,2018.

[6] J. D. Griffin and G. D. Durgin, "Complete link budgets for backscatter-radio and RFID systems," IEEE Antennas and Propagation Magazine, vol. 51, no. 2, pp. 11–25, 2009.

[7] P. Sure and C. M. Bhuma, "A survey on OFDM channel estimation techniques based on denoising strategies," Engineering Science and Technology, an International Journal, vol. 20, no. 2, pp. 629–636, 2017.

[8] C. Knievel and P. A. Hoeher, "Coded sampling bound—how much training is needed for iterative semi-blind channel estimation?" IEEE Transactions on Communications, vol. 62, no. 7, pp. 2422–2431, 2014.

[9] Q. Zhang and Y.-C. Liang, "Signal detection for ambient backscatter communications using unsupervised learning," in 2017 IEEE Globecom Workshops (GC Wkshps). IEEE, 2017, pp. 1–6.

[10] L. Gerhardt, "Pattern recognition and machine learning," IEEE Transactions on Automatic Control, vol. 19, no. 4, pp. 461–462,1974.

The Comparison of BCH Decoding Methods and the Parameters Affecting the BCH Code Performance

Jiajie Huang[†]

School of Information and Control Engineering, Qingdao University of Technology, Qingdao, Shandong Province, 266000, China

tdriver@une.edu

Abstract. In the communication field, channel coding was proposed to increase the data transmission rate and the transmitted data's reliability. Error correction is an essential part of channel coding to improve data reliability. The Bose-Chaudhuri-Hocquenghem (BCH) code is one of the most efficient error-correcting codes. In this paper, a BCH coding method, and several methods to implement BCH decoding are introduced, and these decoding methods are compared with each other. Based on the principles of BCH codes, this study addresses two factors that may impact the performance of BCH codes, the redundancy bit length, and the minimum distance, as well as the relationship between them. This paper composes four flow charts and investigates the redundant bit lengths and minimum distances that may affect the performance of BCH codes during transmission. This paper is potentially beneficial for more extensive applications of BCH codes such as magnetic, optical storage systems, and new generations' digital communication systems.

1. Introduction

The noticeable development of electronic information technology has brought new opportunities and challenges to communication science research. In the process of digital signal transmission in modern communication systems, due to the influence of interference, the symbol waveform may be damaged so that the receiver will receive error information. As an essential part of information security theory, error-correcting code theory plays a decisive role in ensuring the authenticity and reliability of communication. The BCH code is a cyclic code, by which several errors that occur in the data can be corrected. Because of its efficient encoding and decoding algorithms, BCH code has been widely used in communication systems and various electronic devices. BCH code is divided into two types, one is the primitive BCH code, and another one is a non-primitive BCH code. The main difference between the two types of BCH code is that the generator polynomials of primitive BCH codes contain primitive polynomials with the highest degree of m, and the length of the code is $2^m - 1$. In engineering design, BCH codes do not require calculating the generator polynomials since the former scientists have listed the found generator polynomials in a table, so the needed generator polynomials can be found by looking up the table [1].

However, there are different ways to decode BCH code, among which some are simple and some complicated. This paper presents four methods to realize BCH decoding and estimates their advantages and disadvantages. Besides, the performance of the BCH code is worth researching. So, the parameters that may affect BCH code performance are also researched in this paper.

2. Encoding of binary BCH codes

In a digital transmission system, the next step of source coding is channel coding. Channel coding is a way to make communication more reliable. The principle of channel coding is to add check bits, also called redundant bits, after the information bits so that the error correction can be implemented in the receiver [2][3]. So, the BCH code encodes a sequence of bits to a sequence of real symbols. The decoding in the receiver is to detect the errors in the transmission and correct them, then delete the check bits from the received codeword and restore the transmitted sequence [4].

Generally, BCH encoding is simple since the calculation of the generator polynomial is the core of BCH encoding, and the generator polynomials g(x), which former scientists have calculated, are listed in a table. So, the required generator polynomial can be obtained directly by the table. However, decoding BCH code is relatively complicated since different methods can be used to decode BCH code, and each of them may have a different performance. This and the next section will show one general method to encode and several methods to decode, respectively. As shown in Figure 1, the encoded codeword can be obtained [5].

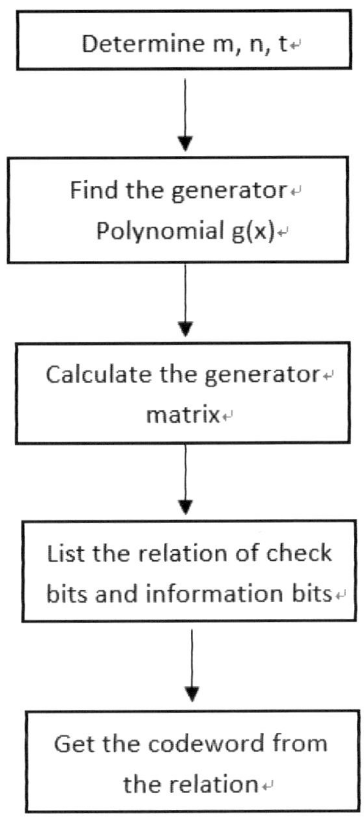

Figure 1. Flow chart of the BCH encoding.

The generator polynomial of the code is expressed as:

$$g(x) = LCM[m_1(x), m_3(x), \dots\dots, m_{2t-1}(x)] \quad (1)$$

where $m_1(x), m_3(x), \dots\dots, m_{2t-1}(x)$ are the minimum polynomials. Because the BCH code is a cyclic code, the generator matrix of the BCH code can be obtained by:

$$G(x) = \begin{pmatrix} x^{k-1}g(x) \\ x^{k-2}g(x) \\ \vdots \\ xg(x) \\ g(x) \end{pmatrix} \tag{2}$$

Then, the generator matrix of the BCH code can be obtained. Based on the generator matrix, the encoded codewords can be determined.

3. Decoding of binary BCH code

Generally, there are several ways to realize BCH code decoding. In this section, four methods to decode BCH codes are illustrated.

3.1 BCH decoding based on Peterson's Algorithm

As shown in Figure 2, there are typically four steps in decoding BCH codes by Peterson's Algorithm [5]. $r(x)$ be the received codeword. Then the error pattern can be expressed as $r(x) = e(x) + c(x)$.

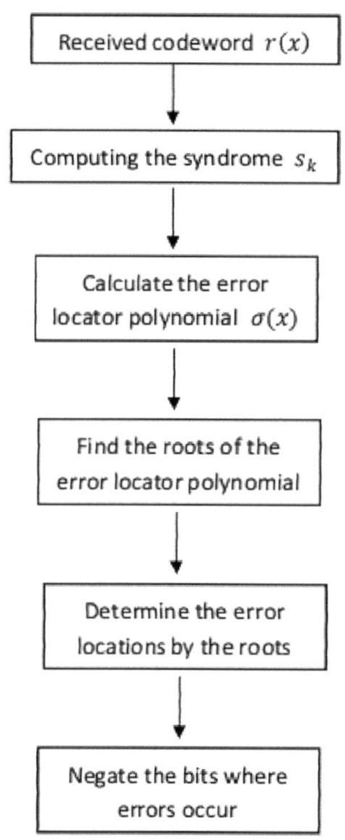

Figure 2. Flow chart of BCH decoding based on Peterson's Algorithm.

The syndrome can be written as

$$s_k = r(\alpha^k) = r_0(\alpha^k)^0 + r_1(\alpha^k)^1 + r_2(\alpha^k)^2 + \cdots + r_{n-1}(\alpha^k)^{n-1} \tag{3}$$

The second step is to calculate the error locator polynomial. Because $c(\alpha^k) = 0$,

$$s_k = r(\alpha^k) = c(\alpha^k) + e(\alpha^k) = e(\alpha^k), k = (1,2, \dots \dots ,2t) \tag{4}$$

Since this paper only analyses binary BCH code, the error values are only 0 and 1, and the syndrome can be written as:

$$s_k = e(\alpha^k) = \sum_{i=1}^{t} x_i^k \tag{5}$$

Where $x_i = \alpha^i$ and x_i is the error position Equation (5) can be simplified as:

$$s_{t+j} + \sigma_1 s_{t+j-1} + \ldots + \sigma_t s_j = 0 \tag{6}$$

Equation (6) is called Newton's Identity. For binary BCH code, the equation can be written as

$$\begin{aligned} &s_1 + \sigma_1 = 0 \\ &s_3 + s_2\sigma_1 + s_1\sigma_2 + \sigma_3 = 0 \\ &s_5 + s_4\sigma_1 + s_3\sigma_2 + s_2\sigma_3 + s_1\sigma_4 + \sigma_5 = 0 \\ &\ldots \end{aligned} \tag{7}$$

3.2 Decoding based on Berlekamp's Algorithm

The high efficiency is the greatest strength of Berlekamp's Algorithm in decoding BCH codes.

3.2.1 Theory of Berlekamp's Algorithm

Berlekamp's Algorithm is the second way to compute the error locator polynomial. This algorithm is an iterative algorithm and has the starting conditions shown in Table 1 [7].

Table 1. Starting Conditions of Berlekamp's Algorithm.

μ	$\sigma^{\mu}(X)$	d_{μ}	l_{μ}	$\mu - l_{\mu}$
-1	1	1	0	-1
0	1	S_1	0	0

Berlekamp's Algorithm is shown in figure 3. After calculating the syndrome, the variables need to be modified as the algorithm requires. An array, which creates the starting conditions table, should be generated to obtain the polynomial. The variables with their values will be contained in the array for one single iteration. $\sigma^m(x)$ and d_m (m =1,2, …, 2t) will be contained in 2 arrays, respectively, in order to implement the arithmetic operations. One of the two arrays will be used to contain the elements $\alpha_1, \alpha_2, \ldots, \alpha_k$ by m-tuples, where m is the power of $GF(2^m)$, and the other one will contain the elements by the powers of α. This algorithm is to iteratively find the error locator polynomial, which satisfies Equation (7). During the process, in each iteration of the algorithm, $\sigma^i(x)$ (i =1, 2, …, 2t) will be calculated. For example, $\sigma^1(x)$, which satisfies the Equation (7), can be calculated in the first iteration, and if $\sigma^1(x)$ also satisfies the second equation of (7), the second iteration can be obtained. Then, it is needed to judge whether $\sigma^1(x)$ equals to $\sigma^2(x)$ or not. If $\sigma^1(x) = \sigma^2(x)$, the algorithm proceeds to the third iteration. If $\sigma^1(x)$ does not equal to $\sigma^2(x)$, a new polynomial $\sigma^2(x)$ which satisfies both the first two equations of (7), is found. The algorithm will not stop iterating until $\sigma^{2t}(x)$ is obtained. Note that, $\sigma^{2t}(x)$ is the error locator polynomial.

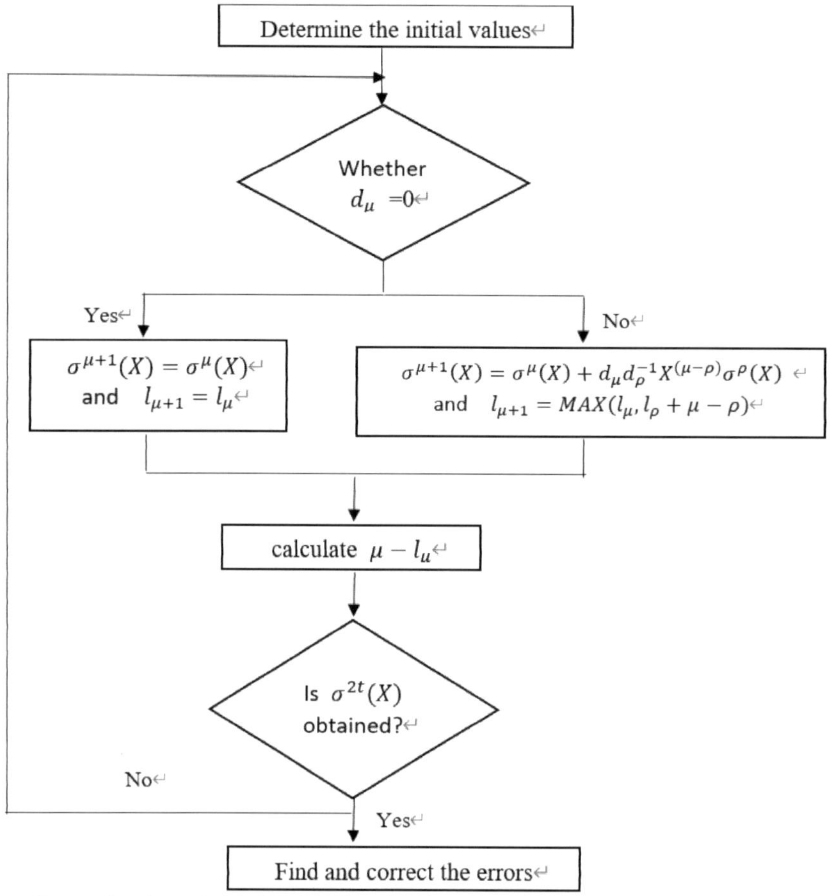

Figure 3. Flow chart of BCH decoding based on Berlekamp's Algorithm.

Since the BCH code that is researched in this paper is binary, the starting conditions of Berlekamp's Algorithm can be modified, as Table 2 shows:

Table 2. Starting Conditions for the modified Berlekamp's Algorithm.

μ	$\sigma^{\mu}(X)$	d_μ	l_μ	$2\mu - l_u$
-1/2	1	1	0	-1
0	1	S_1	0	0

The starting conditions of the simplified Berlekamp's Algorithm have some slight differences from the standard one shown in table 2. But the process of implementing the simplified algorithm and the standard one is the same. It has been said in 2.2.1 that the roots of the polynomial can determine the error locations. The rest job is to negate the bits where errors occur, just like the flow chart shows.

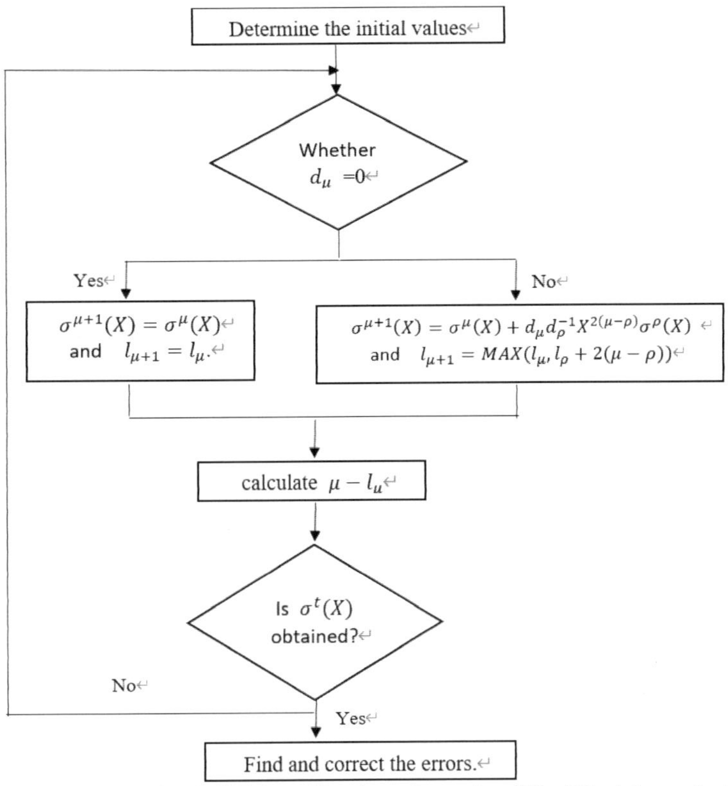

Figure 4. Flow chart of the BCH decoding based on simplified Berlekamp's Algorithm.

3.2.2 Analysis of Berlekamp's Algorithm performance

The error correction capability of the BCH code can affect the performance of Berlekamp's algorithm, since the amount of the syndromes, which need to be computed, is related to the error correction capability. If the BCH code can correct t errors that occur in a codeword, the modified Berlekamp's algorithm, which only needs t iterations from figure 4, saves half the time compared with the common one, which needs 2t iterations [8].

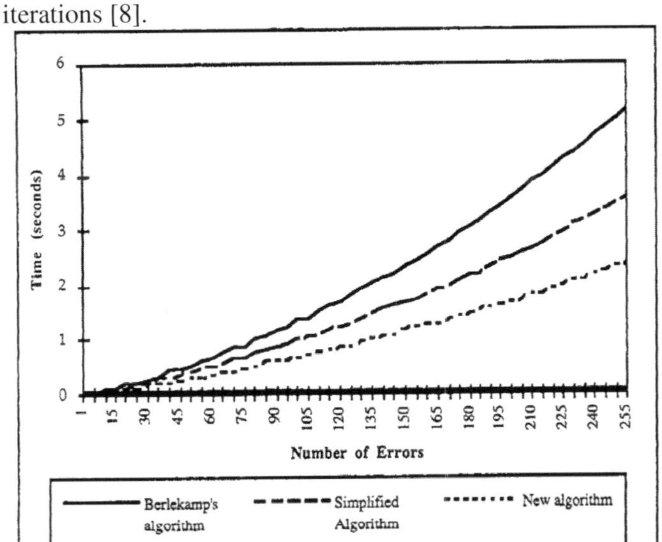

Figure 5. Execution Times for Decoding Binary BCH codes [8].

3.3 Decoding based on the sum-product algorithm

As is known to all, for BCH codes, the more errors exist in the code, the larger complexity the Chase [9] algorithm will have [10].

3.3.1 Theory of sum-product algorithm (SPA)

The cycles with a length of four, which will reduce the performance of decoding performance, in the Tanner graph can be deleted by a certain method [11].

By adding nodes, the performance that is degraded by the cycles can be restored since the cycles with a length of four will be removed by the added variable nodes and check nodes. Note that, in this paper, the removal of cycles with four lengths is represented as RmFC. Now let the variable nodes be v_i, v_j and the check nodes be c_1, c_2. The two kinds of nodes will build one cycle with a length of four. Then, appending the assisting variable node $v_{ij} = v_i + v_j$ and the check node c_{ij} is the core to implement the removal of the cycles with the length of four. After that, the branches that connect with v_i, v_j and c_1, c_2 are removed, and branches that connect with v_i, v_j, v_{ij} and c_{ij} and those who connect with v_{ij} and c_1, c_2 are added. Removing cycles of length four will be accomplished by repeating the above operation.

However, it is irrational only to consider improving the performance of decoding. As this section said before, the complexity of the algorithm should be considered. To solve the contradiction between algorithm complexity and decoding performance, the parity check of echelon canonical form (ECF) is proposed [12]. The central idea of this method is to let the parity check of ECF be the check matrix of removing the cycles with a length of four.

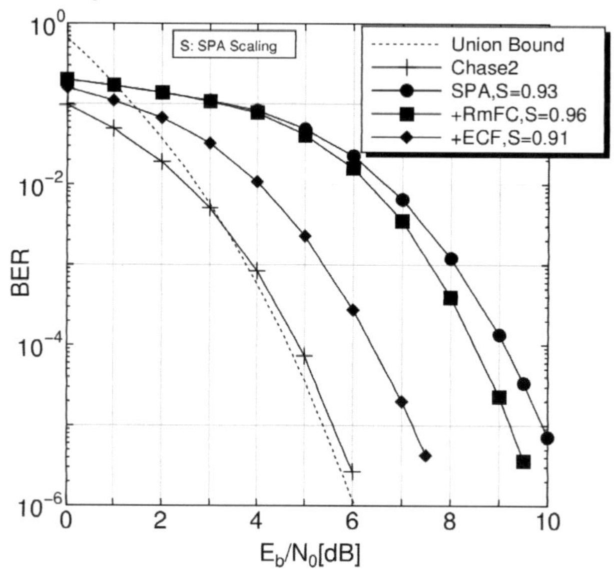

Figure 6. Performance of BCH code in different cases [12].

Figure 6 analyses the performance of the BCH code in three cases. As the figure shows, in the case that the RmFC is used, the performance is noticeably increased compared with the common SPA. What's more, the performance of BCH code with the use of ECF is increased more significantly.

3.4 Tree decoding of BCH code

Tree decoding significantly reduces the time consumption of calculating the error locator polynomial [13]. Sometimes the decoding failures may happen in the receiver, and this decoding method can solve this problem. The core of this method is to compute the discrepancy by the Berlekamp-Massey algorithm [14][15].

Journal of Physics: Conference Series **2384** (2022) 012025 doi:10.1088/1742-6596/2384/1/012025

3.4.1 Theory of Tree Decoding

As mentioned in 2.2.1, the most complicated and time-consuming part of BCH decoding is the calculation of the error locator polynomial. But there are some cases where the roots of the polynomial do not exist, the roots of the polynomial are repeated, or the error correction capability is so poor that it cannot exceed the degree of the polynomial. In these cases, the roots cannot determine the error locations. Consequently, tree decoding is proposed to detect the failures in locating the errors [16]. Figure 7 shows the process of implementing tree decoding.

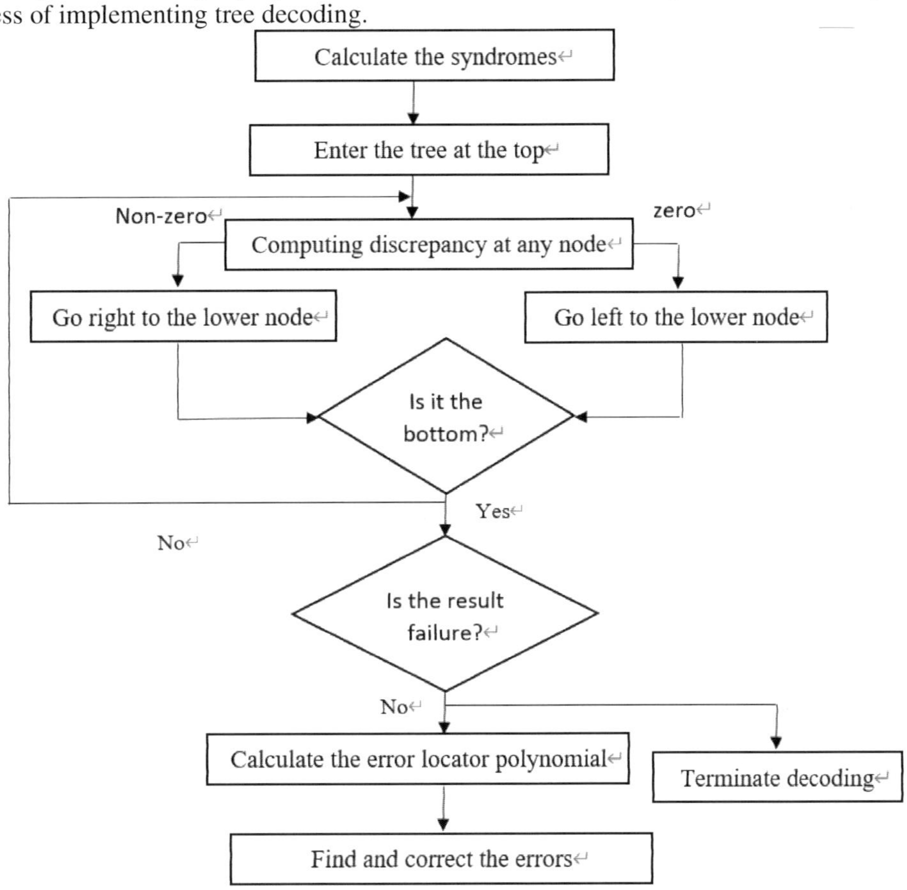

Figure 7. Flow chart of BCH tree decoding.

Figure 8 provides an example of a decoding tree, where T means that the decoding is successful, and F means that a decoding failure exists. In this decoding tree, the leaves will be marked by a word of 4 bits, which can be used to determine the memory location that stores the information of the decoding situation. It is noted that each bit of the word stands for the value of each syndrome, 0 or not 0. Suppose that the syndromes, which have been calculated from the received sequence, are $s_1 = \alpha, s_2 = \alpha^3, s_3 = \alpha^5, s_4 = 0$ and the word is 1110. As the figure shows, the first step is to go right, right, right, and finally left. After that, the result is T, which means the codeword can be decoded successfully.

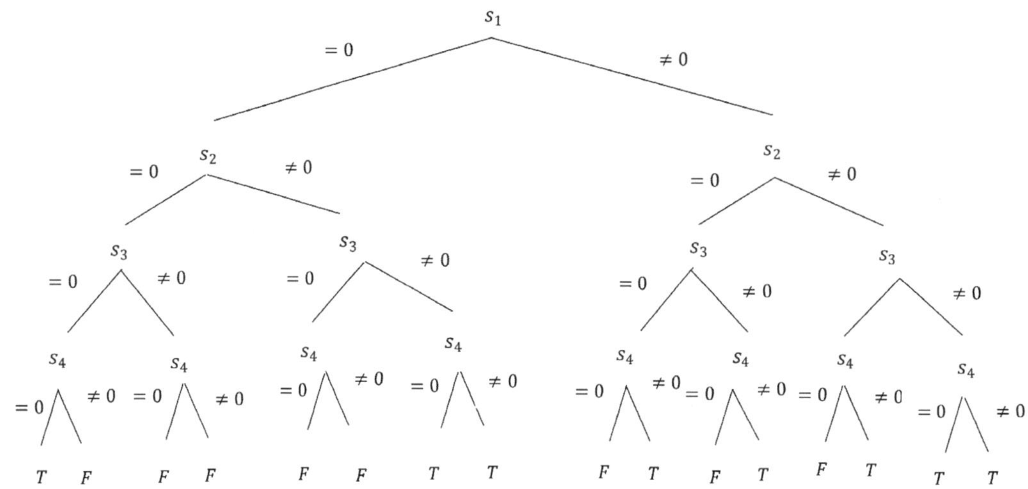

Figure 8. An example of the decoding tree [16].

3.4.2 Advantages and disadvantages of tree decoding

The main advantage of tree decoding is that the decoding tree can detect the failures before calculating the error locator polynomial. And the main disadvantage is that the number of errors to correct t affects the complexity of the formulae at the nodes and leaves. That means the larger the value t will be, the more complicated the formulae will be.

3.5 The difference and relation of the above decoding flow charts.

Figure 2 shows the procedure of Peterson's algorithm, which is the basic way to realize BCH decoding, since the syndrome is the most important part of BCH decoding, this paper shows how to calculate the syndrome in detail. Figures 3 and 4 show the two kinds of BCH decoding based on Berlekamp's Algorithm. Figure 7 shows the process of BCH tree decoding. Figure 2 is the basis of Figures 3, 4, and 7, since, through Figure 2, the syndrome can be easily obtained. Then start the decoding algorithms that are based on the syndrome. There are slight differences between Figure 3 and Figure 4. Figure 4 is the simplified form of Figure 3, and through Figure 3, it requires less time to iterate than the original one.

Table 3. The Comparison of Four Methods of BCH Decoding.

Decoding methods	Advantage	Disadvantage
Peterson's Algorithm Decoding	Easy to calculate syndrome.	Time consumption remains the same
Berlekamp's Algorithm Decoding	Only needs t times iteration.	Related to error correcting capability.
Sum-Product Algorithm (SPA) Decoding	Decreases the complexity of the Chase algorithm.	Performs insufficiently when $t \neq 1\ or\ 0$.
Tree Decoding	Detect decoding failures before calculating polynomial	The complexity of formulae grows with t.

4. Parameters Affecting the Performance of BCH Code

For a given channel, parameters affecting the performance of BCH code include the length of redundant bits (redundancy) and the minimum distance [17].

4.1 Minimum distance

In order to correct t error bits, it requires that minimum distance

$$d_0 \geq 2t + 1. \tag{8}$$

Figure 9(b) shows two code blocks with a distance of 5. If no more than 2 errors occur in code block A or B, the position of A or B will not exceed a circle with a radius of 2 and the original position as the center. The two circles do not overlap with each other. Therefore, it can be decided that, if the received code block falls on the circle with A as the center, it is judged that the received code block is A, and if the received code block falls on the circle with B as the center, it is judged that the received code block is B. In this way, two error bits can be corrected. So, in general, in order to correct t error bits, the minimum distance should not be less than 2t+1.

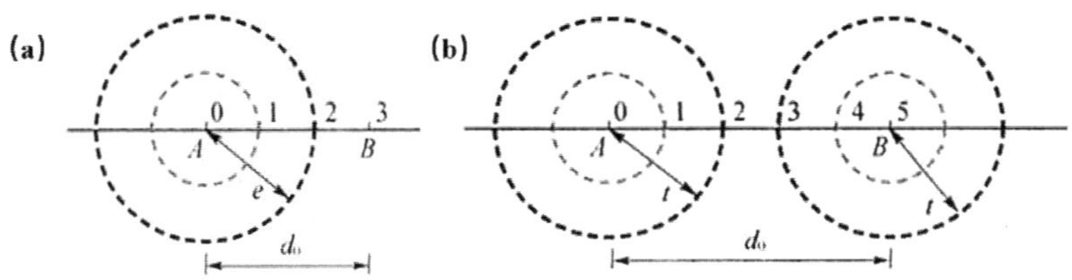

Figure 9: (a) Two code blocks with a distance of 3.
(b) Two code blocks with a distance of 5.

4.2 The length of redundant bits (redundancy)

In order to reduce the number of received error symbols, it is necessary to add check bits to the transmitted sequence. As a result, the transmitted sequence's length grows, and the redundancy increases. If it is necessary to keep the bit rate unchanged, the transmission rate must be increased, which makes the system bandwidth grow. The rise in the system bandwidth will lead to the increased noise power in the system and decreased SNR. The reduction of SNR increases the number of errors in the received sequence.

For a given transmission system, the relation between the transmission rate and $\frac{E_b}{n_0}$ can be expressed as:

$$\frac{E_b}{n_0} = \frac{P_s T}{n_0} = \frac{P_s}{n_0(\frac{1}{T})} = \frac{P_s}{n_0 R_B} \tag{9}$$

where R_B is the bit rate. From Equation (9), if the bit rate grows, the SNR will decrease, and BER will increase.

5. Conclusion

This paper presented four methods or algorithms to implement BCH decoding to get the most effective BCH decoding method. Peterson's Algorithm is the most widely used algorithm, and the main advantage is that when the error correction capacity is low, the effect of high processing speed is more considerable. But the time consumption doesn't change when the number of errors is different, which means this algorithm is not flexible.

The second algorithm is Berlekamp's Algorithm, this only needs to iterate t times instead of 2t times, but the main disadvantage is that it needs some extra time for this algorithm to calculate the new syndromes. The sum-product algorithm (SPA) decoding is a good way to reduce the complexity of the Chase algorithm. However, the drawback of this algorithm is that it requires some additional methods to compensate for the lack of performance improvement, such as ECF. Tree decoding is the last method this paper shows to decode BCH code. The tree decoding can detect the failures of decoding before the calculation of the error locator polynomial. But the complexity of the formulae in the tree will grow as the error correction capacity increases.

This paper also shows two parameters that may affect the BCH code performance, the first one is the length of redundant bits and the second one is the minimum distance. The relationship between the error correction capacity and the minimum distance is $d_0 \geq 2t + 1$. Adding check bits to the transmitted sequence is the core to realize error correction, but the system bandwidth will grow. When the system bandwidth grows, the noise power will increase, increasing the number of sequence errors. So, if the bit rate rises, the SNR will decrease, and BER will increase.

One of the most important parts of modern communication systems is error correction and control. The BCH codes have been widely used in wireless satellite communications, computer networks, and magnetic and optic storage systems. Moreover, BCH codes are also used in CMMB, which is the mobile multimedia broadcasting of China.

Reference

[1] Changxin F and Lina C 2012 *Principles of Communications*. (Publishing House of Electronics Industry) pp 331-354.
[2] Costello D J and Forney G D 2007 Channel coding: The road to channel capacity *Proceedings of the IEEE*, **95(6)** pp 1150-1177.
[3] Shannon C E 1948 A mathematical theory of communication *The Bell system technical journal*, **27(3)** pp 379-423.
[4] Cho J and Sung W 2009 Efficient software-based encoding and decoding of BCH codes *IEEE Transactions on Computers*, **58(7)** pp 878-889.
[5] Yi L 2011 Realization of Decoding Binary BCH Codes Based on Chien's Search Algorithm *Energy Procedia*, **(13)** pp 9678-9682.
[6] Ji X and Song L 2016 Mutual Exclusion Verification of Peterson's Solution in Isabelle/HOL *2016 Third International Conference on Trustworthy Systems and their Applications (TSA)* pp 81-86.
[7] Lin S and Daniel J Costello Jr 1983 *Error Control Coding: Fundamentals and Applications* (New York: Prentice-Hall Inc. Englewood Cliffs) pp 141-159.
[8] Joiner L L and Komo J J 1995 Decoding binary BCH codes *Proceedings IEEE Southeastcon '95. Visualize the Future* pp 67-73.
[9] Chase D 1972 Class of algorithms for decoding block codes with channel measurement information *IEEE Transactions on Information theory*, **18(1)** pp 170-182.
[10] Ogiwara H, Shimamura K and Shohon T 2008 Sum-product decoding of BCH codes *IEICE Transactions on Fundamentals of Electronics, Communications and Computer Sciences*, **91(10)** 2729-2736.
[11] Sankaranarayanan S and Vasic, B 2005 Iterative decoding of linear block codes: A parity-check orthogonalization approach *IEEE Transactions on Information Theory*, **51(9)** pp 3347-3353.
[12] Yedidia J S, Chen J and Fossorier M P 2002 Generating code representations suitable for belief propagation decoding *Proceedings of the annual Allerton Conference on Communication Control and Computing,* **40(1)** pp. 447-456.
[13] Kraft C 1991 Closed solution of Berlekamp's algorithm for fast decoding of BCH codes *IEEE transactions on communications* **39(12)** pp 1721-1725.
[14] BERLEKAMP E 1968 Algebraic Coding Theory *Negacyclic Codes* pp 207-217.
[15] Massey J 1969 Shift-register synthesis and BCH decoding *IEEE transactions on Information Theory*, **15(1)** pp 122-127.
[16] Moore T G and Gulliver T A 1998 Tree decoding of BCH codes *IEEE GLOBECOM 1998 (Cat. NO. 98CH36250)*,**6** pp 3521-3524.
[17] Ding C 2015 Parameters of several classes of BCH codes *IEEE Transactions on Information Theory*, **61(10)** pp 5322-5330.

The Research Process, Application, and the Future Development of Pulse-Position Modulation

Wenzhe Tang[1,†], Sirui Wang[2,†], Yimeng Xu[3,†], Zhouyue Yu[4,*,†]

[1]Automatic Control and Systems Engineering, University of Sheffield, Sheffield, S10 2TN, United Kingdom

[2] School of Electronic Engineering, Xidian University, Xi'an, Shaanxi Province, 710071, China

[3]College of Electronics and Information Engineering, Tongji University, Shanghai, 201804, China

[4]School of Information and Communication, Guilin University of Electronic Technology, Guilin, Guangxi Province, 541214, China

[*]1900210330@mails.guet.edu.cn

[†]These authors contributed equally.

Abstract. With the development of digital communication, pulse-position modulation technology has become an important technology in wireless communication due to its efficient bandwidth utilization and sensitive detection efficiency. This study aims to provide some basic information on PPM. In the trend of rapid development of PPM, researchers have found that PPM plays an irreplaceable role, no matter in daily life or scientific research and development. To arouse people's interest in PPM research, more people are attracted to participate in the study. It began with an introduction to the development process of PPM, and briefly lists the advantages and disadvantages of PPM and the development trend to help readers quickly understand its background knowledge of it. The second part started to work on different modulations based on the previous research on PPM. Various modulation methods are related to each other, and the proposal of each modulation method is based on the shortcomings of existing modulation methods. This study discusses how PPM is used in experiments. Moreover, each experiment's measurement values and corresponding advantages are analyzed through charts, like Bit Error Rate (BER) and Signal to Noise Ratio (SNR) conducted in each experiment. This study draws a conclusion by discussing the application and the broad horizon for future developments mainly used in communication engineering led by light communication. However, in recent years, PPM has been introduced in many other fields, such as detection and mechanic.

1. Introduction

Pulse-Position Modulation (PPM) is a type of signal modulation for optical communications systems that encodes message bits by broadcasting a single pulse in one of the required time shifts, whose transmitted rate is bits per second [1].

PPM's advantages include reduced noise interference and high anti-noise performance. Because of its steady pulse amplitude and width, PPM requires less power [2]. To optimize the quantity of data communicated by each signal block, the PPM compresses the time span over which light pulses occur, leading to abnormal transmission efficiency. On the other hand, the required transmission bandwidth will improve asynchronization between transmitter and receiver. This asynchronization extension is predicted to happen as transmission efficiency improves further. Consequently, the low band-utilization efficiency may restrict transmission capacity when PPM is utilized in a practical system.

The basic principle of PPM is to determine the data information transmitted by the signal according to the time slot position of the pulse. Since 1946 when H. S. Black et al. [3] first proposed this primary term, PPM was mainly used in the military. In 1973, George M. Dillard [4] described a PPM system based on energy detection, and R.A. Cryan; R.T. Unwin discovered that the most efficient modulation format was Pulse Position Modulation (PPM) [5] in 1992. As time went by, the research on PPM by domestic and foreign scientists increased. A. G. Klein, for instance, suggested the decision feedback equalization of pulse position modulation [6] and created the decision feedback equalizer with the minimum mean square error of PPM signal with ISI(cross-code crosstalk) in 2007.

However, the transmission bandwidth significantly impacts how well the modulation technique performs. The multi-pulse position modulation (MPPM) approach was designed to boost accuracy and bandwidth limit to correct this flaw. The MPPM system was initially presented in 1989 by Sugiyama H and Nosu K[7] as a modulation strategy to increase the number of pulses and charge the pulse locations to enhance bandwidth efficiency. In the following 30 years, the evolution of MPPM will make rapid and exciting progress. In 2018, a transmission performance study of wireless optical communication systems based on MPPM modulation was proposed. In 2022, C. Yu found that by analyzing and comparing the amplitude spectra of different shapes after the MPPM [8], triangular and sine waves were better for transmission than rectangular waves, while ladder waves and serrated waves were better suitable under the same transmission conditions.

The future development trend of pulse position modulation is not limited to PPM, but to converting PPM into MPPM, LPPM, and other systems with special functions to satisfy different needs. MPPM features excellent anti-interference, high power utilization, and good frequency band use. With the same transmission efficiency, this technique decreases the bandwidth to roughly half that of standard PPM and increases the band-utilization efficiency of PPM [1]. The difference between LPPM and MPPM is that in MPPM, the PPM signal is composed of pieces of information, but in LPPM, it changes to pieces of information. Both are essential extensions of PPM in global research, but this paper focuses on the relevant content of PPM.

This paper mainly focused on the development process of PPM in recent years and foresaw the future trend of the PPM system. It is primarily divided into four parts. The first part introduces the definition of PPM and its advantages and disadvantages, points out its rapid development worldwide, and analyses its future development trends and prospects. The second part mainly describes the basic principles and international research process of pulse position modulation, which includes and focuses on describing the research progress of PPM, showing the importance of PPM in the research topic. The third part mainly introduces experimental development, which expounds on the experimental methods and results used in the study. The last part presents the application of PPM in daily life and illustrates the absolute position of PPM. The purpose of writing this study is to show the importance of PPM and call on more people to research it. Meanwhile, PPM is currently the mainstream method of digital communication, and it plays a vital role in the future development of technology and 5g research.

2. The basic principles and international research process of pulse position modulation

2.1. Basic principle
Pulse Position Modulations were analyzed and compared with On-Off Keying (OOK) modulation in this part. The advantages and disadvantages of their application in wireless optical communications are compared.

2.1.1. LPPM modulation
This section will concentrate on the LPPM principles. LPPM modulation is translating a binary n-bit

data set to a single pulse signal in a time interval with $2n$ slots. The PPM modulation signal is L bits long, and the data segment's transmission follows the $\log_L^2$ rule. $M = (m1, m2, \Lambda, mn)$ stands for n-bit data collection. l represents the time slot position. Then, the mapping encoding relationship for a single-pulse is as follows:

$$\Phi: \; l = m_1 + 2m_2 + \Lambda + 2^{n-1}m_n \in \{0,1,\Lambda,n-1\} \quad (1)$$

For example, a 4 -PPM modulation, as shown in Table 1:

Table 1. The mapping encoding relationship of 4-PPM modulation

M	l
(0,0)	0
(1,0)	1
(0,1)	2
(1,1)	3

Among them, 0, 1, 2, and 3 correspond to the time slot positions, respectively, as shown in Figure 1.

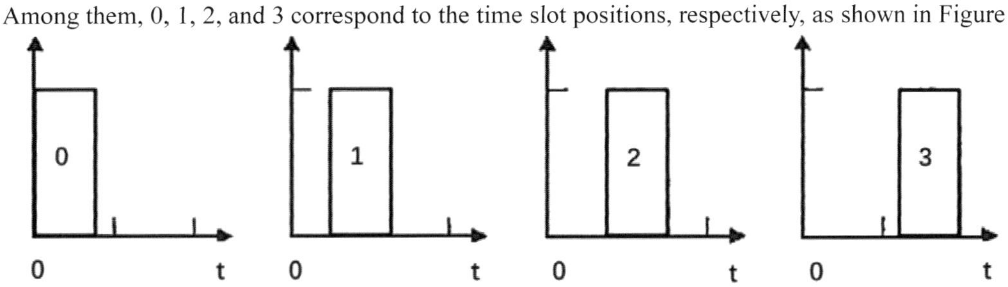

Figure 1. Schematic diagram of L-PPM modulation

The mapping Φ of the Formula (1) is a one-to-one mapping that satisfies the requirements of modulation uniqueness. For a digital baseband signal with a code rate of $R_b = 1/T_b$, its ideal low communication channel bandwidth requirement is $B = R_b/2$. Assuming that the element rate of the OOK signal is $R_b = 1/T_b$, and the bitrate of the L-PPM signal is $R_b' = 1/T_b'$, in the case where the transmission rate is required to be the same, there are: $T_b'L = T_b \log_L 2$, so $B/B' = L/L'$. Generally, $L = 2^n$, n is an integer. It can be seen that L-PPM modulation requires more bandwidth than OOK. However, for an optical communication system that applies L-PPM modulation, its average optical transmission power is P_1/L, and P_1 is the transmitted optical power when the code element is 1. However, the OOK modulated optical system (non-zero code) has the same probability of 1 and 0, and the average optical transmitted power is $P_1/2$. This is particularly advantageous for some handheld devices that use light as a communication carrier, and portable terminals.

One parameter for comparing different modulation methods is the unit transmission rate, i.e. the number of bits transmitted per second per Hz is:

$$\gamma = \frac{R}{B} \; (bit * s^{-1} * Hz^{-1}) \quad (2)$$

where R is the transfer rate (bit $*s^{-1}$); B is the signal bandwidth (Hz).

In optical communication, the laser usually operates in a pulsed state with a pulse duration of τ, and the corresponding signal bandwidth is defined as:

$$B(Hz) = \frac{1}{\tau}(s) \quad (3)$$

For a duty cycle of an OOK modulation r_p, its unit transmittance is:

$$\gamma_{ook} = \frac{1/_T}{1/_\tau} = r_p \quad (4)$$

For single-pulse L-PPM modulation, if the number of slots is $L = 2^n$ and the duty cycle is also r_p, the width of 2^n slots is $T_{ppm} = 2^n * \frac{\tau}{r_p}$. Therefore, the corresponding unit transmission rate

$$\gamma_{ppm} = \left(\frac{n}{T_{ppm}} \Big/ \frac{1}{\tau} \right) = \frac{n * r_p}{2^n} \tag{5}$$

It can be seen from Formulas (4) and (5) that at the same transmission rate, single-pulse PPM modulation requires a higher transmission bit rate than OOK modulation, and the corresponding bandwidth is also large.

2.1.2. DPPM modulation

DPPM was developed from LPPM, it retains the '0' in front of the time slot where the '1' is located in LPPM and the other '0's are removed. So the bits of L-DPPM are indefinite, composed of a string of low levels followed by a high level. This is shown in Figure 2.

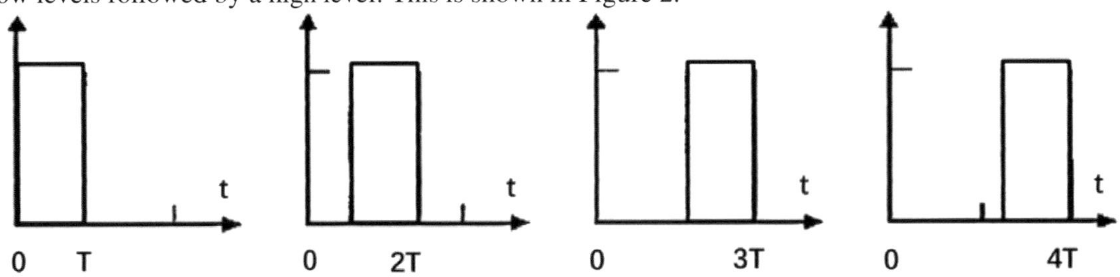

Figure 2. Schematic diagram of DPPM modulation

And Table 2 shows a mapping table between 8-LPPM and the 8-DPPM symbol, the average bandwidth occupied by each DPPM symbol is only about half of the LPPM modulation under the same transmission rate, which can save much bandwidth when transmitting the same amount of information.

Table 2. Mapping among bit groups, 8-LPPM, and 8-DPPM symbols

Signal pulse time slot	Information $m_1m_2m_3$	8-LPPM	8-DPPM
1	000	10000000	1
2	001	01000000	01
3	010	00100000	001
4	011	00010000	0001
5	100	00001000	00001
6	101	00000100	000001
7	110	00000010	0000001
8	111	00000001	00000001

As seen in figure 2 and table 2, one L-DPPM code group transmits the same bits of information as one L-PPM code group, both $\log_L^2$ bits. However, when the transmission rate is the same, DPPM modulation occupies less channel bandwidth than LPPM modulation.

The amount of signal data after DPPM modulation is uncertain, which limits the application of DPPM in some systems. In the following analysis of the transmission rate per DPPM unit, assuming that the probability P (1) of 1 occurrence and the probability of 0 occurrences of 0 in the transmitted information element are equal and that the chance of 1 appearing on either bit time slot of an L-DPPM code group is the same. Then for a DPPM signal with a duty cycle of r_p, its code element rate is $\mathrm{R} = 1/\mathrm{T} = \mathrm{r_p}/\tau$, and the signal bandwidth $\mathrm{B} = 1/\tau$. Because the number of bits of quartiles contained in a code group is indefinite, only the average number of bits of each code group L-DPPM can be derived here. Thus, the period width of a flat code group is:

$$T_{DPPM} = T * \frac{1+2+3+\Lambda+2^n}{2^n} = \frac{2^n+1}{2} T \tag{6}$$

Therefore, the unit transmission rate of DPPM modulation is

$$\gamma_{DPPM} = \frac{N/T_{DPPM}}{1/\tau} = \frac{2nr_p}{2^n+1} \tag{7}$$

It should be noted that this unit transmission rate is only a statistically significant unit transmission rate. The number of occurrences of 1 and 0 may differ over a specific time period, and the output signal number traffic may be time-varying after DPPM modulation.

2.1.3. MPPM modulation
Multi-pulse PPM modulation maps n binary ingress elements to multiple pulses over a time period consisting of M time slots. For example, dual-pulse PPM modulation, as shown in Figure 3.

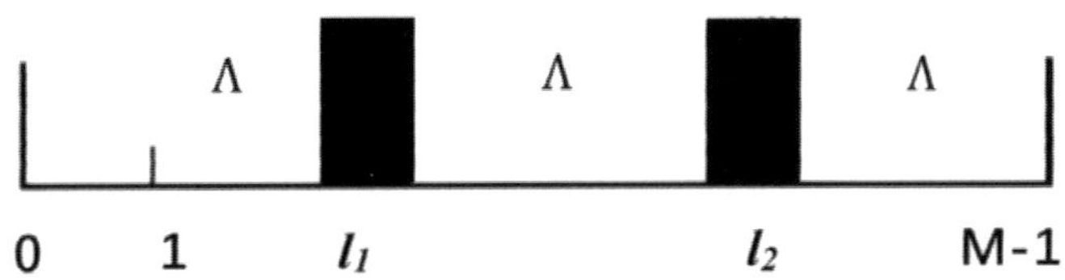

Figure 3. The schematic diagram of MPPM modulation

If it is noted that the 2 pulse positions are l_1 and l_2 respectively, it can be described as:

$$\begin{cases} (l_1, l_2) = \Phi(u_1, \Lambda, u_n) \\ l_1, l_2 \in \{0,1,K,M-1\},\ u_i \in \{0,1\} \end{cases} \tag{8}$$

Since the PPM modulation signal is a time series, the pulse position is also the sequence of time. Therefore, the combination or arrangement of multiple pulses and the types of such arrangements or combinations that characterize their respective communication capabilities are the multi-pulse arrangements and multi-pulse combination PPM. Each pulse has unique characteristics, for example, free selection of level values or pulse widths. It is too complex to implement this modulation method, so it has not been widely used.

For the multi-pulse combined PPM method, the combined type of k-pulse PPM on $\mathrm{I} = 2^{\mathrm{m}}$ time slots are $NC_k(m)$, then there is

$$NC_k(m) = C_l^k = \frac{2^m!}{k!(2^m-k)!} \tag{9}$$

The multi-pulse transmission capacity continues to increase as k increases. Its unit transmittal rate can be written as

$$\gamma_{kPPM} = \frac{r_p \log_2[NC_k(m)]}{2^m} \tag{10}$$

where $[NC_k(m)]$ is the number, whose value is closest to $NC_k(m)$ but smaller than $NC_k(m)$, and $[NC_k(m)] = 2^n$ (n≥m and n is an integer).

2.1.4. Comparison
As can be seen from the above analysis, L-PPM modulation and DPPM modulation have different degrees of bandwidth expansion compared to general OOK modulation. Multi-pulse PPM modulation differs from the first two PPM modulations and choosing the appropriate number of pulses can reduce bandwidth expansion or even no bandwidth expansion and have a higher transmission rate than OOK modulation at the same transmission rate. Their average transmitted optical power has decreased. In addition, L-PPM and MPPM decoding require code group synchronization clocks to determine the start and end of a code group, whereas DPPM does not.

2.2. International research process
Figure 4 briefly presents the historical context of the research on PPM.

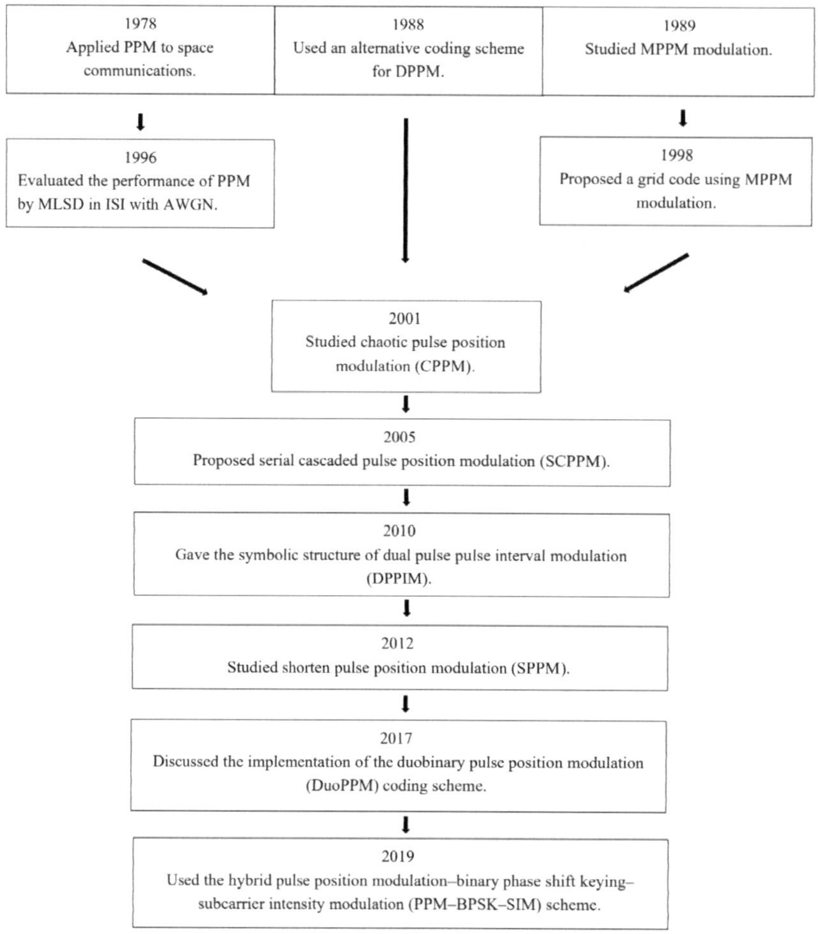

Figure 4. The research process of PPM

As seen in Figure 4, before 2000, the research focused on the initial performance analysis of LPPM, DPPM, and MPPM modulation. Since 2001, The focus of the research began to change, and the method of cascading PPM with other coding methods to improve communication quality. The research content also develops from theoretical derivation and simulation to practical application. As time went by, the study of PPM performance alone is no longer enough to meet a variety of experimental environments. Some researchers started to work on different modulations based on the previous analysis. Various modulation methods are related to each other, and the proposal of each modulation method is based on the shortcomings of existing modulation methods, and this modulation method itself will have some shortcomings. So far, there are nearly twenty PPM methods, and there are still some unsolved problems in these twenty modulation methods. And more modulation methods will be proposed in future studies to improve the performance of communication systems.

In addition, many scientists have worked on analyzing the performance of PPM modulation methods in different environments and comparing them among various methods. In 2016, Correia conducted numerical simulation studies on the propagation performance of OOK, PAM, and PPM modulation in the optical code division multiple access (OCDMA) system [9]. In 2020, Sharma studied a free-space optical communication system using PPM-BPSK-SIM to derive the average bit error rate and closing expression of the hybrid modulation scheme [10]. In 2020, Zaiton studied two single-carrier modulation schemes for visible light communication (VLC), OOK, and PPM's performance. The results show that PPM is superior to OOK in transmission distance and path loss, and OOK is the best choice for achieving the highest data rate [11].

After more than forty years of development, PPM modulation in wireless optical communication has mixed multiple modulation methods. The advantage of PPM is simple to modulate, and the hardware circuit is convenient to implement. However, PPM is a kind of pulse modulation, and its frequency band occupancy is greater than sinusoidal modulation, which is a disadvantage of PPM and is still challenging to overcome at this stage.

3. Experimental development of the PPM system

PPM has become an essential technology for wireless communication and has been systematically investigated due to its high bandwidth utilization and detection sensitivity. The bit error rate, error probability, and signal-to-noise ratio are essential for assessing PPM modulation and demodulation performance. The bit error rate (BER) is one of them. It is the ratio of erroneous bits to total transmitted bits and may be determined using random signals generated at different signal-to-noise ratios (SNRs).

Typically, the PPM system modulation model requires different bits to be used to transmit each signal. The signals are transmitted through PPM under additive white Gaussian noise (AWGN) and are simulated using MATLAB. After averaging, the BER can be obtained, and the characteristics of PPM can be verified by comparing it with other modulation methods.

The block scheme of the MIMO WOC system with QPPM under PPM modulation is shown in Fig. 5. The performance of the MIMO wireless optical communication system with Q-ary pulse position modulation over a spatially correlated log-normal fading channel is analyzed in terms of its uncoded BER and BER channel capacity [12]. Among them, the BER is the crucial variable for this experiment, reflecting the degree of correlation between the number of antennas, the channel, and the system stability. Compared to conventional wireless communication, wireless optical communication (WOC) has a higher transmission rate and more stable security. The spectrum does not need to be authorized or unlicensed, with low power consumption.

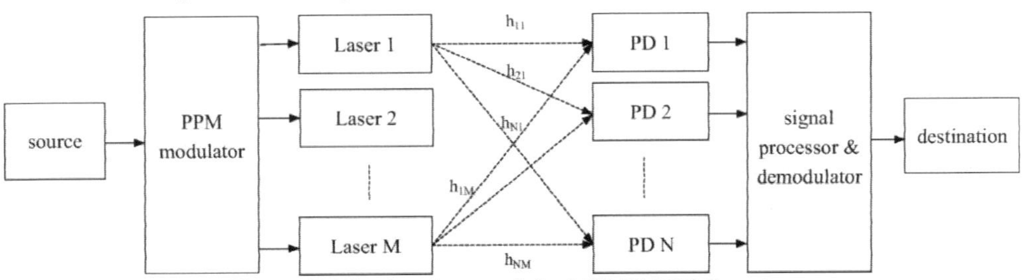

Figure 5. Block scheme of the MIMO WOC system

Figure 6 depicts PPM's detection of the asynchronous rising edge and synchronization of the entire system. This research combines pulse position modulation with Color Shift Keying (CSK) at the rising edge of VPPM modulation, simplifying the synchronization process at the receiver, displaying the increasing trend of the bit rate, and controlling the illumination intensity simultaneously, avoiding the flicker effect of CSK communication systems [13]. The approach allows two extra data bits for signaling in the synchronization scheme, simplifying and reducing power consumption. Furthermore, proven CMOS sensor increases lower BER and improves system performance.

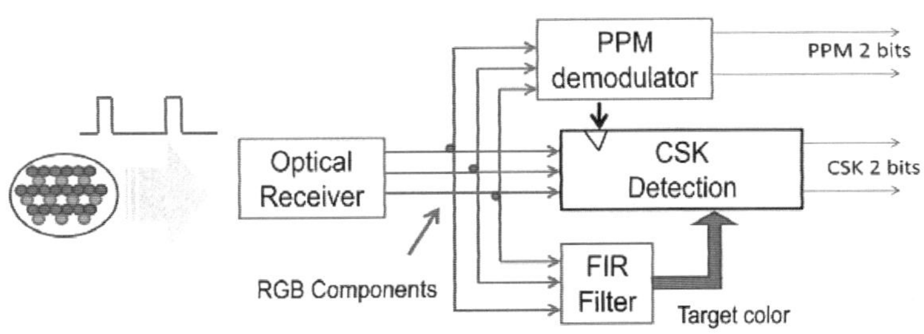

Figure 6. 4-PPM and CSK receiver

Furthermore, PPM can improve transmission system reliability. PPM can be implemented into a single module in the FPGA, as demonstrated in Figure 7, in FPGA experiments [14]. The PPM modulation module's principal function is to convert the original digital signal acquired by timer sampling into a PPM signal. The PPM demodulation module is responsible for converting the collected PPM signal back into the original signal. This system's modulation and demodulation modules are tested using oscilloscopes, FPGA development boards, PC hardware systems, and serial port helpers. The FPGA modulates the data received from the serial port. In the transmission process, the modulated signal is divided into two channels, which are sent to the PPM demodulation module, combined with another modulated signal, and supplied to the PC system via the serial port. To validate the module design, an oscilloscope catches the pin-specific signals by comparing the spectrum and waveform.

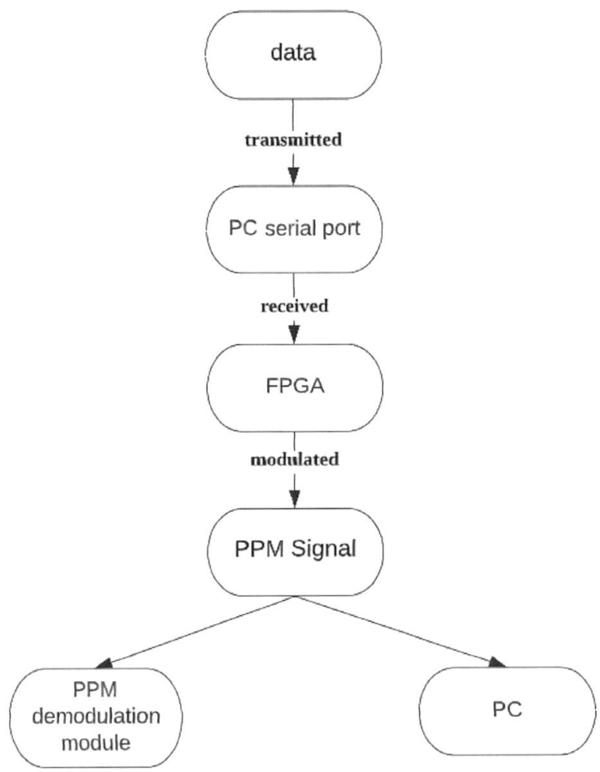

Figure 7. FPGA module

As SNR is a key parameter for experiments close to real life, it is imperative to get the desired SNR value. In a model of a PPM system, by comparing PPM modulation shown in Figure 8. With PWM, a conclusion can be made [15].

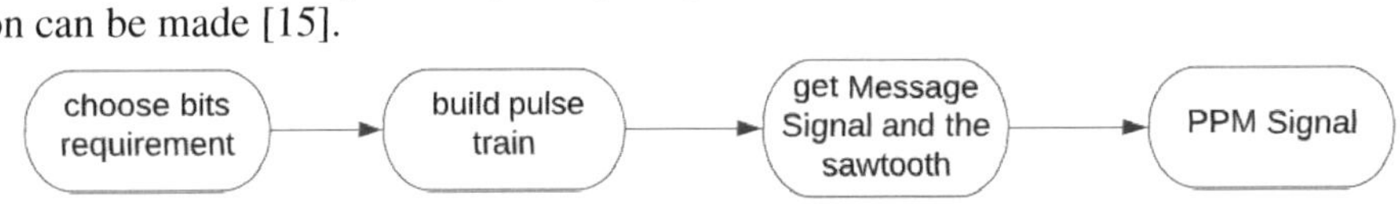

Figure 8. PPM modulation process

Figure 9 (a) illustrates that the PPM error probability will be lower under the same SNR conditions than PWM. It's because the amplitude and width of the pulses stay constant during modulation. In practice, utilizing a system based on PPM is preferable to using PWM when a higher level of immunity to interference is required since PPM has a smaller amplitude of error probability for each SNR value. The study aims to simulate PPM and investigate its features in wireless transmission, finding that PPM exhibits little noise interference. A specific curve of the likelihood of mistakes in PPM is shown in

Figure 9(b). When the SNR increased, the bit error rate's value began to trend lower. It caused the interference to the signal to decline and, in turn, the bit error rate to fall, enhancing the effectiveness of the communication system [15].

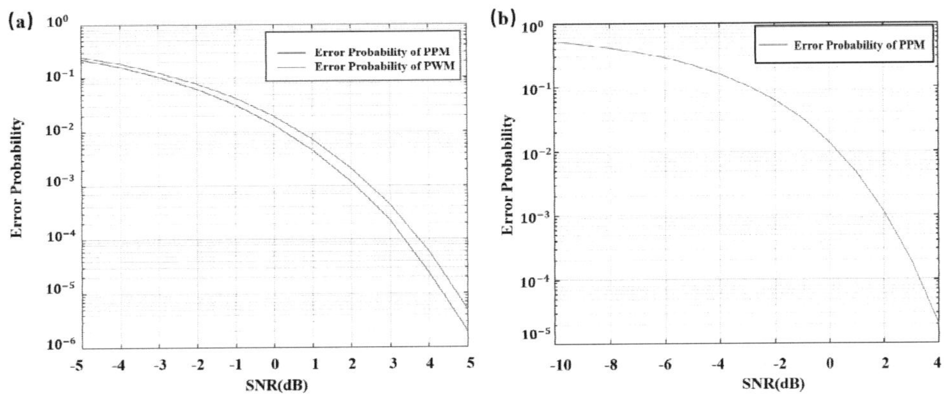

Figure 9. PPM modulation. (a) Error probability of PPM and PWM.
(b) PPM signal transmission in AWGN with correlation demodulation.

A simulated satellite-to-ground experiment was conducted to evaluate the effectiveness of a satellite-to-ground optical communication system [16]. Many parameters should be calculated and considered, such as Rytov variance, wavelength, and aperture diameter of the receiver shown in Figure 10.

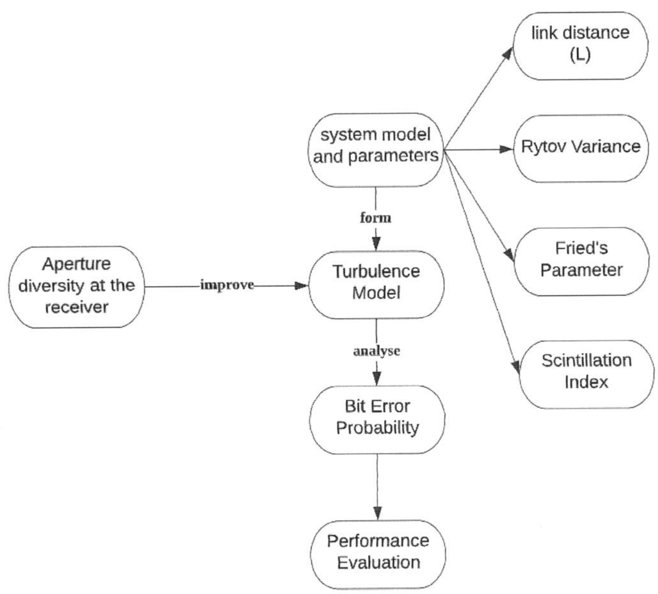

Figure 10. Satellite-to-Ground model and process

The signal-to-noise ratio values acquired in this experiment were necessary to satisfy the desired BER as a function of the Zenith angle. High SNR values result in longer communication times but higher power consumption, necessitating adaptive power regulation. Different zenith angles are acceptable for different SNRs and BERs. The practical SNR value in an optical communication system is close to 30 dB, and the sufficient value of BER is nearly 10^{-9}. As indicated in Figure 11, the communication system can be thought of as beginning at zenith angles of 78°.

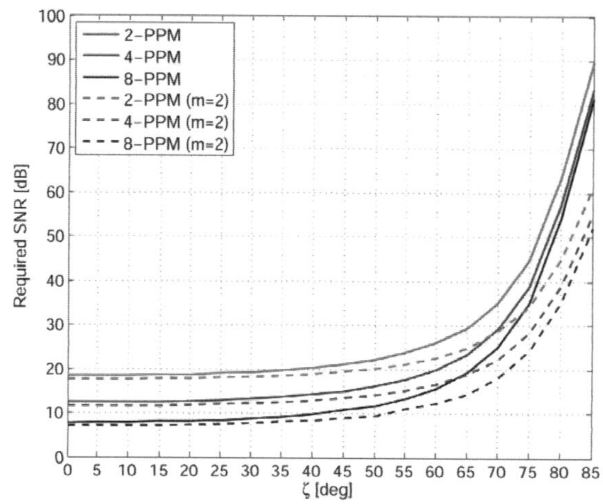

Figure 11. Required SNR values to achieve Pb(e)= 10^{-9}

Thus, according to these experiments mentioned above, precision plays a crucial role in experiments in different aspects. For example, the simulated satellite-to-ground experiment requires data to correspond to the coordinates on the chart. Moreover, PPM can be conducted in other conditions, like optical environments, particularly in color and angular aspects. Conducting PPM at the Zenith angle is an obstacle to achieving a more precise angle. Meanwhile, during experiments, it is a complex process that many intermediate variables need to be achieved before deriving parameters.

Table 3. Comparison of different PPM experiments

Experiments	Complexity	Accuracy	Stability	Application scenarios	Calculation
MIMO WOC system	High	High	Stable	configure the antennas of the MIMO WOC system	Extremely complex
Color Shift Keying communication system	Low	Only general trends	Stable and efficient	Monitor illumination intensity, flickering effects, color detection	Medium
FPGA experiment	Medium	High	Medium	Apply FPGA to the optical system	Slightly complex
Digital communication system	Low	Only general trends	Not mentioned	Improve noise interference	Medium
Satellite-to-ground optical communication system	High	High accuracy in a specific range	Medium in obtaining lower BER	Calculate zenith angle and satellite communication	Complex

4. Applications

The first step to finding out how PPM is applied in other fields is to summarize the history of PPM application from all the information researched. PPM theory has been widely used in many research fields, mainly in communication. In 2013, there was research on the PPM system application in broadband communication [17]. After entering the information research era, PPM theory and its derivative theory (MPPM, random position modulation, etc.) have been applied in various fields, including but not limited to communication technology, since 2019.

PPM has been put into application in a large range of technology development. Firstly, PPM is mainly used in communication, exceptionally light communication. For example, LED communication, atmospheric laser communication, visible light communication, underwater optical communication, and ultra-sensitive real-time space laser communication; Moreover, PPM is also used in detection technology such as underground pipeline detection, laser ranging technology, radar speed range measurement, etc.; Last but not least, PPM can be applied in mechanical technology as well. For example, torque ripple suppression of induction motors for electric vehicles belongs to the mechanical field.

For the Light communication area, there are two main application examples. Firstly, in the LED optical communication system based on CPPM [18], the introduction of PPM or its derivatives (MPPM, random position modulation, etc.) into an optical communication system can reduce the system bit error rate according to the LED optical communication system based on CPPM, as shown in Figure 12.

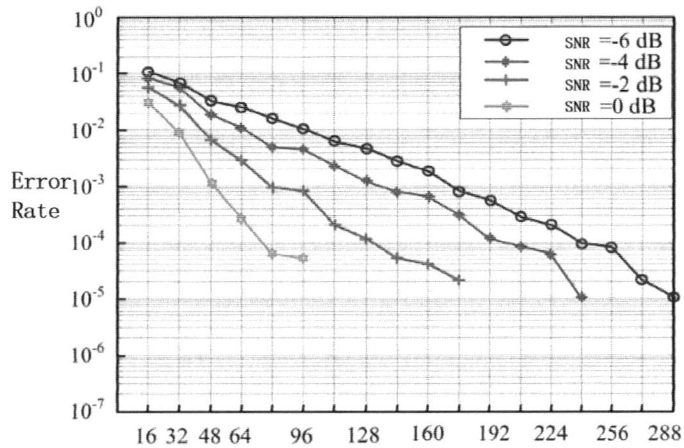

Figure 12. Graph of chaotic sequence spread spectrum code length N and CPPM system error rate at a given SNR

In the Real-Time Space Laser PPM Communication System with Ultra-high Sensitivity [19], PPM offers a higher efficiency power compared with OOK modulation. It can obtain a high peak optical power signal with a high energy-efficiency ratio, strong signal channel resistance to dry disturbance energy, and good reception sensitivity, which is very suitable for the transmission power limited system such as inter-satellite stimulated optical communication.

For the Detection area, the advantages and disadvantages of analog and digital position modulation are compared in the research on laser ranging technology based on pulse position modulation. Digital modulation is chosen as the primary modulation mode in this technology because of its higher precision.

In the Mechanic area, researchers developed their torque ripple suppression of induction motors for electric vehicles by basing random position modulation on the neural network, by which the system can work more efficiently.

Table 4. Applications of PPM

Field of PPM application	Kind of PPM used	Effect of PPM
Light communication	MPPM	Reduce the system bit error rate
	Real-Time Space Laser PPM Communication System	Obtain a high peak optical power signal

Detection technology	Analog and digital position modulation	Digital PPM has higher precision
Mechanic technology	Random position modulation	Develop the efficiency by basing random position modulation on the neural network

5. Conclusion

Given the development of digital communication, the main goal of the current study was to Provide some basic information about PPM. This paper expounds on the importance and future development trend of PPM, the international research process, and specific methods of PPM in detail. Ultimately, its application and broad prospects for future development are discussed. The study contributes to our understanding of the nature and role of PPM. The unique point is that the study manipulates PPM modulation and demodulation experiments in different systems and conditions to determine their effect on PPM's critical advantages. The conclusion is that benefits are not affected by conditions or systems. The shortcoming is that the number of experiments is not large enough to prove the conclusion strictly. As for the outlook for the future, PPM is expected to be more popularized in people's daily life and have a lot better performance as well.

References

[1] Zhou, Z., Liang, B., Cao, Y., & Zhang, M. (2022, January). MPPM Spectrum Analysis Based on PPM. In 2022 14th International Conference on Computer Research and Development (ICCRD) (pp. 356-362). IEEE.

[2] Xizheng, K., Jingyuan, L., Dongsheng, X., & Jiafan, W. (2022). Research progress of pulse position modulation technology in optical wireless communication. Opto-Electronic Engineering, 49(3), 210387-1.

[3] Black, H. S. , Beyer, J. W. , Grieser, T. J. , & Polkinghorn, F. A. . (1947). A multichannel microwave radio relay system. Transactions of the American Institute of Electrical Engineers, 65(12), 798-806.

[4] Dillard, G. M. . (1973). Pulse-position modulation based on energy detection. IEEE Transactions on Aerospace & Electronic Systems, AES-9(4), 499-503.

[5] Cryan, R. A., & Unwin, R. T. . (1992). Optical space communications employing pulse position modulation. IEE Colloquium on Advanced Modulation and Coding Techniques for Satellite Communications. IEEE Xplore.

[6] Klein, A. G., & Duhamel, P. . (2007). Decision-feedback equalization for pulse-position modulation. IEEE Transactions on Signal Processing, 55(11), 5361-5369.

[7] Sugiyama, H., & Nosu, K. . (1989). Mppm: a method for improving the band-utilization efficiency in optical ppm. IEEE Journal of Lightwave Technology, 7(3), 465-472.

[8] Yu, C. (2022, January). Spectrum Analysis of Multi-Pulse Position Modulation under Same Transmission Condition. In 2022 14th International Conference on Computer Research and Development (ICCRD) (pp. 363-367). IEEE.

[9] Correia, D.G., Sales, J.C., Pinto, P.V., Moura, L.P., Ferreira,, A.C., Menezes, J.W., Guimarães, G.F., & Sombra, A.S. (2015). Analysis of the Performance of a PAM/PPM/OOK System Operating with OCDMA, under Nonlinear Optical Effects in Optical Fiber Propagation. Journal of Optical Communications, 37, 233 - 246.

[10] Sharma, K., & Grewal, S.K. (2020). Performance assessment of hybrid PPM–BPSK–SIM based FSO communication system using time and wavelength diversity under variant atmospheric turbulence. Optical and Quantum Electronics, 52, 1-25.

[11] Zaiton, A., Eng, C., & Jasman, F. (2020). Pulse Position Modulation characterization for indoor visible light communication system. Journal of Physics: Conference Series, 1502.

[12] Wang, H., Wang, X., Lynette, K., & Cao, M. (2018). Performance analysis of MIMO wireless optical communication system with Q-ary PPM over correlated log-normal fading channel. Optics & Laser Technology, 102, 153-159.
[13] Raj, F. A. D., Guerra, V., Borges, J. A. R., Torres, J. R., & Perez-Jimenez, R. (2014). Color shift keying communication system with a modified PPM synchronization scheme. IEEE Photonics Technology Letters, 26(18), 1851-1854.
[14] Zhang, G., Wang, X., & Zhao, H. (2020, December). Design of PPM modulation and demodulation module in optical communication. In 24th National Laser Conference & Fifteenth National Conference on Laser Technology and Optoelectronics (Vol. 11717, pp. 224-231). SPIE.
[15] Dai, L., Liu, Z., Meng, S., & Zhao, Y. (2021, August). Research on the Pulse-position Modulation in the Digital Communication System. In 2021 International Conference on Wireless Communications and Smart Grid (ICWCSG) (pp. 8-12). IEEE.
[16] Sahin, S. (2016, July). Improvement of required SNR for satellite-to-ground optical communication via aperture diversity by using PPM. In 2016 10th International Symposium on Communication Systems, Networks and Digital Signal Processing (CSNDSP) (pp. 1-5). IEEE.
[17] Wu N.N, Dou Z & Chen L.L.(2013, April).Application of PAM and PPM in multi-antenna broadband communication system. Journal of Harbin University of Business (Natural Science Edition)(04), 470-473.
[18] Wang F.Y, Hu X.L, Qin L, Guo Y, Xu Y.H & Du Y.X.(2019, December). A LED optical communication system based on CPPM. Optical Communication Technique(12), 7-9.
[19] Wang Z.Z, Ao X.Y, Zhang S.M, Liu C, Yang Q & Xiong L.M.(2022, January).Ultra-High Sensitivity Real-Time Space Laser PPM Communication System Research. Optical communication research(01),18-22.

Research and Simulation of OCDM Switch Layer Based on 2D WDM-Optical Code Switching Structure

Tengfei Dong [1a], Lin Meng[2], Jing Wang[3], Kan Li[4]

[1,2,3,4]Army Infantry Academy, Shijiazhuang 050000, China

[a]15033567859@163.com

Abstract—Optical Communication is widely used in military and civilian backbone network communication, but due to the limitation of "electronic bottleneck", the ability of network nodes to process information and optical layer data transmission capacity is seriously mismatched, which directly causes optical cross connectors in optical nodes. The number of ports carried has skyrocketed. In order to improve the processing speed of network nodes, this paper studies the four-layer multi-granularity switching system of optical fiber, wavelength band, wavelength, and optical code, and finally builds a complete OCDMA system through simulation software, simulates multi-user simulation, and successfully realizes The encoding and decoding process is analyzed, and the performance simulation and analysis of the core switching module of the four-layer multi-layer optical switching system-OCDM switching module are carried out, and the feasibility and superiority of the selected system are verified.

1. Introduction

In the traditional three-level multi-granularity switching system of optical fiber, wavelength band, and wavelength, the adopted wavelength band switching technology only bundles the wavelengths and transmits them, which only reduces the number of switching ports and simplifies the switching management without reducing the wavelength. number, saving resources. Therefore, we study a four-layer multi-granularity model of the fiber, wavelength band, wavelength, and optical code. Multiple users are established on one wavelength, and different address codes are allocated to different users on the same wavelength to distinguish them, which solves the problem of low resource utilization and reduces the number of ports and the use of network equipment. In this paper, the core system model of the four-layer multi-granularity optical switching system is given, and the simulation and comparison of the four-layer multi-granularity switching system with wavelength converters and the four-layer multi-granularity switching system without wavelength converters are carried out.

2. OCDM Translation Layer Switch Fabric

After the light waves carrying multiple optical code words are demultiplexed and output from the corresponding wavelength demultiplexer, the first step is to use the optical signal decoder to decode the user information. This step must use the corresponding address code words. Because a number of different types of decoders are integrated into the system, the received signals can be sent to different decoders, and then the local address code sequence in them can be used to perform correlation operations on the signals. The code word happens to be the same as the code word corresponding to the user information input to the decoder, then the original signal can be recovered correctly, and then the decoded input signal will be transmitted independently because the destination will be different granularity Therefore, what needs to be considered is to input the signal to the converter for conversion, and finally, the signal can be sent to the same outlet as the destination. In order to reduce the probability of collision, we propose a two-dimensional wavelength division-optical code-switching structure, so that the optical code conversion and wavelength conversion can be carried out at the same time. The number of wavelength converters per switching layer is equal to the number of wavelengths of every port [1-2]. As shown in Fig. 1, in order to output at port 1, the data packet $c(\lambda_1, OC_2)$ is converted to a wavelength with sufficient conversion optical code through wavelength conversion, that is, it becomes the data packet $c(\lambda_n, OC_2)$ output at port 1. Through wavelength conversion, the conflict and competition problem at port 1 is solved, which is obviously superior to the one-dimensional wavelength division-optical code-switching structure.

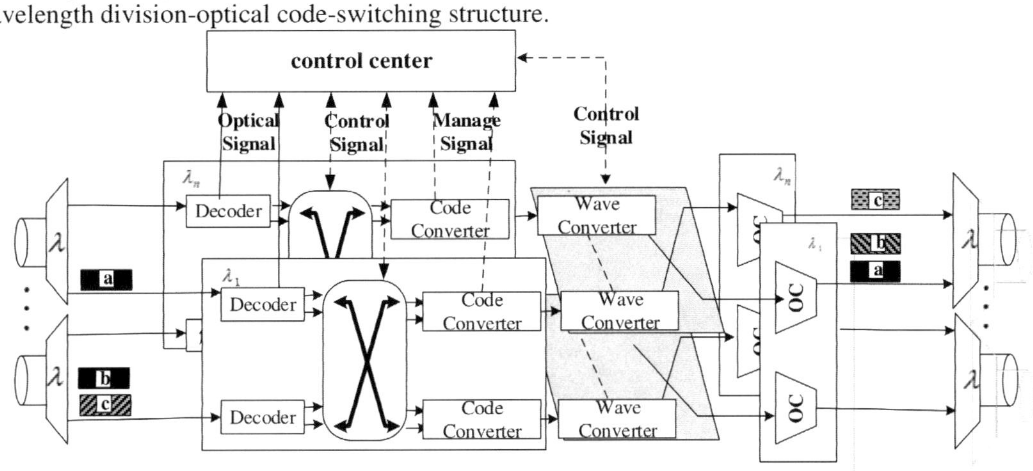

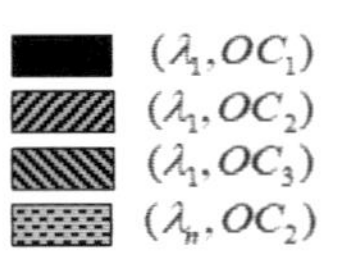

		a	b	c
IN	port	1	n	n
	λ	λ_1	λ_1	λ_1
	OC	OC_1	OC_1	OC_2

		a	b	c
OUT	port	1	1	1
	λ	λ_1	λ_1	λ_n
	OC	OC_1	OC_3	OC_2

Fig.1 Two-dimensional wavelength division-optical code-switching structure

3. OCDM Switching Layer Simulation Implementation

In order to verify the proposed scheme, we used the optical simulation software OptiSystem to simulate the core of the four-layer multi-granularity optical switching system - the OCDM conversion layer [3-5].

3.1 OCDMA Simulation system basic parameter settings

We choose single coincidence code to construct address code and choose fiber Bragg grating codec to complete encoding and decoding. Then we set q =9, d =2, length of code $L = 2k = q - 2d - 1$ =9-4-1=4. In the simulation experiment, these 9 wavelengths $(\lambda_0, \lambda_1, \lambda_2, \lambda_3, \lambda_4, \lambda_5, \lambda_6, \lambda_7, \lambda_8)$ are selected as ITU-T standard wavelengths with 100GHZ interval [4], they are 1546.784nm, 1547.588nm, 1548.392nm, 1549.196nm, 1550nm, 1550.804nm, 1551.608nm, 1552.412nm, and 1553.216nm respectively. The code words of the four users are:

$$\lambda_3\lambda_7\lambda_4\lambda_0,\ \lambda_4\lambda_8\lambda_5\lambda_1,\ \lambda_5\lambda_0\lambda_6\lambda_2,\ \lambda_6\lambda_1\lambda_7\lambda_3 \tag{1}$$

The data propagation rates of the four users are all set to 1.25Gbit/s, and the above-mentioned four single-coincidence codewords are used to design the codec of the fiber grating.

In the process of encoding and decoding, the round-trip time of light passing through the fiber grating array needs to be considered. In general, the next signal bit pulse must wait for the previous signal bit pulse to be completely reflected from the fiber grating array before entering and proceeding. coding. Here we set the total length of the grating array as N, then the total time required for light to pass through the entire grating array is $t = 2N / v$, where $v = c / n$, c is the speed of light in vacuum, n is the effective group refractive index in the fiber, set n =1.4467. Then the time t required to pass through the entire grating array needs to satisfy:

$$t \le T_b = 1/1.25Gbit/s = 0.8ns \tag{2}$$

In the design of the encoder, each grating array takes four gratings, that is, the encoder of each user consists of four FBGs, so three fiber delays are required, each delay:

$$\tau = t/3 = 0.8ns/3 = 0.267ns \tag{3}$$

3.2 Construction of OCDMA Simulation System

Through the setting of the above parameters, combined with the construction principle of the serial structure FBGs codec, a 4-channel single-channel 1.25Gbit/sOCDMA simulation system is constructed[6]. Since the codeword of user 1 is n, the center wavelengths of the four fiber gratings of user 1 should be set to 1549.196nm, 1552.412nm, 1550nm, and 1546.784nm respectively. Figure 2 shows the structure of the encoder of tag 1. The Bragg reflection spectrum width was set to 0.3 nm, and the reflectance was 99%.

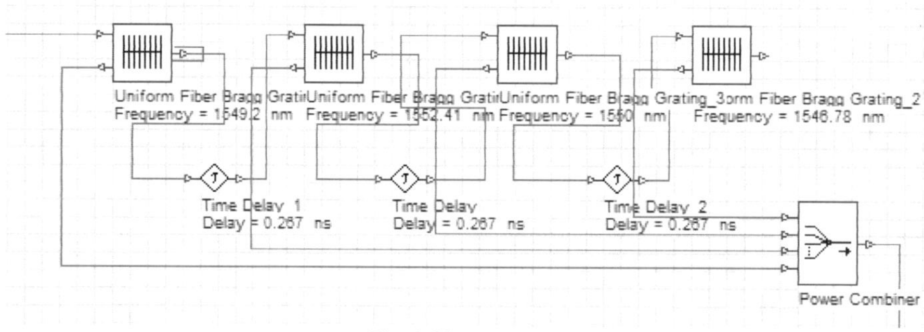

Fig.2 User 1 Encoder

Similarly, the center wavelengths of users 2, 3, and 4 are set in sequence. After the four users are encoded, they are coupled and superimposed by the coupler, and then transmitted to the decoder for decoding. Here we only decode user 1.

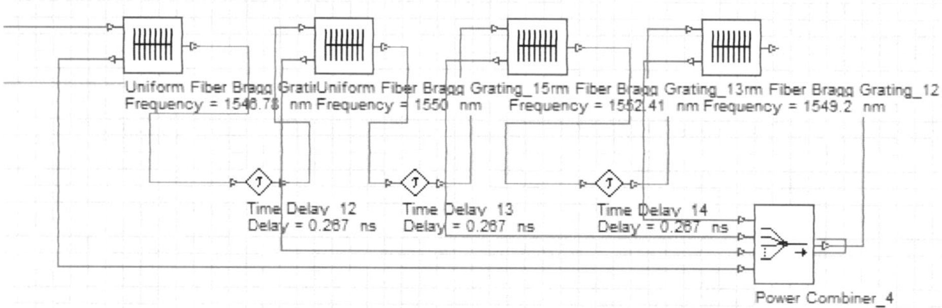

Fig.3 User 1 Decoder

Figure 3 is the structure diagram of the user 1 decoder, which is also composed of four fiber gratings, the delay $\tau = 0.267ns$ between each grating, the Bragg reflection spectrum width is still 0.3nm, and the reflectivity is still 99%. Different from the user 1 encoder, the center wavelength of each fiber grating of the decoder is just opposite to that of the encoder, that is, the center wavelengths of the four fiber gratings are 1546.784nm, 1550nm, 1552.412nm, and 1549.196nm respectively.

Journal of Physics: Conference Series **2384** (2022) 012027 doi:10.1088/1742-6596/2384/1/012027

3.3 OCDMA System Simulation and Analysis

It is assumed that the bit sequence of user 1 is 0101101110, the bit sequence of user 2 is 1110001110, the bit sequence of user 3 is 1100110011, and the bit sequence of user 4 is 0001000111. Figure 4 is the optically encoded signal of user 1. After the four chips of the encoder are respectively delayed by 0.267ns, the encoded signal is formed by coupling and superposition by the coupler[7-8]. The superimposed signal obtained by the superposition of the four-channel coded signals is shown in Fig. 5.

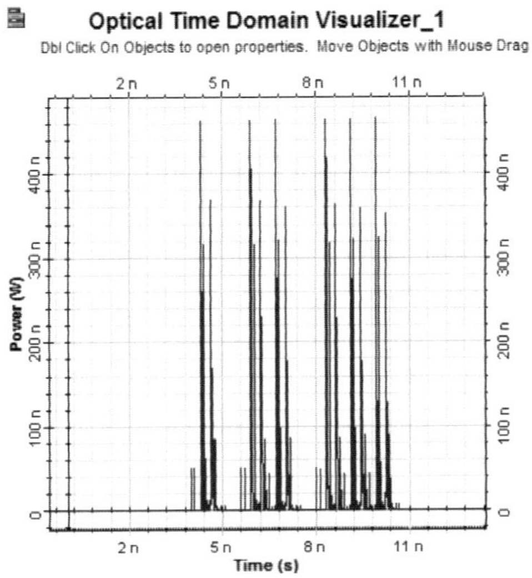

Fig.4 User 1 encoded signal

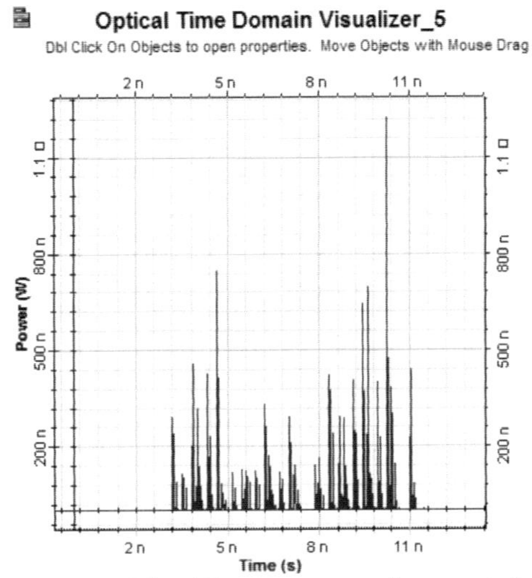

Fig.5 Four-user superimposed signal

In this simulation, the influence of external noise is not considered, and only the signal of user 1 is decoded and restored. After the four chips of the decoder are respectively delayed by 0.267ns, they are coupled and superimposed by the coupler to form the decoding The signal is shown in Figure 5. From the recovered signal diagram, we can see that although the decoding can be successful, there must be multiple access interference in the system, that is, during the encoding and decoding process, the user information will be interfered with by other undesired user information and superimposed. Eventually, each user's information will be doped with other users' information, causing noise to accumulate.

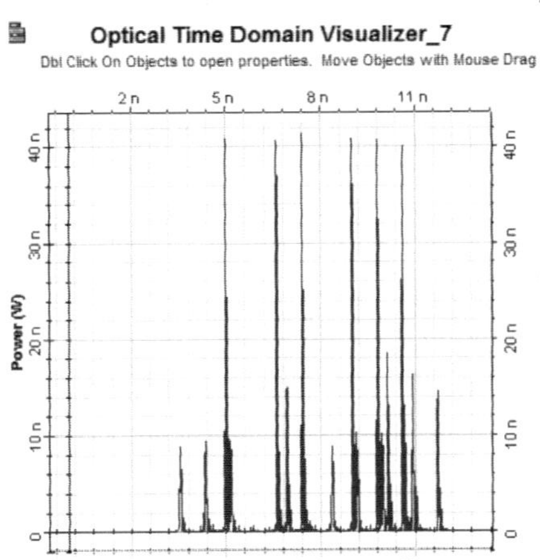

Fig.6 User 1 decoded signal

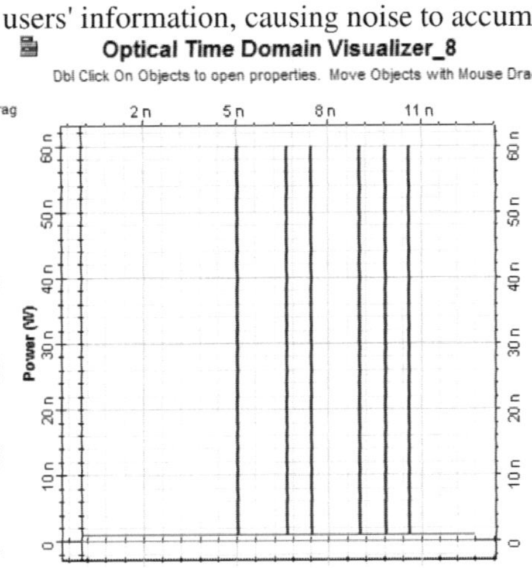

Fig.7 User 1 decodes the signal after limiting

In order to effectively reduce the multiple access interference and expand the user capacity of the OCDMA system, we added an optical hardware limiter (Optical Hard Access, OHL) after the coupler of the user 1 decoder. Different from the filter, the optical hardware limiter performs optical processing on the signal and effectively filters out the superimposed noise. T

Figure 7 shows the decoded signal of user 1 after being limited by the optical hardware limiter, from which we can see that the decoded signal after the limiter eliminates noise interference, and the original pulse signal of user 1 can be successfully decoded from it, which is 0101101110.

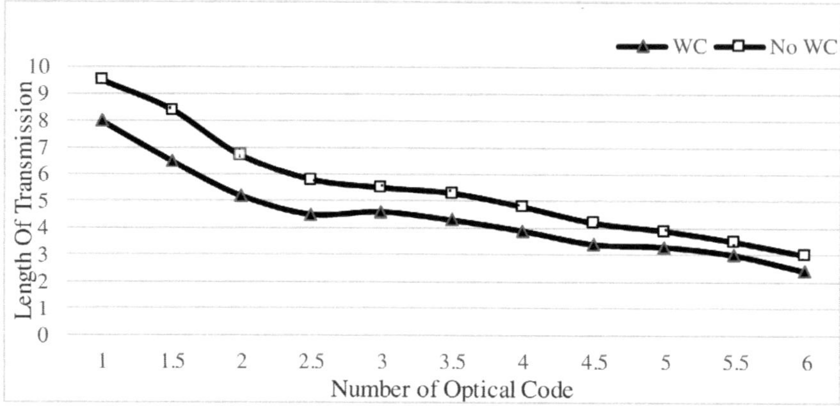

Fig. 8 Transmission performance of different optical codes

The OCDM conversion module has the function of wavelength conversion. Through wavelength conversion, the problem of conflict and competition at the output port is solved. From Figure 8, we can see that no matter how many optical code users are carried by each wavelength, the network topology path length of the system after adding the wavelength converter to the OCDM conversion module is smaller than that of the system without adding the wavelength converter. The reason for this is that the addition of the wavelength converter reduces the competition at the output port, thereby reducing the path of the system network topology during routing, reducing the time, and improving efficiency.

4. Conclusion

In this paper, the core switching layer of the four-layer multi-granularity switching system, the OCDM switching layer, is deeply analyzed. In view of the port competition problem in the one-dimensional wavelength division-optical code-switching structure, a two-dimensional wavelength division-optical code-switching structure is proposed. The OCDMA system is built by OptiSystem, and the single coincidence code with better performance is selected as the address code, the selection of optical address code, and the optical codec. The user is expected to implement the codec successfully through simulation, thus verifying the feasibility and superiority of the scheme.

References

[1] Wang Y, Cao X. Multi-granular optical switching: A classified overview for the past and
[2] future[J]. Communications Surveys & Tutorials, IEEE, 2012, 14(3): 698-713.
[3] Huang S W, Baba K, Murata M, et al. Evaluation of OCDM-Switching and Code Convertiton for All-Optical End-to-End Path Provisioning in Multi-granularity Networks[J]. IEEE Network, 2006.
[4] Jiang J, Wu D, Fan P. General Constructions of Optimal Variable-Weight Optical Orthogonal Codes [J]. IEEE Trans. Inf. Theory., 2011, 57(7): 4488-4496.
[5] Zhang Z, Tian C, Petropoulos P, et al. Distributed-Phase OCDMA Encoder–Decoders Based on Fiber Bragg Gratings[J]. Photonics Technology Letters, IEEE, 2007, 19(8): 574-576.
[6] Tao S, Miao T, Xin-Lao W, et al. Performance of all fiber optical current transducer in optical transmis sion based on optisystem[C]//Optoelectronics and Microelectronics (ICOM), 2013 International Conference on. IEEE, 2013: 1-4.
[7] Matsumoto R, Kodama T, Shimizu S, et al. 40G-OCDMA-PON System With an Asymmetric
[8] Structure Using a Single Multi-Port and Sampled SSFBG Encoder/Decoders[J]. Journal of Lightwave Technology, 2014, 32(6): 1132-1143.

A feature extraction method for XML documents based on PCA

Zhiwen Yu [1,a*], Ruifeng Zhao [1], Kaiwen Zeng [1], Wenjie Zheng [1], Hao Liu [2] and Ting Yang [2]

[1.] Guangdong Grid Co, China

[2.] School of Electrical and Information Engineering, Tianjin University, China

[a.] yzw_javelin@163.com

Abstract. XML document stores the information of the new power system source load interaction. It has the characteristics of self-description, extensibility, structure, and content, which makes it widely used. Improving the method of extracting elements from XML documents is very helpful in solving the problem of distributed object operation measurement in the power grid. To classify or analyze XML documents better, based on the theoretical analysis of principal component analysis and the study of the text representation model, this paper proposes two methods: vectorization representation of XML documents and further feature extraction. The experiment result shows that the method of all path feature extraction for XML document can represent the main feature of XML document effectively, and is an important work for latter handling XML documents of power grid efficiently

1. Introduction

Currently, China Southern Power Grid (CSG) usually uses the CIM/XML model regulated by the IEC61970-301 standard to describe equipment entity objects, equipment attributes, and equipment connection relations related to power system operation. The file formed after the CIM model is built in XML language is called CIM/XML model. This model describes various data of power system operation.

XML is a semi-structured data description language with the characteristics of self-description, extensibility, structure, and content, making it widely used[1]. Especially in recent years, in the face of massive XML data, how to find useful information and cluster or classify XML information makes XML data mining one of the most active branches in database research, development, and application[2]. Data preprocessing is an important step before data mining.

At present, the common data preprocessing methods include data cleaning, data integration, data transformation, and data reduction. These data processing technologies are used before data mining, which greatly improves the quality of data mining patterns and reduces the time required for actual mining. XML document is a kind of text with structural information. The information representation of the XML document and the selection of its feature items is a preprocessing step to be solved before data mining and information retrieval of the XML document. Guided by the theoretical basis of principal component analysis, this paper puts forward the methods of full path feature vector extraction and edge set vector extraction of the XML document, which paves the way for the subsequent processing of the XML document[3].

2. Representation of XML documents

2.1. Common text representation models

CSG use a common information model (CIM) to describe all kinds of data of power system operation, including the attributes of power grid equipment and the connection relationship between equipment. The CIM model is finally stored in XML files. The information described by an XML document is a kind of text with structural information, which is also the essential difference between an XML document and a plain text document. At present, the research on text representation mainly focuses on the selection of text representation models and the selection of feature word selection algorithms. There are many common text representation models, including the Boolean model, vector space model, probability model, and so on.

The most famous vector space model is proposed by Gerard Salton. It is the most widely used text representation model at present. Its basic idea is to take each feature as one dimension of the feature space coordinate system and use the vector of the feature space to represent the text. In the vector space model, each text is formalized as a vector in the n-dimensional space, so the text can be abstracted as $V(d) =(w(t_1),\cdots,w(t_2),\cdots, w(t_m))$, where t_i is the feature item, m is the dimension of the text space, $w(t_i)$ is a function, and its basic function is to calculate the weight III of the feature item t_i in the text vector[4]. Here, we use a vector space model to represent XML documents.

2.2. Vector representation of XML documents

XML documents of CSG are stored in a tree structure. The commonly used representation methods of XML documents include element representation, edge set representation, path representation, and so on. Here we give some specific definitions and descriptions, as shown below.

Definition 1:XML document tree. $DT= (V, \Sigma, tag, root)$, where V is the finite set of nodes, Σ is the finite set of labels: the function tag: $V\rightarrow\Sigma$ defines the labels of nodes. The root is the only node in V, tag(root) =r is called the root of DT. It is specially noted here that in the bilge tree structure, the attributes of elements are not specially distinguished. The attributes of elements are also treated as ordinary elements and the text content of leaf nodes in the tree [5]. Figure 1 shows two XML document trees DT_1 to DT_2.

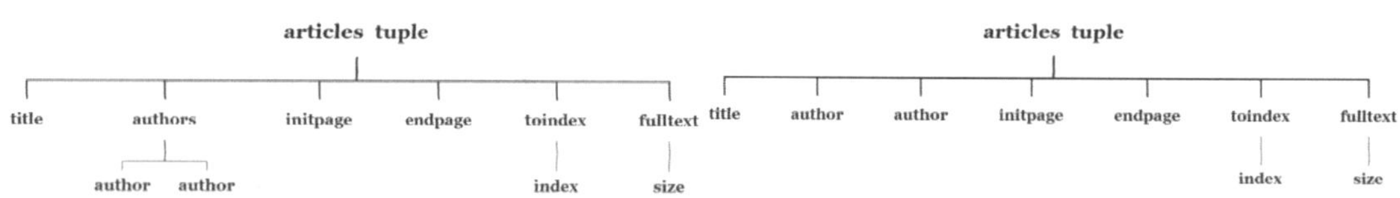

Figure 1.XML document tree DT_1 and DT_2

Define 2 Full paths. In the XML document tree DT, the full path of node v_i is defined as an ordered set of labels from root to v_i. Where, $v_i \in VDT$,i=1,2,⋯, m. An XML document is modeled through path information, which reflects the hierarchical and structural information of XML.

Define 3 Edge set. In the XML document tree DT, the edge set E_i is defined as the finite edge set from all v_i nodes to v_j nodes in the tree, which is recorded as $D\_edgeDTEI = \{v_i, v_j\}$, where $v_i \in VDT, i,j \in [1, m]$, and $i < j$.The edge set of XML is used to model XML, which better reflects the structural information of XML.

3. Feature extraction based on PCA

The main function of feature extraction is to reduce the number of processing without damaging the core information of the text, so as to reduce the dimension of vector space, simplify the calculation, and improve the speed and efficiency of document processing. Document feature selection has a very important impact on the research of document content filtering and classification, clustering, automatic summarization, user interest pattern discovery, knowledge discovery, and so on. Usually, the scoring value of each feature is calculated according to a feature evaluation function, then these features are sorted according to the scoring value, and several features with the highest scoring value are selected as feature words, which is feature extraction[6].

3.1. PCA-based feature extraction process

Principle component analysis is a non-parametric pattern recognition method that aims to represent high-dimensional data in low-dimensional space. PCA is a very effective dimensionality reduction processing technology. It is a statistical analysis method based on the principle of K-L transformation. The processing process of this method is as follows: first, assume that there are n documents, each document has p variables, and use these p variables to describe or represent the document. In this way, n documents and p variables of each document together form a data matrix X of $n*p$ order

$$X = \begin{bmatrix} x_{11}x_{12}\cdots x_{1p} \\ x_{21}x_{22}\cdots x_{2p} \\ \cdots\ \cdots\ \cdots\ \cdots \\ x_{n1}x_{n2}\cdots x_{np} \end{bmatrix} \tag{1}$$

Facing the data matrix X, when p is large, it will be very troublesome to investigate or study problems in this space. We hope to find a method to reduce the dimension; that is, the original more variable indicators can be replaced by fewer indicators after dimension reduction, and the original main information is saved after dimension reduction. At the same time, they are independent of each other. Here, $x_1, x_2 \cdots x_p$ are defined as the original indicator variables, and the new indicator variables are $z_1, z_2, \cdots z_m$ $(m<=p)$

$$\begin{cases} z_1 = l_{11}x_1 + l_{12}x_2 + \cdots + l_{1p}x_p \\ z_2 = l_{21}x_1 + l_{22}x_2 + \cdots + l_{2p}x_p \\ \cdots \\ z_m = l_{m1}x_1 + l_{m2}x_2 + \cdots + l_{mp}x_p \end{cases} \tag{2}$$

Here, the rules for determining the coefficient Lij are described as follows: ① z_i and z_j ($i \neq j$; i,j=1,2,...,m) are independent of each other; ② z_1 is the one with the largest variance among all linear combinations of $x_1, x_2, \cdots, x_p$, and z_2 is the one with the largest variance among all linear combinations of $x_1, x_2, \cdots, x_p$ not related to z_1, z_m is the largest variance among all linear combinations of $x_1, x_2, \cdots, x_p$ that are not related to $z_1, z_2, \cdots, z_{m-1}$. Then the new indicator variables $z_1, z_2, \cdots$, and z_m are respectively called the first, second, ⋯, and m-th principal components of the initial indicator variables $x_1, x_2, \cdots, x_p$.

It can be seen from the above analysis that the essence of principal component analysis is to determine the load L_{ij} (i=l, 2,⋯, m. j=1,2,⋯, p) of the original variable x_j (j=1, 2,⋯,p) on the principal components z_i (i=1, 2, ⋯, m). It can be proved mathematically that they are the eigenvectors corresponding to m larger eigenvalues of the correlation matrix.

4. Experimental results

Based on the theoretical analysis of PCA and the two vector representation methods of XML documents, combined with the characteristics of the XML document of CSG, we have done the following experiments: In the experiment, the vector representation of XML documents in the test set is realized by visual C++. Here, two methods are used to represent XML documents. One is to extract full path information from all test documents and then vectorize. The other is extracting the edge set information from all test documents and using the same vectorization. The process of feature vector extraction and dimension reduction is realized by MATLAB programming[7]. Ten samples are taken from each of the five types of test documents for the above test set to form 50 XML test documents.

After vectorization, feature extraction is performed. The comparison data of the specific experimental results are shown in Figure 2. The abscissa in Figure 2 is the dimension, and the ordinate represents the cumulative contribution rate.

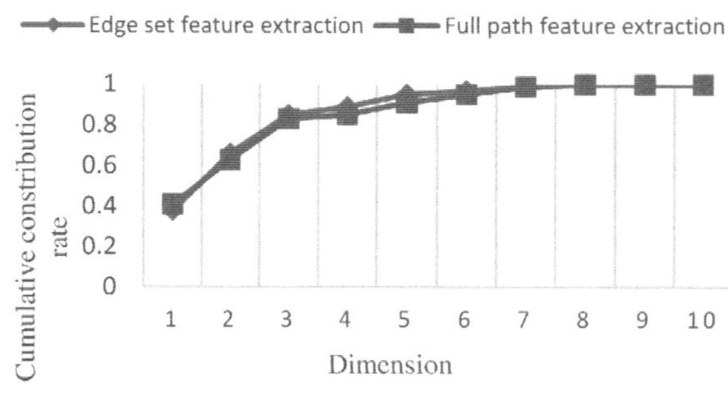

Figure 2. full path feature extraction and edge set feature extraction

From the analysis of the experimental results, the initial dimension of the full path method for representing XML documents is 49. After PCA dimensionality reduction, when the dimension is 4, the cumulative contribution rate reaches 88.7%. This result shows that when the dimension is greater than or equal to 4, the initial XML document information can be completely represented; When using an edge set to represent an XML document, the initial dimension is 62. When the dimension is 4, the cumulative contribution rate reaches 85.6%. This result shows that when the dimension is greater than or equal to 4, the initial XML document information can be completely represented. The experiment further proves that the method of using a full path to represent XML documents can better describe XML documents. After PCA dimensionality reduction, the XML documents can be described with a small number of features, and the effective dimensionality reduction is realized, which paves the way for the subsequent classification or clustering of XML documents.

5. Conclusion

This paper proposes a PCA-based feature vector extraction method for XML documents, including full path feature vector extraction and edge set vector extraction for power grid data. From the experimental results, we can know that the full path feature vector extraction method can well describe the XML information, effectively reduce the dimension, solve the problem of distributed object operation measurement in the power grid, and establish a good foundation for further research on the automatic classification and clustering of XML. It also solves the problem of analyzing distributed object operation measurement data for CSG. However, XML documents are a combination of structure and content, and it is necessary to combine the specific situation of the grid data. Simply studying the structure or content can not fully represent the information of XML documents, so the feature extraction method needs to be further improved.

Acknowledgments

China Southern Power Grid science and technology project（GDKJXM20210063）

References

[1] B. Hazela, V. Singh, K. K. Agarwal, P. Singh, and A. Kumar, "A review on elaborative approach for the transmutation of XML entities for recommendation based system," 2017 International Conference on Intelligent Sustainable Systems (ICISS), 2017, pp. 1092-1095,

[2] HUANG Hua, DAI Jiangpeng, WANY Yi, et al. Graph database based construction and network topology of CIM/E for power grid[J]. Automation of Electric Power Systems, 2019, 43(22): 122-129(in Chinese).

[3] LI Guangye, LI Wei, WANG Lixia, et al. Research on grid topology search engine based on graph database[J]. Information Technology, 2019, 43(5)： 155-158(in Chinese).
[4] Yousuke Watanabe, Hidetaka Kamigaito, Haruo Yokota. Similarity search for office XML documents based on style and structure data[J]. International Journal of Web Information Systems,2013,9(2).
[5] Zhen Zhao, Zongmin Ma, Li Yan. An Efficient Classification of Fuzzy XML Documents Based on Kernel ELM[J]. Information Systems Frontiers,2019,23.
[6] Yue Zhao, Ye Yuan, Guoren Wang,Amaury Lendasse. Keyword Search over Probabilistic XML Documents Based on Node Classification[J]. Mathematical Problems in Engineering, 2015,2015(Pt.9).

Research on Transmission Characteristics of 5G Signal for Indoor Substation Based on Ray Tracing

Chengbin Huang[1], Wei Huang[2], Hua Jiang[1], Jia Li[1], Li cong[1], Xiaoyan Tian[3,4]

[1]State Grid Jilin Province Electric Power Company Information Community Company, Changchun 130000

[2]State Grid Changchun Power Company, Changchun 130021

[3]Department of Electronics and Communication Engineering, North China Electric Power University, Baoding 071003

[4]Any letter should be addressed to the author Xiaoyan Tian, 2952543151@qq.com

Abstract: The 5G networks can be used to enhance the transmission ability of substation power services. However, because of the complex and diverse environment of substations, such as the presence of a large number of obstacles of different materials and electrical equipment indoors, we need to conduct wireless coverage modeling and transmission quality assessment. In this paper, the 5G signal transmission characteristics in the indoor environment of substations for the established indoor environment and equipment layout are analyzed using methods such as road loss statistical analysis and 3D ray tracing. It can quickly evaluate the 5G indoor coverage quality of substations and provide technical support for planning a reasonable and efficient 5G network.

1. Introduction

A large number of electrical equipment with different specifications in the substation room can cause strong reflection and scattering and produce obvious multipath effects. Moreover, the electromagnetic environment of substations is very complex[1]. Wireless sensor units placed in substations are subject to varying degrees of electromagnetic interference. This effect can cause communication errors or packet loss and may cause communication interruptions[2]. Therefore, the signal coverage and transmission quality must be studied using fading statistical models, ray-tracing techniques, etc.

The literature [3] presents a proposed maximum inner product (MIP) method to extract path loss parameters for 110 kV and 220 kV primary substations. The maximum inner product of the test data is calculated to obtain the loss index n and the standard deviation σ, and then the path loss model is established. In literature [4], a virtual multi-antenna array channel measurement system for 5G millimeter wave band 28GHz is used in high voltage substations to perform channel measurements for LOS/NLOS scenarios, respectively. In this paper, the path loss values of direct, reflection, diffraction, and transmission are calculated for each path by ray tracing technique and path loss equations.

Ray tracing is an important technique for studying the propagation characteristics of radio waves in mobile communications and can be used to identify all possible ray paths between transceivers and transmitters in multipath channels. For obstacle bypassing, GO theory is complemented by introducing bypassing rays, namely GTD (Geometric Theory of Diffraction) and UTD (Uniform Theory of Diffraction)[5]. However, unlike the public network environment, the indoor environment of substations is not only affected by doors, windows, and walls, but also by the presence of electrical devices. So it is necessary to conduct research on indoor 5G signal coverage in substations.

For the indoor environment of the substation, this paper investigates the analysis and calculation method of 5G signal loss and coverage quality using a combination of a statistical model and 3D ray tracing. Considering the indoor electrical equipment with different specifications in the substation, firstly, Line of Sight (LOS)/Non-Line of Sight (NLOS) classification is performed for the indoor 2D environment, and secondly, indoor 3D channel evaluation and wireless network coverage prediction are performed based on ray tracing model, and then suggestions for indoor network coverage planning in the substation are given.

2. Ray Tracing Path Analysis Method

There are two methods of implementing ray-tracing technology: the emission ray method based on the forward algorithm and the image method based on the reverse algorithm[6]. This paper focuses on the mirror method based on the reverse algorithm, using the mirror principle for ray path determination. The algorithm is based on the mirroring theory, which builds a virtual source point hierarchy tree by the forward search to establish visible polygons, prisms, and field points, and reverse search to establish ray propagation paths to achieve the prediction of electromagnetic wave field intensity in a three-dimensional environment[7].

For the direct path, it can be obtained by directly connecting the lines based on the coordinate information of the sending and receiving ends. For the reflected path, the predicted reflected path is transformed into a spatial geometry problem according to the mirroring principle of the inverse algorithm. The specific process is as follows.

1) Arrange and combine all possible path combinations according to the number of reflections needed. 2) For any path arrangement combination case, determine the transmitting point about the first reflective plane mirror point, the second reflective plane mirror point, and so on. 3) Connect the last reflective plane mirror point to the receiving point. At the same time, connect this reflective point with the mirror points of the previous step one by one to obtain all reflective surface points. 4) Determine whether each reflective point is in the plane, and if a point is not in the plane, there is no reflective path that satisfies the reflective count for this combination case. 5) Connect all reflective points that meet the emission point, all reflective points, and the receiving point to obtain the reflective path.

3. Substation indoor wireless network channel estimation

3.1 Channel modeling based on ray-tracing techniques

The environment in which wireless communication in substation systems is located is complex, and there are many secondary equipment and cabinets indoors. When signals propagate in substations, they are inevitably affected by the reflections, scattering, and refraction of this equipment and buildings, resulting in large-scale and small-scale fading. By building a typical structural model of the substation, the actual substation structure is abstractly constructed to comprehensively describe the location of different large-scale equipment and buildings in the substation.

Figure 1 shows the 3D layout of the substation interior, and the cube figure indicates the electrical equipment. As shown in Figure 2, in the indoor layout of the substation, the white part indicates the LOS area, the gray part indicates the NLOS area, the "+" indicates the emitter location, and the red circle indicates the user (UE) location, and the solid black line indicates the ray tracing path.

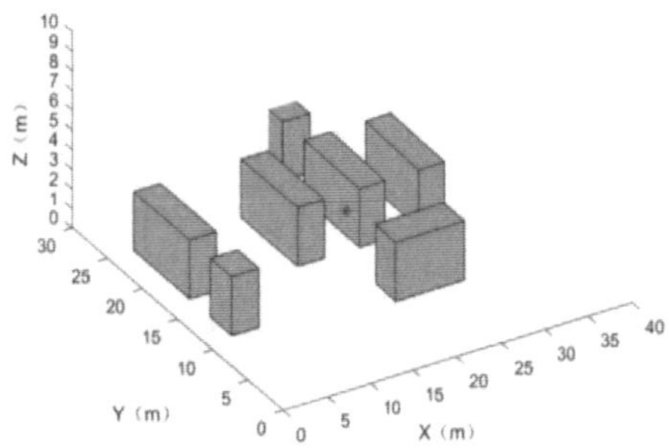

Figure 1 Indoor 3D view of the substation

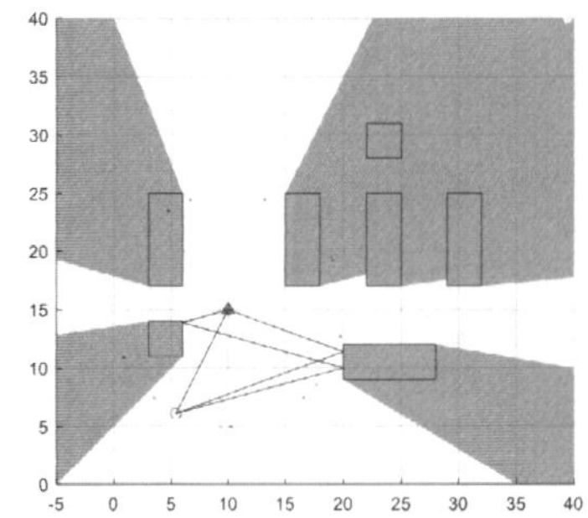

Figure 2 Indoor LOS/NLOS division and ray tracing

The transmitting and receiving points of rays are in an indoor environment. And passing through the electrical equipment, the signal transmission from the transmitter to the user may occur many times of reflection, diffraction, and transmission. Considering the three path cases, the specific calculation of network coverage based on ray tracing technology is as follows.

1） Direct path

The direct path can be considered as signal propagation in the LOS region, so the propagation loss of each direct signal is calculated $Lfs(f,d_{3Di})(dB)$ [8]:

$$Lfs(f,d_{3Di}) = \begin{cases} Lfs1(f,d_{3Di}) \\ Lfs2(f,d_{3Di}) \end{cases} \tag{1}$$

Among them,

$$Lfs1(f,d_{3Di}) = 20\lg(f) + 22\lg(d_{3Di}) + 28.0 \quad (10m \le d_{2Di} \le d'_{BP}) \tag{2}$$

$$Lfs2(f,d_{3Di}) = 20\lg(f) + 40\lg(d_{3Di}) + 28.0 - 9\lg(d'^2_{BP} + (h_{BS} - h_{UT})^2) \quad (d'_{BP} \le d_{2Di} \le 5\text{km}) \tag{3}$$

Where $Lfs(f,d_{3Di})$ denotes the ith path loss value of the ray propagating in the LOS region, and f(MHz) is the signal frequency, d_{2Di}(Km) is the horizontal transmission distance between the base station and the mobile user of the ith ray in the Los area, d_{3Di}(Km) is the transmission distance between the base station and the mobile user of the ith ray in the Los area,h_{BS} is the effective height of the base station antenna, and h_{UT} is the effective height of mobile user antenna.

2） Reflection, bypass, transmission path

Due to the presence of electrical equipment, propagation path loss is the total sum of attenuation for reflection, diffraction, and transmission. $Lrtd(f,\theta_{ji})(dB)$:

$$Lrtd(f,\theta_{ji}) = \sum_j Lr(f,\theta_{ji}) + \sum_k Lt(f,\varphi_{kl}) + \sum_m Ld(f,\omega_{ml}) \tag{4}$$

Where $Lrtd(f,\theta_{ji})$ is the total attenuation of reflection, bypassing, and transmission caused by ith ray passing through electrical equipment. $\sum_j Lr(f,\theta_{ji})$ is the attenuation of the ith ray after j times reflection, and θ_{ji} is the angle of incidence. $\sum_k Lt(f,\varphi_{kl})$ is the attenuation of the ith ray after k times transmission, and φ_{kl} is the angle of incidence. $\sum_m Ld(f,\omega_{ml})$ is the attenuation of the ith ray after m times bypassing, the ω_{ml} is the angle of incidence. If the attenuation of reflection, transmission, and diffraction of the ray reaches the preset value, the field strength of the ray is considered to be minimized at this time and can be ignored, and ray tracing is stopped.

3)Total propagation loss per ray $Los(f,d_{3Di})(dB)$:

$$Los(f,d_{3Di}) = Lfs(f,d_{3Di}) + Lrtd(f,\theta_{ji}) \tag{5}$$

4) Calculate the average loss of the propagation path of the N rays arriving at the receiving point $PL(dB)$

$$PL = 10\lg\left(\frac{1}{N}\sum_{i=1}^{N}\left(10^{\frac{Los(f,d_{3Di})}{10}}\right)\right) \tag{6}$$

5) Calculate the wireless signal strength of a single transmitting antenna at the receiving point $Ps(dBm)$:

$$Ps = Pt - PL + Gt + Gr \tag{7}$$

where Pt is the antenna gain of the transmitting antenna port, andGt is the antenna gain of the transmitting antenna, andGr is the antenna gain of the receiving antenna.

6) Calculate the total signal strength of all antennas in the room at the reception point$P(dBm)$:

$$P = 10lg\sum 10^{\frac{Ps}{10}} \tag{8}$$

3.2 Simulation of indoor wireless network coverage in substations

The band planning of the power wireless network and the band planning of industrial IoT are not exactly specified, and the potential bands are 230MHz, 700MHz, 1.8GHz, 2.4GHz, 3.35GHz, and 5.8GHz [9]. Also, considering the different materials of different electrical equipment in the substation, the impact on the signal transmission will be different, so the simulation parameters need to be defined in advance, as shown in Table 1.

Table 1 Parameter presets.

Parameters	Preset value
Signal Frequency f/GHz	2.4, 3.6
Antenna Gain Gr, Gt	50
Transmitter Power Pt/dBm	15
Electrical equipment dielectric constantε	1~10
Height of electrical equipment /m	3
Transmitter height /m	5

Based on the Matlab simulation platform, the signal strength coverage maps are obtained for different transmitting point locations and frequencies. As shown in Figure 3 and Figure 4, the two plots represent the wireless signal strength coverage map when the signal frequency is taken as 3.6GHz. It can be seen that the signal strength is greater in part closer to the Tx. At the same time, Figure 5 shows signal strength coverage when the frequency is 2.4 GHz.

Also, the existence of different dielectric constants of electrical devices, which are made of different materials, leads to different degrees of attenuation of the signal. Figure 6 shows that without changing the position of Tx and ensuring that Rx is in the same position, The signal frequency f is transformed into the wireless signal strength curve at 1.8GHz, 2.4GHz, 3.35Ghz, 3.6Ghz, and 5.8GHz, respectively. And the farther away from Tx, the signal propagation path through the electrical equipment, the greater the signal attenuation and the smaller the signal strength.

As a result, the best coverage of the wireless network can be determined by setting Tx at different locations, and wireless network planning can be carried out to meet the communication performance requirements of the substation.

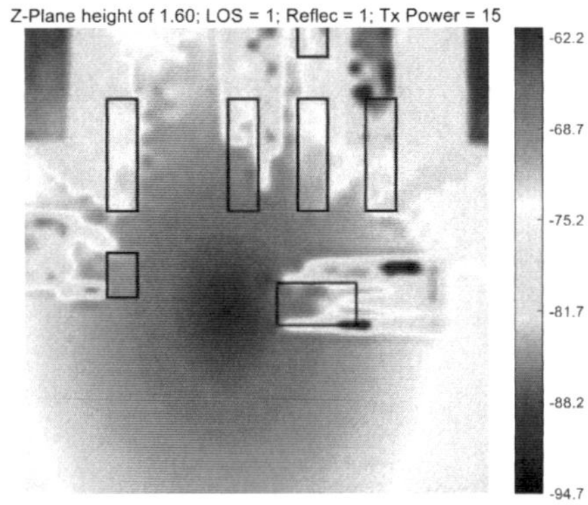

Figure 3 Tx at the center (f=3.6GHz)

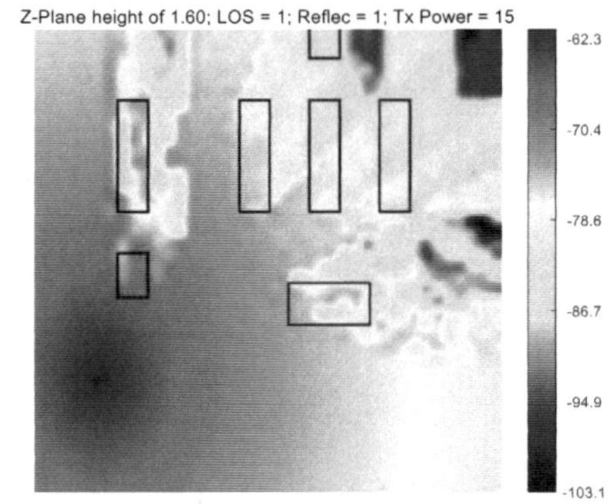

Figure 4 Tx at the lower left (f=3.6GHz)

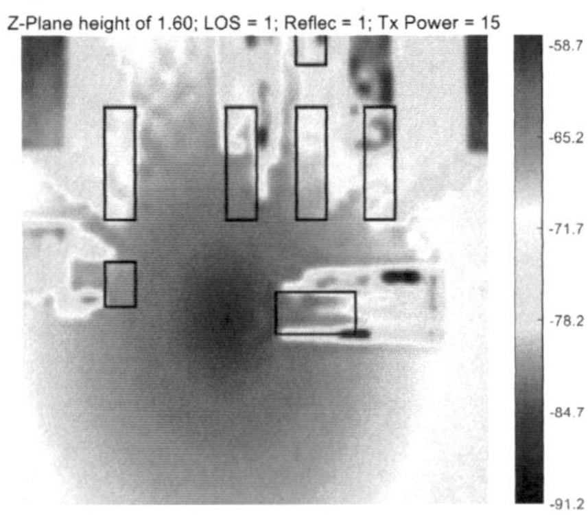

Figure 5 Tx at the center (f=2.4GHz)

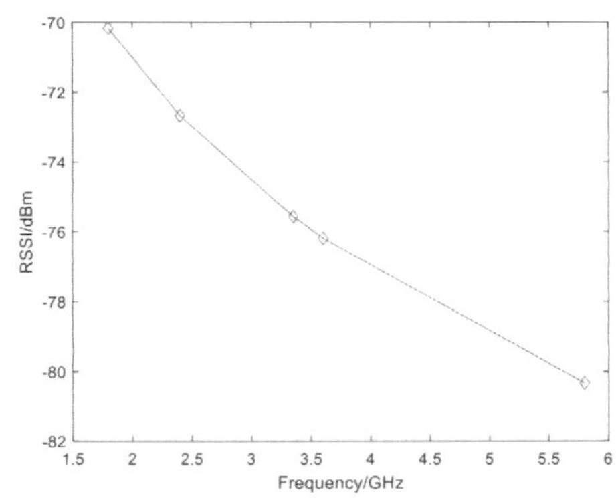

Figure 6 Curve of Rx signal strength versus frequency at the same position

4. Conclusion

5G indoor wireless network coverage is one of the key points in the future 5G development. In this paper, we have studied the wireless network coverage of substations based on ray tracing technology, and from the simulation results, we can predict the wireless signal coverage of substation start-up better. In addition to considering the dielectric constants of different electrical devices, the difficulty of this paper is the need to consider the complex electromagnetic environment of substations, and further work is needed to refine the difficulty and guide the application in practice.

Acknowledgment

The paper is supported by the 2022 science and technology project of State Grid Jilin Province Electric Power Company with the name "Research on 5G Transmission Characteristics and Access Ability of Substation (2022-35)".

References

[1] Jianfei Ji, Yifei Yang, Yubo Yuan and Lihui Wang, 2014. Experimental analysis of EMC immunity for field installed intelligent equipment of intelligent substation[C]. In: 2014 China International Conference on Electricity Distribution (CICED). Shen Zhen, China. vol. 41, no. 3, pp. 998-1007.

[2] Qiang Lee , Hongxin Zhang , Duan Li , Yang Lu, Jianqi Li, Yinghua Lu ,11 May 2017. The modeling about path loss of wireless channel in primary substation [C].In: 2016 2nd *IEEE International Conference on Computer and Communications,* Cheng Du, Chian.

[3] Qiang Li, Hongxin Zhang, Tianyi Zheng, Xueli Wang, Yinghua Lv, (April 2019) Maximum inner product method for extracting the path loss parameters in primary transformer substations[J]. *China Communications*, Volume:16, Issue: 4, Page(s): 120-129.

[4] Fu Z, Cui H, Geng S, et al, 2019. 5GMillimeter Wave Channel Modeling and Simulations for a Hight-Voltage Substation [C]. *2019 IEEE Sustainable Power and Energy Conference (iSPEC)*. IEEE, 2019: 1822-1826.

[5] Bo Lv. 2015. Research on multi-floor environment ray tracing algorithm and the influence of building geometry modeling on ray tracing algorithm [D]. Xi'an University of Electronic Science and Technology.

[6] Ling H, Chou R C, Lee S W. (1989) Shooting and bouncing rays: Calculating the RCS of an arbitrarily shaped cavity[J].*IEEE Antennas and Propagation*, 37(2): 194-205.

[7] Zhaokui Li, Lidong Wang, Cong Li, Xinshao Zhou. (2012) An improved ray tracing algorithm using mirror theory[J]. *Journal of Wuhan University*, 37(7): 784-788.

[8] 2019.Study on channel model for frequencies from 0.5 to 100 GHz (Release 14): 3GPP TR 38.900 [S/OL]. [2019-01-06].ftp://3gpp.org/specs.

[9] Guangyun Zhu. 2021, Wireless channel measurement and characterization in substation scenarios [D]. Beijing Jiaotong University.

Electric field characteristics of shared towers and electric field distribution characteristics near 5G base stations

Huan Huang[1,a], Yan Hong Xiao[1], Bin Li[2], Lei Liu[2], Siwu Yu[1], Xianyin Mao[1], Chi Peng[3], Hao Du[3]

[1] Electric Power Research Institute Guizhou Power Grid Company., Ltd Guiyang, China

[2] State Key Laboratory of HVDC, Electric Power Research Institute, China Southern Power Grid line 3-City Guang Zhou, China

[3] Guizhou Power Grid Company Ltd Guiyang, China

[a]email: zjq0709@yeah.net

Abstract. With the continuous promotion of domestic 4G network construction and the gradual arrival of 5G networks, the requirements of mobile communication networks on capacity and coverage are constantly improving, and the number of communication base stations is also increasing. Therefore, the "shared tower" with the function of a communication base station added to the existing high-voltage transmission line tower is becoming a new resource-sharing mode. There is no systematic study on the installation location of the shared tower base station on the high-voltage tower, which is not conducive to the application and promotion of "shared tower". In this paper, the finite element simulation model of the tower installed with the base station is built. The electric field distribution near the tower with the base station, the electric field distribution characteristics under different base station installation quantities and different positions of the base station are mainly studied, and the electric field distribution near the surface of the base station is obtained. Combined with the electrical safety distance limit of communication equipment and iron tower, the influence of the installation location and quantity of the base station on the electric field near the base station is evaluated.

1. Introduction

With the development of 5G mobile communication technology, the construction of 5G communication base stations will be spread on a large scale. In the future, the number of 5G base stations will be several times that of 4G base stations. The in-depth development of the shared tower mode will greatly support the investment and construction of 5G mobile communication base stations. At present, there is much research on the electric field distribution near the high-voltage tower. In literature 1, the power frequency electric field distribution on the ground around the 500 kV transmission line is studied, and the influence of the tower on the nearby electric field environment is analyzed [1]. Li Y et al. and Li C et al. [2] [3] have carried out electric field simulation analysis of live working on 800kV transmission lines and studied the distribution of electric field on the surface of the human body working on high-voltage transmission lines [2-3]. Literature 4 studies the electric field distribution of 110kV composite towers [4], and literature 5 studies the electric field distribution near typical towers of 500kV double circuit lines on the same tower, providing a reference for live working [5]. Yang L et al. [6] studied the electric field distribution of 750kV wine cup type composite cross-arm towers under different design schemes, providing guidance for tower design [6]. Yang L et al. [7] studied the electric field distribution around the corner tower of 750kV transmission line [7]. In Literature 8, the electric field simulation of the metal tools of the ±1100kV transmission line is carried out to study the corona characteristics [8]. Literature 9 studies the electric field distribution of 110kV double-circuit AC T-connector towers, which provides a reference for tower design [9]. In Literature 10, the electric field distribution in the vicinity of buildings under ultra-high voltage transmission lines is studied [10]. In [11], the safety distance of UAV inspection for ±500kV DC transmission line is studied [11]. Wang L et al. [12] studied the influence of line patrol helicopters on the electric field distribution of 500kV AC transmission lines [12]. Most of the existing research uses the method of electric field simulation to study the electric field distribution of the tower. The transmission tower with a base station needs to consider the electric field distortion characteristics of the base station to better set the installation location to meet the insulation requirements.

In this paper, the electric field simulation model of the tower with the base station is built, the electric field distribution near the tower with the base station is studied, and the influence of different mounting positions and numbers of the base station on the electric field distribution is analyzed, and the electric field distribution near the surface of the base station is obtained, which provides theoretical guidance for the design of "shared tower".

2. Simulation model establishment and simulation setup of the tower with a base station

The actual application field photos of the shared tower are shown in Figure 1. The application of the shared tower has not been widely promoted, and the actual field measurement is difficult to carry out. Therefore, this paper chooses to use mature commercial electric field simulation software to study the electric field distribution of the shared tower, so as to determine the erection method and location of the base station. In this paper, a 110 kV AC shared tower is selected for finite element modeling, and the established model is shown in Figure 2. The tower height is 23 m, and the height of phase A, B, and C conductors above the ground is 18 m, 15.5 m, and 11 m, respectively. In the simulation, the height of the base station above the ground is set to 10.5 m, 9 m, and 7.68 m respectively.

The model ignores the parts such as insulators and fittings that have little influence on this analysis. The electric field distribution is analyzed by COMSOL software, the mesh generation model is shown in Figure 3. The main material of the tower was set as the built-in material "Soft Iron with Losses", while the wire was the built-in material "Copper". The calculation domain is set to an air domain eight times the size of the model.

Figure 1 The actual application field photos of the shared tower

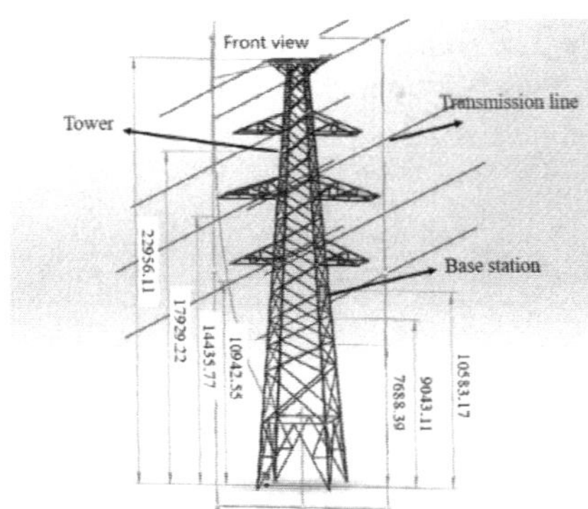

Figure 2 110kV Shared tower model

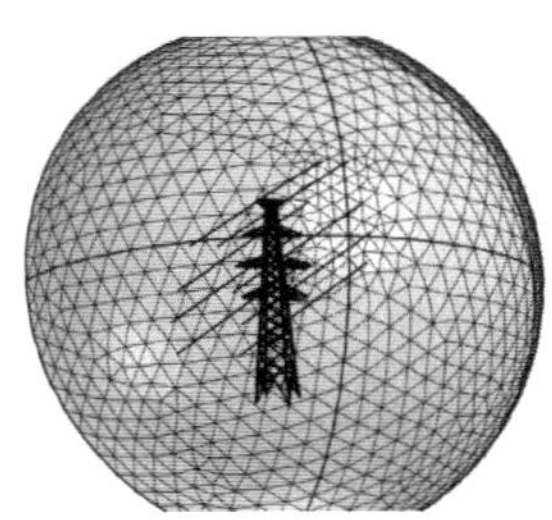

Figure 3 Finite element modeling diagram of electric field distribution analysis of transmission tower

The electric field simulation study of the above model is a typical electrostatic field problem, and its satisfying equations and boundary conditions are shown in Equations 1 and 2.

$$\nabla^2\varphi = \frac{\partial^2\varphi}{\partial x^2} + \frac{\partial^2\varphi}{\partial y^2} + \frac{\partial^2\varphi}{\partial z^2} = 0 \quad (1)$$

$$\left.\begin{aligned} \varphi\big|_{\Gamma_1} &= f_1(x,y,z) \\ \frac{\partial\varphi}{\partial n}\bigg|_{\Gamma_2} &= f_2(x,y,z) \end{aligned}\right\} \tag{2}$$

Where Γ1 is the boundary of the three-phase conductor and Γ2 is the boundary of the calculation domain.

3. Simulation results and analysis of electric field distribution

3.1. The situation with different phase sequences

In this paper, 110kV AC Double Circuit Tower is used as the simulation model, so the influence of different phase sequences of double circuit conductors on the electric field distribution should be considered. There are 6 kinds of phase sequence conditions, and the electric field strength of each phase sequence condition can refer to table 6-1 below for the electric field strength at 1.5m above the ground.

TABLE 1. Electric field intensity at 1.5m above the ground

Phase sequence	Maximum field strength / (kV/m)	Phase sequence	Maximum field strength / (kV/m)
abc-abc	1.236	abc-bca	0.643
abc-acb	0.825	abc-cab	0.670
abc-bac	1.208	abc-cba	0.517

It can be seen from the table that the reverse phase sequence (abc-cba) has the smallest ground field strength and the conditions are ideal, while the in-phase sequence (abc-abc) has the largest ground field strength and the most severe conditions. Considering the influence on the electric field distribution after mounting the base station, this paper studies the electric field distribution near the base station under the different phase sequences. Figure 4 is a detailed diagram of the electric field intensity near the communication base station under the condition of phase sequence abc-abc. At this time, the electric field intensity near the base station is 25000V/m, which is the case with the largest modulus. It is consistent with the rules in Table 1. Therefore, in the subsequent simulation research, this paper is carried out under the condition of phase sequence abc-abc.

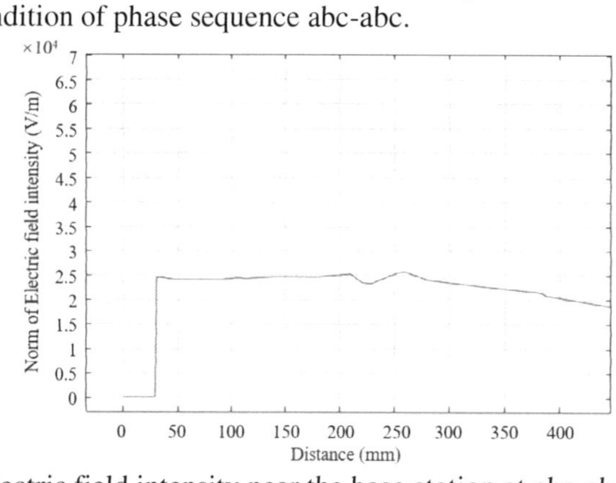

Figure 4 the electric field intensity near the base station at abc-abc phase-sequence

3.2. The situation with different mounts

Figures 5 and 6 are the norm of electric field intensity with one and two (symmetric) base stations mounted at the intermediate position defined above (about 9m above the ground). Obviously, the potential contour of the case of loading two base stations in the same section is more symmetrical than that of loading one base station, and because of the shielding effect of the shared tower, the distribution of electric field lines between the tower materials is sparse, and the potential value is almost is 0.

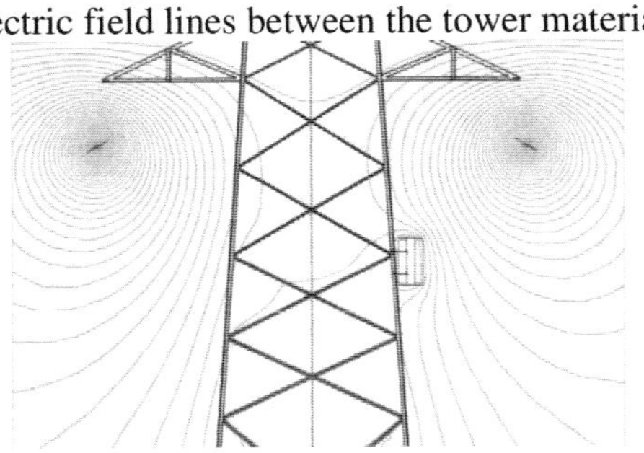

Figure 5 The potential contour of the tower with one base station

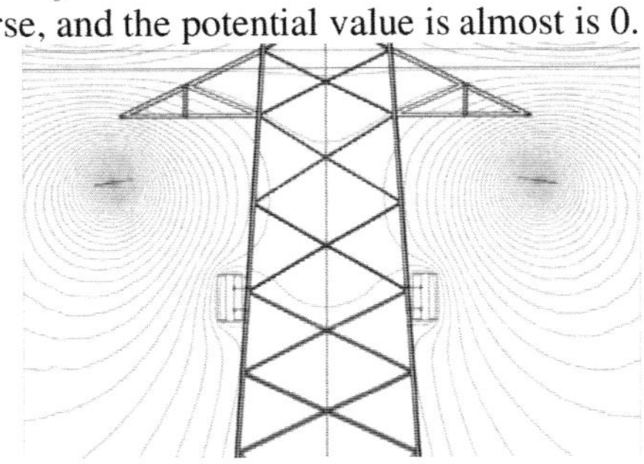

Figure 6 The potential contour of the tower with two base stations

3.3. The situation under different mount locations

Figure 7 is the potential contour map at a higher mounting position (10.5m above the ground), and Figure 8 is the potential contour map at a lower mounting position (7.68m above the ground). Comparing the potential contour map in the middle of Figure 5, it can be seen that the base station is closer to the wire, the potential contour near the base station is denser, the base station is farther from the wire, and the potential contour near the base station is sparser.

Figures 9 and 10 show the distribution of the electric field between the base station and the C-phase (the phase closest to the base station) conductors when a base station is loaded at higher and lower positions, respectively. The magnitude of the electric field strength is 40,000 V/m at the upper position and 20,000 V/m at the lower position. The results show that the distance between the base station and the wire leads to a severe situation of the electric field strength, but due to the increased height, it may be beneficial to the signal propagation of the base station under certain conditions. The greater distance from the wire results in a numerically lower electric field strength, but the reduction in height may be detrimental to the signal propagation of the base station.

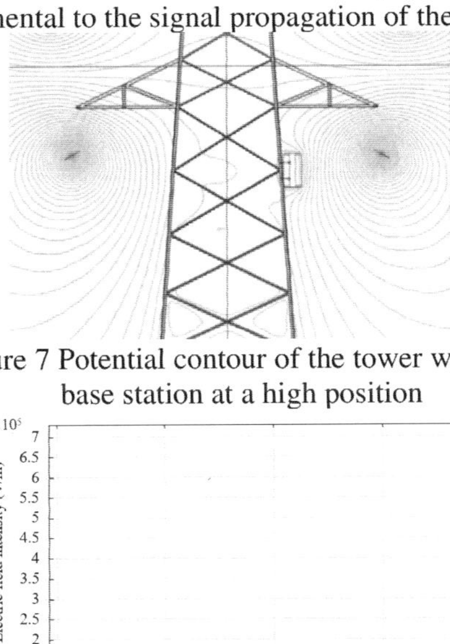

Figure 7 Potential contour of the tower with one base station at a high position

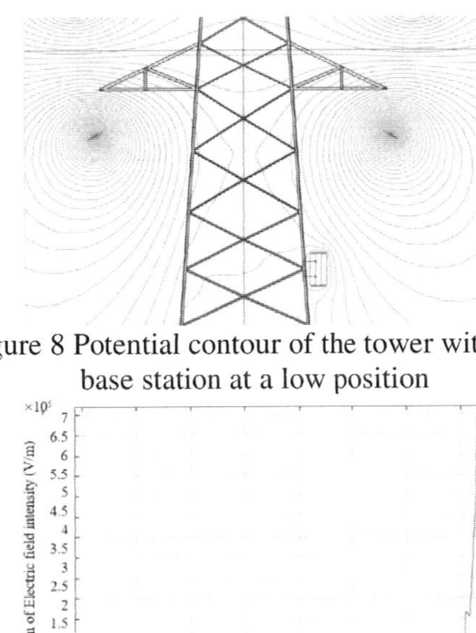

Figure 8 Potential contour of the tower with one base station at a low position

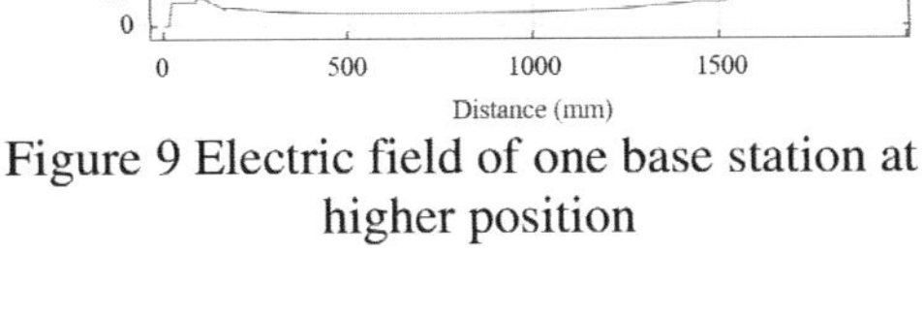

Figure 9 Electric field of one base station at a higher position

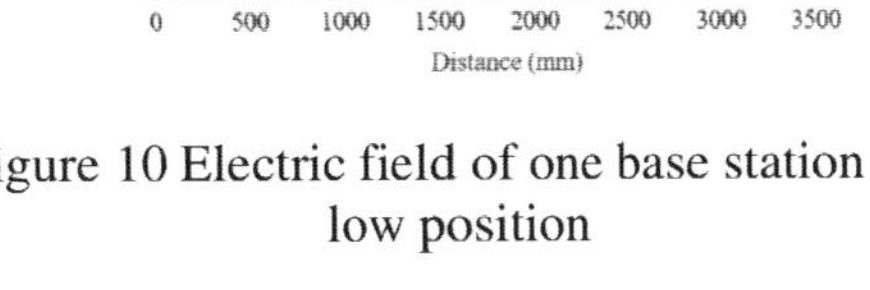

Figure 10 Electric field of one base station at a low position

3.4. The electric field near the base station surface

Since the three-dimensional structure of the base station is similar to that of a cube, the electric field strength experienced at the intersection of the cube's surface and the center of the wider surface will be different. This paper extracts the electric field strength near the surface of the communication base station, as shown in Figure 11 below. In Figure 11, the abscissa is the straight line parallel to the height of the base station (the height of the base station is 795.0mm) on the base station surface starting from the vicinity of the lower surface boundary and ending at the vicinity of the upper surface boundary, and the origin is near the edge of the lower surface of the base station. The electric field strength at the corners of the upper and lower surfaces of the base station is higher than that at the center of the base station surface, while the electric field strength at the corners of the lower surface is smaller than that at the corners of the upper surface. This is because the lower surface is farther from the wire than the upper surface and is not facing due to the direction of the wire. It can be seen that for the communication base station installed on the shared iron tower, more attention should be paid to the evaluation of the electric field strength at the corners of the base station.

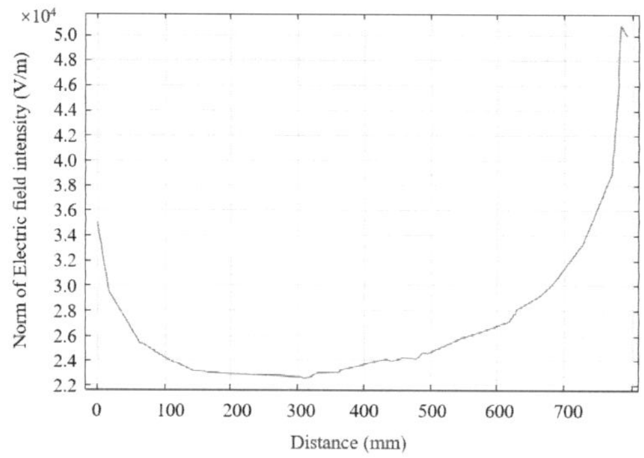

Figure 11 Electric field strength near the surface of the base station

4. Conclusions

In this paper, a typically shared tower is selected to conduct finite element electric field simulation analysis, focusing on the analysis of the influence of multiple factors such as the mounting position of the communication equipment, the number of mounting, and the influence of different phase sequences on the conductors mounted on the tower on the electric field distribution and electric field strength, and conclusion are as below:

1. Under different phase sequence conditions, the same-phase sequence (abc-abc) has the largest field strength to the ground, and it is also the case where the field strength near the base station is the largest for each phase sequence condition of the communication base station installed on the shared tower.
2. If different numbers of communication base stations are mounted on the same side, or the communication base stations are mounted asymmetrically, the electric field near the base station will not change greatly due to the asymmetric mounting and the symmetrical mounting. This is mainly because the main material of the tower has a shielding effect.
3. When the base station is mounted at different heights, the electric field near the base station also changes accordingly. Generally speaking, the higher the mounting position of the base station, the closer it is to the wire and the greater the electric field intensity around the base station. It is necessary to weigh the signal coverage and the influence of the electric field on itself according to the actual situation and the parameters of the communication base station to determine the better one. Mount height.
4. The electric field intensity at the corners of the upper and lower surfaces is higher than that at the center of the base station surface, and the electric field intensity at the corners of the upper surface is greater than that of the corners of the lower surface. Therefore, more attention should be paid to the evaluation of the electric field strength at the corners of the base station, especially the corners of the upper surface (the surface closest to the wire).

Acknowledgments
This work is supported by the Science and Technology Project of China Southern Power Grid Co., Ltd. (066600KK52190068).

References
[1] Li Y, Fan Y, Xu L. (2013) Simulation Analysis on Power Frequency Electric Field at Ground Surface Near EHV Power Transmission Tower. Power System Technology, 37(03): 782-787.
[2] Liu C, Kong X. (2019) Simulation Analysis on Body Surface Electric Field of Live Operating Personnel on ±800 kV Power Transmission Lines. Guangdong Electric Power, 32(07): 142-148.
[3] Pan J, Luo R, Pan Q, (2017) Simulation Analysis of Electric Field on Live Working of ±800 kV Transmission Line. High Voltage Apparatus, 53(11): 197-202.
[4] Wang C, Yue P, Zheng N, Jiang W, Xia Z, Chu J, Yang J, Lan L. (2015) Electrostatic Field Simulation of 110 kV Composite Tower. Insulating Materials, 48(04):40-44+51.
[5] Chen B, Lan L, Wang Y, Long G. (2018) Simulation analysis of electric field on entering equal potential of 500 kV double circuit transmission line towers for live working. Engineering Journal of Wuhan University, 51(05): 432-436+464.
[6] Yang L, Tan H, Wang J, Peng Z. (2009) Electric Field Computation and Corona Test for Composite Cross-Arms in 750kV Cup Type Tower. Power System Technology, 42(06):2009-2015.
[7] He X, Zhu Y, Zhang M, Kong M. (2019) Simulation on Electric Field of Operators Around Corner Tower of 750 kV Transmission Line. Guangdong Electric Power, 32(07): 106-111.
[8] Mao Y, Deng T, Wei J, Niu H, Zhang X. (2019) Electric Field Characteristic and Corona Performance of ±1 100 kV Power Fittings for Transmission Lines. High Voltage Engineering, 45(04): 1137-1145.
[9] Yang W, Yang F. (2020) Simulation analysis of the electric field of 110 kV double circuit AC T-tower. Journal of Electric Power Science and Technology, 35(05): 138-143.
[10] Wang Q, Luo Y, Yang F, Chu X. (2013) Three-dimensional Simulation of Electric Field Around Building Under EHV Transmission Lines. High Voltage Apparatus, 49(10):1-6.
[11] Liu Z, Du Y, Chen Y, Ma J, Wu X, Yao J. (2019) Simulation and Experiment on the Safety Distance of Typical ±500 kV DC Transmission Lines and Towers for UAV Inspection. High Voltage Engineering, 45(02):426-432.
[12] Wang L, Yu D, Chen F, Wan S, Bian X, Chen L. (2013) Simulation of Effect on 500 kV AC Transmission Line Electric Field of Patrol Helicopters. High Voltage Engineering, 39(01): 201-209.

Journal of Physics: Conference Series **2384** (2022) 012031 doi:10.1088/1742-6596/2384/1/012031

Time Domain Measurement System for Electric Field of Radio Interference Generated by Corona Discharge

Huan Huang [1,a], Yan Hong Xiao[1], Bin Li[2], Lei Liu[2], Siwu Yu[1], Xianyin Mao[1], Chi Peng[3], Hao Du[3]

[1] Electric Power Research Institute Guizhou Power Grid Company., Ltd Guiyang, China

[2] State Key Laboratory of HVDC, Electric Power Research Institute, China Southern Power Grid line 3-City Guang Zhou, China

[3] Guizhou Power Grid Company Ltd Guiyang, China

[a]email: zjq0709@yeah.net

Abstract. When corona discharge occurs around the wire, it will be accompanied by the movement of space charge, and then generate corona discharge effects such as radio interference and audible noise. In order to reveal time domain characteristics of the electric field of radio interference generated by corona discharge, a set of time domain measurement systems for radio interference, including oscillator antenna, sampling resistance, and data acquisition system, is developed. Based on the characteristics of corona-generated radio interference and antenna theory, the size of the antenna, the equivalent circuit model, and the selection of sampling resistance are discussed carefully to satisfy the requirements for the measurement of corona-generated radio interference. To verify the effectiveness of the measurement system, an experimental platform with a single corona source is built, and the time domain waveforms of corona current and radio interference are simultaneously measured. Besides, the attenuation law of radio interference is obtained by using a time domain measurement system. The results show that the amplitude of the measured voltage of radio interference decreases exponentially with distance. The time domain measurement system presented in this paper can lay a solid foundation for the time-domain evaluation of the radio interference level generated by corona discharge.

1. Introduction

The radio interference induced by the corona discharge is one of the electromagnetic field environment factors during the design of the transmission lines [1]. Thus, the investigation of the characteristics and prediction method for radio interference may have a certain guiding significance.

At present, some scholars have conducted experimental and theoretical research on radio interference of transmission lines [2-6]. Due to the complexity and randomness of the corona discharge, the corona current, i.e., the source of the radio interference, cannot be directly calculated. Thus, the investigations on radio interference were mainly based on the empirical formulas obtained from measured data from actual transmission lines or corona cages, et al. The empirical formulas usually lack universality and are highly dependent on environmental conditions.

Besides, due to the complexity of corona discharge and many factors affecting the radio interference of DC lines, it is still the main technical means to study the characteristics of radio interference through experiments. However, the existing research mainly focuses on the measurement of single frequency points (0.5MHz or 1MHz) or multiple frequency points of radio interference. This measurement method belongs to the frequency domain measurement method [7-11]. The measurement results reflect the comprehensive effect of radio interference over a period of time, but it is difficult to reflect the intrinsic characteristics of radio interference [12].

In recent years, to master the characteristics of DC line radio interference sources, many scholars have conducted in-depth theoretical experiments on corona current pulse characteristics and influencing factors. Reference [13] established an accurate corona current pulse time-domain test system, and investigated the influence of different electrode structures and environmental factors. Reference [14] established the conversion relationship between corona current pulse parameters and steady-state corona current, and proposed the conversion relationship between corona cage radio interference and line radio interference. These works show that the characteristics of corona current can be fully mastered by means of time domain measurement, which can lay a foundation for obtaining accurate radio interference prediction methods. However, few scholars have studied the time-domain test system and test method of line radio interference.

This paper uses an array antenna to build a time-domain measurement system for the electric field strength of radio interference. Combined with the characteristics of corona discharge radio interference, the antenna size and structural parameters are systematically analyzed by using antenna theory, and a DC single-point discharge experimental platform is built. Based on the measurement system, the time-domain attenuation characteristics of single-point corona discharge radio interference are obtained.

2. Time-domain measurement system for radio interference

2.1 The components for the measurement system

As shown in Fig. 1, the main components of the radio interference electric field strength time-domain measurement system designed in this paper include array antenna, measurement resistance, grounding plate, voltage probe, and high-speed data acquisition system. The array antenna is the receiving antenna to convert the radio interference electric field intensity signal generated during corona discharge into a voltage signal. The time-domain antenna is connected with the measuring resistor, and the voltage at both ends of the resistor is collected by the voltage probe. The output signal of the voltage probe is collected by the high-speed data acquisition system and finally stored and displayed on the computer.

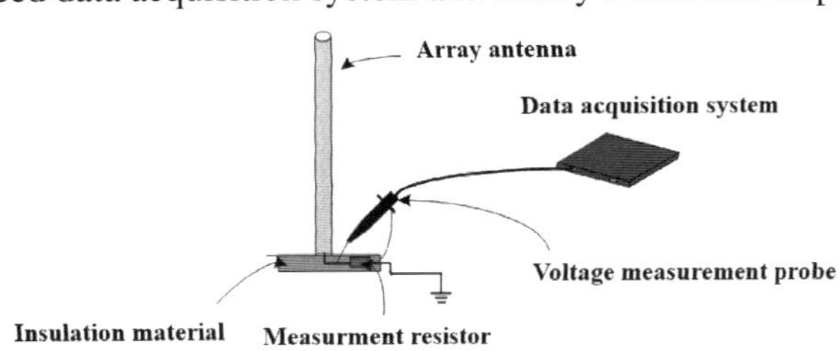

Figure 1 A schematic diagram of the connection of the time domain measurement system

Existing research shows that the maximum frequency of the line corona discharge radio interference signal is lower than 30 MHz. In order to achieve effective collection of the radio interference signal, the effective test bandwidth of the voltage probe used in the measurement system is 0-250 MHz, and the sampling frequency of the acquisition card is set to 200 MS /s.

In order to verify the effectiveness of the voltage probe and the acquisition system, a typical double pulse signal waveform [12] is generated by the signal generator and output to the matching sampling resistor. The voltage probe measures the voltage at both ends of the sampling resistor, and the measured waveform is displayed by the data acquisition card and the computer. The experimental arrangement and wiring diagram are shown in Fig. 2. The comparison between the output signal of the signal generator and the acquisition signal of the voltage probe and the acquisition card is shown in Fig. 3. It can be seen that the output waveform is consistent with the acquisition waveform, which can verify the effectiveness of the voltage probe and acquisition system.

Figure 2 Experimental arrangement for the validity of voltage probe and sampling system

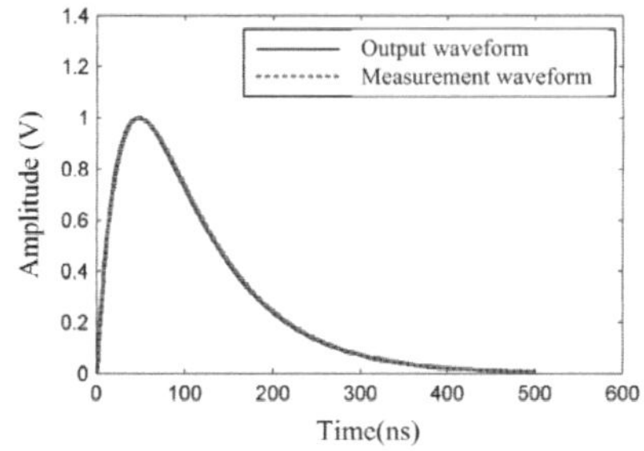

Figure 3 Comparison of the experimental results for the validity of voltage probe

2.2 Determination of array antenna size

The frequency band range of corona current generated by corona discharge is 0-30 MHz. Considering the mirror effect of the ground plate, only half of the length of the designed dipole antenna needs to be less than 50 cm; that is, $l_e = h/2 < 50$cm. Using Ansoft HFSS software, the antenna gain characteristics are simulated and analyzed to determine the length of the time domain dipole antenna. The maximum antenna gain under different lengths is shown in Table 1. From the simulation results, the antenna gain is the maximum when the length is 40 cm. The longer the antenna length is, the greater the effective height of the antenna is. Therefore, the length of the antenna is set as 40 cm.

Table 1 Antenna gain size under different antenna lengths

Length	Antenna gain
20cm	-3.52dB
25cm	-3.71dB
30cm	-3.68dB
35cm	-3.55dB
40cm	-3.38dB

2.3 Equivalent circuit model of the test system

The antenna parameters of an electrically small antenna can be considered to be basically independent of frequency. The electrically small antenna can be equivalent to a capacitor C_a independent of frequency [15], which can be expressed as

$$C_a = \frac{2\pi h \varepsilon_0}{k - 2 - \ln 4} \quad (1)$$

where, $k=2\ln(2h/a)$, a is the radius of the antenna, and ε_0 is the vacuum dielectric constant.

Fig. 4 is the equivalent circuit diagram of the antenna time domain measurement system. In the figure, $u_i(t)$ is the induced voltage of the antenna, Z_a is the impedance corresponding to the equivalent capacitance C_a of the antenna, R_l is the measured resistance, and $u_L(t)$ is the measured voltage signal. According to the induction principle of the dipole antenna, the induced voltage received by the antenna is

$$u_i = h_e E_i \quad (2)$$

$$h_e = \frac{h(\mathrm{k}-1)}{k - 2 + \ln 4} \quad (3)$$

where E_i is the component of the incident electric field in the antenna polarization direction, and h_e is the effective length of the antenna.

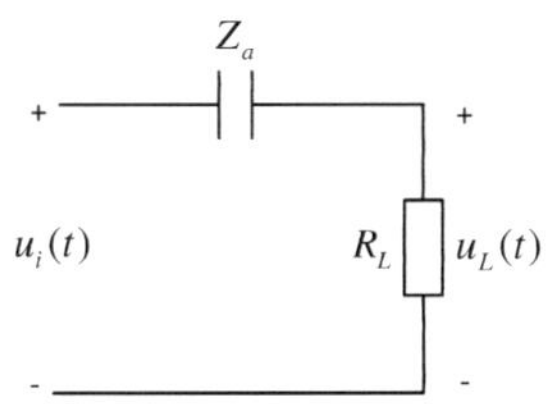

Figure 4 The circuit model of the time domain measurement system

2.4 The choice of the measurement resistance

The measurement resistance shall ensure that the measurement system shall meet the bandwidth requirements of corona discharge radio interference as much as possible and have a sufficient standing wave ratio. In addition, the measurement resistance shall ensure that the time-domain pulse signal of radio interference measurement will not overlap. The duration of a single corona current pulse is 250ns. In order to prevent pulse aliasing, the time constant of the measuring circuit should be less than or equal to the duration of the corona current pulse, i.e., $RC_a \leq 250$ns. Thus, the resistance should be $R \leq 10.7$kΩ.

In order to explore the influence of the measurement resistance on the measurement bandwidth, it is necessary to analyze the standing wave ratio of the measurement system. The standing wave ratio reflects the impedance matching degree between the antenna and the feeder [16-17]. The standing wave ratio is 2.0. The corresponding reflectivity is 8.16%. The calculation formula of the standing wave ratio is

$$SWR = \frac{1+K}{1-K} \tag{4}$$

$$K = \frac{R-r}{R+r} \tag{5}$$

where SWR is the standing wave ratio, K is the reflection coefficient, R and r are the output impedance and input impedance, respectively

In this paper, the effective bandwidth range with standing wave ratio below 2.0 under 2kΩ, 4kΩ, 6kΩ, 8kΩ, and 10kΩ resistors is analyzed by HFSS simulation software. The bandwidth range of standing wave ratio below 2.0 under different measured resistances obtained by simulation is shown in Table 2. Since there are few frequency components of corona current above 10MHz, 0-10MHz is taken as the effective bandwidth range of corona current in this paper. It can be seen that with the resistance increase, the effective bandwidth with the standing wave ratio below 2.0 becomes larger. Therefore, to ensure the measurement system's sensitivity, the resistance value of 10kΩ is selected as the measurement resistance of the measurement system.

TABLE 2 Effective bandwidth ratio under different measurement resistors

Measurement resistors	Effective bandwidth for standing wave ratio below 2.0	Effective bandwidth ratio
2kΩ	7.3MHz-10MHz	27%
4kΩ	6.1MHz-10MHz	39%
6kΩ	4.5MHz-10MHz	55%
8kΩ	2.8MHz-10MHz	72%
10kΩ	1kHz-10MHz	99%

3. Validation of time domain measurement system

In order to verify the radio interference electric field strength time-domain measurement system, a DC single point corona discharge radio interference time-domain measurement platform is built in the laboratory. The schematic diagram of the experimental platform is shown in Fig. 5. In the experiment, an insulator is used to erect a smooth conductor with a diameter of 8mm, and a small ball with a radius of 2.5 mm is set in the middle of the conductor to simulate single point corona discharge. Positive polarity high-voltage DC power is applied to the conductor, and both ends of the conductor are grounded through high-voltage coupling capacitance and matching resistance to ensure that corona current will not be distorted in the propagation process.

In the experiment, the current probe is used to collect the corona current generated by the corona discharge of the wire in real time, and simultaneously collect the time-domain voltage signal of radio interference. The frequency range obtained by the effective measurement of the current probe is 1.5k-200MHz, and the adoption rate is set to 200 MS/s.

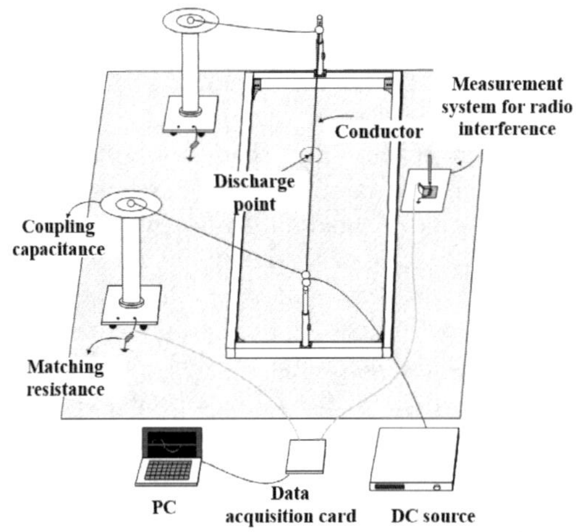

Figrue 5 Schematic diagram of time domain measurement platform for radio interference

Fig. 6 (a) shows the time domain waveforms of corona current and measured voltage measured when the voltage is 80kV, and the partial enlarged diagram of a single pulse is shown in Fig. 6 (b). It can be seen from Fig. 6 that both corona current and radio interference measurement voltage exhibit double exponential pulse characteristics and have a one-to-one correspondence in time domain waveform.

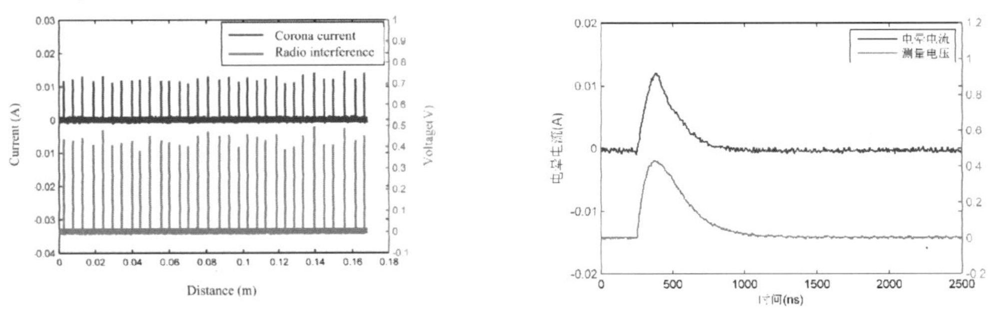

(a) Measured waveforms (b) Local enlargement waveform

Figure 6 The measured waveforms of corona current and radio interference

4.Attenuation characteristics of time domain pulse of radio interference electric field intensity

On the basis of the experimental measurement platform shown in Fig. 5, the wire position is kept unchanged, and the radio interference time-domain measurement system is placed at 1m, 1.2m, 1.4m, 1.6m and 1.8m away from the projection distance of the wire. The change law of the radio interference measurement voltage with the measurement point position are measured and statistically analyzed.

Then, the amplitude of radio interference voltage pulse signal under different measurement distances is extracted. The statistical results of radio interference voltage pulse amplitude under different voltages are shown in Fig. 7. Due to the randomness of corona discharge, the pulse amplitude has certain randomness. Fig. 7 shows the error diagram of pulse amplitude, that is, the arithmetic mean value of the measured pulse amplitude is the midpoint, and half of the length of the error line is the standard deviation of the pulse amplitude.

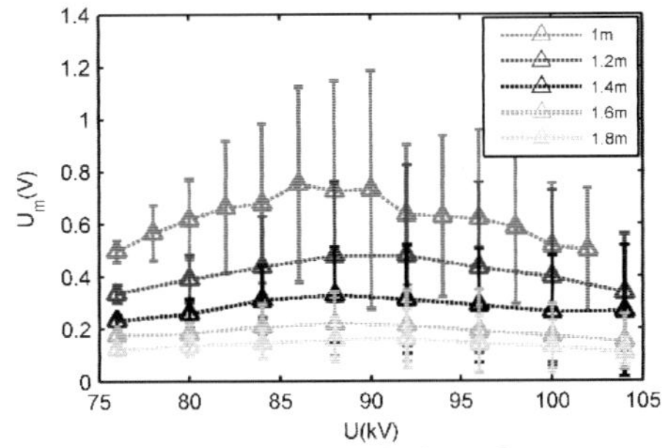

Figure 7 The statistical results of amplitude of ratio interference at different measuring distances

It can be seen from Fig. 7 that the amplitude of radio interference pulse will decrease with the increase of the distance between the measurement points. In order to further analyze the attenuation law of radio interference, the measured voltage amplitude of radio interference under different measurement distances is made, as shown in Fig. 8. Fig. 8 also shows the results of fitting the measurement results with exponential function, and the fitting expression is shown in formula (6).

$$u_{\mathrm{m}} = 4.8e^{-1.9l} \tag{6}$$

where u_{m} is the pulse amplitude of the radio interference, and l is the distance between the array and conductor. It can be It can be seen from equation (7) that the pulse amplitude of radio interference measurement voltage exhibits exponential attenuation characteristics with spatial distance.

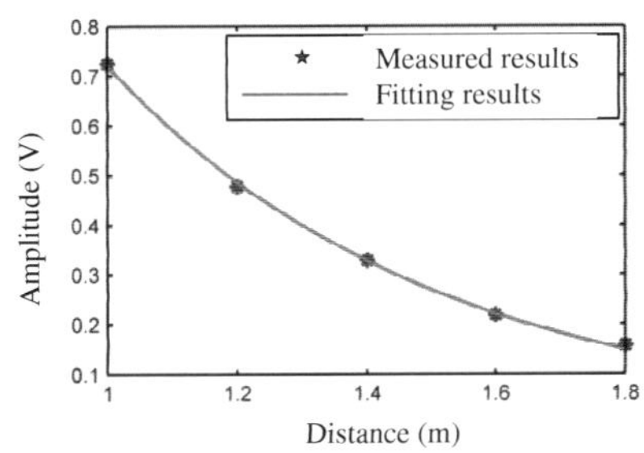

Figure 8 The fitting figure of the amplitude of measured radio interference

5. Conclusions

In order to measure and obtain the time-domain characteristics of the radio interference electric field intensity generated by corona discharge, a time-domain test system is designed and built for the radio interference electric field intensity of corona discharge. Besides, the attenuation characteristics of the radio interference time-domain waveform are analyzed. The main conclusions of this paper are as follows.

1) The time-domain measurement system of radio interference electric field intensity built in this paper includes a dipole antenna, sampling resistor, grounding metal plate, voltage probe, and high-speed data acquisition system. In the case of small size, the dipole antenna can be equivalent to a capacitor, and the RC loop equivalent radio interference equivalent circuit can be used.

2) Through the simulation of the measurement system, the influence of antenna size and measurement resistance on antenna gain and measurement bandwidth is analyzed, and the basic parameters of the time-domain measurement system of radio interference electric field intensity are determined.

3) The attenuation characteristics of a single point corona discharge radio interference are measured by the time domain measurement system of radio interference electric field strength. The measured radio interference pulse amplitude attenuates exponentially with distance.

Acknowledgments
This work is supported by the Science and Technology Project of China Southern Power Grid Co., Ltd. (066600KK52190068).

References
[1] P. Sarma. Maruvada. (2011) Corona in Transmission systems: Theory, Design, and Performance, Johannesburg, South Africa: Eskom Holding.
[2] Moreau M. R., Gary C. H. (1972) Predetermination of the Radio-Interference Level of High Voltage Transmission Lines I-Predetermination of the Excitation Function. IEEE Transactions on Power Apparatus and Systems, 91(1): 284-291
[3] Adams G. E. (1956) The Calculation of the Radio Interference Level of Transmission Lines Caused by Corona Discharges. IEEE Transactions on Power Apparatus and Systems-Part III, 75(3): 411-419
[4] A. J. Otto, H. C. Reader. (2010) Wideband and narrowband HVDC conductor corona test methods for radio noise prediction. IEEE Trans. Power Del, 25(4): 2950-2957.
[5] Zhao L, Cui X, Xie L, Lu J, He K, Ju Y. (2017) Altitude correction of radio interference of HVDC transmission lines Part II: measured data analysis and altitude correlation. IEEE Transactions on Electromagnetic Compatibility, 59(1): 284-292.
[6] Nakano Y, Sunaga Y. (1989) Availability of corona cage for predicting radio interference generated from HVDC transmission line. IEEE Transactions on Power Delivery, 5(3):1436-1442.
[7] Liu Y, Zhou L, Liu Y, Yuan H, Ji L, Wang Q. (2016) Analysis on the spectrum characteristic of corona current and its relationship with radio interference on UHVDC transmission line. IEEE Transactions on Dielectrics and Electrical Insulation, 23(6):3336-3345.
[8] Annestrand S. A. (1972) Radio interference from HVDC converter stations. IEEE Transactions on Power Apparatus & Systems, PAS-91(3):874-882.
[9] G. E. ADAMS. (1956) The calculation of the radio interference level of transmission lines caused by corona discharges. Power Apparatus and Systems. Part III. Transactions of the American Institute of Electrical Engineers, 75(3): 411-419.
[10] Radio interference characteristics of overhead power lines and high voltage equipment Part 2: Methods of measurement and procedure for determining limits. CISPR 18-2, 1986.
[11] Kanda M. (2002) Standard probes for electromagnetic field measurements. IEEE Transactions on Antennas & Propagation, 2002, 41(10):1349-1364.
[12] Li X, Li D, Chen B, Cui X, Lu T, Li Y. (2018) Measurement and analysis of time-domain characteristics of corona-generated radio interference from a single positive corona source. Physics of Plasmas, 20(5): 043510
[13] Cui X, Liu Y, Lu T, Wang Z, Li X and et al. (2015) Wideband measurement technology of original corona currents. Proceedings of the CSEE, 35(7): 1818-1827.
[14] Zhang B, Wang W, He J. (2017) Theoretical study on radio interference of HVDC transmission line based on cage tests. IEEE Transactions on Power Delivery, 32(4): 1891-1898.
[15] Lin C. (2002) Antenna Engineering Handbook. Electronic Industry Press.
[16] Makarov S. N, Puzella A. (2007) Scan Impedance for an infinite dipole array: Hansen's formulas compared with Ansoft HFSS simulations. IEEE Antennas & Propagation Magazine, 49(4):143-156.
[17] Bennett W. S. (1986) Properly applied antenna factors. IEEE Transactions on Electromagnetic Compatibility, 28(1):2-6.

ILP-based multiperiod flow routing and scheduling method in time-sensitive network

Jin Wang[1,*], **Liang Zhou**[1], **Li Tian**[1]

[1]State Grid Hubei Electric Power Research Institute, Wuhan, China

*Corresponding author. E-mail: syywangjin@foxmail.com

Abstract. Time-sensitive network (TSN) provides end-to-end data transfer with very low latency and high reliability on the basis of Ethernet. It is suitable for time-sensitive applications and will be widely used in scenarios such as autonomous driving and industrial Internet. When the existing time-sensitive flow routing and scheduling methods do not consider the shared bandwidth of the flow triggered by different cycle times, the possibility of conflict is different, which leads to the inaccuracy of the index to measure the scheduling bottleneck. This paper analyses the different cycle time-sensitive flow between the possibility of conflict, puts forward a kind of applicable to passes based on integer linear programming cycle time-sensitive traffic scheduling algorithm, according to the characteristics of the cycle time-sensitive flow carries on the preliminary classification and sorting, and developed a series of scheduling constraints, implementation is sensitive to time flow of packet scheduling. The experiment can be conducted from the three perspectives of average end-to-end delay, maximum load, and flow scheduling. The proposed routing method was compared with the existing benchmark routing algorithm evaluation. The results show that the proposed method can significantly improve the capacity of the network sensitive to time flow, reduce load scheduling by optimizing the network bottleneck at the same time, and improve the schedulability.

1. Introduction

With the continuous development of network communication, many application scenarios, such as industrial IoT, automotive, aviation electronic network, electric power communication network, etc. [1], the network of real-time and deterministic demand increasing, traditional Ethernet has not completely satisfied the requirements of the corresponding field, is the introduction of new technology to solve urgently. In order to meet the real-time requirements, real-time communication technologies represented by EtherCAT, Profinet, and TTE are applied in scenarios with high real-time requirements. These technologies are based on Ethernet, but at the same time, they have some special mechanisms, and they are not compatible with each other, which limits the further development of real-time Ethernet. In this context, the IEEE802 TSN task group proposed the concept of a time-sensitive network, intending to build a unified data link layer protocol, which could operate identically in all fields through standardization and provide real-time data transmission [2].

Compared with the traditional Ethernet network, a time-sensitive network is a new deterministic network transmission technology, which mainly provides clock synchronization, flow reservation, traffic scheduling and shaping, network configuration, and other functions [2], so as to guarantee the QoS of time-sensitive data flow. The protocols include IEEE 802.1-AS-REV [3] (Timing and Synchronization), IEEE 802.1-QBV [4] (Enhancements for Scheduled Traffic), and IEEE Multiple protocols such as 802.1QBU [5] (Frame Preemption).

Journal of Physics: Conference Series **2384** (2022) 012032 doi:10.1088/1742-6596/2384/1/012032

Flow scheduling is the core of the deterministic transmission mechanism. By selecting a specific scheduling algorithm, all switches determine each data frame transmission time slot at the exit, ensure that all frames without conflict in turn on the export link in transmission, at the same time satisfy the requirement of the flow each time delay and bandwidth, etc., and make the different categories of business requirements coexist on the same network. Ieee802.1qbv's gated Time-aware Shaper (TAS) defines a gating mechanism based on queue transmission selection. Each transmission queue is associated with a transmission gate with two states: open and closed. The queue door opens, and if there are frames to be transmitted in the queue, they can be transmitted. If the queue gate is closed, the queue cannot be transmitted. The status of the transmission Gate is controlled by the Gate Control Lists (GCLs), which periodically schedule the traffic of each queue and control the opening time of each queue Gate. The gating mechanism is shown in Figure 1.

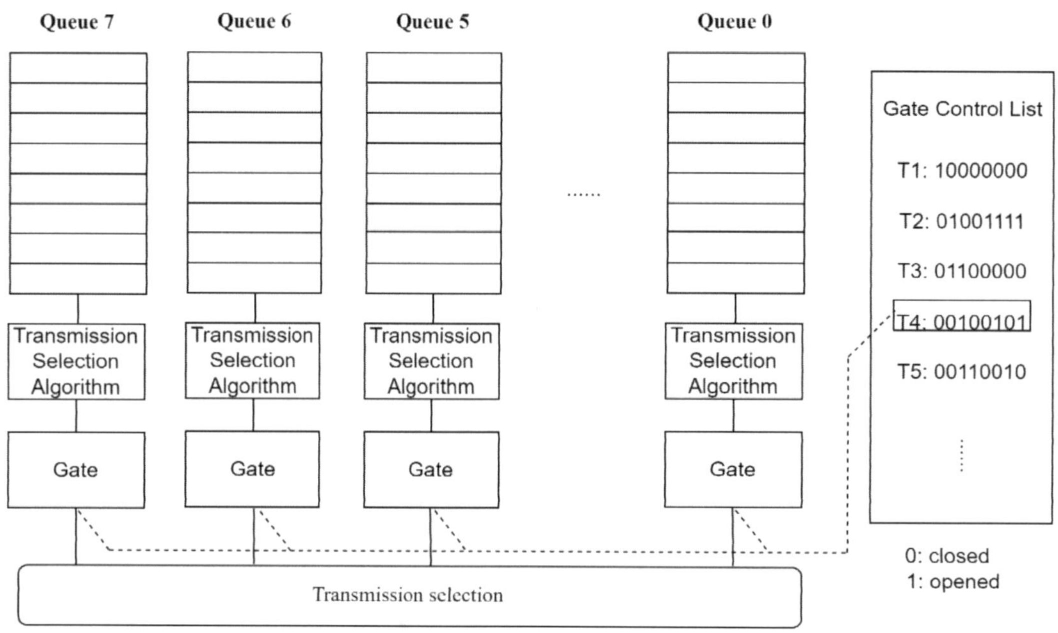

Figure 1 IEEE802.1 Qbv gating mechanism

The establishment of GCL is the core of TAS scheduling, and how to set its parameters reasonably has become the key to solving the scheduling problem. However, since the dataflow scheduling standard was introduced, it only stipulated some criteria of dataflow scheduling, but did not specify the specific method of scheduling and routing. This makes the implementation method of scheduling and routing based on dataflow scheduling standards become a research hotspot. Scheduling of data streams specifies which frames to send and when to send, and ensures that frames do not collide with each other. The scheduling problem of data flow is proved to be a non-deterministic Polynomial (NP) complete problem [6]. The methods to solve this kind of problem include scheduling algorithms based on search and scheduling algorithms based on direct solutions. Search-based scheduling algorithms are generally heuristic algorithms, such as tabu search algorithm and genetic algorithm [7-9]. The search-based scheduling algorithm approximates the optimal solution or the suboptimal solution by a certain search strategy in the solution space and stops the search within a certain time limit. The solution efficiency is high, but the solution quality is not guaranteed. Satisfiability Modulo Theories (SMT) [10, 11] or Integer Linear Programming are generally used as the scheduling algorithms for direct solutions. ILP's mathematical method directly solves the scheduling problem [12-14]. Methods based on SMT or ILP define the scheduling problem by putting forward a set of first-order logic constraints, select different optimization objectives, and search the whole solution space through the solver to find the optimal solution, but the solution time is long and suitable for small-scale networks.

However, the types and quantities of time-sensitive flows in the current industrial Internet are becoming more and more complex, and a large number of flows with different cycles passing through the same path are prone to conflict, even leading to unscheduling. The existing methods do not consider the problem of how to combine flows of different periods, and there is still a lot of room for improvement in schedulability. In this paper, different cycle time-sensitive flow between the possibility of conflict, puts forward a kind of applicable to passes based on integer linear programming cycle time-sensitive traffic scheduling algorithm, according to the characteristics of the cycle time-sensitive flow carries on the preliminary classification and sorting, and developed a series of scheduling constraints, implementation is sensitive to time flow of packet scheduling.

2. System model

However, the types and quantities of time-sensitive flows in the current industrial Internet are becoming increasingly complex, and many flows with different cycles passing through the same path are prone to conflict, even leading to unscheduling. In this paper, the routing and scheduling problem of time-sensitive networks is defined as finding a feasible network schedule for time-triggered network infrastructure and traffic pattern F to maximize the number of schedulable data flows under the constraints of the scheduling model. The time-sensitive network topology is modeled as an unweighted directed graph $G = (V, E)$. Where $V = \{v_1, v_2, \cdots, v_V\}$ is the set of network nodes, which consists of End stations (ES) and switches (SW). $E = \{e_1, e_2, \cdots, e_E\}$ denotes the set of interconnected directed edges. Each link $e = [v_i, v_j] \in E$ contains a long link delay from the node v_i to node v_j and a processing delay l_e. $|V|$ and $|E|$ are the number of nodes and links, respectively. The bandwidth capacity of each link e is C_e. The maximum traffic transmitted by a link in a period T is defined as the periodic traffic, which is represented by $C_e T$. The existing methods do not consider the problem of how to combine flows of different periods, and there is still a lot of room for improvement in schedulability. In this paper, different cycle time-sensitive flow between the possibility of conflict, puts forward a kind of applicable to passes based on integer linear programming cycle time-sensitive traffic scheduling algorithm, according to the characteristics of the cycle time-sensitive flow carries on the preliminary classification and sorting, and developed a series of scheduling constraints, implementation is sensitive to time flow of packet scheduling.

In a time-sensitive network, we consider a set of TSN flows $F = \{f_{1,} f_2, \cdots, f_n\}$ with delay and jest constraints. Each flow $f_k \in F$ is represented by tuples $\{src_k, dst_k, prd_k, siz_k, ddl_k, td_k\}$, Where src_k represents the source node of f_k, dst_k indicates the destination node of f_k, prd_k represents the period of f_k, siz_k represents the frame size of f_k. The real-time performance of time-triggered flow shows that the transmission of a message has a deadline, and sending the message to the destination node before the deadline is considered a secure transmission. The transmission deadline of f_k is denoted by ddl_k. td_k indicates the transmission time of f_k on the link. In order to ensure real-time performance, it is necessary to ensure that f_k is transmitted from src_k to dst_k within ddl_k.

The application data in a real-time network is generally divided into two kinds: the real-time data controlled by the system, that is, TSN data flow, and the best-effort data flow used for auxiliary services. The TSN flow has a fixed period and size and must reach its destination node within its cutoff time and satisfy the jitter constraint. In contrast, BE data streams do not require strict time guarantees. Different time-sensitive flows have different transmission cycles. In order to achieve unified scheduling of time-sensitive flows and meet their real-time and deterministic requirements, the least common multiple of all TSN flow cycles, namely the super period HP, is taken as the scheduling cycle of TSN flows in the network, as shown in Equation (1).

$$hp = lcm(prd_1, prd_2, \cdots, prd_n) \quad (1)$$

Two variables are introduced to explain the routing and scheduling problems. Binary decision variable $r_{[v_i, v_j], f_k}$ indicates whether the link $[v_i, v_j] \epsilon E$ has been taken up by $f_k \in F$.

$$r_{[v_i,v_j],f_k} = \begin{cases} 1: & f_k \; is\ routed\ over\ [v_i, v_j] \\ 0: & otherwise \end{cases} \tag{2}$$

In order to describe the scheduling problem, a real decision variable $t_{[v_i,v_j],f_k}$ is introduced in $[0, prd_k]$ to represent the scheduling result. If f_k through $[v_i, v_j]$ (that is, $r_{[v_i,v_j],f_k} = 1$), $t_{[v_i,v_j],f_k} \in [1, prd_k]$ is flow f_k in link $[v_i, v_j]$ sent on time, otherwise, $t_{[v_i,v_j],f_k} = 0$, it follows from above that schedule S is uniquely determined by the following tuple:

$$S = \{< r_{[v_i,v_j],f_k}, t_{[v_i,v_j],f_k} > | \forall [v_i, v_j] \in E, \forall f_k \in F\} \tag{3}$$

3. Proposed method

3.1 Multiperiod time-sensitive flow classification and sorting

Time-sensitive flows in networks have different transmission cycles. When multiple time-sensitive flows with different transmission periods are combined through the same link, to ensure that there is no time conflict between the flows, different flow combinations lead to a different bandwidth utilization of the link. In order to improve the bandwidth utilization of the link without time conflict, reasonable packet transmission should be carried out as far as possible. Time-triggered flows with different periods are defined as heterogeneous flows, and all links are assumed to be homogeneous and have the same characteristics.

Literature [15] shows that if a flow frame period and other flow frame cycle are co-prime, then the flow is different from any other frame cycle flow through the same link will conflict; namely, the flow path of the cycle is different from other frames can be separated flow; otherwise, it will cause the failure of scheduling, so should be the first to choose for this kind of flow routing. If the frame period of a flow is the factor of the frame period of other flows, the greatest common divisor of the frame period of the link is the frame period of the flow when this flow and other flows pass through the same link. Therefore, routes should be selected for this type of flow first. The flow with different periods should be classified before the flow scheduling.

On the other hand, considering that the greatest common divisor of the frame period of all flows on the link is not changed, the route should be selected for the time-triggered flows with a smaller frame period first. If the later added stream frame period is smaller than the previous one, it is likely to cause the greatest common divisor to decrease. Therefore, the time trigger flow should be sorted from smallest to largest. The classification and sorting algorithms of multi-periodic flows are as follows.

Algorithm 1: flows classification and sorting

Input: initial set of time-sensitive flows F, and set of flow period Prd_{set}
Output: the ordered set of flows by flow classification F_{order}

```
lcm_pre = LCM(Prd_set)
for f_k ∈ F do
  remove prd_k from Prd_set
    lcm_next = LCM(Prd_set)
    if lcm_next==lcm_pre/prd_k then
      class_k = 1
    else if lcm_next == lcm_pre
      class_k = 2
    else
      class_k = 3
    end
    add prd_k into Prd_set
end
F_order ← sort F by ascending order according to class first, then prd
```

3.2 Time-sensitive flow scheduling based on ILP

ILP is a widely accepted and mature method that can theoretically be used to solve all problems that can be modeled by integer decision variables. This paper presents a scheduling scheme based on ILP (integer Linear Programming). After the system model is established, the routing and scheduling problem is planned into an ILP model under the consideration of various link constraints and delay requirements, so as to select the best communication route for TSN flows and allocate traffic reasonably, so as to meet the data transmission delay requirements of TSN flows. The general form of the integer linear programming problem is as follows.

$$\begin{aligned} &min\ z = C^T X \\ &s.t. \begin{cases} GX \le h \\ AX = b \end{cases} \end{aligned} \tag{4}$$

Where A, G and C^T are constant matrices, and X is a variable matrix. Each element in each matrix is an integer, and the variable value range can be restricted to the whole integer field, or can be restricted to take 0 or 1. Equation (4) contains an objective function and a set of equality or inequality constraints. The solver traverses the entire solution space to find the optimal solution for the objective function.

In the process of data transmission, due to transmission conflict and forwarding delay in network communication, if there is no delay constraint, scheduling cannot provide a real-time guarantee. To ensure its real-time performance, scheduling should satisfy certain constraints, which are called time-sensitive constraints. Time-sensitive constraints include routing constraints, path-related constraints, no contention constraints, delay constraints, and link load constraints. The constraints of flow f_a are as follows.

3.2.1 Routing constraint. The routing constraint describes the transmission continuity of the network flow and ensures that the flow f_k reaches the destination node dst_k from the source src_k through the switch.

$$\sum_{\forall v_j|[src_k,v_j]\in E} r_{[src_k,v_j],f_k} = 1, \sum_{\forall v_i|[v_i,src_k]\in E} r_{[v_i,src_k],f_k} = 0 \tag{5}$$

Equation (5) ensures that the unicast flow f_k is emitted from the source node src_k and src_k does not receive f_k.

$$\begin{aligned} \forall v_s \in V\backslash(src_k \cup dst_k)&: \sum_{\forall v_i|[v_i,v_s]\in E} r_{[v_i,v_s],f_k} \le 1 \\ &\sum_{\forall v_j|[v_s,v_j]\in E} r_{[v_s,v_j],f_k} - \sum_{\forall v_i|[v_i,v_s]\in E} r_{[v_i,v_s],f_k} = 0 \end{aligned} \tag{6}$$

Equation (6) mainly defines the constraints of the intermediate link. The first inequality indicates that each switch can only receive the flow f_k from at most one port. Due to the unicast model, each switch sends and receives the flow f_k equal times.

3.2.2 Path order constraint. Path-related constraints describe the transmission timing constraints of a flow. Since a flow can only be forwarded to the next switch after it has reached the current switch, the path-related constraints for flow f are as follows.

$$\forall f_k \in F, \forall[v_x, v_y], [v_y, v_z] \in E: t_{[v_y,v_z],f_k} - t_{[v_x,v_y],f_k} \ge td_k + M \times (r_{[v_y,v_z],f_k} - 1) \tag{7}$$

Where M is a sufficiently large integer, it is commonly used to represent conditional constraints in integer linear programming.

3.2.3 Contention constraint. In order to achieve zero congestion transmission, it is required that all allocated time slots do not overlap. For any two streams f_m and f_k, you cannot send any flow to the same link at the same time; otherwise, a conflict will occur. Since a link can transmit only one data stream at a time, the resource limit is as follows.

$$\begin{aligned} &\forall f_m, f_k \in F(m \ne k), \forall[v_i, v_j] \in E, r_{[v_i,v_j],f_m} + r_{[v_i,v_j],f_k} \ge 2: \\ &t_{[v_i,v_j],f_m} \ge t_{[v_i,v_j],f_k} + td_k \text{ or } t_{[v_i,v_j],f_k} \ge t_{[v_i,v_j],f_m} + td_m \end{aligned} \tag{8}$$

3.2.4 Delay constraint. The delay constraint states that the time from send to receive of a stream should be less than the upper limit of the delay, so the delay constraint is as follows.

$$\sum_{\forall v_i|[v_i,dst_k]\epsilon E} t_{[v_i,dst_k],f_k} - \sum_{\forall v_j|[src_k,v_j]\epsilon E} t_{[src_k,v_j],f_k} \le ddl_k - td_k \quad (9)$$

3.2.5 Link Load Constraint. Link load constraint describes that a load of TSN flow in each link between switches should be less than a certain threshold. Therefore, the link load constraint is as follows.

$$\forall v_i, v_j \epsilon V, [v_i, v_j] \epsilon E: \sum_{f_k \epsilon F} L_{[v_i,v_j],f_k} \le L_{max} \quad (10)$$

Where, $L_{[v_i,v_j],f_k}$ represents the bandwidth utilization of the flow f_k on the link $[v_i, v_j]$, and L_{max} is the maximum bandwidth utilization of the link, which is set as 20% in this paper. This constraint ensures that the maximum load of TSN flows on inter-switch links does not exceed 20%, avoiding the extreme case where some links are overloaded while the rest are idle.

3.2.6 Optimization objective. Time-sensitive flow scheduling is a problem of scheduling periodic tasks. When switching between queues with different priorities, there is a guard band, which causes a waste of bandwidth. In order to reduce the number of handoffs between queues and improve the overall bandwidth utilization, the scheduling length is used as the optimization objective. The scheduling length refers to the latest transmission completion time of all flows. The objective function is set to minimize the sum of the sending time of the last frame of each flow on the last link, as shown below: The time-sensitive constraint only ensures that the scheduling is feasible but not necessarily optimal. Therefore, the algorithm in this paper designed the following optimization objectives for bandwidth utilization.

$$min\ (\sum_{f_k \in F} t_{[v_{last},dst_k],f_k}) \quad (11)$$

3.3 Routing and scheduling algorithm

Given a TSN network G and a flow set F, the output is a mapping $S[r, t]$ of routing and transmission time slots. Firstly, the flow classification sorting function is called to classify and sort the flow set F so that the path can be shared better for conflict-free routing. One stream group is processed at each iteration to generate ILP constraints for G_i, and ILP functions search for feasible solutions satisfying time-sensitive constraints. The link through which the current flow group passes is deleted from graph G because the next flow group cannot use the link contained in the current group path. If the scheduling result is null, the search continues for routes to the next set of flows. The scheduling algorithm is as follows.

Algorithm2: Multiperiod flow routing and scheduling algorithm

Input: network G, flow set F
Output: schedule S_v

```
G ← flows_classification_and_sorting(F), C← ∅
for i ∈ {1,2,3} do
    C← generate time sensitive constraints(5)-(11) for G(i)
    S_f ← ILP(C)
    if S_f ≠ ∅ then
        S_v ← S_v ∪ S_f
        G ← G − S_f
    end
end
return S_v
```

4. Experimental validation and analysis

4.1 Experiment setup

Our proposed mechanism is implemented in a Python-based framework. The Python toolkit NetworkX was used to model the network topology and generate different traffic patterns. Linear programming software Gurobi 9.0 was used to implement and solve the scheduling scheme based on ILP. The computing environment is Windows 10 PC, the processor is Intel i7-7700k quad-core 4.20 GHz, and the memory size is 32 GB.

In order to facilitate the observation of test results, this paper assumes the network structure as shown in Figure 2. The network structure uses five switches, and each switch is connected to 3 terminals and consists of 22 full-duplex links. The link transmission speed is assumed to be 1 Gbps, and the switch processing delay is set to 20μs.

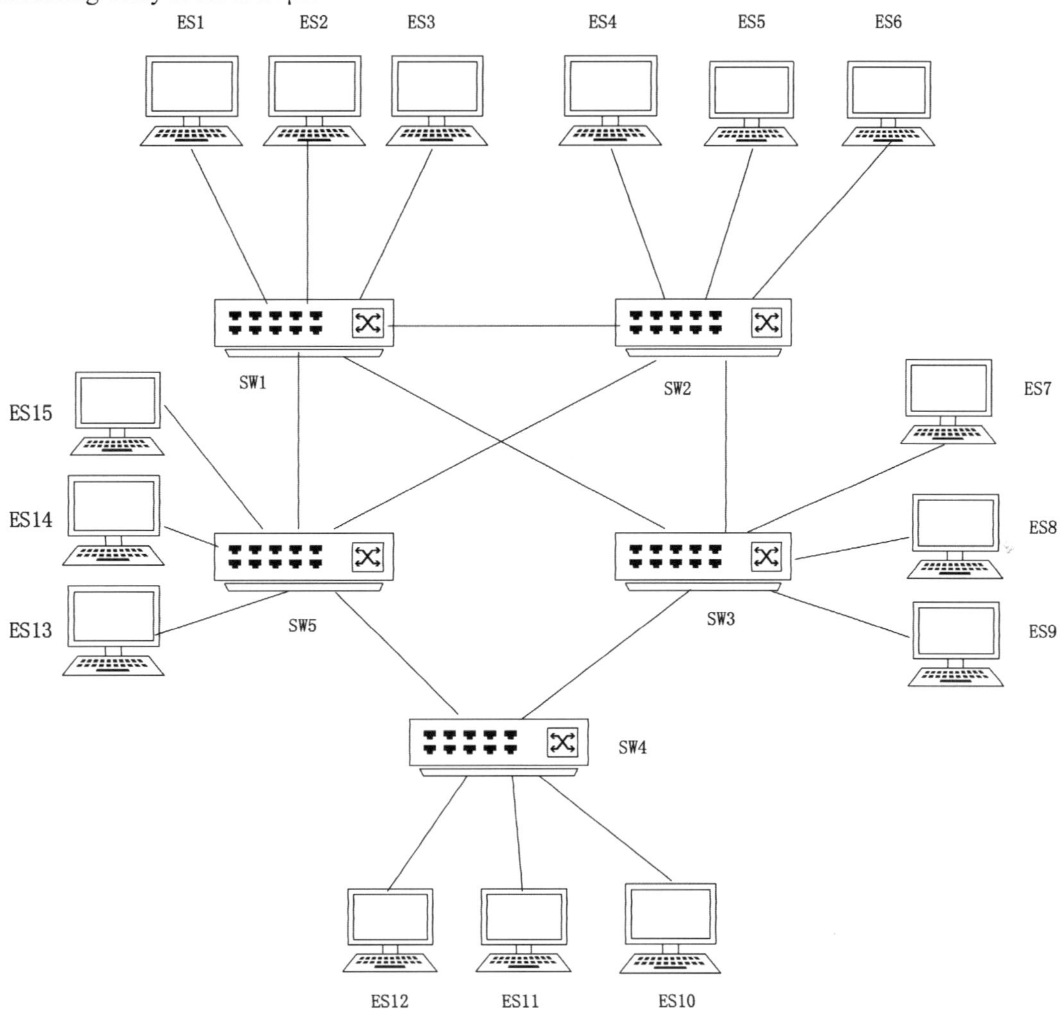

Figure 2 Experimental network topology

Four types of control flows are set in the experiment to form time-sensitive flows, and the specific flow characteristics are shown in Table 1. In the simulation, the proportion of time-sensitive traffic is 25%, and the maximum allowable delay (MADB) is the maximum delay requirement after end-to-end traffic transmission. TSN is an emerging technology, and the shortest path forwarding mechanism [16] is a common path optimization method. The shortest path routing optimization method is taken as the comparison object of the algorithm in this paper.

Table 1 Setting of flow parameters

Flow ID	Size (byte)	Prd(μs)	MADB((μs))	Proportion
F1	10	80	200	25%
F2	20	60	300	25%
F3	30	30	150	25%
F4	40	40	400	25%

4.2 Performance Metrics
Three indexes are defined to evaluate the proposed method: end-to-end average delay, schedulability, and maximum link load ratio. The average end-to-end delay D_{avg} is the sum of the time for each traffic to reach the destination and then divided by the total number of traffic N, as shown in Equation (12). Schedulability [17] Sched refers to the ratio between the number of traffic N_s that can be successfully scheduled and the number of traffic N that needs to be scheduled, and is used to evaluate the impact of path selection on subsequent traffic scheduling. The calculation method is shown in Equation (13). The maximum link load ratio U_{max} is the ratio between the amount of data that passes through a unit of time and the maximum amount of data that can pass through a link. It is an important index to measure whether an algorithm can guarantee the load balancing of network links.

$$D_{avg} = \frac{\sum_{i=1}^{N} ET_i}{N} \tag{12}$$

$$Sched = \frac{N_s}{N} \tag{13}$$

4.3 Average end-to-end delay
Figure 3 shows the comparison of the average end-to-end delay. The proposed method has a lower average end-to-end delay than the shortest path routing method. The algorithm in this paper classifies and sorts the flows of different cycles, which reduces the possibility of conflicts between flows, so that a better transmission path can be obtained, and the average end-to-end delay will be relatively low.

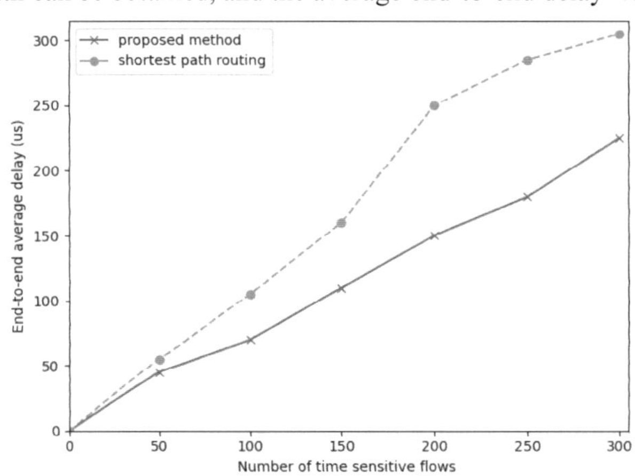

Figure 3 Average end-to-end delay of time-sensitive flows

4.4 Maximum link load
The usage of link load determines the quality of the scheduling result. The proposed algorithm ensures that if the load on a link exceeds 20%, the link will be bypassed, and other links will be selected. That is, the load on all links is guaranteed to be less than 20%. Figure 4 compares the maximum load of links

using the shortest path scheduling algorithm and using the algorithm in this paper to generate scheduling results.

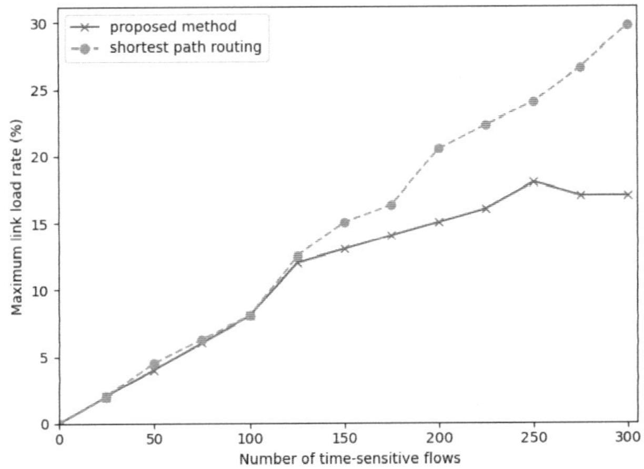

Figure 4 Maximum link load rate

4.5 Schedulability

With the increase of the number of flows, the load in the network increases, the number of shared links between flows will inevitably increase, and the possibility of conflicts will also increase. Therefore, the schedulability of all routing algorithms will inevitably decrease with the increase in the number of flows. Figure 5 shows that when the number of flows is consistent, the proposed algorithm classifies flows of different periods compared with the shortest path optimization algorithm, which increases the associability between flows and has high schedulability.

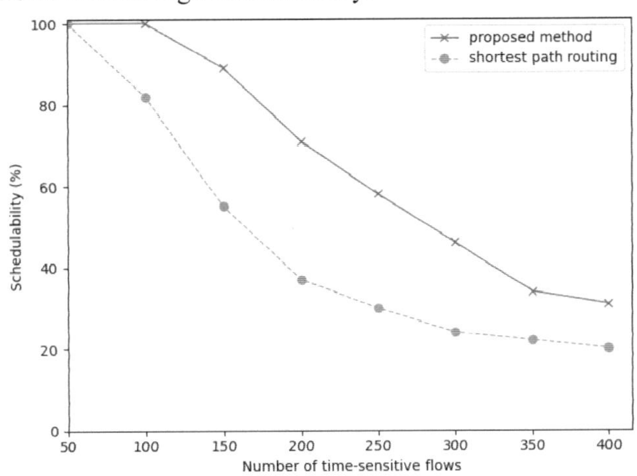

Figure 5 Schedulability comparison of time-sensitive flows

5. Conclusion

Based on the study of IEEE802.1 Qbv agreement and the analysis of the different cycle time-sensitive to flow between the possibility of conflict, this paper puts forward a kind of applicable to passes based on integer linear programming cycle time-sensitive traffic scheduling algorithm. According to the characteristics of the cycle, we carried out the preliminary classification and sorting of time-sensitive, developed a series of scheduling constraints, and implemented packet scheduling for time-sensitive flow. Although the solution time of the ILP-based method is generally longer than that of the heuristic algorithm, the search of solution space is more sufficient, and the evaluation of schedulability is more accurate. The experiment is conducted from three perspectives: the average end-to-end delay, the maximum load, and flow scheduling. The proposed routing method was compared with the existing benchmark routing algorithm evaluation. The results show that the proposed method can significantly improve the capacity of the network sensitive to time flow, reduce load scheduling by optimizing the network bottleneck at the same time, and improve the schedulability.

Acknowledgment
This paper is supported by the Science and Technology Project of State Grid Corporation of China (Project No. 5108-202222039A-1-1-ZN).

References

[1] Lucia Lo Bello and Wilfried Steiner. A perspective on IEEE time-sensitive networking for industrial communication and automation systems. Proceedings of the IEEE, 2019. **107**(6): 1094-1120.
[2] Norman Finn. Introduction to time-sensitive networking. IEEE Communications Standards Magazine, 2018. **2**(2): 22-28.
[3] IEEE Standards Association. IEEE Standard for Local and Metropolitan Area Networks—Timing and Synchronization for Time-Sensitive Applications in Bridged Local Area Networks. IEEE Std, 2011. **802**.
[4] IEEE Standards Association. IEEE Standard for Local and Metropolitan Area Networks—Bridges and Bridged Networks—Amendment 25: Enhancements for Scheduled Traffic. Amendment to IEEE Std, 2016. **802**: 1-57.
[5] IEEE. IEEE standard for local and metropolitan area networks—bridges and bridged networks—amendment 26: frame preemption: 802.1 Qbu-2016. 2016.
[6] Ken W Tindell, Alan Burns, and Andy J. Wellings. Allocating hard real-time tasks: an NP-hard problem made easy. Real-Time Systems, 1992. **4**(2): 145-165.
[7] Yang Wang, Jidong Chen, et al. A time-sensitive network scheduling algorithm based on improved ant colony optimization. Alexandria Engineering Journal, 2021. **60**(1): 107-114.
[8] Frank Dürr and Naresh Ganesh Nayak. *No-wait packet scheduling for IEEE time-sensitive networks (TSN)*. In *Proceedings of the 24th International Conference on Real-Time Networks and Systems*. 2016.
[9] Maryam Pahlevan and Roman Obermaisser. *Genetic algorithm for scheduling time-triggered traffic in time-sensitive networks*. In the *2018 IEEE 23rd international conference on emerging technologies and factory automation (ETFA)*. 2018. IEEE.
[10] Ramon Serna Oliver, Silviu S Craciunas, and Wilfried Steiner. *IEEE 802.1 Qbv gate control list synthesis using array theory encoding*. In the *2018 IEEE Real-Time and Embedded Technology and Applications Symposium (RTAS)*. 2018. IEEE.
[11] Qing Li, Dong Li, et al. A simple and efficient time-sensitive networking traffic scheduling method for industrial scenarios. Electronics, 2020. **9**(12): 2131.
[12] Eike Schweissguth, Dirk Timmermann, et al. *ILP-based routing and scheduling of multicast real-time traffic in time-sensitive networks*. In the *2020 IEEE 26th International Conference on Embedded and Real-Time Computing Systems and Applications (RTCSA)*. Gangnueng, Korea (South), 2020: 1-11
[13] Zaiyu Pang, Xiao Huang, et al. Flow scheduling for conflict-free network updates in time-sensitive software-defined networks. IEEE Transactions on Industrial Informatics, 2020. **17**(3): 1668-1678.
[14] Pozo F, Rodriguez-Navas G, Hansson H. *Schedule reparability: Enhancing time-triggered network recovery upon link failures*. 2018 IEEE 24th International Conference on Embedded and Real-Time Computing Systems and Applications (RTCSA). IEEE, 2018: 147-156.
[15] Kai Huang, Jingkang Wu, et al. A period-aware routing method for IEEE 802.1 Qbv TSN networks. Electronics, 2020. **10**(1): 58.

[16] Eike Schweissguth, Peter Danielis, et al. *ILP-based joint routing and scheduling for time-triggered networks*. In the *Proceedings of the 25th International Conference on Real-Time Networks and Systems*. 2017: 8-17.

[17] Johannes Specht and Soheil Samii. *Synthesis of the queue and priority assignment for asynchronous traffic shaping in switched Ethernet*. In the *2017 IEEE Real-Time Systems Symposium (RTSS)*. 2017: 178-187.

User multi-scenario energy storage configuration and operation strategy

Bin Zhou [1], Zhiqiang Zhou [1 *], Yunfei Zheng [1], Qing Gong [2]

[1] State Grid Hubei Electric Power Co. Ltd, Wuhan Hubei 4300770, China

[2] Power China Hubei Electric Engineering Co. Ltd, Wuhan Hubei, 430040, China

Corresponding author's e-mail: 625309960@qq.com (Zhiqiang Zhou)

Abstract. With the transformation of China's energy consumption side, users' energy demand will become more diversified and comprehensive, and the coupling degree of energy consumption systems such as cold, heat, and electricity will deepen day by day. Among them, the energy storage system will play an important role in comprehensive energy service. Using energy storage equipment on the users' side can significantly improve the utilization rate of renewable energy, realize peak shaving and valley filling, and thus reduce the energy cost of the system. User-side residential and commercial scenarios have different load characteristics and face different energy storage configuration requirements. Therefore, this paper studies the operation optimization strategy of multi-scenario energy storage configuration on the user side, studies the definition and constraints of the state of charge of electric energy storage for residential areas, and establishes the configuration and operation model of electric energy storage.

1. Introduction

Energy storage is a national strategic emerging industry, an important part of the energy Internet, and an important supporting technology, which has the technical advantages of quick response, two-way monitoring, strong environmental adaptability, short construction period, and so on. The application of large-scale energy storage will affect energy conversion and grid structure and have a significant impact on the power structure. [1]. As for the user side, since 2017, it has been the most active application field of energy storage in China. More and more energy storage project developers are starting from electricity tariff management to build a comprehensive energy service system with energy storage as the core, and further exerting the role of energy storage in the value-added distribution of electricity and building a smart energy model has become a hot topic for discussion [2]. The energy storage on the user side can promote peak shaving and valley filling, and promote distributed generation. In the long run, the energy storage on the user side can reduce electricity demand, improve power quality and enhance the reliability of the power supply. This paper considers the operation optimization strategy of multi-scenario energy storage configuration on the user side, studies the operation optimization strategy of energy storage configuration in residential or commercial areas on the user side, and considers the operation strategy of energy storage under demand response management and energy cascade utilization at the same time, which is of great significance to the configuration and operation of energy storage on the user side in the future [3].

2. Overview of user-side energy storage

2.1 Development and application status of energy storage

At present, all countries are actively exploring the application of user-side energy storage, establishing corresponding energy storage according to the characteristics of user-side energy storage, making use of peak-valley electricity price difference to make profits, improving the utilization efficiency of user-side energy, ensuring the stability of power grid, improving the quality of user-side energy consumption, and establishing energy storage in industrial areas, commercial areas, ports, hospitals and other places to relieve the pressure of peak power grid [4]. Common energy storage types of users are electric energy storage and cold and hot energy storage. The configuration of electric energy storage devices needs to consider its capacity and power, and the cost model of the whole life cycle of energy storage should be considered in the configuration of energy storage, so as to establish the configuration model of energy storage to maximize the benefits [5].

2.2 Research on Type and Configuration of Energy Storage at the User Side

Physical energy storage is divided into pumped energy storage, compressed air energy storage, and flywheel energy storage, electrochemical energy storage is divided into lithium battery energy storage, lead-acid battery energy storage and flow battery energy storage, and electromagnetic energy storage is divided into superconducting energy storage, capacitive energy storage, and supercapacitor energy storage. A comparison of electric energy storage modes is shown in Table 1:

Table 1 Comparison of energy storage modes of meter electricity

Project	Classification	Principle	Advantages and disadvantages
Electric energy storage	Physical energy storage	Pumped energy storage and compressed air Energy storage, flywheel energy storage	The advantages of environmental protection and greenness require special geographical conditions and sites, and the construction has great limitations.
	Electrochemical energy storage	Lithium battery, lead-acid battery Pool energy storage and flow battery energy storage	Different electrochemical energy storage technologies have their own advantages and disadvantages, and the construction site is not affected by climate and geographical factors.
	Electromagnetic energy storage	Superconducting energy storage and capacitor storage Energy, super capacitor energy storage	High energy consumption

Heat storage technology can be roughly divided into sensible heat storage, latent heat (phase change) storage, and chemical reaction storage according to different heat storage modes of heat storage materials. The comparison of heat storage modes is shown in Table 2. Cold storage technology also uses the sensible heat or latent heat characteristics of substances to store cold energy and releases the cold energy for the demand side when needed. In the aspect of energy storage operation strategy, according to the characteristics of energy storage, the influence of uncertainty on energy storage operation strategy is studied, and the economy of energy storage operation is considered to be solved by the optimization method. From the perspective of economic optimization, the investment cost and operation cost of energy storage are considered, and the configuration model is established.

Table 2 Comparison of table heat energy storage modes

Project	Classification	Principle	Advantages and disadvantages
Thermal energy storage	Sensible heat energy storage	Its unique specific heat capacity is used to store and release the heat generated by temperature change.	Has stable physical and chemical properties and good thermal conductivity.

	Latent heat storage	By using the coexistence of two stages of the material itself, when one state changes to the other, the absorption and release of heat energy can be realized.	With high energy density, compact system, and low heat loss, it is suitable for long-term heat storage.
	Chemical reaction heat storage	Make use of chemical reactions. Thermal energy storage	The energy density of stored energy is high, and it has good reversibility, high heat storage density, and a fast reaction rate.

3. Study on Optimal Operation Strategy of Energy Storage Configuration in Residential Area 3

3.1 Overview of energy storage configuration in residential areas

User-side energy storage can promote energy production and consumption, realize multi-energy cooperative operation, and complete the transfer of various energy sources in time and space. Through the optimal scheduling of energy storage and discharge, the production and consumption of energy can be decoupled, and the consumption ratio of renewable energy can be increased. Energy storage can reduce the peak load of the system and reduce the capacity investment of equipment in energy transmission and distribution links. Energy storage can also provide auxiliary services for the energy system so that the energy system can continue to run safely and reliably. Energy storage can generate electricity at any time, so as to reduce the peak load, reduce the peak-valley difference of the power system and improve the operating efficiency of the power system.

3.2 Operation strategy model of electric energy storage configuration in residential areas

To allocate electric energy storage in residential areas, consider dividing the annual load of residential areas into three categories: winter, summer, spring, and autumn. Among them, residential areas use heat and electric energy in winter, electric energy and cold energy in summer, and electric energy in spring and autumn. The expenditure of the energy storage device on the user side should include the initial investment cost, annual operation and maintenance cost, and replacement cost of the electric energy storage device. The minimum cost expression on the user side is shown in Formula 1:

$$\min C = C_1 + C_2 + C_3 + C_4 \tag{1}$$

In formula 1, C1 is the initial investment cost of the energy storage device on the user side, C2 is the annual operation and maintenance cost of the energy storage device on the user side, C3 is the replacement cost of the energy storage device on the user side, and C4 is the expenditure of purchasing electric energy. Assuming that the life of the energy storage device is T years, the benchmark discount rate is i0, and the energy storage device is updated and replaced N times in the life cycle, the cost of each stage in the life cycle is analyzed.

(1) Initial investment cost, the initial investment cost is mainly related to the capacity and power of the electric energy storage device, and its formula is shown in Formula 2:

$$C_1 = (C_e E + C_p P)\frac{i_0(1+i_0)^T}{(1+i_0)^T-1} \tag{2}$$

In Formula 2, C1 is the initial investment cost of the electric energy storage device, Ce is the investment coefficient of electric energy storage cost per unit capacity, Cp is the investment coefficient of electric energy storage cost per unit power, P is the rated power of electric energy storage device, and E is the capacity of electric energy storage device.

Quantity.

(2) Annual operation and maintenance cost, as shown in Formula 3:

$$C_2 = C_{ex} E \frac{i_0(1+i_0)^T}{(1+i_0)^T-1} \tag{3}$$

In formula 3, C2 is the annual operation and maintenance cost of the electric energy storage device, Cex is the operation cost of the unit power electric energy, and E is the capacity of the electric energy storage device.

(3) Replacement cost. If the actual running time of the electric energy storage device is less than the expected time, the electric energy storage device needs to be replaced. The expression of the cost is shown in Formula 4:

$$C_3 = aC_1\frac{i_0(1+i_0)^T}{(1+i_0)^T-1} = \alpha(C_{eE} + C_pP)\frac{i_0(1+i_0)^T}{(1+i_0)^T-1} \quad (4)$$

Where C3 is the replacement cost of the energy storage device, and α is the replacement coefficient of the energy storage device.

(4) Expenditure on purchasing electric energy

The expenditure for purchasing electric energy is the cost, and its formula is shown in Formula 5:

$$C_4 = \sum_{i=1}^{8760}(c_{price,e}(t)W_{buy,e}(t)) \quad (5)$$

In Formula 5, cprice,e(t) is the price of unit electric energy purchased in the T period, and Wbuy,e(t) is the amount of electric energy purchased in the T period.

4. Study on the Optimal Operation Strategy of Energy Storage Configuration in Business District

4.1 Load Characteristics and Equipment of Business District

Commercial load mainly refers to urban commercial centers and commercial hotels, etc. Its electricity consumption characteristics and laws are quite different from those of traditional high-energy-consuming factories and mines and labor-intensive manufacturing enterprises, which are mainly manifested in the strong correlation between its electricity consumption characteristics and daytime work and rest, seasonal changes, etc.

(1) Air conditioning model: Air conditioning is the main refrigeration equipment in commercial buildings, which can convert electric energy into cold energy. The relationship between its output cold power and input electric power is shown in Formula 6:

$$C_{ac} = COP_cP_{ac} \quad (6)$$

In 6, Pac is the electric power input by the air conditioner, in kW; Is the cold power output by Cac, in kW; COPc is the energy efficiency ratio of air conditioning refrigeration. The air-conditioning system is composed of the refrigeration unit, water pump, air-conditioning box fan with terminal heat exchange, fan coil unit, etc. Among them, the electric-cooling conversion coefficient of the refrigeration unit is about 5.6 ~6 (that is, 5.6~6kWh of cold energy can be produced by consuming 1kWh of electric energy). As the terminal equipment such as the water pump also consumes electric energy, the comprehensive energy efficiency ratio of the air-conditioning system is about 3.5.

(2) Electric boiler model: Electric boiler can convert electric energy into heat energy by resistance heating or electromagnetic induction heating, and the relationship between the output thermal power and the input electric power is shown in Formula 7:

$$H_{eb} = \eta_{eb}P_{eb} \quad (7)$$

7, Peb is the electric power input by the electric boiler, in kW; Heb is the thermal power output by the electric boiler,

Unit kw; Eb is the electrothermal conversion efficiency of an electric boiler (about 90%-95%). The energy conversion equipment in the commercial area is shown in Figure 1:η

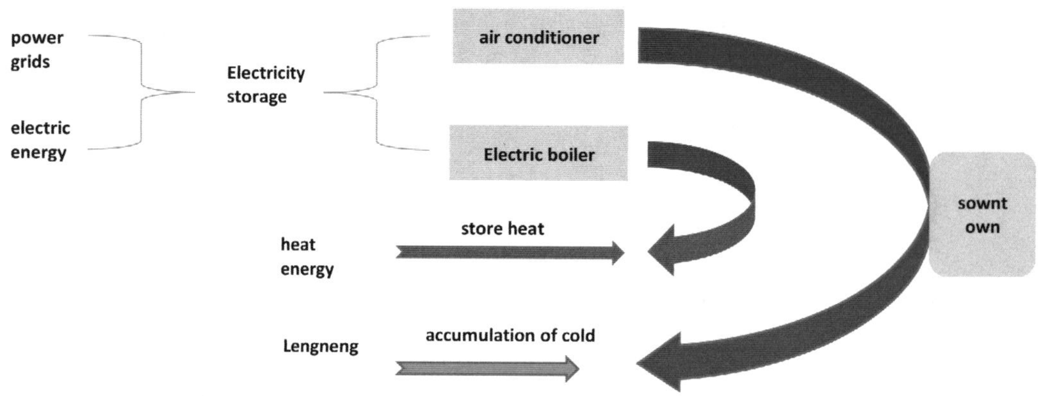

Figure 1 Energy conversion equipment in the business district

4.2 Characteristics of Cold, Thermoelectric and Energy Storage in Business District

According to the model of electric energy storage in residential areas, it can be seen that lithium battery energy storage is a more suitable type of electric energy storage for users. The storage modes of heat energy are mainly divided into three types: sensible heat, latent heat, and chemical reaction heat storage. Sensible energy storage stores heat or cold by increasing the temperature of a solid or liquid medium, latent energy storage stores energy by melting and evaporating substances or changing the state at a certain temperature, and chemical reaction storage stores energy by combining reversible chemical reactions with heat. A comparison of three heat storage modes is shown in Table 3.

The heat storage technology of phase change uses the latent heat of material to realize energy conversion. The process of phase change can be divided into four types: solid-solid phase change, solid-liquid phase change, solid-gas phase change, and liquid-gas phase change. Although solid-gas phase transition and liquid-gas phase transition have great latent heat in a phase transition, it is difficult to control in practical application because a large amount of gas will be produced in the process of phase transition, and the volume of material change is large. A heat storage boiler can continuously and stably provide the required heat energy to users by converting intermittent renewable energy or valley electricity with a low utilization rate into heat energy and storing it in composite phase change materials. The regenerative electric heater uses low-valley electric heating for heating, and stores a part of the heat in the heat storage material. During the peak period of electricity consumption, the refrigerator is turned on or not, and the stored cold energy is used for supply.

Table 3 Comparison of three heat storage methods

Project	Sensible heat storage	Latent heat storage	Chemical heat storage
Heat storage material	Water, rocks, soil, etc.	Organic matter and inorganic matter	Metal chloride, hydrogenation Things, oxides, etc.
Heat storage technology	Heat storage in pools, caves, and underground water/soil.	Composite phase change material, ice storage	Adsorption/absorption heat storage
advantage	Has little impact on the environment, cheap heat storage materials, a simple and easily controlled system, and high reliability.	The energy density is high, A constant temperature can be obtained.	The energy density is high, and the system Compact, low heat loss, suitable Combined long-term heat storage
disadvantage	Low heat storage density, huge system, large heat dissipation, large initial investment, and affected by geology.	Poor thermal stability, crystallization, Corrosion problem, expensive heat storage materials	Heat storage materials are expensive.

4.3 Optimal allocation model of energy storage in the business district

To allocate cold, heat, and electrical energy storage in the business district, consider dividing the annual load of the business district into three categories: winter, summer, spring, and autumn. Among them, thermal energy and electric energy are used in commercial areas in winter, electric energy and cold energy are used in commercial areas in summer, and electric energy is used in commercial areas in spring and autumn. Typical daily electric heating and cooling load data: Take a hotel with an area of 20000m² as the research object. The hotel has electric load demand all day, and the hourly change trend of each typical daily electric load is basically the same: 8:00~22:00 is the peak period of electricity consumption, and electricity is mainly used for lighting and power. Lighting includes lighting in different areas, and power includes kitchen power and other power. In other time periods, the hotel still needs to meet the requirements of power and lighting in some areas, and the electric load is maintained at a certain level, which is about 40%~50% of the peak load. Hotels have heat load demand all day in winter, and the heat load characteristics of typical heating days are as follows: 0:00~5:00 is the hot valley period, and there is no heat demand in restaurants and other areas, so the heating power is low. From 6: 00 to 17: 00, the heat load is relatively stable, about 45% to 50% of the peak load. After 18:00, the temperature dropped. In order to meet the indoor heating demand, the hotel's heat load increased and reached its peak load. Phase change energy storage is selected for hot energy storage and cold energy storage, and lithium battery is selected for electric energy storage, as shown in Table 4:

Table 4 Configuration optimization results of cold, heat and electricity energy storage in the business district

		Sowntown
Configuration result	Electrical energy storage capacity (MWh)	3.67
	Thermal energy storage capacity (MWh)	8.58
	Cold energy storage capacity (MWh)	10.83
Equivalent annual cost (ten thousand yuan)	Energy storage cost	101.02
	The cost of electric energy	396.61
	total cost	497.62

The allocation capacity of hot energy storage and cold energy storage in commercial areas is small, mainly because there are still cold and hot loads during the night electricity price valley. Electric boilers and air conditioners must first meet the demand for cold and hot energy at night, and the surplus power will refrigerate heat and store energy in cold and hot energy storage.

5. Conclusion

Based on the background of users, this paper studies the energy storage allocation and operation strategies in different scenarios such as residential areas and commercial areas, and obtains the following results: Aiming at residential areas, comprehensive residents, including office workers, children, and elderly users, are selected, and the minimum cost is taken as the objective function. With various types of energy conversion balance constraints, optimization equations are established to calculate the allocation and operation strategies of different types of electrical energy storage. In terms of energy storage configuration, it is more appropriate for residents to choose lithium batteries and lead-acid batteries for energy storage. Lithium battery is the most economical electric energy storage for residents. In terms of energy storage operation strategy, electric energy storage is in a two-charging and two-discharging working state, which reduces the cost of electricity purchase through low storage and high arbitrage. According to the commercial scene, this paper analyzes the energy-related information and energy conversion efficiency of each equipment of commercial users, calculates the operation strategy of various types of equipment and energy storage after the load aggregator implements price-based demand response, and users adjust their own energy consumption according to the electricity price. In order to evaluate the economy of energy storage, the evaluation index of energy storage is established, and the economic benefits brought by cold, thermal, and electric energy storage are studied. The results show that the energy storage subsidy policy can reduce the initial investment of the project and shorten the investment return period.

Acknowledgments
The authors would like to show sincere thanks to those techniques who have contributed to this research.

References

[1] Martin-Ramos, J. A., Pernía, A. M., Díaz, J., Nuño, F., & Martínez, J. A. (2008). Power supply for a high-voltage application. *IEEE Transactions on power electronics*, 23(4), 1608-1619..
[2] Heydt, G. T., Liu, C. C., Phadke, A. G., & Vittal, V. (2001). Solution for the crisis in electric power supply. *IEEE Computer Applications in Power,* 14(3), 22-30.
[3] Gutnik, V., & Chandrakasan, A. P. (1997). Embedded power supply for low-power DSP. *IEEE Transactions on Very Large Scale Integration (VLSI) Systems*, 5(4), 425-435.
[4] Akhmatov, V., Knudsen, H., & Nielsen, A. H. (2000). Advanced simulation of windmills in the electric power supply. *International Journal of Electrical Power & Energy Systems,* 22(6), 421-434.
[5] Li, C., Zhang, S., Zhang, J., Qi, J., Li, J., Guo, Q., & You, H. (2018). Method for the energy storage configuration of wind power plants with energy storage systems used for black-start. *Energies,* 11(12), 3394.

Journal of Physics: Conference Series **2384** (2022) 012034 doi:10.1088/1742-6596/2384/1/012034

Separation Structure Speech Separation Model Fusing Conformer and NBC

Zhou Qiang, Cheng Xiangyang*, Ding Tian

School of Computer Science and Engineering, Wuhan Institute of Technology, Hubei Wuhan 430205

*Corresponding author: 3426205919@qq.com

Abstract. The quality of speech separation affects the entire speech technology ecosystem. Aiming at the problems of low utilization of local feature information, insufficient convergence speed, too many calculation parameters and too long calculation time in blind source separation in view of Transformer dual-path cyclic neural network, a blind source separation model based on fusion of Conformer and NBC (Narrow-band Conformer NBC) is proposed. First, for the problems of low utilization of speech local feature information and insufficient convergence speed, the Transformer block is replaced by the Conformer in the block of the dual-path recurrent network. It can improve the utilization and convergence speed of local features. Secondly, the NBC block is used to replace the Transformer block in the inter-block loop. The NBC block simplifies the calculation of vector similarity and vector aggregation, reduces lots of parameters and calculation cost, and reduces the computational complexity of the model. In the experiments on WSJ0-2mix[7] dataset and WHAM[13] dataset, contrast with other models, the convergence speech is faster and the blind source separation effect is better.

1. Introduction

Speech separation is single or multiple signal sources extracted from a noisy environment. The rich auditory nervous system of people can well separate and collect the desired signal from the mixed signal sources. How to use a computer to simulate human signal separation is known as the cocktail party problem[1]. Speech separation is the pre-processing of many speech-oriented technologies such as speech translation, speech synthesis and speech synthesis.

Speech separation has been intensively studied for decades, resulting in many traditional methods: computer auditory scene analysis(CASA[2]), non-negative matrix factorization(INF[3]) and probabilistic models[4]. But these models can only be used in closed loudspeakers and have big limitations in practical applications. Anytime deep learning has been successful in various fields, deep learning has also been introduced in the field of speech separation. Deep learning studies separate unknown mixtures, overcoming the barriers of traditional methods. In the current research stage, deep learning techniques for single-channel speech blind source signal separation can be divided into two categories: end-to-end time domain methods and time-frequency domain (Time-Frequency T-F[5]) methods. The time-frequency domain method is to establish the frequency domain signal characteristics through short-time Fourier transform, use the T-F[5] method to separate the signal, and finally reconstruct the source waveform through inverse. The time-frequency domain methods mainly include[5, 6]. The end-to-end time-domain method mainly uses the waveform mixed in the original stage. The source signal includes multiple signal sources and noise. The mixed signal is separated by using an encoder-mask network-decoder model. This separation method improves the speech upper limit of separation. End-to-end time-domain methods include[7, 8, 9, 10]. The input mixed signal is efficiently encoded in the encoder, the separator estimates the optimal mask to separate the mixed signal, and the decoder reconstruct the speech signal through the corresponding mask in the time domain. The separation model relies heavily on the learned masking strategy in Conv-TasNet[7]. A relatively complete set of comprehensive filters is learned from the input signal data to calculate each speech source in each learning domain. The subsequent DPRNN[8] carried out long-term modeling, and the separation effect was greatly improved. By dividing and combining the input signal sources, blocks of the same size are formed for processing, and different feature extraction methods are used for processing within and between blocks. Using RNNs is limited by recurrent connections, especially in global processing steps. The DPTNet[9] model formed by integrating the transformer into the separation pipeline is better than the previous DPRNN model, but the RNN is still used in the network model, which effectively negates the parallelization ability of the pure attention model. SepFormer [10] replaces RNNs with a transform dimensional multi-scale pipeline consisting of complex transformers that simultaneously learn corresponding short-term dependencies and long-term scale dependencies. However, the utilization rate of the local features of the DPTNet[9] model is not high. Due to the simultaneous processing of corresponding short-term dependencies and long-term scale dependencies, the network model calculation is more complicated and the calculation parameters increase.

In response to this problem, this paper proposes a new model DPCNBCN (Dual-Path Conformer NBC Network), which fuses the Conformer block and the NBC block into the DPTNet[9] model. This network model adopts the dual-path framework of DPTNet[9]. The Conformer block is used for processing within the block, and the NBC block is used for processing between blocks. The RNN is replaced with a converter that learns short-term and long-term dependencies. The local information features are effectively used in the block, and the processing is more efficient. For small blocks, the inter-block processing is effectively simplified and the convergence is faster.

2. DPCNBCN NETWORK MODEL

The model proposed in this paper adopts an encoder-masked network-decoder network construct. From figure 1, the encoder uses convolutional layer processing, uses normalization processed and then divides the signal into blocks of the same size, uses the Conformer block for processing within the block, and uses the NBC block for inter-block processing after connection transformation. The mask network construction predicts the mask of each speech source signal, and the decoder reconstruct the speech signal through the corresponding mask in the time domain.

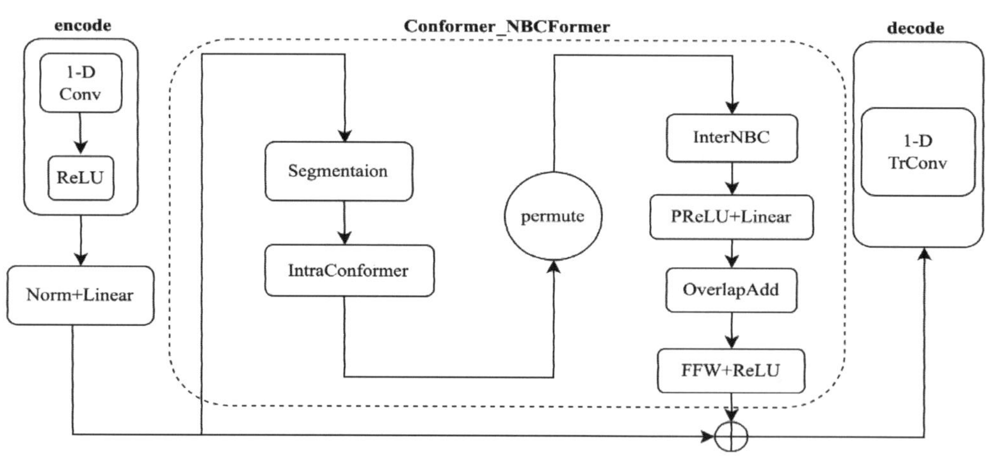

Figure 1. DPCNBCN Network Model

2.1. Encode

The time-domain mixed signal $x \in R^{T}$ is input to the encode and processed using a convolutional layer as in Eq. 1:

$$h = \mathrm{Re}LU(conv1d(x)) \tag{1}$$

The processed signal $h \in R^{F \times T'}$ is subjected to a layer normalization operation and processed using a linear layer, and F represents the dimension.

2.2. Split and combine

The mixed signal $h \in R^{F \times T'}$ is divided into blocks of size C, the divided signals overlap by 50% on the time axis, and the first and last blocks are not long enough to be zero-filled. Combining into blocks $K \in R^{F \times C \times N_C}$, F represents the dimension, C represents the divided block length, N_C represents numbers of divided blocks.

2.3. Mask Network

The dual-path recurrent neural network is used as a framework, which is processed by the Conformer in the block. The multi-scale pipeline performs the conversion and processes the NBC blocks between the blocks. The Eq. 2:

$$h'' = NBC(P(Conformer(K))) \tag{2}$$

$Conformer(\bullet)$ represents the Conformer module, $P(\bullet)$ represents the multi-scale converter, $NBC(\bullet)$ represents the NBC module, and the whole process is repeated N times.

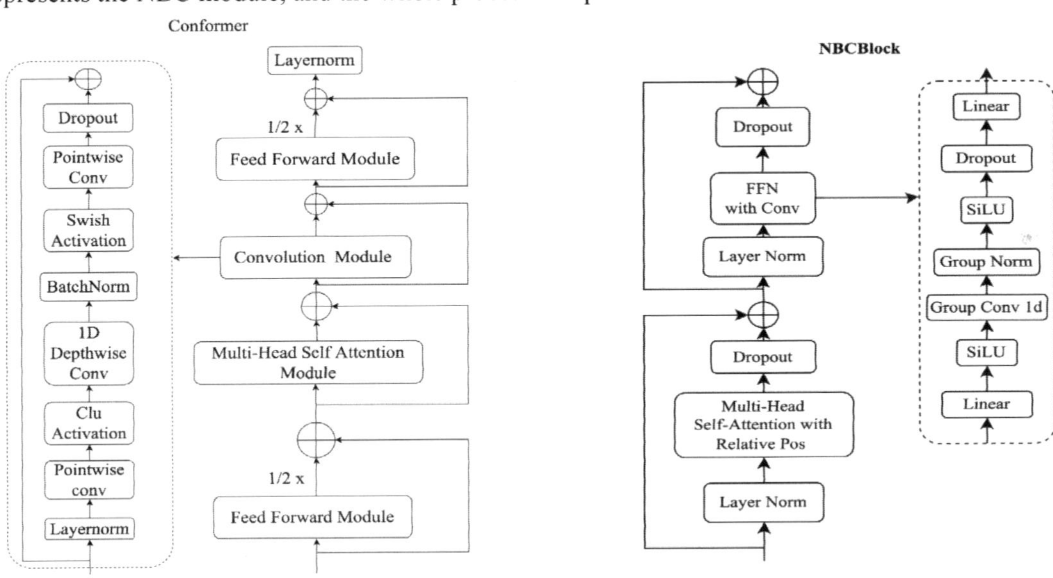

Figure 2. (left) Conformer Module and (right) NBC Block

The Conformer[11] module uses CNN to effectively utilize local features, and combines convolutional neural networks and converters to model the portion dependencies and global dependencies of mixed audio signal sequences in a parameter-efficient manner. The specific modules are shown in figure 2(left). Two feedforward layers are used to sandwich the multi-head attention layer and the convolution block to form the entire Conformer module. The multi-head attention layer in the model in this paper uses a relative-position multi-head attention mechanism[12], which can capture longer-term dependencies and solve the context fragmentation problem. The specific formula is as Eq. 3:

$$\begin{aligned} w' &= x + \frac{1}{2} FFM(x) \\ w'' &= w' + \mathrm{R}\, PSA(w') \\ w''' &= w'' + Conv(w'') \\ \tilde{x} &= Layernorm(w''' + \frac{1}{2} FFM(w''')) \end{aligned} \tag{3}$$

$FFM(\bullet)$ represents the feedforward layer module, $RPSA(\bullet)$ represents the relative position of the multi-head attention mechanism layer, and $Conv(\bullet)$ represents the convolutional block.

The NBC module is an effective simplification of the Conformer module. When the Conformer module processes the global features, there are too many parameters, low feature utilization, and too long a calculation time. The Conformer module is simplified, and most of the features are retained. At the same time, the calculation time is greatly reduced. The specific NBC module is shown in Figure 2 (right). The relative multi-head attention layer and feed-forward convolution layer are used to extract features and reduce calculation time effectively. The specific formula is as Eq. 4:

$$\begin{aligned} x_i' &= x_{i-1} + dropout(RPSA(Layernorm(x_{i-1}))) \in R^{H_1 \times T} \\ x_i'' &= SiLU(Linear(Layernorm(x_i'))) \in R^{H_2 \times T} \\ x_{i,j}''' &= SiLU(GroupNorm(GroupConv(x_{i,j-1}'''))) \in R^{H_2 \times T} \\ x_i &= dropout(Linear(dropout(x_{i,L_2}'''))) \in R^{H_1 \times T} \end{aligned} \tag{4}$$

In Eq. 4, the hidden unit is increased to H_2 and the group of convolutional blocks is used to capture local temporal information. Each block consists of a 1D convolutional layer, a normalization layer, and a $SiLU$. Using 1D convolutional layers along the time series. $GroupConv(x_{i,j}''')$ represents the j-th group of convolution block processing in the i-th group of modified Confirmer modules. J is in the range of 1 and L_2.

A mask is learned for each learned source through a 2-D convolutional layer, and the mask is finally converted back to a sequence using overlapping addition.

2.4. Decode

The model of decode adopts a transposed convolutional layer with strides and kernels corresponding to the model of encode. The output h of the encode into the decode and the k-th mask m_k in the mask network. The formula of the decoder is as Eq. 5:

$$\hat{s}_k = conv1d - transpose(h * m_k) \tag{5}$$

$\hat{s}_k$ represents the estimated source of network model separation.

3. Experiments and Analysis

The experimental datasets are WSJ0-2mix[7] and WHAM! [13] dataset to evaluate network models. The WSJ0-2mix[7] dataset is from more than 30 hours of original training speech data and more than10 hours of original validation speech data from the WSJ0 corpus database[14] to randomly mix the two speakers and select different speakers original speech from the WSJ0 corpus database[14] training set si_tr_s to randomly SNR mixing, SNR is set from -5dB to 5dB. The test set is generated with a mixture of WSJ0[7] validation set from si-dt-05 document and evaluation set from si-et-05 document. The training and test sets were created using different speakers with sampling rate is set to 8kHz. WHAM![13] dataset is to randomly add urban noise and random sounds to the WSJ0-2mix[7] dataset and perform speech separation in a noisy environment with a sampling rate of 8kHz.

In the experiments, the encoder the stride factor size is 8 samples, kernel size is set to16 samples, and contains 256 convolution filters. The decoder's kernel and stride factor are the same as the encoder configuration. The length of the block in the experimental treatment is C=250 with 50% signal overlap and zero padding. 8 layers of transformers are used within and between blocks, the intra-block processing is repeated N=2 times, and the Conformer block uses 8 parallel attention heads and a 1024-dimensional position feedforward network. The NBC block uses 8 parallel attention heads and a 1024-dimensional feedforward network, and this model has about 23 million parameters. The optimizer uses the Adam algorithm with a learning rate of $15e-5$, after epoch15, if the results obtained for testing on the training set did not change over three consecutive epochs, the learning rate is done by halving. The L2 norm of the gradient is set to 5. By replacing the invariant loss with a scale-invariant signal-to-noise ratio (SI-SNR[11]). The system trains for 100 epochs, about 2 hours per epoch on a single NVIDIA 3090ti GPU with 24G of video memory.

3.1. Experimental Results

The best result of the DPCNBCN model experiment on the WSJ0-2mix[7] dataset is that the SI-SNR[11] is 21.6dB and the SDR is 21.8dB. The DPCNBCN model is compared with better models, as shown in Table 1:

Table 1 Experimental comparison of DPCNBCN model and literature model on WSJ0-2mix

Model	SI-SNR	SDR	Stride
Tasnet[7]	10.8	11.1	20
ConvTasnet[7]	15.3	15.6	10
DeepCASA[14]	17.7	18.0	1
Dual_PathRNN[8]	18.8	19.0	1
Sudo rm -rf[15]	18.9	n.a.	10
DPTNet[9]	20.2	20.6	1
Wavesplit[10]	21.0	21.2	1
Wavesplit+DM[10]	22.2	22.3	1
SepFormer[10]	20.4	20.5	8
DPCNBCN	**21.6**	**21.8**	**8**

Table 2 Ablation experiments of DPCNBCN on WSJ0-2mix

SI-SNR	N	Conformer	NBC	Heads	DFF	PosENC
21.6	2	8	8	8	1024	yes
21.2	2	4	4	16	2048	yes
21.0	2	4	4	8	2048	yes
20.9	3	4	4	8	2048	yes
20.7	2	4	1	8	2048	yes
20.2	2	3	3	8	2048	no

At the same time, an ablation experiment was performed on the DPCNBCN model on the WSJ0-2mix[7] validation set, and different parameters were quantitatively analyzed in Table 2.

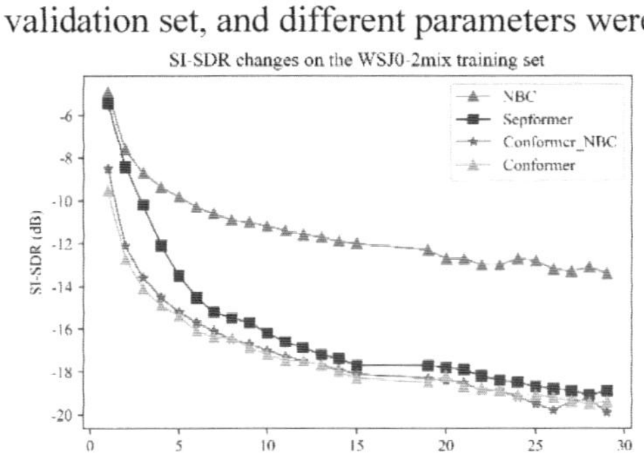

(a)

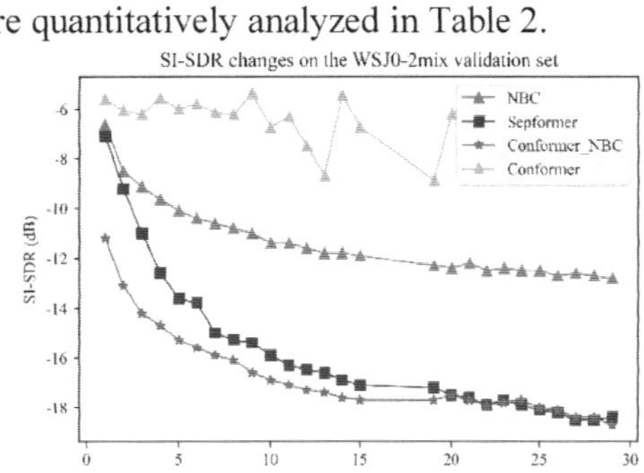

(c)

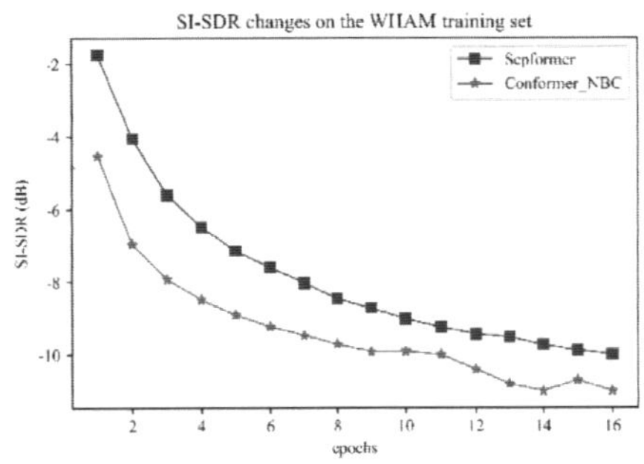

(b)

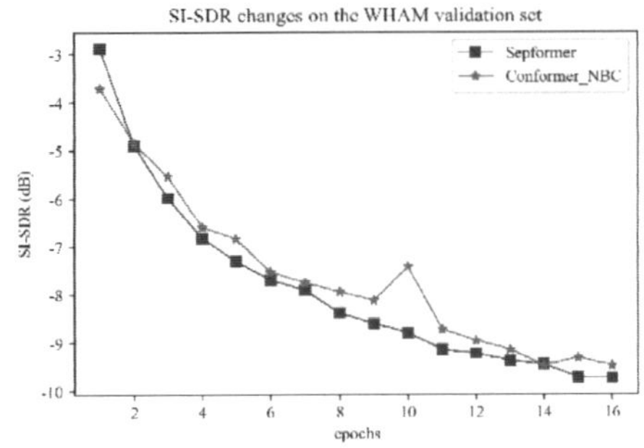

(d)

Figure 3 (a) SI-SDR[10] variation of each model on the WSJ0-2mix[7] training set
(c) SI-SDR[10] variation of each model on the WSJ0-2mix[7] test set
(b) Sepformer[10] and DPCNBCN SI-SDR[10] variation on the WHAM training set
(d) Sepformer[10] and DPCNBCN SI-SDR[10] variation on the WHAM test set

In figure3(a), the red line indicates that the NBC module is used both within and between blocks; the blue line show the experimental results of Sepformer[10]; the green line show the experimental results of the DPCNBCN model; the yellow line indicates that the Conformer module is used both within and between blocks. It is obvious from the figure that the DPCNBCN model converges faster in the process of training the model, and the separation effect is better. In figure3(c), it is obvious that the convergence of the Conformer module is not obvious within and between blocks. The DPCNBCN model converges faster than the Serformer[10] model. In figure3(b) and (d), in the WHAM[13] dataset, the DPCNBCN model converges faster than the Sepformer[10] model, but in the test set, the Sepformer[10] model and the DPCNBCN model converge at the same speed. The synthesized speech has noise, but the noise is eliminated by separation. The separated speech does not contain noise, and the model has better robustness.

4. Concluding remarks

This paper proposes a separation model DPCNBCN that integrates the Conformer module and the NBC module. Using the dual-path framework of DPTNet, Coformer is used within speech blocks and NBC is used between blocks, replacing the RNN with a Transformer that learns short-term and long-term dependencies. The DPCNBCN model converges faster, takes less training time, and has better separation. There is also good separation in noisy mixed sources. It not only has a better performance effect on the blind mixed voice source, but also has better effect on the separation of other mixed sources.

References

[1] Bronkhorst A W. The cocktail party phenomenon: A review of research on speech intelligibility in multiple-talker conditions[J]. Acta Acustica united with Acustica, 2000, 86(1): 117-128.
[2] Wang D L, Brown G J. Computational auditory scene analysis: Principles, algorithms, and applications[M]. Wiley-IEEE press, 2006.
[3] Schmidt M N, Olsson R K. Single-channel speech separation using sparse non-negative matrix factorization[C]//Interspeech. 2006, 2: 2-5.
[4] Virtanen T. Speech recognition using factorial hidden Markov models for separation in the feature space[C]//Interspeech. 2006.
[5] Hershey J R, Chen Z, Le Roux J, et al. Deep clustering: Discriminative embeddings for segmentation and separation[C]//2016 IEEE international conference on acoustics, speech and signal processing (ICASSP). IEEE, 2016: 31-35.
[6] Yang G P, Tuan C I, Lee H Y, et al. Improved speech separation with time-and-frequency cross-domain joint embedding and clustering[J]. arXiv preprint arXiv:1904.07845, 2019.

[7] Luo Y, Mesgarani N. Conv-tasnet: Surpassing ideal time–frequency magnitude masking for speech separation[J]. IEEE/ACM transactions on audio, speech, and language processing, 2019, 27(8): 1256-1266.
[8] Luo Y, Chen Z, Yoshioka T. Dual-path rnn: efficient long sequence modeling for time-domain single-channel speech separation[C]//ICASSP 2020-2020 IEEE International Conference on Acoustics, Speech and Signal Processing (ICASSP). IEEE, 2020: 46-50.
[9] Chen J, Mao Q, Liu D. Dual-path transformer network: Direct context-aware modeling for end-to-end monaural speech separation[J]. arXiv preprint arXiv:2007.13975, 2020.
[10] Subakan C, Ravanelli M, Cornell S, et al. Attention is all you need in speech separation[C]//ICASSP 2021-2021 IEEE International Conference on Acoustics, Speech and Signal Processing (ICASSP). IEEE, 2021: 21-25.
[11] Gulati A, Qin J, Chiu C C, et al. Conformer: Convolution-augmented transformer for speech recognition[J]. arXiv preprint arXiv:2005.08100, 2020.
[12] Dai Z, Yang Z, Yang Y, et al. Transformer-xl: Attentive language models beyond a fixed-length context[J]. arXiv preprint arXiv:1901.02860, 2019.
[13] Wichern G, Antognini J, Flynn M, et al. Wham!: Extending speech separation to noisy environments[J]. arXiv preprint arXiv:1907.01160, 2019.
[14] Liu Y, Wang D L. Divide and conquer: A deep CASA approach to talker-independent monaural speaker separation[J]. IEEE/ACM Transactions on audio, speech, and language processing, 2019, 27(12): 2092-2102.
[15] Tzinis E, Wang Z, Smaragdis P. Sudo rm-rf: Efficient networks for universal audio source separation[C]//2020 IEEE 30th International Workshop on Machine Learning for Signal Processing (MLSP). IEEE, 2020: 1-6.

A New Bandwidth Reconfigurable Antenna

Yiping Zang ,Weishan Li, Jie Ma*, Jiangjun Yang, and Wenjun Xiong

Xi'an Institute of Electromechanical Information Technology, Xi'an, China

*875129145@qq.com

Abstract—A bandwidth reconfigurable monopole antenna with disc-shaped is presented in this article. This presented antenna is integrated with a disc-shaped monopole antenna and a cascaded bandpass structure, and four different bandwidth states can be adjusted by the bandpass structure. By switching the diode's on-off state integrated with the bandpass structure, this presented monopole antenna can switch quickly among four different bandwidth states whose fractional bandwidth is range from 16% to 55% centered at around 1.8 GHz. The experiment also shows good and stable electromagnetic properties of this presented antenna in different working bandwidths.

1. Introduction

Currently, needs for reconfigurable terminal devices are continuously growing in order to satisfy multifunctional communication systems. Additionally, in next-generation wireless communication, many new reconfigurable antennas will be extensively applied owing to their quick switching property on the basis of different application backgrounds. Reconfigurable antenna technologies, such as frequency reconfiguration, polarization reconfiguration, radiation pattern reconfiguration, and so on, can flexibly control its characteristic. For instance, antennas with tuned frequency are investigated in literature [1]-[4]. Antennas with polarization reconfigurable are proposed by adopting a parasitic pixel layer in [5]-[6]; Antennas with radiation pattern reconfigurable are studied in [7]-[8]. Compared to the adjusted frequency, polarization, and radiation pattern, scarce research is investigated on adjusting the antenna's operating bandwidth. However, more and more requirements come out for the controlling of users' numbers by the working bandwidth in a local or secure communication system. Therefore, a disc-shaped microstrip monopole antenna with reconfigurable bandwidth will be studied in this research.

In this article, a new antenna with bandwidth reconfigurable is proposed by incorporating a reconfigurable compact bandpass structure based on short-circuited coupled lines and short-circuited stubs. Through turning the diode's on-off state which integrated into the bandpass structure, the antenna's operating state can switch quickly among four different bandwidths with similar center frequencies. The present antenna has four bandwidth states whose relative bandwidth is range from 16% to 55% centered at near 1.8 GHz.

2. Antenna Design

Firstly, a microstrip disc-shaped monopole antenna depicted in Fig. 1(a) is introduced. The monopole antenna is simulated by simulation software HFSS. The calculation results of the software are displayed in Fig. (b) and Fig. (c). The antenna's operating band can cover from 1.2GHz to 3GHz shown in Fig. 1(b), and the radiation characteristic at 1.8GHz is plotted in Fig. 1(c).

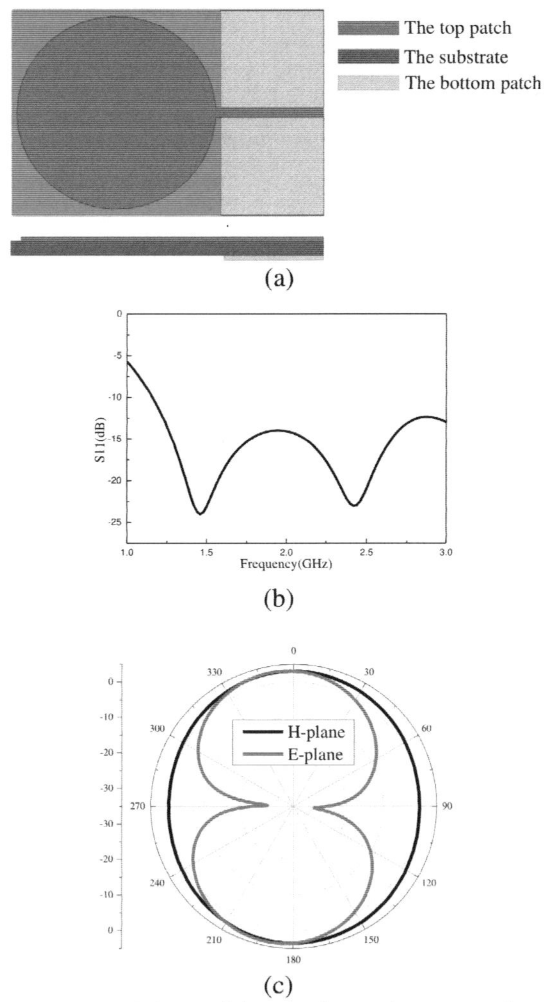

Fig. 1. (a) Simulation structure of the traditional microstrip monopole antenna, (b) Reflection coefficients, (c) Radiation pattern at 1.8GHz.

A new antenna with bandwidth reconfigurable based on the monopole antenna mentioned above is designed, displayed in Fig.2(a). Reconfigurable antenna is devised using RO4730 (thickness=1.542mm, dielectric constant ε_r=2.98, tanδ= 0.0027). The antenna contains a disc-shaped monopole antenna and a reconfigurable bandpass structure. As seen from the side view of the presented reconfigurable antenna, the T-shaped metal ground is etched on bottom of the plate. Bandpass structure is composed of an input microstrip line whose impedance is 50Ω, short-circuited coupled lines, the biasing circuits for PIN diodes' on-off, and short-circuited stubs with a $\lambda/4$ length (λ is a wavelength at the center frequency). Specific parameter are provided in Table 1. The PIN diode can be regarded as a series of the capacitor (capacitance C=0.025pF) and resistor (resistance R=4).

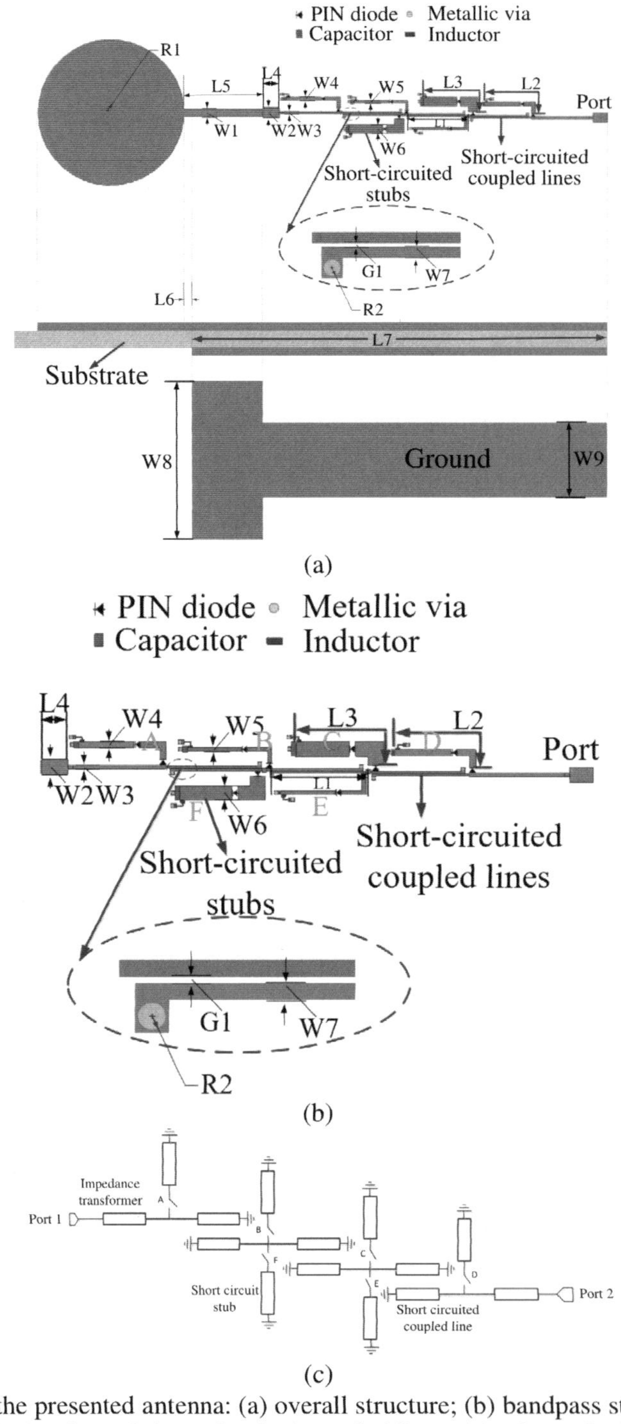

Fig. 2. Schematic of the presented antenna: (a) overall structure; (b) bandpass structure based on short-circuited stubs and short-circuited coupled lines; (c) schematic diagram.

For practical implementation, depending on the odd-even mode theory, the bandwidth of the passband structure could be designed and altered in [9]-[10] by turning the short-circuit stub's characteristic impedance. Therefore, by selectively switching the short-circuited stub's on-off or altering the short-circuit stub's impedance, the bandwidth can be changed. The performances of the bandpass structure have been verified by HFSS software [11]. Fig. 3 depicts the simulated results of passband structure at different states: (i) all PIN diodes are OFF; (ii) only PIN diodes are embedded in stub "B" and "E" are ON; (iii) only PIN diodes are embedded in stub "C "and "F" are ON; (iv) all PIN diodes are ON.

TABLE 1 DIMENSIONS OF THE PROPOSED ANTENNA (UNIT: MM)

Parameters	W1	W2	W3	W4	W5	W6	W7	W8	W9	R1
Values (mm)	3.82	0.95	1.60	1.00	3.82	3.80	0.50	84.00	19.50	30.00
Parameters	G1	L1	L2	L3	L4	L5	L6	L7	R2	
Values (mm)	0.20	24.10	25.50	27.60	6.70	33.10	0.62	172.8	0.40	

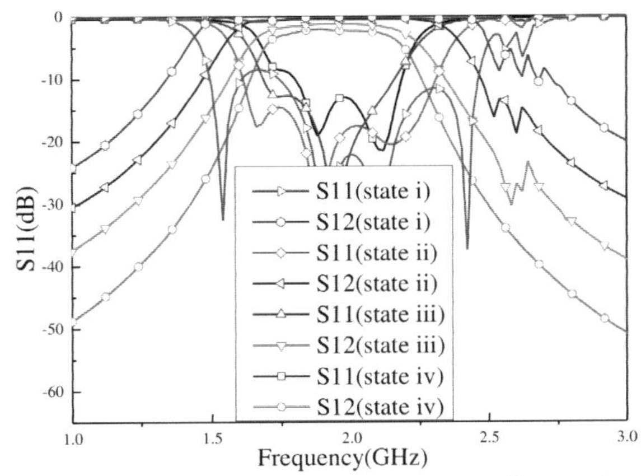

Fig. 3. Simulated results of the bandpass structure at four various situations.

As depicted from Fig. 3, state (i) corresponds to the widest bandwidth, state (ii) corresponds to the second widest bandwidth, state (iii) corresponds to the third widest bandwidth, and state (iv) corresponds to the narrowest bandwidth. It is clearly seen that the designed compact bandpass structure has a good function of reconfigurable bandwidth through switching PIN diodes' on-off.

After completion of the bandpass structure's design, the whole antenna depicted in Fig. 2(a) is devised and simulated. The presented antenna's simulated result of reflection coefficient for four different bandwidth states is depicted in Fig. 4. The presented antenna's bandwidth will be changed by switching the corresponding PIN diode's on-off state. To further study the presented antenna's working mechanism, the surface current on the reconfigurable antenna at four different bandwidth states is displayed in Fig.5. From the figure, the different stubs are effectively controlled by turning PIN diodes' on-off, and four stubs' working state corresponds to four different bandwidth states, so the bandwidth can be adjusted by selectively switching short-circuited stub's on-off.

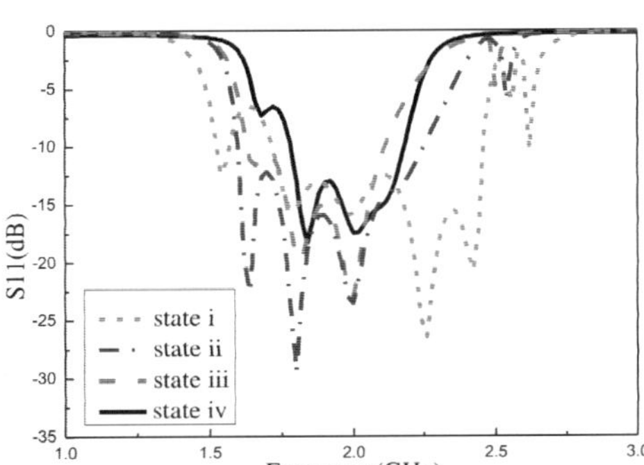

Fig.4. Simulated results of the presented antenna at four various situations.

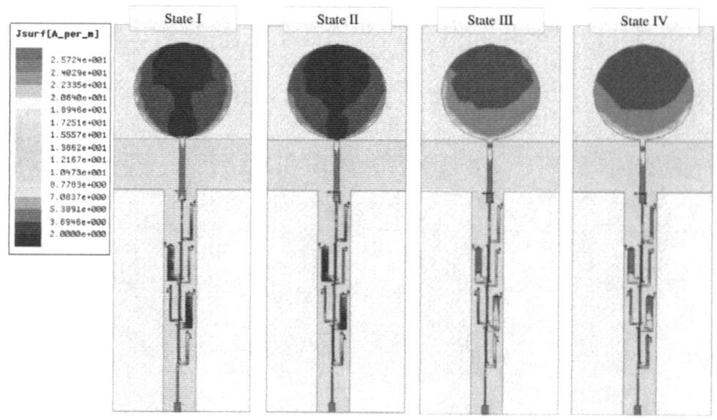

Fig. 5. Surface current of the presented antenna under four different bandwidth states

3. Measured Results and Analysis

In order to verify the idea, the presented antenna prototype depicted in Fig. 6 is fabricated. The PIN diode model MA-COM/MA4FCP300 is chosen to switch stub's on-off.

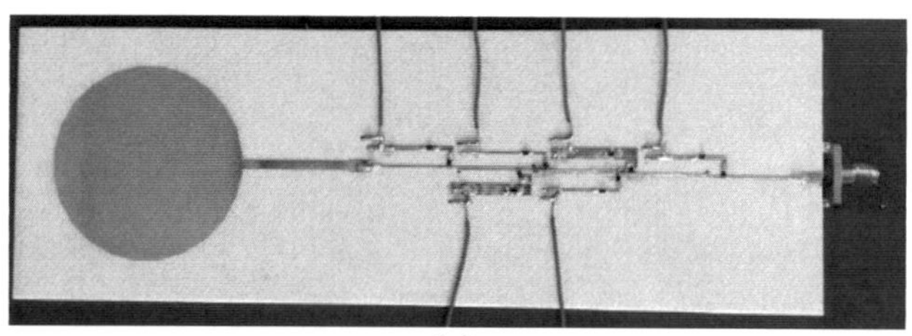

(a)

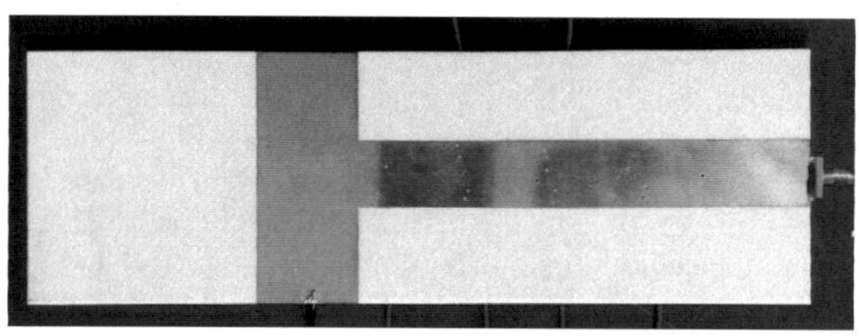

(b)

Fig.6. The manufactured antenna: (a) Top View; (b) Bottom View.

The measured results of reflection coefficients at four different states are shown in Fig. 7. Based on the above analysis, all PIN diodes are switched OFF, and the widest bandwidth is obtained. When PIN diodes on stub "B" and "E" are ON, the second widest bandwidth is obtained. When PIN diodes on the stub "C" and "F" are ON, the third widest bandwidth is obtained. When all PIN diodes are ON, the narrowest bandwidth is obtained. From the measured data, the designed antenna has four different bandwidth states, whose fractional bandwidth ranges from around 16% to 55%.

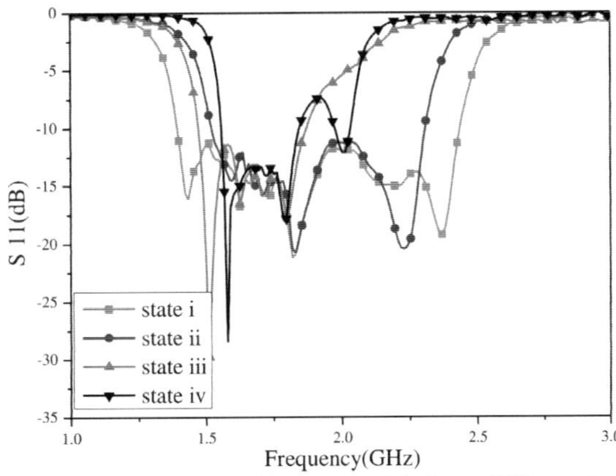

Fig. 7. The measured reflection coefficients with four different bandwidth states.

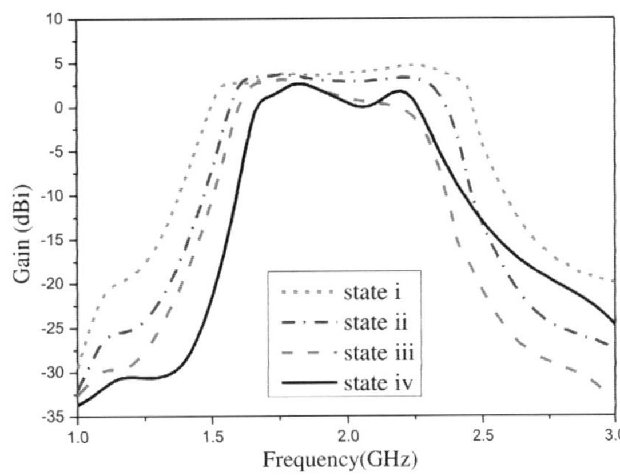

(a)

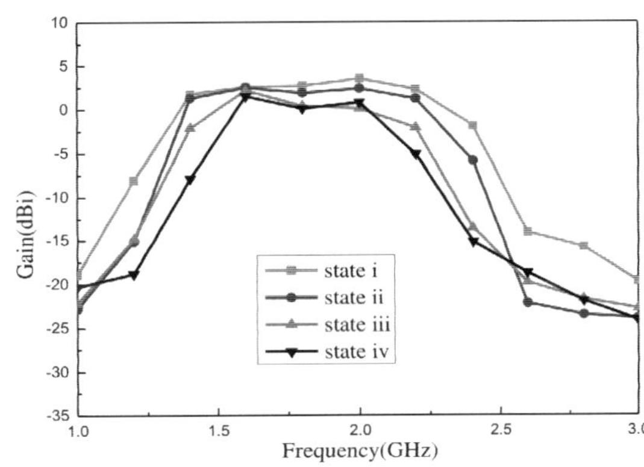

(b)

Fig. 8. The gains of the antenna with four distinct bandwidth states: (a) simulated results; (b) measured results.

From Fig. 4 and Fig. 7, the simulated result is consistent with the measured results except for a small frequency shift and deteriorated impedance match. Moreover, better out-band suppression of the proposed antenna is obtained. The discrepancies mainly come from the tolerances of the substrate's dielectric constant (±1.6% error specified by the manufacturer), the machining error, and the tolerances of the PIN diode.

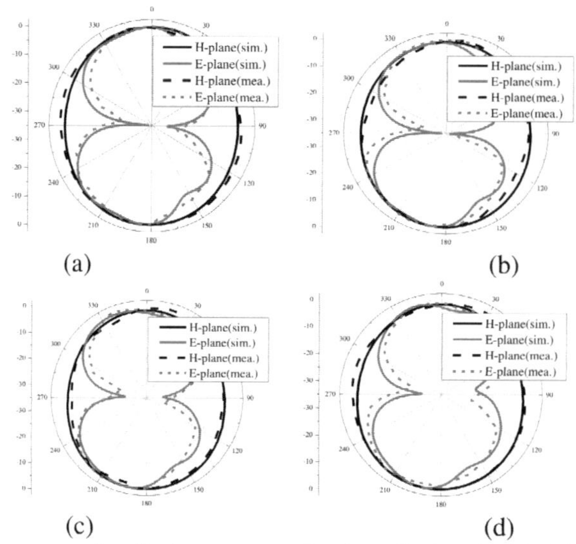

Fig. 9. Simulated and measured radiation patterns for four distinct bandwidth states at 1.8GHz: (a) state (i); (b) state (ii); (c) state (iii); (d) state (iv).

The calculated and measured gains of the proposed antenna at four various bandwidth states are displayed in Fig. 8 (a) and Fig.8(b). From Fig.8, stable gains in working bands are obtained. Calculated and measured radiation characteristic at four different situations at 1.8GHz are displayed in Fig. 9 (a), (b), (c), and (d). Better consistency in calculated and measured results is displayed in Fig. 9. Moreover, the H-plane and E-plane of radiation pattern are stable and similar to classical monopole antenna at four different working states, which shows that the radiation characteristic of the presented antenna are relatively stable, even though short-circuit stub, short-circuited coupling line, and PIN diodes are added. More flexible reconfigurable bandwidth states can be extended by introducing more short-circuit stubs, short-circuited coupling lines, and PIN diodes.

4. Conclusion

A disc-shaped microstrip monopole antenna with reconfigurable bandwidth has been presented. The switch of short-circuited stubs are used to vary the antenna among four distinct states with fixed center frequency, so the four bandwidth states of the antenna are flexibly changed through tuning the on-off situation of six PIN diodes. Moreover, the introduction of short stubs and PIN diodes in the proposed antenna has almost no influence on the antenna's performance.

References

[1] M. S. Alam, and A. Abbosh, "A compact reconfigurable antenna with wide tunable frequency and 360° beam scanning," IEEE Antennas Wireless Propag. Lett., vol.18, no.1, pp.4-8, 2019.

[2] S. Lee, S. I. H. Shah, H. L. Lee, and S. Lim, "Frequency-reconfigurable antenna inspired by origami flasher," IEEE Antennas Wireless Propag.Lett., vol.18, no.8, pp.1691-1695, 2019.

[3] S. Subbaraj, M. Kanagasabai, M.G. N. Alsath, et al. "A compact frequency-reconfigurable antenna with independent tuning for hand-held wireless devices," IEEE Trans. Antennas Propag., vol. 68, no. 2, pp. 1151-1154, 2020.

[4] G. Jin, C. Deng, Y. Xu, J. Yang, and S. Liao, "Differential frequency-reconfigurable antenna based on dipoles for sub-6 GHz 5G and WLAN applications," IEEE Antennas Wireless Propag.Lett., vol.19, no.3, pp.472-476, 2020.

[5] D. Rodrigo, B. A.Cetiner, and L. Jofre. "Frequency, radiation pattern and polarization reconfigurable antenna using a parasitic pixel layer," IEEE Trans. Antennas Propag., vol. 62, no. 6, pp. 3422-3427, 2014.

[6] W. Lin, and H. Wong, "Polarization reconfigurable wheel-shaped antenna with conical-beam radiation pattern," IEEE Trans. Antennas Propag., vol. 63, no.2, pp. 491-499, 2015.
[7] G. H. Huff, and J. T. Bernhard, "Integration of packaged RF MEMS switches with radiation pattern reconfigurable square spiral microstrip antennas," IEEE Trans. Antennas Propag., vol.54, no.2, pp. 464-469, 2006.
[8] S.-J. Shi, and W.-P. Ding, "Radiation pattern reconfigurable microstrip antenna for WiMAX application," Electron. Lett., vol.51, no.9, pp. 662-664, 2015.
[9] G. L. Matthaei, L. Young, and E. M. T. Jones, Microwave Filters, Impedance Matching Networks and Coupling Structures. Norwood, MA: Artech House, 1985.
[10] A. Miller, and J. S. Hong, "Cascaded coupled line filter with reconfigurable bandwidths using LCP multilayer circuit technology," IEEE Trans. Microw. Theory Tech., vol. 60, no. 6, pp. 1577-1586, 2012.
[11] High Frequency Structure Simulator (HFSS). Ansoft Corp., Pittsburgh, PA.

Low profile filtering antenna with high selectivity without extra circuit

Weishan Li, Yiping Zang, Jie Ma*, Jiangjun Yang, and Wenjun Xiong

Xi'an Institute of Electromechanical Information Technology, Xi'an, China

*875129145@qq.com

Abstract—A novel microstrip filtering antenna with high selectivity is presented. A shorting SIR stub and a square patch with a rectangular ring slot are designed to produce a couple of radiation nulls near the passband, respectively. The bandwidth (3.44GHz-3.94GHz) is wider than that in a traditional antenna. In the presented antenna, the selectivity of the right side of passband is 30.3 dB, and the selectivity of the left side of passband is 18.2dB without an extra circuit. The proposed antenna consists of three layers of stacking structure connected tightly up and down, so it has an easy fabrication structure. Flat gain, compact size 0.49λ*0.49λ*0.07λ, and higher front and back ratio (>20dB) are obtained than the conventional antenna.

1. Introduction

Microwave components' requirement for high integration and multi-functions increased day by day with wireless communication technology's rapid development. In order to realize various functions, the antenna is integrated with different passive components like filter, diplexer, phase shifter, and power divider at RF Front-End. This series integration method is very convenient to use, but it leads to a larger size at RF Front-End and an increased loss such as the integration of antenna with filter in series. Moreover, the operating band's performance of the filtering antenna will be disturbed owing to the mismatch at the joint.

Filtering antennas are analyzed by adopting different approaches [1]-[9]. The first technique method adopted in [1]-[3] is the combination of an antenna with a filtering network. In [1], a filtering antenna with a wideband and circularly polarized characteristic is designed. A single band filtering antenna [2] with a low profile and wideband are designed and analyzed. However, the selectivity of these antennas is not very satisfactory, and the front and back ratio of these antennas is lower compared with the antenna whose ground is distributed on the whole structure's bottom. Filtering antenna with harmonic restriction and high selectivity is represented in [3]; nevertheless, the bottom distribution of the filtering network will limit the integration with other devices, which are located at the antenna's bottom and could affect the performance of the antenna via the filtering network. The second technique applied in [4][5] is the combination of an antenna with a feeding structure. Parasitic patch [4] and dielectric resonator [5] are used as radiation structures; moreover, new filtering feeding structures and various slots' shapes are employed to realize the filtering function. Furthermore, these antennas have high gain and low profile. Nonetheless, owing to the feeding structure being situated at the bottom of the ground, the back side radiation will be enhanced, and thus the front and back ratio is not very high. The third technique employed in [6] is a multi-layers coupling method based on the filter's design principle. The filtering antenna with high selectivity and dual beam [6] is designed, but the size of the ground is not very small. The fourth method adopts multiple shorting pins to produce radiation null to satisfy the

 2384 (2022) 012036 doi:10.1088/1742-6596/2384/1/012036

filtering ability [7]. High gain and filtering ability are achieved in [7] by loading shorting pin, U-shaped slot, and stack patch, but the whole dimension of the antenna is a little bigger. The fifth method is a single-layer coupling method based on the theory of passband filter in [8]. Both antennas have a high gain and low profile, but the bandwidth is not very wide. A filtering antenna with a novel feeding scheme and low profile is designed in [9], but it has a little bigger size and narrower bandwidth.

2. ANTENNA CONFIGURATION AND MECHANISM

2.1. Antenna Configuration

Fig.1 displayed the filtering antenna's whole structure. The high selectivity filtering antenna contains a top patch, middle patch, feeding structure, two small U-shaped patches, two square parasitic patches, big U-shaped patch, two square shorting patches, two long metal pins, two short shorting pins, ground, and three layers substrate, and these metal structures are printed on the surface of these substrates whose permittivity are 2.65 and height is 2 mm. The top patch printed on the top patch is a square patch with an etched square groove in the center. The middle patch is a square patch with a square annular shape slot in the center. The feeding structure, square shorting patch, big U-shaped patch, small U-shaped patch, and square parasitic patch are etched on the bottom layer substrate. Two square shorting patches, small U-shaped patches, and square parasitic patches are surrounded by the big U-shaped patch symmetrically located on the feeding structure. The ground is distributed on the bottom dielectric slab. Two long metal pins distributed on outside of the big U-shaped patches connect middle patch to the ground through the middle layer substrate and bottom layer substrate. Two short metal pins connect the square shorting patch to the ground through the bottom layer substrate. Three layers of substrate connect closely to each other from top to bottom. The filtering antenna's detailed sizes are displayed in Table I.

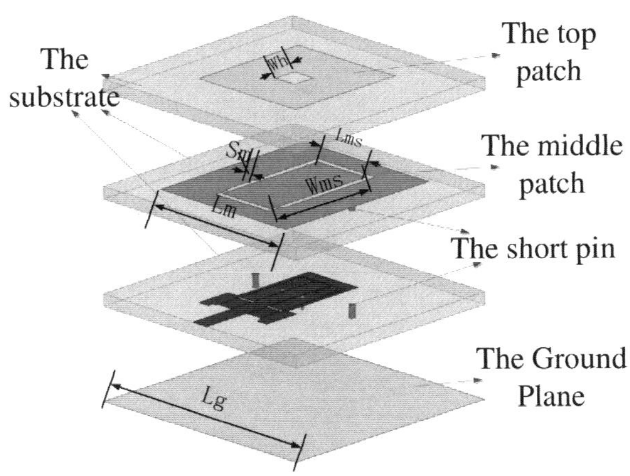

(a)

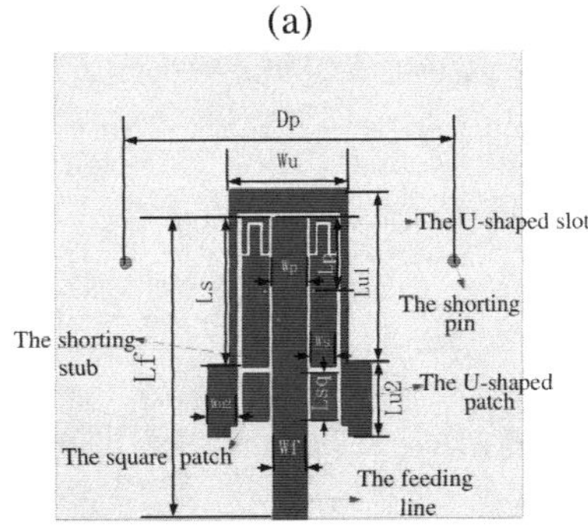

(b)

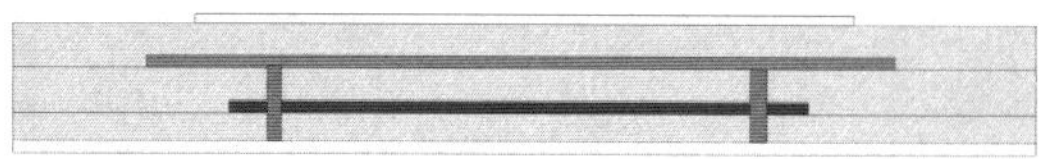

(c)

Figure. 1. Structure of the presented antenna: (a) exploded view; (b) top view of the bottom substrate; (c) side view.

2.2. Antenna Mechanism

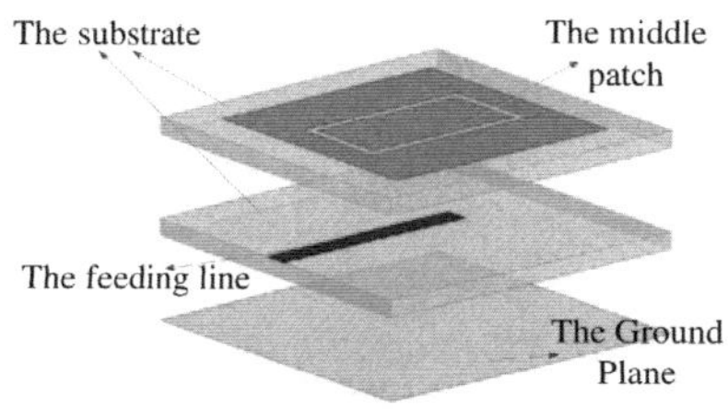

Figure. 2. Exploded view of the antenna I.

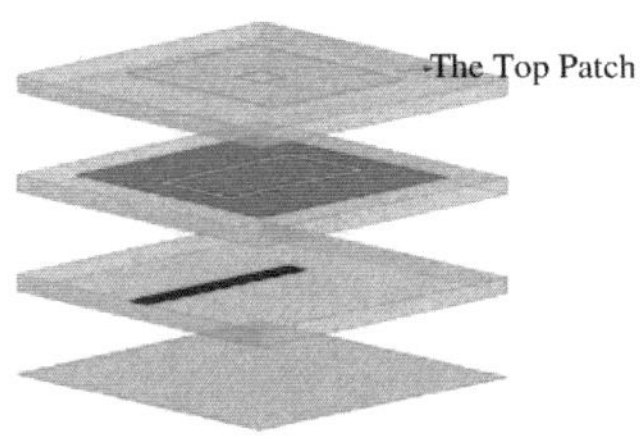

Figure .3. Exploded view of the antenna II.

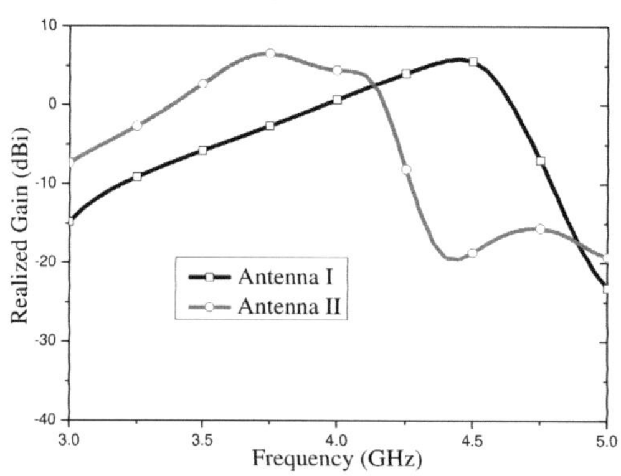

Figure. 4. Simulated gain of the antenna I and antenna II

Firstly, antennas I and II, whose structures are depicted in Fig.2 and Fig.3, are designed. The only difference between antenna I and II is that an extra top patch on the top substrate is added in antenna II. The results are displayed in Fig.4. In antenna II, the gain is more flatter than antenna I, and one radiation null is generated.

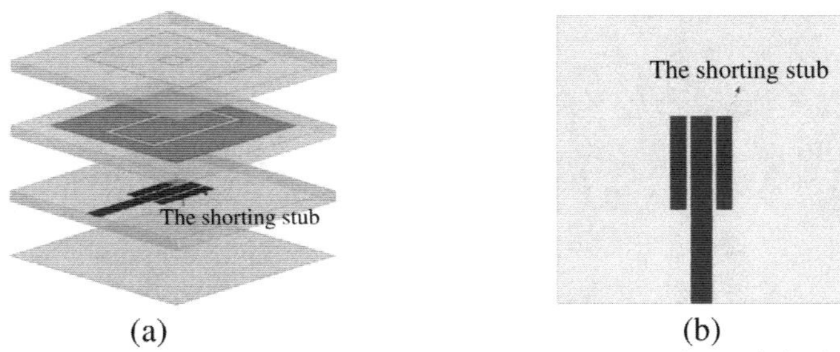

Figure. 5. Structure of the antenna III: (a) exploded view; (b) top view of the bottom substrate.

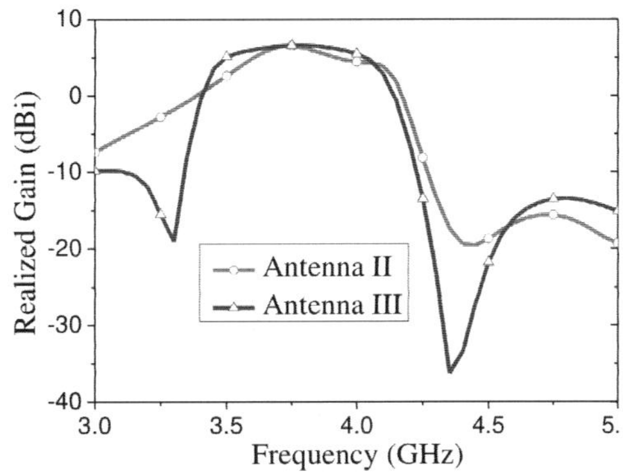

Figure. 6. Simulated gain of antenna II and antenna III.

Secondly, antenna III, whose structure is shown in Fig. 5, is devised to raise the selectivity on the left side of the passband. In antenna III, two shorting stubs situated on both sides of the feeding line are added based on antenna II's structure. The shorting stubs as a parallel resonator can suppress the energy's radiation at the resonant frequency. The gain results are shown in Fig. 6. In antenna III, a radiation null is obtained compared with antenna II from Fig.6. Two radiation nulls' generation enhances the selectivity on both sides of the passband, but the gain far away from the radiation null is a little big.

Thirdly, antenna IV, which is depicted in Fig. 7, is designed to strengthen the suppression on the right side of the passband. Two shorting pins which link the middle patch to the ground are added in antenna IV based on antenna III. The simulated gain result is displayed in Fig. 8. Another radiation null is generated compared with antenna III, and the gain in the passband and the radiation null are little affected by these two shorting pins.

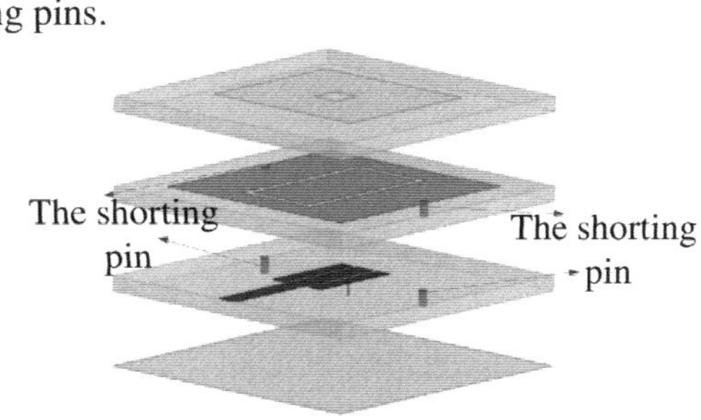

Figure. 7. Exploded view of the antenna IV.

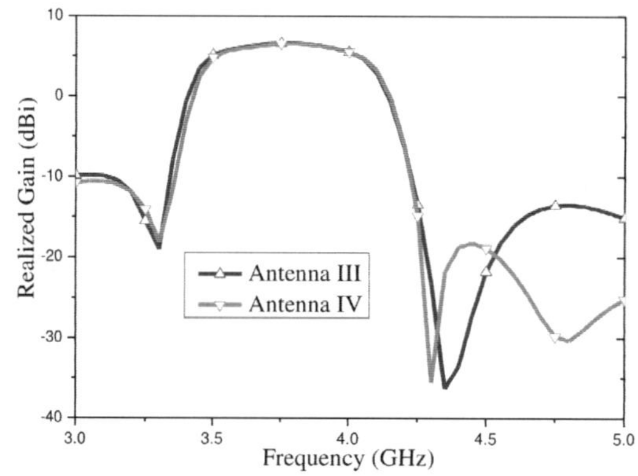

Figure. 8. Simulated gain of antenna III and antenna IV.

Fourthly, antenna V, whose structure is displayed in Fig. 9, is devised to raise the suppression on the left side of the passband. In antenna V, a U-shaped patch and two square patches distributed around the shorting pin line are added based on antenna IV. The result of the gain is depicted in Fig. 10. Another radiation null is introduced compared with antenna IV, and the gain operating band and the radiation null on right side of passband is little influenced by the U-shaped patch. Moreover, antenna V at around 3.7 GHz is not working based on the traditional definition of impedance bandwidth (S11< -10dB) displayed in Fig.11.

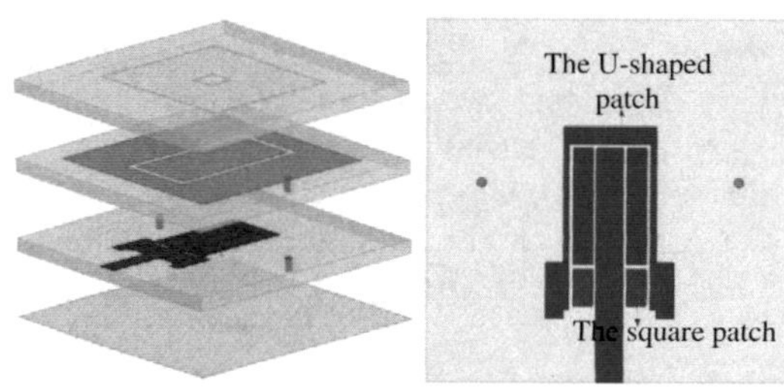

Figure. 9. Structure of the antenna V: (a) exploded view; (b) top view of the bottom substrate.

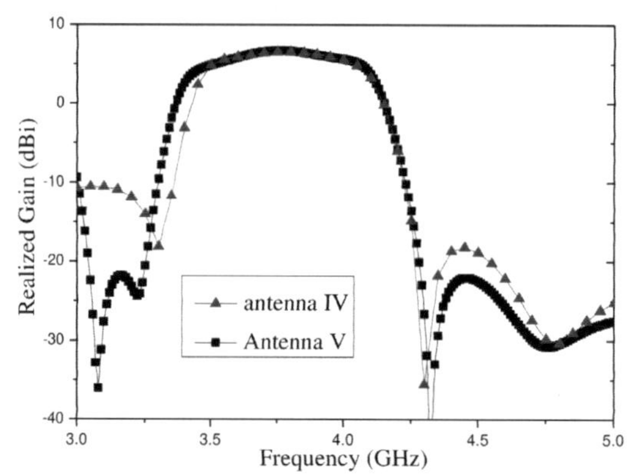

Figure. 10. Simulated gain of the antenna IV and antenna V.

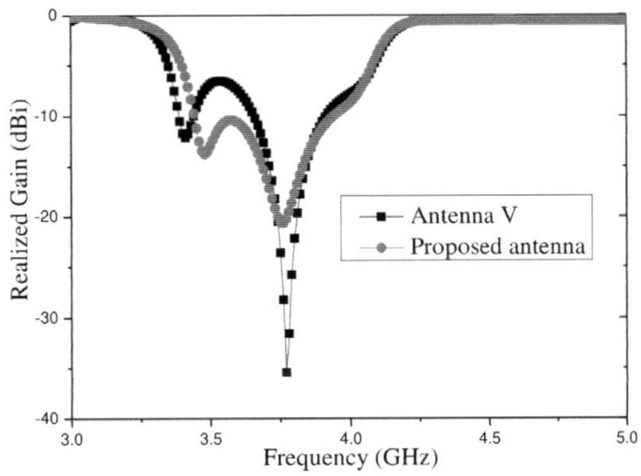

Figure. 11. Simulated reflection coefficient of the antenna IV and antenna V.

Finally, the filtering antenna, which can work at around 3.7 GHz, is designed. In the presented antenna, the input impedance will be appropriately changed through altering the size of the U-shaped groove on the shorting stub. The result depicted in Fig. 11 can reveal that the presented antenna can work at around 3.7GHz. Although introducing the U-shaped slot, the gain and the radiation null cannot be influenced significantly.

TABLE 1 DIMENSIONS OF THE PROPOSED ANTENNA

Parameters	Lg	Lm	Ls	Lms	Lf	Lu1	Lu2	Lsq	Lp
Values(mm)	40	30	13	10	26	15	6.2	4.1	6.5
Parameters	Wh	Wms	Wu	Wf	Wp	Ws	Wu2	Dp	Sm
Values(mm)	4	20	10.1	2.9	4.1	2.6	2.5	28	1

3. RESULTS AND DISCUSSION

The detailed size is listed in Table 1. HFSS is employed to simulate all models during the design process.

Fig. 12 depicts the simulated results of the gain. It can be observed that the simulated average gain is 6.4dBi in operating band. Two radiation nulls emerged at 3.1GHz and 3.3GHz on the left of the passband, which will raise the selectivity of the left side of band, while two radiation nulls emerged at 4.34GHz and 4.68GHz will raise the selectivity of the right side of band.

Fig.13 displays the calculated radiation characteristics of the designed antenna at 3.75GHz. The radiation patterns are stable, and a higher front and back ratio (>20dB) is obtained. The 3-dB beamwidth at H-plane and E-plane is ~90°and ~88°.

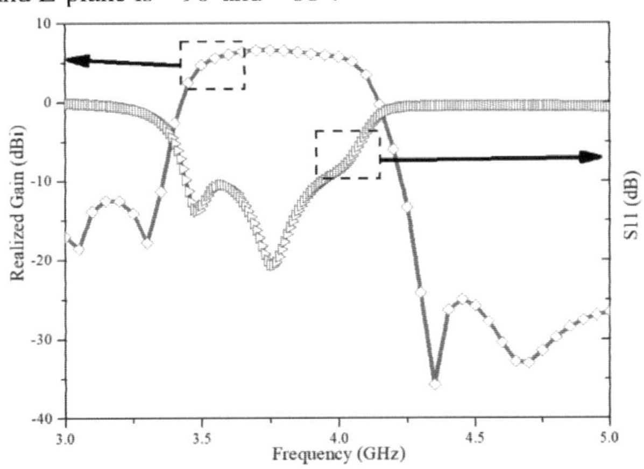

Figure. 12. The simulated S11 and gain of the designed antenna

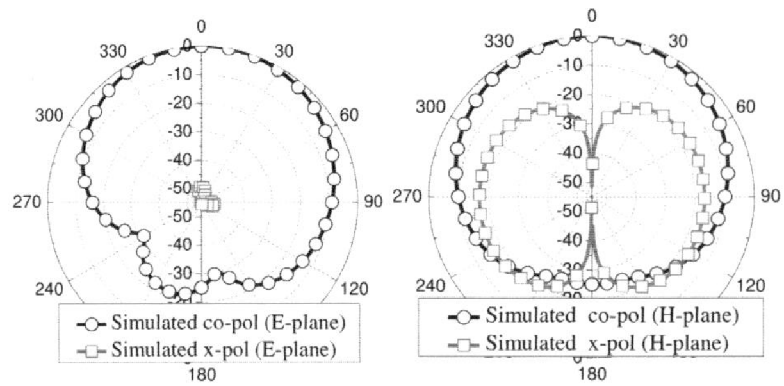

Figure. 13. Simulated radiation patterns of the antenna in E-plane and H-plane at 3.75GHz.

4. Conclusion

A multilayer filtering antenna with high selectivity and low-profile is presented. A shorting SIR stub and a square patch with a rectangular ring slot are designed to produce a couple of radiation nulls near the passband, respectively. The wide bandwidth (3.44GHz-3.94GHz) is wider than that of a traditional antenna. In the presented antenna, the selectivity of right side of band is 30.3dBi, and the selectivity of the left side of band is 18.2dBi without an extra circuit. The proposed antenna consists of three layers of stacking structure connected tightly up and down, so it has an easy fabrication structure. A small size ($0.49\lambda*0.49\lambda*0.07\lambda$) is acquired. Moreover, the gain in operating band is flat. A higher front and back ratio (>20dB) is obtained compared with a traditional antenna in the designed antenna.

References

[1] Z. H. Jiang, D. H. Werner, "A compact, wideband circularly polarized co-designed filtering antenna and Its application for wearable devices with low SAR" IEEE Trans. Antennas Propag., vol.63, no.9, pp. 3808-3818, Sep. 2015.

[2] C. X. Mao, S. Gao, Y. Wang, B. Sanz-Izquierdo, Z. Wang, F. Qin, Q. X. Chu, J. Li, G. Wei, and J. Xu, "Dual-band patch antenna with filtering performance and harmonic suppression" *IEEE Trans. Antennas Propag.*, vol.64, no.9, pp. 4074-4077, Sep. 2016.

[3] C-X. Mao, S. Gao, Y. Wang, Z. Wang, F. Qin, B. Sanz-Izquierdo, and Q-X, Chu, "An integrated filtering antenna array with high selectivity and harmonics suppression" *IEEE Trans. Microw. Theory Techn.*, vol.64, no.6, pp.1798-1805, Sep. 2016.

[4] Y. M. Pan, P. F. Hu, and X. Y. Zhang, "arasitic Patch Antenna with Filtering Response," *2016 IEEE 5th Asia-Pacific Conference on Antennas and Propagation (APCAP)., Auckland, pp. 315-316, 2016.*

[5] P. F. Hu, Y. M. Pan, X. Y. Zhang, and S. Y. Zheng, "A Compact Filtering Dielectric Resonator Antenna With Wide Bandwidth and High Gain" IEEE Trans. Antennas Propag., vol.64, no.8, pp. 3645-3651, Aug. 2016.

[6] X-Y. Zhang, D. Xue, L-H. Ye, Y-M. Pan, and Y. Zhang, "Compact Dual-Band Dual-Polarized Interleaved Two-Beam Array With Stable Radiation Pattern Based on Filtering Elements" IEEE Trans. Antennas Propag., vol.60, no.5, pp. 4566-4575, Sep. 2017.

[7] X. Y. Zhang, W. Duan, and Y-M. Pan, "High-Gain Filtering Patch Antenna Without Extra Circuit" IEEE Trans. Antennas Propag., vol.63, no.12, pp. 5883-5888, Dec. 2012.

[8] C-K Lin and S-J Chung, "A Filtering Microstrip Antenna Array" IEEE Trans. Microw. Theory Techn., vol.59, no.11, pp.2856-2863, Nov. 2011.

[9] Y. Zhang, X. Y. Zhang, and Y-M Pan, "A compact, wideband circularly polarized co-designed filtering antenna and Its application for wearable devices with low SAR" IEEE Trans. Antennas Propag., vol.65, no.8, pp. 3808-3818, Aug. 2017.

Slice resource allocation based on Comprehensive Utility

Zhentao Shen

Beijing Fibrlink Communications Co., Ltd.

shenzhentao66@163.com

Abstract. Due to the increasing demand for quality of service in power communication services quality.smart grids require network slicing technology to provide differentiated services. However, the existing algorithms lack in-depth research on the comprehensive service, particularly in power services and system accessibility. To address these issues, this paper proposes an integrated utility-based resource allocation scheme. Under the premise of meeting user needs, the priority of slices is comprehensively considered to optimize system utility and access rate. By comparing to local optimization algorithms, the results show that the new algorithm can maximize the overall user utility of the system and optimize the slice access rate simultaneously.

1. Introduction

Along with the fast development of new technologies for smart grids, The range of electrical services is becoming increasingly diverse, leading to differing individualised service requirements, such as communication speed, delay, reliability, etc. It is very important to dynamically and flexibly allocate network resources according to business requirements. Network slicing technology for different industries, as an important technology of the 5G mobile communication system, can realize network isolation in physical or logical terms, provide Corresponding personalised services for different businesses, and build the corresponding virtualization according to the diverse needs of different businesses. However, Because of the increasing demand for limited radio frequency resources from various services, how to make full use of the limited spectrum resources, maximize user experience and system benefits, and meet the individual needs of multiple services needs to be solved urgently.

In References [1]-[3], the prospect and possibility of network slicing applied to the smart grid are analyzed, as well as the bi-directional business flow and of smart grids, and the open issues of slicing technology for grid applications is discussed. Reference [4] proposes an opportunity-based spectrum sharing allocation, which effectively improves the utilization of wireless resources. Reference [5] aims at maximizing network benefits and proposes a genetic algorithm resource allocation strategy to solve network slicing. In Reference [6], mobile cloud computing (MCC) technology minimizes joint energy consumption. Reference [7] proposes a network slice resource allocation scheme based on the online auction. The slice resource allocation problem is transformed into a multi-service online bidding problem to ameliorate the efficiency of the network while ensuring that different service needs are met. References [8-9] consider improving the profit of wireless networks in resource allocation, which formulate optimization problem as Mixed Integer Non-Linear Problem (MINLP) to use population-based to reduce the complexity of the algorithm. References [10-11] consider reliability, and quality-of-service (QoS), where QoS includes optimization goals such as delay, rate, and throughput. Reference [12] proposes a resource allocation algorithm based on utility maximization, but it belongs to a local optimization algorithm.

At present, most literature does not fully consider smart grid business scenarios. In typical power business scenarios such as differential protection, precision timing and load control, it is critical to guarantee low latency and high reliability business requirements, wireless resource reservation helps to solve this problem to a certain extent. Excessive reservation will result in a waste of resources. Therefore, a reasonable and effective resource allocation algorithm needs to be adopted to solve the differentiated power business needs. This paper proposes a resource allocation scheme that maximizes the comprehensive utility of the system and optimizes the system utility and user access rate in a flexible way under the premise of satisfying the user's QoS requirements.

2. System Model and Assumptions

2.1. System Model

Considering the resource allocation of the uplink cellular network, it is assumed that the total bandwidth of the radio resources is expressed as B, wherein a resource block includes 12 carriers, a total of 1 time slot, i.e., 0.5ms, with a resource block as a unit. The set of users k = {1,2,..., K}. Meanwhile, considering the rate slice (Enhanced Mobile Broadband, eMBB) and the delay constrained slice (ultra-reliable low-latency communication,uRLLC), this paper assumes that a user belongs to a slice.

2.1.1. Rate equations:

In the rate constraint slice, the user's minimum rate is $v_k^{\min}$, assuming that the user's rate on a resource RB is expressed as in equation (1).

$$\mathrm{r}_{k,BS}^{\mathrm{RB}} = B_{rb} lb(1+\frac{ph_{k,BS}}{N_0 B}) \tag{1}$$

2.1.2. Minimum number of resources for rate slice users equations:

In the rate constraint slice, the minimum number of resources required by user k is

$$N\mathrm{v}_k^{\min} \geq \frac{v_k^{\min}}{\mathrm{r}_{k,BS}^{RB}} \tag{2}$$

2.1.3. Delay equations:

In the delay constraint slice, the maximum delay requirement of the user is τ_k^{max}. According to the queuing model, the arrival rate of the user's data packet is λ_k, and the length of the data packet is L_k bits. The average delay calculation formula is in equation (3).

$$\tau_k^{max} = \frac{1}{\frac{r_k^{\min}}{L_k} - \lambda_k} \tag{3}$$

2.1.4. Minimum number of resources for delay slice users equations:

In the delay constrained slice, the minimum number of resources required by user k is

$$N\tau_k^{\min} \geq \frac{L}{\mathrm{r}_{k,BS}^{RB}\tau_k^{\max}} + \frac{\lambda_k L_k}{\mathrm{r}_{k,BS}^{RB}} \tag{4}$$

Since the service requirements of the access users are different for each slice, the QoS definition of each slice is different.

2.1.5. Utility function in the rate slice:
If the actual rate of the user after obtaining the minimum number of resources is r_k, the user utility function in the rate slice is defined as in equation (5).

$$U_k^m = \begin{cases} \dfrac{1}{1+e^{-b_0(r_k - v_k^{\min})}} & r_k \geq v_k^{\min} \\ 0 & otherwise \end{cases} \tag{5}$$

where b_0 represents the gradient value of the rate utility function. The steeper the gradient of the function, the more sensitive it is to the velocity.

2.1.6. Utility function in the delay slice:
If the actual delay of the user is τ, the user utility function in the delay slice is defined as in equation (6).

$$U_k^d = \begin{cases} 1 - \dfrac{1}{1+e^{-b_1(\tau - \tau^{\max})}} & \tau \leq \tau^{\max} \\ 0 & otherwise \end{cases} \tag{6}$$

where b_1 represents the gradient value of the delay utility function. The steeper the gradient of the function, the more sensitive it is to the time delay.

2.1.7. Access rate equations:
Define the access rate of the system under slice s.

$$A_s = \frac{\sum x_{k,s}}{Num_s} \tag{7}$$

2.1.8. Comprehensive utility equations:
W is defined as the comprehensive utility of the current user, the comprehensive user service utility, and the system access rate.

$$w = \beta U_k + (1-\beta)A_s \tag{8}$$

where β is the weight of user QoS utility, U_k and A_s respectively represent user utility and user access rate.

2.1.9. The problem mode equations:
Therefore, the problem model can be expressed as follows.

$$\max \sum_{s=1}^{S}\sum_{k=1}^{K} \alpha_s (\beta U_k x_{k,s} + (1-\beta)\frac{x_{k,s}}{Num_s}) \tag{9}$$

$$\text{s.t.} \sum_{s=1}^{S} x_{s,k} \leq 1 \; \forall s \in S, \forall k \in K$$

$$x_{k,s}(1 - x_{k,s}) = 0 \; \forall s \in S, \forall k \in K$$

$$\sum_{s=1}^{S}\sum_{k=1}^{K} x_{k,s}(N_V + N_\tau) \leq B \; \forall s \in S, \forall k \in K$$

Journal of Physics: Conference Series **2384** (2022) 012037 doi:10.1088/1742-6596/2384/1/012037

$$\tau_k \le \tau_k^{\max} \ \forall k \in K$$

$$v_k \ge v_k^{\min} \ \forall k \in K$$

where α is the slice weight factor, the first constraint indicates that each user is connected to only one slice, the second constraint indicates the user access state, which can only take a value of 1 or 0, and the third constraint indicates that wireless resources blocks' number is less than system bandwidth, the fourth and fifth constraints represent the constraints of delay users and rate users in different slices, respectively.

2.2. Algorithm Description

The above model is a combinatorial optimization 0-1 planning problem, which can be optimized using the knapsack algorithm(KA). The knapsack algorithm decision-making problem can be visualized as a knapsack of a given fixed size and quality, and the most valuable items are selected to be loaded into it. In a set of items, each item has its own mass W and value v, which is limited within the range of the total mass of the backpack, and select the optimal combination.

For the above problems, consider the limited bandwidth resources as a backpack with limited capacity, the corresponding resources allocated by the user as the quality of the user, the system utility realized by the user as the value of the backpack items, the bandwidth allocation problem is transformed into a backpack problem, how to choose Allocate resources to maximize the Comprehensive system utility.

The algorithm flow is as follows in Figure 1.

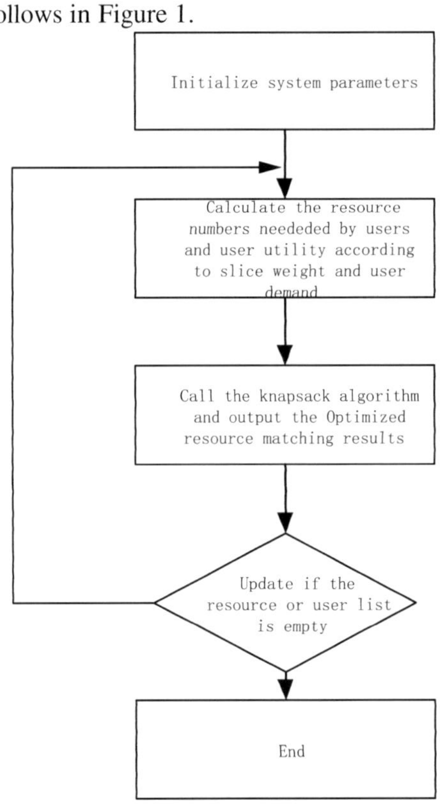

Figure 1. Knapsack algorithm process

3. Simulation and Numerical Results

In the simulation verification algorithm phase, Consider rate constraints and delays constrained two types of users are randomly distributed in a 500m radius cell. The simulation parameters used are shown in Table 1.

Table 1. System Simulation Parameters

Simulation parameters	Values
Bandwidth	20MHz
Number of slices	2
Amount of users	100
Users per slice	50
Transmit power for cellular users	23dBm
Path Loss Model	128. 1 + 37. 6 lgd
Noise spectral density	-174 dBm / Hz
Minimum rate constraint	100Mb/s
Maximum Delay Constraint	1ms
Packet length	100kb
Packet arrival rate	150packet/s

In terms of algorithm validity comparison, considering the delay slice and rate slice scenarios, the proposed algorithm and the maximization principle resource allocation algorithm GA in the literature [12] and the round-robin algorithm (Round-Robin, RR) are considered. Set b0, b1, α, and β to be 0.8, 0.01, 0.5, and 0.5, respectively, as shown in Figure 2. When the users' number is less than 40, due to sufficient resources, the user needs can be met, and the system utility is basically the same, but as the users' number gradually increases, the user utility value of the KA algorithm is obviously greater than that of the GA algorithm and the RR algorithm. This is because the KA algorithm is based on global optimization, which improves the QoS utility of the system users.

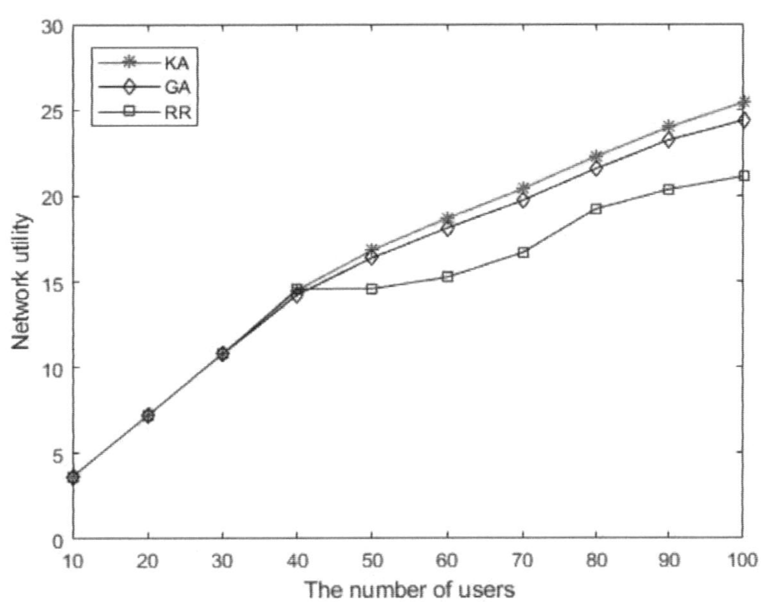

Figure 2. Network utility for different resource slicing schemes.

Considering the user access rate, the influence of the uRLLC slice weight factor α on the access rate is considered separately, and β is set to 0. As α gradually increases, the uRLLC slice access rate gradually increases, but this is at the expense of eMBB slice access. Therefore, on the premise of ensuring the low-latency and high-reliable service access rate, setting an appropriate α weight can improve efficiency of wireless resources, while ensuring the low-latency service user access rate.

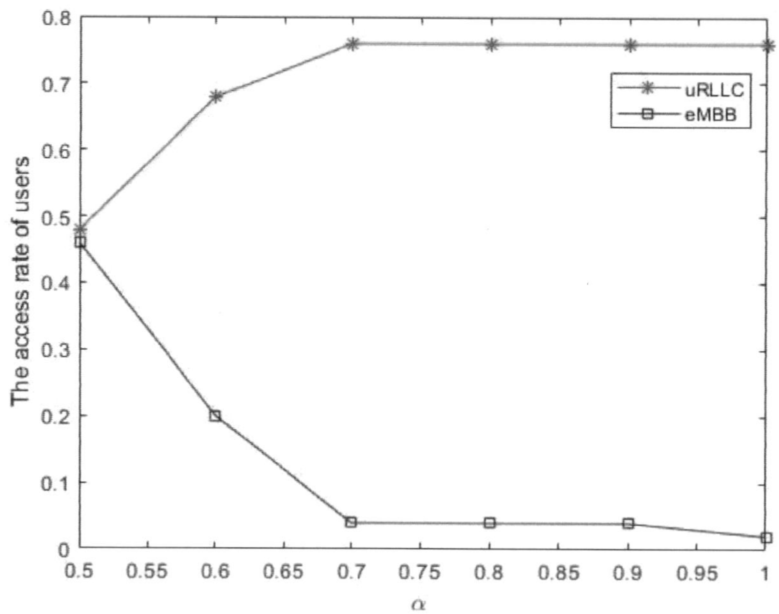

Figure 3. The access rate of users under different slice weights.

4. Conclusions

For the purpose of satisfying differentiated user business requirements, a resource allocation optimization model is constructed to seek to maximize the utility of the system, and propose a solution strategy by introducing the knapsack algorithm. Finally, the proposed algorithm is simulated by comparing with existing algorithms. The results show that the algorithm we proposed can improve the system utility in the context of ensuring certain user requirements and can improve the access rate of corresponding slice users in a targeted manner.

Acknowledgment

This research was supported by the Independent Technology Projects of Beijing Fibrlink Communications Co., Ltd: 5G-based power communication terminal research and development and application management improvement research.

References

[1] Dorsch N, Kurtz F and Wietfeld C 2018 On the economic benefits of software-defined network and network slicing for smart grid communications *J.Netnomics.* 19(1-2): 1-30

[2] Ahmad A, Rehmani M, Tempine H 2017 IEEE Access Special Section Edition: Optimization for Emerging Wireless Networks: IoT, 5G, and Smart Grid Communication Networks *J. IEEE.Access.* 5(1): 2096-100

[3] Saleem Y, Crespi N, Rehmani M.H, et al. 2019 Internet of Things-aided Smart Grid: Technologies, Archives, Applications, Prototypes, and Future Research Directions *J. IEEE Access.* 7(2): 62962-3003

[4] Liu X, Li M, Song M, et al. 2016 Wireless virtual network embedding based on Spectrum sharing allocation *J.IEEE.ICCSE. Nagoya.* 670-75
[5] Li. J, Su F, Lu Y, et al. 2003 A Heuristic Genetic Algorithm for Solving Resource Allocation Problems *J. Knowledge & Information Systems.* 5(4): 503-11
[6] Wang K, Yang K, Magurawalage C S 2016 Joint energy minimization and resource allocation in C-RAN with mobile cloud *J.IEEE Transactions on Cloud Computing.* 6(3): 760-70
[7] Liang L, Yanfei W, Gang F, Xin J and Yunjian J 2019 Online Auction-Based Resource Allocation for Service-Oriented Network Slicing *J.IEEE Transactions on Vehicular Technology.* 8063-74
[8] JIA Q, XIE R, HUANG T, et al. 2017 Efficient caching resource allocation for network slicing in 5G core network *J.IET Communications.* 11(18): 2792-99
[9] Sunday O, Olabisi E 2018 Profit-aware resource allocation for 5G sliced Networks C.EuCNC 43-9
[10] Arjun A, Gustavo D, Sanjay S 2018 Joint Scheduling of URLLC and eMBB Traffic in 5G Wireless Networks *J. IEEE.INFOCOM.* 1970-78
[11] TANG L, ZHANG X, XIANG H, et al. 2017 Joint resource allocation and caching placement for network slicing in fog radio access networks *J. IEEE.SPAWC.* 1-6
[12] Mingyue L 2021 Research on resource allocation algorithm for 5G network slices in smart grid *J.North China Electric Power University.* 17-26

Embedded remote monitoring system based on NBIOT

ZiChen Guo[1], YiFan Liu[1], Fang Lu[1*]

[1]School of Electrical and Information Engineering, Zhengzhou University, Zhengzhou, China

*Corresponding author: ieflu@zzu.edu.cn

Abstract. With the rapid development of information technology, the remote control has become an important means to improve production efficiency and intelligence level. In this paper, an embedded remote control system based on NBIoT technology is designed and produced. Based on STM32F427 as the control core, the system can collect the current and voltage of each branch and perform data analysis, data processing, and branch on-off control. It connects multiple smoke sensors to realize the fire warning function. Data transmission is realized through NBIoT technology and the OneNet platform, and users can realize remote control through APP. Multiple built-in communication interfaces can be extended for different scenarios. It has a good effect on the test and can meet the requirements of remote control. It has high application value in industrial control, smart city, and other fields.

1. Introduction

With the development of information technology, remote monitoring systems have been more and more widely used in industrial production, smart home, building automation, smart grid, and other fields. In recent years, with the rapid development of microprocessor and Internet of Things technology, the performance and application of embedded systems have developed rapidly, and remote monitoring systems based on embedded platforms is an important development direction [1]. An industrial remote control system based on CC320 is designed [2], which uses a PID algorithm for liquid level control and transmits data through wifi. An elevator monitoring system based on GPRS and an embedded system is introduced, which can diagnose elevator faults by time and send an SMS alarm [3]. An embedded monitoring device based on a Linux system was developed, and the data acquisition and transmission were tested, and good results were obtained [4]. A low-cost GSM-based home appliance remote intelligent control system was developed and designed, which can remotely set parameters such as temperature and humidity, and control the switch state of home appliances. has important practical significance [5].

In this paper, a low-cost embedded remote control system is designed and produced. The internal integrated sensors can collect the voltage, current, temperature, and other parameters of each branch, and upload them to the OneNet platform through the NBIOT module, and users can view them through the APP. parameters and remote control. The built-in LoRa wireless communication module is used to collect multi-node smoke sensor data to realize the fire warning function. After testing, the system runs stably and is easy to use.

2. System design

2.1. System architecture
The structure of the system is shown in Figure 1. The system can be simply viewed as composed of intelligent terminals, sensor nodes, and cloud platforms.

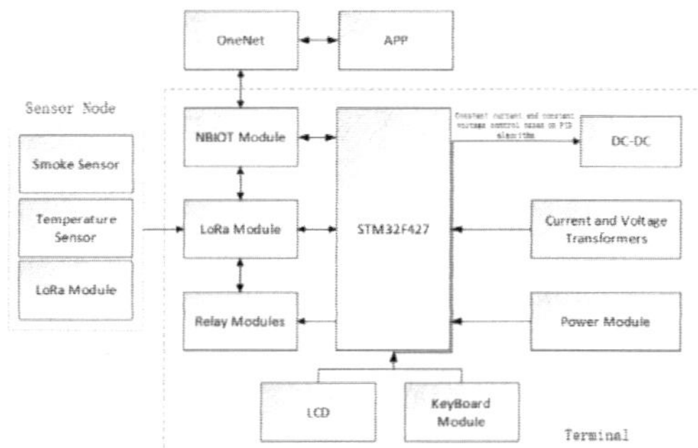

Figure 1. System architecture.

As can be seen from Figure 1, the intelligent terminal part is mainly composed of a core control module, a power supply module, a human-computer interaction module, a voltage and current detection module, a relay, and a communication module. Sensor nodes include temperature and smoke sensors, as well as LoRa modules that transmit data. In this paper, STM32F427 is selected as the main control chip of the system. Its operating frequency is 180MHZ, and it has common communication interfaces such as IIC, SPI, CAN, and USART, which can meet the needs of the normal operation of the system in complex environments. The smart terminal is shown in Figure 2.

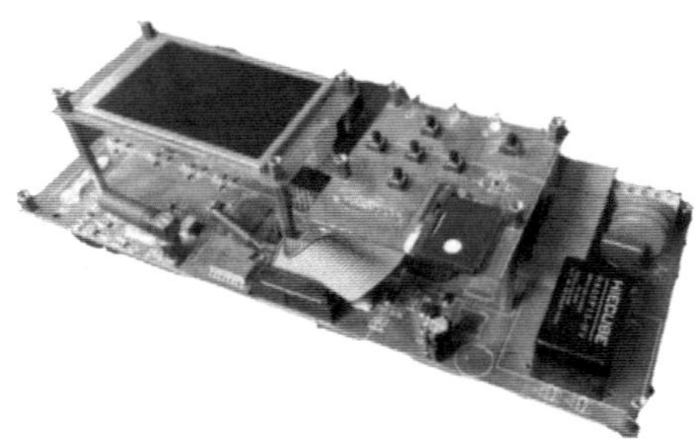
Figure 2. Physical view of the terminal.

2.2. Design of power module
The power module of the system divides the input circuit into twelve branches, which are connected to the relay module. The HA05P1 module converts 220V/380V AC into 36V DC and then performs multi-stage step-down processing to drive the functional modules. This paper adopts a leakage protection circuit based on M54123, which can quickly turn off the power protection device when leakage occurs. Taking into account the possible interruption of the power supply during the working process, the auxiliary power supply part is designed, which is composed of four super capacitors connected in series. The model is CXHP2R750R, the rated voltage is 2.7V, and the capacitance value is 5F. When the system is running normally, the supercapacitor module remains fully charged. When the system is powered off, the supercapacitor supplies power for a short time and sends the information to the cloud platform.

2.3. Design of communication module
The complex working environment makes it difficult to deploy wired communication. Therefore, reliable wireless communication is required to collect sensor information from each node and upload the collected information to the cloud platform. Currently, the commonly used wireless communication

methods for the Internet of Things include NBIOT, LoRa, Zigbee, etc. NBIOT uses narrow bandwidth for communication (180kHZ), it can directly deploy LTE network and have excellent LPWA characteristics [6]. In this paper, NBIoT technology is selected for the communication between the cloud platform and the terminal device, and at the same time, it receives user instructions to realize remote control. This function is implemented by the BC26 module. For the communication between the smoke sensor and the terminal device, LoRa wireless communication technology is used. Compared with NBIoT technology, LoRa has smaller power consumption and lower cost but a low transmission rate and requires a gateway, which is inconvenient to deploy [7]. Therefore, in the design of this paper, LoRa is only used for the connection between sensor nodes and terminals, not for connecting to the cloud platform. Multiple nodes with LoRa modules and smoke sensors are deployed at various locations in the building and transmit data to smart terminals.

2.4. Human-computer interaction and measurement module
The human-computer interaction module of the terminal part is mainly composed of LCD and buttons. In order to facilitate development and post-debugging, a 4.7-inch LCD screen and five independent buttons are installed in the intelligent terminal. Through the preset program, the threshold current, threshold voltage, and relay state can be set through the buttons. At the same time, the current voltage, temperature, and sensor node data of each branch can be obtained. Among them, the measurement of AC current and voltage is realized by a current transformer and voltage transformer, and alarm information is sent when the measured data is higher than the set threshold.

3. Software Design

3.1. Main Program Design
The main program flow of the terminal is shown in Figure 3, which is written by Keil Vision5.14.0.0. The leakage protection and alarm functions are directly performed by the hardware circuit without software detection. The operating system used by the intelligent terminal is FreeRTOS. FreeRTOS is an open-source real-time operating system that can be ported to a variety of microprocessors [8]. Compared with bare metal development, the use of the FreeRtos system can reduce the difficulty of task scheduling of complex functions and improve the portability of code. According to the specific functions of the system, the following six key tasks are established: system initialization task, signal acquisition, analysis task, LoRa module transceiver task, NBIoT module transceiver task, relay module control task, LCD display, and control task.

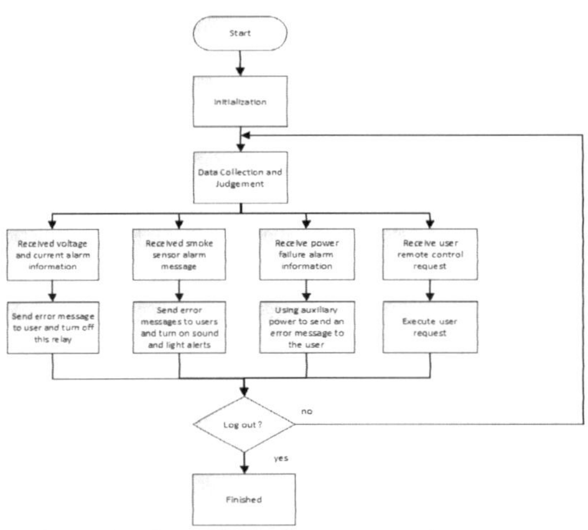

Figure 3. Flow chart of system main program.

3.2. Design of Communication Module Program

Due to resource constraints, IoT often uses lightweight protocols, such as MQTT and CoAP. MQTT is a publish/subscribe-based protocol, and its advantage is that it can provide reliable information transmission services for remotely connected devices through low bandwidth or unreliable network transmission with very low power consumption [9]. In this paper, the MQTT protocol is used to realize the message push between the terminal device and the OneNet platform. Figure 4 describes the flow chart of the terminal collecting and uploading data and receiving the request to control the device.

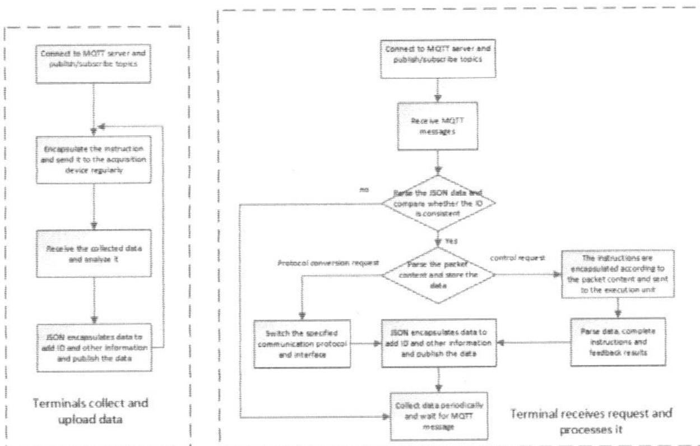

Figure 4. Flow chart of the communication program.

4. Functional Test

In order to verify the function of the system, we conducted a preliminary test on the function of the system. The test results are shown in Figure 5 and Figure 6. During the experiment, the equipment was connected to 220V AC and connected to other electrical appliances.

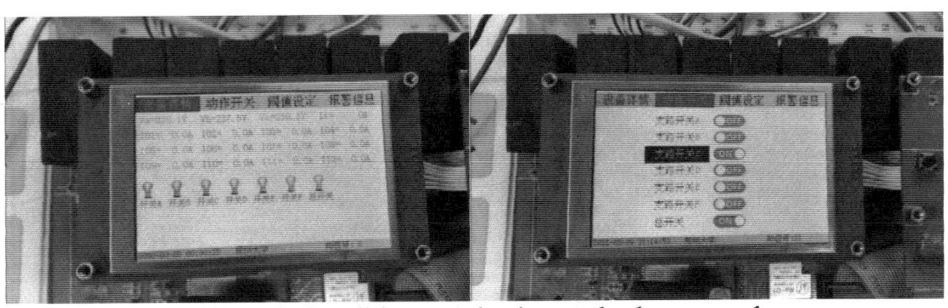

Figure 5. Condition monitoring and relay control.

Figure 6. Threshold setting and warning information display.

In addition to the test of the terminal part, we also tested the function of the sensor node and the user's remote control. In this paper, a lit cigarette is placed close to a smoke sensor in the test site to simulate a fire scenario. Figure 7 shows the alarm information of the APP and the effect of the remote control.

Figure 7. Threshold setting and warning information display.

5. Conclusion

This paper designs an embedded remote monitoring system that can perform current and voltage detection, multi-node smoke and temperature detection based on LoRa technology, and realize the information transmission between the communication terminal and the cloud platform through the NBIoT module. In functional tests, the system performed well. It can feedback on the current state of the system in real time and realize the remote control of the terminal through the mobile APP, which has certain application value in the fields of smart home, industrial control, and smart grid. In the next step, we will conduct more tests and optimize the hardware circuit and communication algorithm and hope it will be put into use.

Acknowledgments

This work was supported by the School of Electrical and Information Engineering of Zhengzhou University. ZiChen Guo and YiFan Liu contribute equally to this work.

References

[1] Jen-Hao Teng, Chin-Yuan Tseng and Yu-Hung Chen 2004 *IEEE Region 10 Conference TENCON 2004* vol 3 (Thailand: Chiang Mai) pp. 566-569

[2] A. L. Sangeetha, N. Bharathi, D. G. M. Blesson, S. V. Girish and A. B. Ganesh 2016 *2016 International Conference on Wireless Communications, Signal Processing and Networking (WiSPNET)* (India: Chennai) pp. 729-732

[3] E. Dawei and Xiaobin Zhai 2011 *2011 International Conference on Computer Science and Service System (CSSS)* (China: Nanjing) pp. 2695-2698

[4] Xiang-hai Li and Jian-hui Liu 2010 *2010 International Conference on Computer Application and System Modeling (ICCASM 2010)* vol 3 (China: Taiyuan) pp. 299-302

[5] G. M. S. Mahmud Rana, A. A. Mamun Khan, M. N. Hoque and A. F. Mitul 2013 *2013 2nd International Conference on Advances in Electrical Engineering (ICAEE)* (Bangladesh: Dkaka) pp. 291-295

[6] S. K. Routray, K. P. Sharmila, E. Akanskha, A. D. Ghosh, L. Sharma and M. Pappa 2021 *2021 Third International Conference on Intelligent Communication Technologies and Virtual Mobile Networks (ICICV)* (India: Tirunelveli) pp. 393-398
[7] A. Zourmand, A. L. Kun Hing, C. Wai Hung and M. AbdulRehman 2019 *2019 IEEE International Conference on Automatic Control and Intelligent Systems (I2CACIS)* (Malaysia: Selangor) pp. 324-330
[8] L. Jiang, Z. Zeng and Y. Liu 2021 *2021 IEEE 3rd International Conference on Frontiers Technology of Information and Computer (ICFTIC)* (USA: SC, Greenville) pp. 579-583
[9] B. Mishra and A. Kertesz 2020 "The Use of MQTT in M2M and IoT Systems: A Survey" *IEEE Access* **8** 201071-201086

Journal of Physics: Conference Series **2384** (2022) 012039 doi:10.1088/1742-6596/2384/1/012039

Research on Intelligent Anti-misoperation of Operation Ticket in Urban Distribution Network

Peng Zhang [1], Yutao Xu [1*], Xi Yue [1], Changjun Li [2], Xiaohui Yang [1]

[1]State Grid Urumqi Power Supply Company, Urumqi Xinjiang 830000, China

[2]State Grid Xinjiang Electric Power Co., Ltd, Urumqi Xinjiang 830000, China

*905011351@qq.com

Abstract. The traditional anti-mischeck mode of regulation and control operation relying on the experience of dispatchers can no longer meet the needs of daily work. To solve the problems of poor actual operation effect and low anti-misoperation efficiency of the traditional anti-misoperation method due to the large number of tasks of the current distribution network, this paper optimizes the traditional anti-misoperation method and takes the distribution network of M power supply company as an example to design a new intelligent anti-misapplication method of operation ticket. First of all, this paper uses intelligent distribution network maintenance technology to overhaul the distribution network operation ticket application server in an all-round way. Based on real-time analytical constraints, combined with the power supply demand of urban distribution networks, this paper designs an intelligent anti-misoperation constraint mechanism for operation tickets. The experimental analysis shows that after the application of the proposed anti-error method, the qualified rate of the operation order scheduling tasks is more than 95. 6% and the error rate is less than 1. 1%, which shows that the anti-error method proposed in this paper has significant advantages and high feasibility.

1. Introduction

With the continuous expansion of the power-grid scale and the deepening of technologies such as the smart grid and UHV, China's power system has entered the stage of a large power grid [1]. With the rapid development of urban distribution networks and distribution systems in China, there are more and more power dispatching tasks and operation ticket dispatching tasks. The distribution network operation ticket system, as an important part of urban distribution network dispatching and operation management, has a greater impact on the stable construction and development of the urban distribution network [2].

Due to the influence of human factors and various uncertain factors, the dispatching security and dispatching efficiency of power grids at all levels cannot be effectively guaranteed in the dispatching of distribution network operation tickets, which leads to certain security risks and hidden dangers in the operation of distribution network [3]. The electrical misoperation in the power system includes mistakenly manning the live interval, closing with ground wire, closing the switch by mistake, pulling the knife switch with load, hanging the ground wire with electricity, etc. These are all frequent faults that threaten the safe operation of power equipment, which may lead to large-scale blackouts, destruction of power equipment, the collapse of system oscillation, and even casualties. It has brought serious losses to enterprises and society. To realize the safe and reliable operation of power equipment and prevent the occurrence of misoperation accidents, although the power system compulsorily implements that high-voltage switchgear must be equipped with anti-misoperation devices and takes various management and technical measures to prevent the occurrence of misoperation, it still fails to fundamentally eliminate such accidents.

The traditional distribution network operation ticket is mainly filled by the staff, and the relevant contents of the distribution network operation are filled in the operation ticket module corresponding to the production management system, which provides the basic guarantee for the stable operation of the urban distribution network [4]. However, in the actual operation of the dispatching task of the operation ticket, the traditional way of filling in the operation ticket still has great shortcomings, and the filling speed is slow, especially for some atypical operation tickets, which are prone to different degrees of logical errors, thus causing the failure of the power production management system and reducing the efficiency and quality of urban distribution network operation [5]. In addition, the operation ticket issued by dispatching operation is often not closely combined with the actual situation, and only depends on the work experience of dispatching staff and the surface judgment of maintenance operators, which will lead to the lack of sufficient basis for the issuance of operation ticket. This will not only cause the problem of waste of resources but also can not guarantee the rationality of the operation order, which greatly affects the orderly development of power grid dispatching operations.

Based on this, the scientific and reasonable anti-misoperation method of the distribution network is very important. In this paper, based on the traditional anti-misoperation method, a new intelligent anti-misoperation method is designed, which contributes to the rapid development of urban distribution networks.

2. Design of intelligent anti-misoperation method for operation ticket

2.1. Overhaul power distribution network operation ticket application server

In this paper, an intelligent anti-misoperation method for urban distribution network operation tickets is designed. Firstly, the intelligent distribution network maintenance technology is used to formulate the operation ticket maintenance constraint plan, and the distribution network operation ticket application server is maintained in an all-round way. As an important part of the urban distribution network and power system operation, the distribution network operation ticket application server is responsible for storing the relevant information of operation tickets and application tickets [6]. The composition structure of the operation ticket application server is shown in Figure 1.

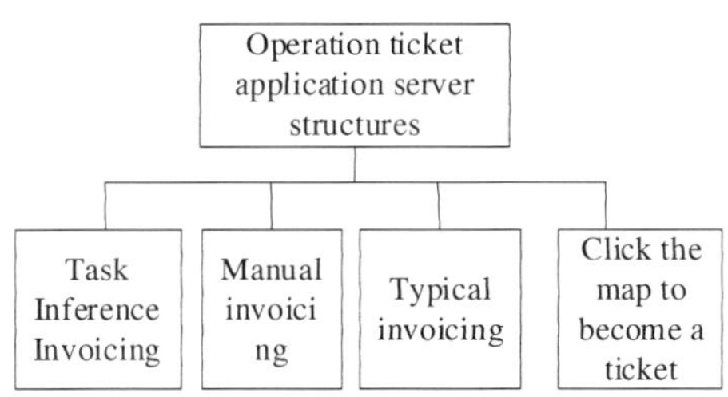

Figure 1 Composition structure of power distribution network operation ticket application server

As shown in Figure 1, the power distribution network operation ticket application server is composed of four parts, namely, point map ticket generation, power distribution network operation ticket task reasoning invoicing, manual invoicing, and typical ticket invoicing. The operation status of the urban distribution network can be directly reflected by the point map ticket. Through the point selection and application instructions of the server, the dispatching task of the distribution network operation ticket can be automatically analyzed, and the complete operation ticket can be issued. The task reasoning and invoicing function of the operation ticket has a power grid topology model and a five-prevention check system. By safely checking each operation mode of the operation ticket of the distribution network, the intelligent and automatic reasoning mode is adopted to generate an accurate and complete operation ticket of the distribution network, which is highly executable [7].

For the maintenance of task reasoning and invoicing of operation order, the whole process and comprehensive maintenance of task reasoning and invoicing function of operation order should be carried out according to the relevant contents of the knowledge base in the field of urban distribution networks. The manual invoicing of operation tickets is mainly aimed at unconventional operation tasks, and the requirements for filling in information such as the date of filling in the operation ticket, the name of the operator, the operating unit, and the operating equipment are relatively high, in which errors are prone to occur in the process of copying and copying the operation instructions of the operation ticket [8]. In order to improve the quality and efficiency of the manual operation order, it should be improved from the perspective of the operation order dispatcher, and the operation object and instructions of the operation order should be detected.

For typical ticket issuing, start with the typical ticket management module of the production management system, comprehensively overhaul the search function and retrieval conditions of the module, and judge whether the operation ticket function is in good condition [9]. Through the maintenance of the distribution network operation ticket application server, the basic guarantee is provided for the intelligent error prevention of the urban distribution network operation ticket.

2.2. Design of intelligent anti-misoperation constraint mechanism for operation ticket

After the power distribution network operation ticket application server is overhauled, an intelligent anti-misoperation constraint mechanism of the operation ticket is designed in an all-round way, so that the actual operation quality of the urban power distribution network and the task scheduling efficiency of the operation ticket are improved. In the intelligent anti-misoperation constraint mechanism of the operation ticket, the real-time analytical constraint is mainly used to adjust the connection mode of the power grid and optimize the configuration of electrical equipment in the distribution network according to the power supply demand of the urban distribution network. In general, in the task scheduling process of urban distribution networks, due to the different operation modes of primary electrical equipment, there are different requirements for the operation sequence and operation procedures of task scheduling of operation tickets [10]. At this time, the perfect intelligent anti-error constraint mechanism of the operation ticket can effectively avoid the hidden danger of accidents caused by the wrong operation sequence and rules of task scheduling of operation tickets, and reduce the possibility of accidents in urban distribution networks. Based on the reminder requirements of the urban distribution network operation ticket, the process of the intelligent anti-misoperation constraint mechanism of the operation ticket is designed, as shown in Figure 2.

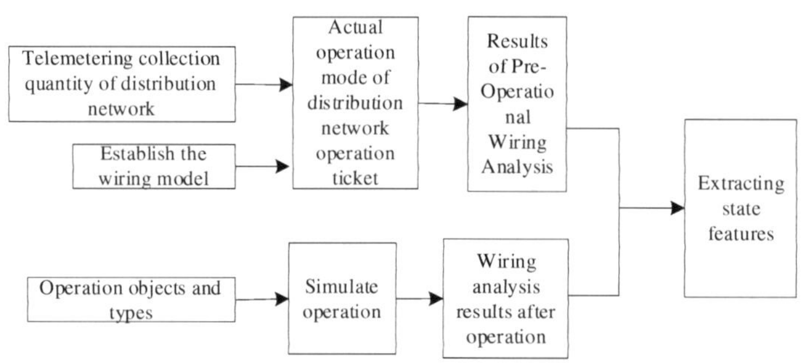

Figure 2 Design flow of intelligent anti-misoperation constraint mechanism for operation order

As shown in Figure 2, multi-sensors are used to measure the telemetry acquisition quantity of the urban distribution network, establish the connection model of the urban distribution network, and obtain the actual operation mode and operation characteristics of the city distribution network. Simulate the operation mode of distribution network operation tickets according to the type and operation object of distribution network operation tickets [11]. A comprehensive analysis of the wiring results of the distribution network is conducted before and after the operation of the distribution network operation order, and the average utilization coefficient of the operation of the distribution system and the number of electric equipment dispatched by the operation orders are calculated. The formulas are:

$$k = \frac{\sum_{i=1}^{m} P_{av}}{\sum_{i=1}^{m} P_e} \tag{1}$$

$$n = \frac{\left(\sum_{i=1}^{m} P_e\right)^2}{\sum_{i=1}^{m} P_{le}^2} \tag{2}$$

Where k represents the average utilization coefficient of distribution network operation; n represents the number of electric equipment dispatched by operation order; P_{av} represents the active power of distribution network operation; P_e represents the reactive power of distribution network; P_{1e}^2 represents the dispatching task amount of operation order. The calculation results are collected and processed, and the state characteristics of the task scheduling of the urban distribution network operation ticket are extracted to complete the comprehensive design of the intelligent anti-misoperation constraint mechanism of the operation ticket.

3. Experimental analysis

Based on the above content, this paper designs the overall process of the intelligent anti-misoperation method of the urban distribution network operation ticket. On this basis, in order to further verify the effectiveness of the intelligent anti-misoperation method designed in this paper, the following experiments are carried out. In this experiment, the distribution network of an M power supply company in a city is selected as the research object. The task scheduling system at all levels of the distribution network of the power supply company is gradually improved. The unified management is carried out through the production management system, and the operation status of the dispatching operation ticket is reflected in the dispatching terminal in real-time. Relatively speaking, the anti-misoperation technology of the power grid is relatively mature. However, in the process of high-speed operation of the distribution network, there are still some deficiencies in operation ticket scheduling.

On this basis, the intelligent anti-misoperation method of distribution network operation ticket designed in this paper is applied to the distribution network of the M power supply company. First of all, a microcomputer anti-misoperation device is installed in the substation to effectively prevent the misoperation of the primary electrical equipment from the source. The electric lock is connected in series, the electrified device of the bus and the line is connected to the circuit of the power distribution network, and the anti-misoperation function of the power distribution network is realized by locking the circuit. A schematic diagram of the anti-misoperation wiring of the power distribution network of M Power Supply Company is shown in Figure 3.

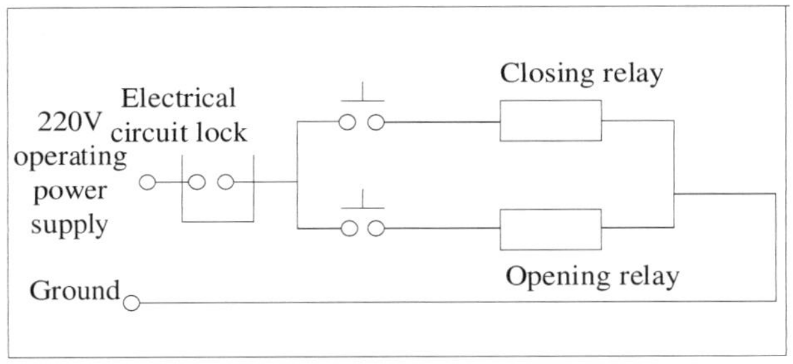

Figure 3 Schematic diagram of electric anti-misoperation wiring of power distribution network of M power supply company

As shown in Figure 3, it is the overall structure of the electric anti-misoperation wiring of the distribution network of M Power Supply Company. An analog drawing board is added to the distribution network to reflect the real-time operational status of the equipment and find the dangerous points of misoperation on the transmission line. A distribution network operation ticket operation and maintenance department is established to overhaul the operation status of the operation ticket application server by using the basic data in the electrical equipment database of the power supply company. From the two different dimensions of electrical business direction and control direction, the intelligent anti-error control point of operation ticket is deployed to carry out the anti-error and wiring management of the operation ticket. A communication connection is established between the operation ticket application server and the production management system to extend the power distribution line so that the power distribution network operation ticket achieves the goal of cooperative control of power transmission error prevention and power distribution error prevention. In order to verify the feasibility of the method designed in this paper more intuitively, the experimental method of comparative analysis is used to compare the intelligent anti-misoperation method designed in this paper with the traditional anti-misoperation method. Set 6 different groups of invoicing scheduling tasks, and mark them as MRTL01, MRTL02, MRTL03, MRTL04, MRTL 05, and MRTL06 respectively. Using the finite element analysis software, respectively determine and calculate the qualified rate of each group of invoicing and dispatching tasks after the application of the two operation order anti-misoperation methods, and the calculation formula is:

$$P = \frac{M_a}{M} \times 100\% \tag{3}$$

Where P represents the qualified rate of operation ticket issuing and dispatching tasks; M_a represents the number of qualified operation tickets executed; M represents the number of operation tickets executed. The comparison results of the qualified rate are shown in Table 1.

Table 1 Comparison of qualified rate of dispatching tasks of two operation order error prevention methods

	The method of this paper		Traditional method	
Schedule the task	Pass rate	Error rate	Recall ratio	Error rate
MRTL01	95.6%	0.5%	89.4%	6.4%
MRTL02	97.2%	0.4%	88.6%	6.8%
MRTL03	96.8%	0.5%	82.3%	6.2%
MRTL04	96.4%	0.2%	89.7%	5.1%
MRTL05	97.3%	1.1%	88.6%	4.9%
MRTL06	97.9%	0.9%	88.3%	6.7%

According to the comparison results in Table 1, among the two distribution network operation ticket anti-misoperation methods, the intelligent anti-misoperation method designed in this paper has a qualified rate of more than 95.6% and an error rate of less than 1.1% for each group of operation ticket issuing and dispatching tasks. Compared with the traditional anti-misoperation method, the operation ticket dispatching task execution effect is more advantageous.

4. Conclusion

A distribution network has the characteristics of a complex structure and heavy workload of dispatching, operation, maintenance, overhaul, and transformation. With the continuous improvement of power supply reliability requirements of the distribution network, the huge risks caused by misdispatching and misoperation must be effectively avoided in each link of the distribution network operation process. Therefore, it is necessary to reduce the possibility of misscheduling and misoperation, so as to improve the overall safety and production efficiency. To sum up, in order to improve the shortcomings of the traditional anti-misoperation method of urban distribution networks, this paper makes an optimization design on the basis of the traditional anti-misoperation method and proposes an intelligent anti-misoperation design.

Through the research of this paper, the quality and efficiency of intelligent anti-misoperation of the urban distribution network operation ticket are effectively improved. The security risks in the distribution network operation ticket are reduced. The relevant data of the distribution network operation tickets are updated in real-time, and the transmission efficiency of information in the operation process is significantly improved, which has great research significance.

References

[1] Sun G, Huang W, Yi J 2022 Research and application of intelligent operation ticket generation method based on data interaction between distribution network technical support platform and distribution network automation master station. China high tech, (06): 29-30.

[2] Cai J, Lin Y, Wu Z 2022 Error prevention method of abnormal motion sensing and adaptive correction of interconnection switch in the distribution network. Electrical technology, 23 (03): 92-97.

[3] Lu Z, Wang S, Chai J 2022 Application Research of electric system operation order ticket system based on artificial intelligence recognition. Electrical applications, 41 (03): 49-55.

[4] Song Z, Leng L 2022 Research on the application of new intelligent operation ticket system. Journal of electric power, 37 (01): 84-90.

[5] Yin C, Cheng R 2021 Design of intelligent management and control system for networked dispatching of a distribution network based on topology error prevention. Electronic design engineering, 29 (19): 146-149 + 154.

[6] He Z, Li Y, Wang N 2021 Research and application of digital "two tickets" of distribution network considering multi-source data interaction. Electrical applications, 40 (08): 56-60.

[7] Chen W 2021 Research on application optimization of dispatching operation tickets under the mode of regulation integration. Smart city, 7(04): 66-67.

[8] Xu Y, Zhu S, Liu Q, Intelligent operation ticket integrated system based on first-order predicate logic. Electrical Automation, 43 (01): 4-9.

[9] Cai X, Qi Y 2020 Construction and practice of intelligent error prevention system for power grid dispatching operation based on artificial intelligence. Power big data, 23 (04): 16-23.

[10] Liu Y, Zheng M, Bai Y 2020 Research on design and application of OMS dispatching intelligent ticketing system in the distribution network. Electrotechnics, (02): 86-89.

[11] Dai X, Huang G 2014 Research on Maloperation Prevention System of Distribution Network. Electric Power Security Technology, 64-66.

Journal of Physics: Conference Series **2384** (2022) 012040 doi:10.1088/1742-6596/2384/1/012040

Multi-path QoE Routing Algorithm for wireless mesh network Based on Q-Learning

Tang Jie[*]**, Xu Yan, Che Yali, Liu Xuhui**

School of Electronic and Information Engineering, Lanzhou Jiaotong University, No.88 Anning West Road, Lanzhou, 730070, China

[*]1250608134@qq.com

Abstract. To address the problem of poor quality of experience(QoE) of wireless mesh network users in the current video streaming service, a wireless mesh network multi-path QoE algorithm based on Q-learning(QLMPQ) is proposed. The selection strategy is used to find the optimal master-slave path to improve the reliability of data transmission. Virtual simulation experiments are carried out under the conditions of different concurrent flows to verify the feasibility and accuracy of the algorithm. The experimental results show that compared with Q-Routing, QLAODV, and AODV algorithms, this algorithm can significantly improve the performance of wireless mesh networks.

1. Introduction

With the popularization of the 5G communication network, the rich multimedia content of today's Internet has proliferated. The traffic of video services has exceeded 50% of the mobile Internet traffic, so the future communication network must provide effective support for the rapidly growing media traffic[1]. Currently, most works of literature use network quality of service (QoS)[2] to represent the ability of the network to utilize equipment and resources to provide better service performance for network communication. But obviously, this indicator is not suitable for today's mobile Internet scenarios dominated by multimedia content. To this end, some studies have proposed the quality of user experience (QoE), which measures the user's subjective perception of equipment, network, application, and service quality[3,4].

A wireless mesh network (WMN) is a new type of high-speed, high-capacity broadband wireless network, in which its nodes have both routing and forwarding functions, and can dynamically self-organize and adaptively configure multi-hop wireless distributed networks[5]. The biggest benefit of this structure is that if the nearest AP is congested due to excessive traffic, the data can be automatically re-routed to a neighboring node with less traffic[6,7]. But the network topology The diversity increases the difficulty of optimizing QoE quality.

In this regard, reinforcement learning algorithms can be used to solve the above problems. In recent years, reinforcement learning has been applied to various wireless networks, such as wireless ad hoc networks, wireless sensor networks, and cognitive wireless networks[8]. The routing algorithm mainly studies the reward function according to the optimization target. The researchers design different reward functions in the direction of the optimization target based on the reinforcement learning algorithm and train the routing algorithm to meet the routing performance index through the reward mechanism. Reference [9] proposes a reinforcement learning routing algorithm that comprehensively considers distance and energy loss. Reference [10] proposes a distributed, energy-efficient reinforcement learning routing algorithm for wide-area wireless mesh Internet of Things (LPWAN)

Journal of Physics: Conference Series **2384** (2022) 012040 doi:10.1088/1742-6596/2384/1/012040

networks. In [11], the author proposed a distributed reinforcement learning routing algorithm RLBDR in the multi-gateway WMN scenario.

Based on the above analysis, this paper proposes a multi-path QoE algorithm (Q-Learning Multi-path QoE Routing Algorithm, QLMPQ) for WMN based on Q reinforcement learning. The Q-learning algorithm of Q-learning algorithm finds paths to complete data transmission; this paper proposes a multi-path selection strategy, selects the optimal main route according to the state value function table, calculates the correlation between other paths and the main path, selects the path with the lowest correlation as the secondary path by setting different numbers of concurrent streams to conduct virtual computing experiments, and analyzes the performance of WMN using QLMPQ, Q-Routing, QLAODV, and AODV algorithms.

2. WMN QoE model

2.1. Mathematical model of WMN

This paper adopts the Cartesian Product of Graph (CPG) model[12] to discuss and study the paths of multiple concurrent flows in WMN. The network structure under multiple concurrent flows can be used $O=\{M,I,\Omega,\mu,(b_i,d_i)\}$ to represent.

- M=(N, L) represents the topology diagram of the wireless network. Among them, N represents the node set of the network, and L represents the link set of the network. r is the transmission range of the network node, $n\in N$ and d represents the adjacent distance between the two nodes, if $d\le r$, indicates that there is a link l between two adjacent nodes, $l\in L$.
- I is the wireless interference relationship between nodes on M.
- $\Omega=\{c_1,c_2,\cdots,c_k\}$ is the set of available orthogonal channels on the network topology map, where k is the number of available orthogonal channels.
- μ is the number of node interfaces.
- (b_i,d_i) is a set of multiple source node and target node pairs.

2.2. Video Streaming QoE Model

This paper adopts the video stream distortion model of [13], first evaluates the impact of the loss of a slice of a video frame, and then calculates the relationship between the video frame structure and the occurrence of slice loss. The total distortion D of a video stream is calculated as follows:

$$D=\frac{\sigma_S^2\left(\gamma_q{}^{\delta+1}-(\delta+1)\gamma_q+\delta\right)}{\delta\left(1+\gamma_q\right)^2} \tag{1}$$

σ_S^2 is the distortion caused by slice loss; δ is the number of P frames in each image group; γ_q is the attenuation factor ($\gamma_q<1$) of this model, which represents the number of spatial filtering of the video in the encoding/decoding process.

Extending the model to scenes with different frame loss, D pairs are averaged, and the total distortion is calculated as:

$$\overline{D}=s\overline{n}P_eLD \tag{2}$$

s is the number of slices of each video; $\overline{n}$ represents the burst degree of loss (represents the Bernoulli packet loss model, which is burst packet loss); P_e is the message error rate; L is the number of messages per frame.

The mean squared error is mapped to a peak signal-to-noise ratio(PSNR) metric using the following equation:

$$PSNR = 10\log_{10}\frac{255^2}{\overline{D}} \tag{3}$$

Finally, map PSNR to QoE score[14]:

$$QoE_{video} = 4.5 - \frac{3.5}{1+\exp\left(x_1\left(PSNR - x_2\right)\right)} \tag{4}$$

According to the literature [14], when PSNR=40dB, the video quality is the highest; when PSNR=20dB, the video quality is the lowest. Therefore, the slope of the mapping curve is set as x1=0.5, and the center point is set as x2=30.

3. Q Reinforcement Learning Algorithm Research

Reinforcement Learning (RL) is a computational learning method, which is a process in which an agent continuously interacts with the environment and adjusts its strategy based on experience to maximize the long-term cumulative value of all rewards.

3.1. Q-learning algorithm

Q-Learning algorithm is a classic algorithm in reinforcement learning, which uses the temporal difference to solve reinforcement learning control problems. In the Q-Learning algorithm, the Q value Q(s, a) is defined as selecting behavior a in state s and following the estimate of the optimal policy. Therefore, the ultimate goal of the Q-Learning algorithm is to find the optimal policy by estimating its Q value.

In the Q-learning algorithm, the agent maintains and updates a state-value function table with a capacity of |S|*|A|, where each Q value is $Q(s_t,a_t)$:

$$Q(s_t,a_t) = E_\pi\left[\sum_{t=0}^{+\infty}\gamma^t r_{t+1} \mid S_t = s_t, A_t = a_t, \pi\right] \tag{5}$$

Performing action at in state st at time t, and the optimal action strategy that the agent can choose is the action corresponding to the maximum Q value in the table $a_t = \arg\max_a Q(s_t,a_t)$. The probability is Transferred to the next state st+1, and the immediate reward rt+1 for this action is harvested. At this time, the optimal Q value of the subsequent state is obtained according to the $\varepsilon - Greedy$ strategy to iterate the state value function table of the state st, so it is also An offline(off-policy) algorithm. The update formula is:

$$Q_{t+1}\left(s_t,a_t\right) \leftarrow \left(1-\alpha\right)Q_t\left(s_t,a_t\right) + \alpha\left[r_{t+1}\left(s_t,a_t\right) + \gamma\max_\alpha Q\left(s_{t+1},a\right)\right] \tag{6}$$

$\alpha \in \left(0,1\right]$ where represents the update rate of the estimated value. As Q-learning continues to iteratively learn the sample data, approaching the optimal action decision, the decision action will gradually form a policy rule that focuses on the action with the largest Q value.

3.2. Routing Algorithm Design

First, the number k of orthogonal channels is obtained. The CPG network model divides the network topology into k virtual network layers. The network topology of each layer is the same, but the orthogonal channels belonging to each node are different. For multiple channels and interfaces, the identity of the node is transformed into k nodes with different channels in the virtual network layer, and the layered nodes are still regarded as the same node during virtual computing. Then, when assigning available channels to each path, it is necessary to layer by layer check the availability of the channel under the current network layer; if the current link on the path under the current network layer can satisfy the interference relationship I of the link coexistence condition in the same channel, then it is needed to assign the channel to which the current network layer belongs to the link.

In the QLMPQ algorithm designed in this paper, in the path-finding stage, the Q-Learning algorithm in reinforcement learning is used to find the given request node. And the reward function designed according to the video streaming QoE metric is used to update the state value function Table. In the path s-election stage, the multi-path selection strategy is used to select the master-slave path according to the updated state value function table for data transmission.

3.2.1. Design of Route finding algorithm

At this stage, the Q-Learning algorithm is mainly used to iteratively update the state value function table. The following is the pseudo-code of this algorithm.

Algorithm1 Q-learning route finding algorithm

Input: number of iterations T, state s, action a, learning rate α , discount coefficient γ , exploration rate ε

Output: state value function table

```
initialize s and state value function table
for T do
    using a greedy strategy, choose action a in state s to determine the next hop node, and
obtain a new state s' and reward value r
    update Q(s,a)
    s=s'
    if s' is the destination node
        end the current iteration
    else
        continue to find the destination node
    end if
end for
```

$r_t(s_t,a_t)$ designed according to the video streaming transmission in this paper is the reward value of the action a_t selected by the agent according to the state s_t at time t, and the specific value is determined by two nodes. The QoE score and the available network resources between them are determined. The available network resources refer to the unoccupied channels. The more available network resources on a path, the more the average available resources of the nodes on the path. The network load is lighter. In this scheme, the reward value is set as:

$$r_t(s_t,a_t)=QoE(sender\_node)+QoE(receiver\_node) \\ -C_m(sender\_node,receiver\_node)+C_e(sender\_node,receiver\_node) \tag{7}$$

Among them, QoE(sender_node) and QoE(receiver_node) represent the QoE value of the sending node and the receiving node respectively, Cm(sender_node,receiver_node) represents the number of all channels between the sending node and the receiving node, and is also the virtual layer of the CGP model The number of layers, Ce(sender_node, receiver_node) represents the number of all occupied channels between the sending node and the receiving node. The node with more available resources has a higher reward value, and vice versa.

3.2.2. Design of multi-path selection algorithm

After obtaining the updated state value function table, the main path needs to be selected according to the maximum reward value and the maximum Q value in the table. However, when the main path has problems such as channel allocation or uneven load, which causes the QoE quality to fail to meet the standard, it needs to switch to the secondary path In this regard, a calculation method of the path correlation degree is designed. According to this method, the degree of correlation between multiple communicable routing paths obtained by the routing algorithm is specifically quantified, and the selection is based on the relationship between the correlation degree. The required multipath. In order to clearly reflect the difference between the three types of multipath routing, the specific calculation method of the correlation is as follows:

Assuming that there are X nodes shared in the two paths and Y links shared. The correlation calculation formula is as follows:

$$P_c = u(N-1) + vL + 1 \tag{8}$$

Among them, v>u>1, v represents the weight of the uncorrelated node relevance of the link; u represents the weight of the shared link.

Algorithm2 Multi-path Finding Algorithm

input: network architecture O, state value transfer table
output: the compatible path set P

```
for the set of node pairs (bi,di)
    select the main transmission path for the routing request node pair according to the state
value transfer table
    select the slave path for the master path according to the multi-path selection strategy
    store the master path and slave path in the compatible path set P
    for the compatible path set P do
        for path link set L do
            if there are available channels between nodes on the main path
                at time slot t, the link is assigned a channel
            else if there are available channels between the nodes of the slave path
                at time slot t, the link is assigned a channel
            else
                no channel is allocated and the link is placed at the next time slot t+1
            end if
        end for
    end for
end for
```

4. Simulation and result analysis

4.1. Simulation environment

In order to evaluate and compare the performance of the algorithm proposed in the paper, NS3 network simulation software is used to simulate the routing protocol proposed in the paper, and the NS3.28 version is selected. The simulation environment of the experiment is Ubuntu18.04 operating system, Intel I5-6300 processor, 8 GB memory. The simulation scene is an area of 1000m*1000m, the number of nodes used in the simulation is 8*8, each node is configured with three wireless interfaces, three channels are allocated, and the nodes are randomly distributed in the area.

the performance of this algorithm is the best when the parameters of the reinforcement learning algorithm are set as learning rate, discount coefficient, and greedy strategy exploration rate.

4.2. Analysis of Simulation Results

The simulation experiment mainly studies the effects of different numbers of concurrent streams on the overall network throughput, packet loss rate, and transmission delay of QLMPQ, Q-Routing, QLAODV, and AODV algorithms.

The simulation test data packet sending rate is set to 300kbps, and the number of concurrent streams is increased from 2 to 12. Figures 1 and 2 respectively show the trends of the overall network throughput and average end-to-end delay as the number of streams increases.

It can be observed from Figure 1 that as the number of concurrent flows in the network increases, the overall network throughput of each comparison algorithm is on the rise. Intuitively, the network throughput of QLMPQ is always better than other algorithms. The reason is that QLMPQ is based on the use of multi-path selection. The strategy's reinforcement learning route discovery mechanism and its QoE metric design reward function for the advantages of video streaming, so the QLMPQ throughput performance is optimal.

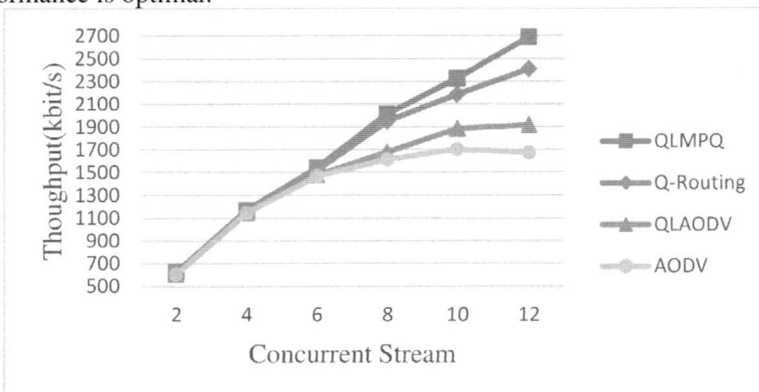

Figure1 Comparison of overall network throughput under different numbers of concurrent flows

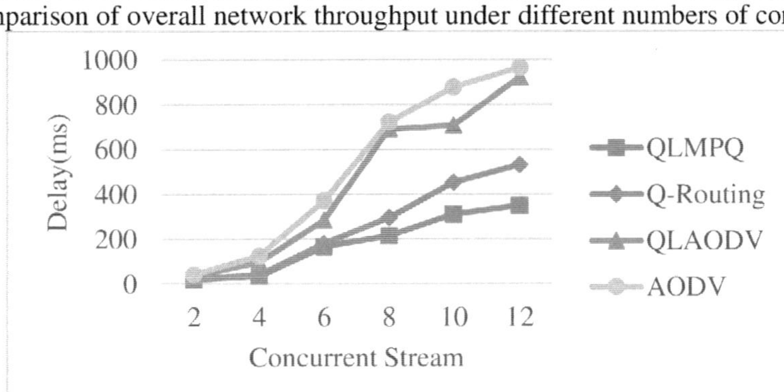

Figure2 Comparison of the average end-to-end delay of the network under different numbers of concurrent flows

As seen from Figure2, as the number of concurrent flows in the network increases, the average end-to-end delay of each comparison algorithm shows an upward trend, which is similar to the trend of packet loss rate. The average end-to-end delay of QLMPQ is always better than that of other routes. The algorithm reflects its route discovery mechanism based on reinforcement learning using a multi-path selection strategy and the advantages of the QoE metric reward function for video streaming, and the simulation results are in line with expectations.

5. Summary

This paper mainly studies a reinforcement learning routing algorithm suitable for WMNs. By analyzing the WMN, an adaptive distributed intelligent routing algorithm is proposed. Considering the QoE index of multimedia transmission, a reinforcement learning algorithm is designed. The reward function is learned, master-slave routing is flexibly selected through a multi-path strategy, and finally a routing algorithm that can comprehensively consider routing performance is learned. The NS-3 simulation software is used to compare QLMPQ, Q-Routing, QLAODV, and AODV algorithms. From simulation results, it is shown that QLMPQ comprehensively considers various routing indicators of the WMN, effectively improves the network throughput, and reduces the average end-to-end delay of the network.

References

[1] Systems C. Global mobile data traffic forecast update, 2009-2014[R]. [S.l.]: Cisco, 2010.

[2] Kourtesis D, Alvarez-Rodríguez J M, Paraskakis I. Semantic-based QoS management in cloud systems: current status and future challen-ges[J]. Future Generation Computer Systems, 2014, 32(2): 307-323.
[3] Lin C, Hu J, Kong X Z. A Review of Models and Evaluation Methods of User Quality of Experience(QoE)[J]. Journal of Computers, 2012, 35(1): 1-15.
[4] Takahashi A, Hands D, Barriac V. Standardization activities in the ITU for a QoE assessment of IPTV[J]. IEEE Communications Magazine, 2008, 46(2): 78-84.
[5] [Akyildiz I F, Wang X D, Wang W L. Wireless Mesh Networks: A survey Computer Networks, 2005, 47(4): 445-487.
[6] Lin H, Ma J F. A Secure Routing Protocol Based on Cross-layer Reputation Mechanism in WMNs[J]. Journal of Xidian University: Natural Science Edition, 2014, 41(1): 116-123.
[7] Fang Y L, Li F M, Wu P, et al. Multicast Routing Protocol for Wireless Mesh Networks[J]. Journal of Software, 2010, 21(6): 1308-1325.
[8] Yao M M, Cao Z M, Huang Q S, Shan Z L. Q-Learning Routing and Connection Scheduling Based on Traffic Patterns[J]. Journal of South China Normal University(Natural Science Edition), 2021, 53(04): 107-114.
[9] Kiani F, Amiri E, Zamani M, et al. Efficient intelligent energy routing protocol in wireless sensor networks[J]. International Journal of Distributed Sensor Networks, 2015, 11(3): 618072.
[10] Liu Y, Tong K, Wong K. Reinforcement learning based routing for energy sensitive wireless mesh IoT networks[J]. ELECTRONICS LETTERS 22nd August, 2019, 55(17): 966-968.
[11] Boushaba M, et al. Reinforcement Learning Based Routing in Wireless Mesh Networks[J]. Wireless Networks, 2013,19(8): 2079-2091.
[12] Cao Z M, Xiao W J, Peng L. A mesh chain graph model for MIMO scheduling in IEEE802.16 WMN[C]//Proceedings of the 2010 Second International Conference on Computer Modeling and Simulation. Sanya, Hainan, China: IEEE, 2010: 547-551.
[13] Tao S, Apostolopoulos J, Guérin R. Real-time monitoring of video quality in IP networks[J]. IEEE/ACM Trans on Networking, 2008, 16(5): 1052-1065.
[14] Gauss S, Müller T, Roll T, et al. Objective video quality assessment of mobile television receivers[C]//IEEE International Symposium on Consumer Electronics(ISCE 2010). IEEE, 2010: 1-6.

Energy management optimization strategy of virtual power plant based on deep reinforcement learning

Xiu Ji[1], Cong Li[1*], Jiarui Wang[2], Yan Wang[1], Faming Hou[1] and Shuanghao Guo[1]

[1]National Local Joint Engineering Research Center for Smart Distribution Grid Measurement and Control with Safety Operation Technology, Changchun Institute of Technology, Changchun, China

[2]State Grid Jilin Electric Power Co., Ltd. Electric Power Research Institute

*Correspondence: licongcoo@126.com

Abstract. The large-scale integration of new energy into the grid in the future will have an impact on the dispatching operation of the power grid. As a potential controllable resource, a controllable load is gradually being tapped, and the application of virtual power plants and artificial intelligence provides a solution. It can effectively perform resource aggregation scheduling and energy management optimization. Therefore, this paper proposes an energy management optimization strategy for virtual power plants based on deep reinforcement learning. The strategy first establishes three types of controllable load models and energy storage models and then combines the deep reinforcement learning Double-DQN algorithm with the internal model of the virtual power plant to construct an environment, action, and reward functions, and finally, conducts simulation and result analysis. The calculation example shows that this strategy can realize the optimal scheduling of the virtual power plant and improve the demand side response, and the experimental model in this paper has a higher yield curve compared with other learning algorithms, which verifies the effectiveness and rationality of the strategy. It is significant to inspire and promote the high-quality development of a green energy economy with new ideas of energy management models.

1. Introduction

Green, low-carbon, and high-quality development is the main theme of current development. A virtual power plant is a special power plant that aggregates and synergistically optimizes distributed resources such as distributed power, energy storage, controllable loads, and electric vehicles through information technology and software systems. Power plant construction and supporting power grid construction have great economic advantages [1-3].

The energy management strategies of virtual power plants have been intensively studied by predecessors, most of which are combined with optimization algorithms or heuristic algorithms. Reference [4] proposes a generalized optimization model, which is transformed into a mixed integer linear programming (MILP) to calculate the optimal scheduling of prosumer assets while minimizing costs by taking into account DR requests and end-user comfort requirements. Reference [5] applies the optimization method to the energy management strategy of virtual power plants and proposes an adaptive optimal robust optimization method to optimize resource allocation. Reference [6] applied the heuristic algorithm to study the optimal energy management strategy of renewable energy, energy storage, and controllable load under a virtual power plant. Reference [7] employs an enhanced particle swarm optimization algorithm and a commercial solver to solve a formulated two-stage model to optimize the scheduling of virtual power plants. The heuristic algorithm can obtain the corresponding optimal solution under the given conditions, but there are many limitations, and the generalization learning ability of the algorithm is insufficient. With the gradual development of artificial intelligence technology, some artificial intelligence algorithms have also been applied to solve this problem. AI algorithms solve more selective problems. Through continuous "learning," they can make more appropriate choices when faced with a piece of information. They have certain advantages in over-optimization algorithms and heuristics [8-9]. Reference [10] studied the response characteristics of distributed energy resources and used the Q-learning algorithm to solve the operation optimization problem of the VPP auxiliary service market and achieved good results. In [11], a fuzzy Q-learning method is proposed for the electricity market model with the presence of renewable resources. The proposed method can model a series of continuous multi-dimensional renewable energy penetration electricity markets with relatively few iterations.

In summary, this paper takes the virtual power plant as the research entry point, establishes an internal model of the virtual power plant with a controllable load model and an energy storage model, and proposes an optimal energy management strategy based on deep reinforcement learning. The algorithm uses the experience playback and parameter freezing mechanism to improve the performance of the deep reinforcement learning algorithm, reduce the possibility of over-learning, and avoid convergence failure. Finally, the results of the optimization strategy are verified, and the convergence of other different algorithms is compared.

2. Build Aggregate Models

2.1. Controllable load model

This load is suitable for air-conditioning and water heater systems, and it is usually controlled within a certain range without affecting customer satisfaction. A virtual power plant cuts some or all of it based on supply and demand. It needs to meet equation (1).to ensure that the temperature is within a certain acceptable range.

$$k(t)=\begin{cases}1, & \mathrm{T}_{in}(t)\le T_{\min}\\ 0, & T_{in}(t)\ge T_{\max}\\ k(t-1), & T_{\min}\le \mathrm{T}_{in}(t)\le T_{\max}\end{cases} \tag{1}$$

Loads such as washing machines and dishwashers have certain working time requirements. Combined with electricity price information and according to the operating time range of such loads, the time when the load starts to run can be reasonably planned to optimize the power consumption management.

$$c(t)=\begin{cases}1, & \mathrm{T}_{z}(t)<T_{set}\\ 0, & T_{z}(t)\ge T_{set}\end{cases} \tag{2}$$

The electric vehicle charging load model is related to the initial charging time, the rated charging power of the vehicle battery, the initial state S_{OC} of the battery, and the final full charge requirement. When there is no demand response control signal, when the state of charge S_{OC} of the electric vehicle does not meet the power requirement, it will be in the continuous charging stage until it reaches the full charge capacity requirement. The charging characteristics of electric vehicles in the period t are:

$$S_{OC}(t)=S_{OC}(t-1)+\gamma\frac{p_{ev}(t)\Delta t}{Q_B} \tag{3}$$

Journal of Physics: Conference Series **2384** (2022) 012041 doi:10.1088/1742-6596/2384/1/012041

$$S_{AC}(t)=\begin{cases}1, & T_{in}(t)<T_{\max}\\ 0, & T_{in}(t)\geq T_{\max}\end{cases} \tag{4}$$

In the formula: $S_{OC}(t)$ represents the state of charge of the battery at time t; $p_{ev}(t)$ is the actual charging power of the electric vehicle at time t; p_{ev} is the rated charging power of the electric vehicle; Q_B is the rated battery capacity; γ is the battery charging efficiency; $S_{ev}(t)$ is the charging state of the battery of the electric vehicle at time t (a value of 0 means power off to stop charging; a value of 1 means the charging state is powered on);

2.2. Energy storage model

Electric energy storage is an important part of the micro-energy grid, which can alleviate the uncertainty of renewable energy and load in the micro-energy grid to a certain extent. This paper takes the battery as an example to establish an electric energy storage model. The state of charge of the battery is related to the state of charge of the previous period and the charging power of the battery, which satisfies the following differential equation.

$$\Phi(t)=\Phi(t-1)+\frac{P_{BC}(t)\Delta t-P_{BD}(t)\Delta t}{W_B} \tag{5}$$

In the formula: $\Phi(t)$ is the state of charge of the battery at time t； $P_{BC}(t)$ $P_{BD}(t)$ are the charging and discharging power of the battery in the t period, respectively, kW； W_B the maximum capacity of the battery； Δt In order to prolong the service life of the battery as much as possible, the battery energy storage constraints and battery charge and discharge power constraints are specified as follows:

$$\eta_{B.\min}<\Phi(t)<\eta_{B.\max} \tag{6}$$

In the formula: $\eta_{B.\min}$ and $\eta_{B.\max}$ are the maximum energy storage coefficient and the minimum energy storage coefficient of the battery.

Battery charge and discharge power constraints：

$$\begin{cases}0<P_{BC}(t)\Delta_t<W_B\eta_{BC}\\ 0<P_{BD}(t)\Delta_t<W_B\eta_{BD}\end{cases} \tag{7}$$

In the formula: η_{BC} and η_{BD} are the maximum charge and discharge rates of the battery, respectively.

3. Deep Reinforcement Learning algorithm

DQN Compared with Q-learning, DQN mainly converts the Q function into a deep neural network through the value function approximation method. This article uses Double DQN. The specific steps are shown in Figure 1.

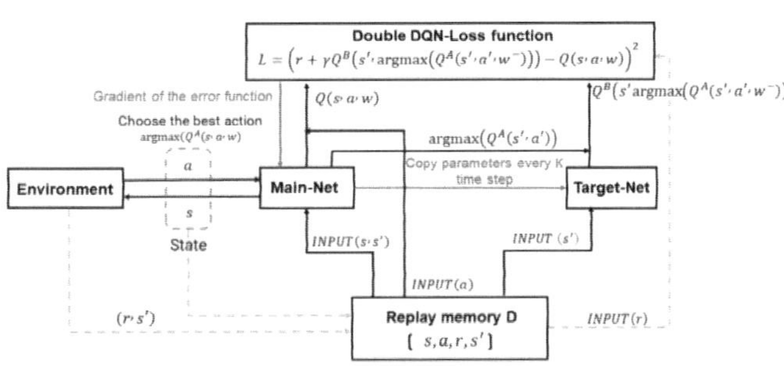

Figure 1. Double-DQN algorithm schematic diagram.

1) Initialize the two networks Q^A and Q^B and the state s.

2) Q^A and Q^B alternately perform evaluation calculation and action output.

3) Input the state s into Q^A and Q^B respectively, get the action a output by the two networks respectively, migrate to the respective state s' according to the respective actions, get the respective reward value r, and select two networks to update with equal probability.

4) If Q^A is selected, define $a^* = \arg\max Q^A(s', a)$ and update it as follows:

$Q^A(s,a) = Q^A(s,a) + \alpha(s,a)[r + \gamma \max Q^B(s', a^*) - Q^A(s,a)]$

5) If Q^B is selected, define $b^* = \arg\max Q^B(s', a)$ and update it as follows:

$Q^B(s,a) = Q^B(s,a) + \alpha(s,a)[r + \gamma \max Q^A(s', b^*) - Q^B(s,a)]$

6) $s \leftarrow s'$ Repeat the process of 2-6 until the number of training sessions is reached.

The advantage brought by Double DQN is to use max to evaluate when learning the evaluation and to use another weight set without max evaluation to do the action when selecting the output action, thus reducing the over-estimation to a certain extent. It can also be said that this method of decoupling action selection and action estimation with two networks can avoid the accumulation of overestimated positive noise to a certain extent.

The choice of state space is related to the final effect of deep reinforcement learning. Since the training of the agent is carried out in the modeled environment, the agent only observes part of the state in the real scene. For the above model, the information provided by the environment to the agent is wind, photovoltaic power output, virtual power plant bidding power (total demand state), time-of-use electricity price, controllable load state, and energy storage state of charge.

$$S = [P_W(t), P_v(t), L(t), \Gamma(t), \mathrm{H}_{1-3}(t), \Phi(t)] \tag{8}$$

In the formula: $P_W(t), P_v(t)$ is the power output in the t time period, kW, $L(t)$ is the load demand of the load in the t time period, $\Gamma(t)$ is the time-of-use electricity price in the t time period, $\mathrm{kW{\bullet}h}$, $\mathrm{H}_{1-3}(t)$ is the state of the controllable load, that is, whether to participate in the regulation of the virtual power plant, $\Phi(t)$ is the charge of the electric energy storage in the t time period.

After the agent observes the state information of the environment, according to its own policy set, the Double-DQN algorithm in this paper selects an action in the action space A. According to the dispatching strategy based on the complete absorption of wind and light in this paper, the controlled devices are controllable load, electric vehicle, and energy storage. Each controllable load can receive two actions of open and disconnect. It can receive three actions of charging, discharging, and idle. The total action space of the system is the Cartesian product of the three device action spaces, and the total action space is defined as follows: $A = [A_L \mathrm{x} A_C \mathrm{x} A_{ESS}]$.

The reward function designed in this study mainly includes: whether the bidding quantity balance can be achieved during the operation of the virtual power plant. If it exceeds or lacks, it will be calculated according to the formula. The virtual power plant will punish the over-discharge or over-charge of the energy storage battery.

$$C = c_p \sum_{i=1}^{T} (|d_{p,i}| + |d_{b,i}|) \tag{9}$$

In the formula: c_p is the penalty for the unit difference of power, Yuan/ $\mathrm{kW{\bullet}h}$; $d_{p,i}$ and $d_{b,i}$ are the unbalanced power of the power bus and the over-discharge or over-charge of the battery during the i period, $\mathrm{kW{\bullet}h}$.

In response to the incentive mechanism, there will be a greater reward for the controllable load operation regulation during peak electricity prices and peak electricity prices, and there will be rewards for underestimated electricity prices and normal price controllable load operation, that is,

Journal of Physics: Conference Series **2384** (2022) 012041 doi:10.1088/1742-6596/2384/1/012041

when the electricity output is insufficient, or the electricity price is high, it can be turned off. Then we control the load and turn on the load at other times.

$$\sum_{i=1}^{T} E = \begin{cases} \dfrac{p_{l1}}{p_l} \times D \\ \dfrac{p_{l2,3}}{p_l} \times D \end{cases} \quad \text{t} \in \textit{peak electricity price} \tag{10}$$

When the internal balance of the virtual power plant is unbalanced, it also needs to purchase electricity, so the amount of electricity purchased also becomes one of the reward functions. The following standards are defined:

$$O = \frac{P_{in}}{L_D} * G \tag{11}$$

In the formula, P_{in} indicates the purchase of electricity from an external grid, L_D indicates the load. G indicates the reward coefficient. The less electricity purchased, the greater the reward value.

The total reward function is: $R = C + E + O$.

4. Experimental Analysis and Verification

The specific parameters of the internal equipment of the virtual power plant model used in this paper are shown in Table 1.

Table 1. Abbreviation and working parameters of each equipment.

Type	*Abbreviation*	Parameter	*Value*
Air conditioner + Water heater	L1	Rated power KW	6
Washing machine + Dishwasher	L2	Rated power KW	2
Electric car	L3	Rated power KW	4
Energy storage	B/BD/BC	Maximum charging power	0.2
		Maximum discharge power	0.4
		Maximum energy storage coefficient	0.9
		Minimum energy storage coefficient	0.2

The total control capacity of the controllable load is 40,000 households. The capacity of energy storage battery is a battery with a capacity of 200 kW•h, and the initial state of charge of the energy storage device is set to 20%. The figure below shows a typical daily virtual power plant scenario, with forecasted daily load, forecasted renewable energy output, and time-of-use electricity price curve. In Figure 2. LD is the predicted load; PV and WT are the output of photovoltaic and wind turbines, respectively.

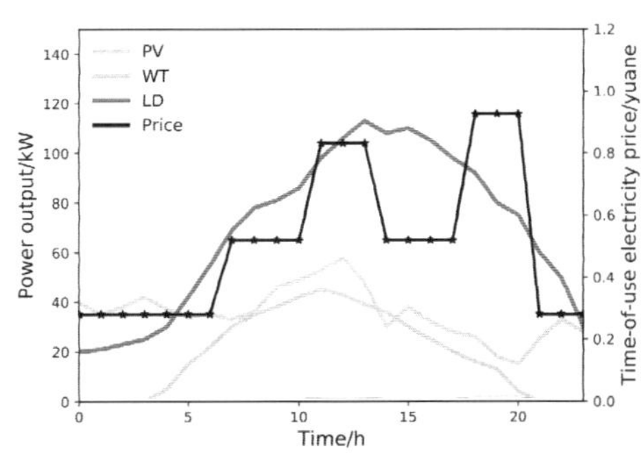

Figure 2. Virtual power plant scene diagram.

In this example, the output, load, and time-of-use electricity prices of renewable energy are all based on the forecast data in Figure 2. The specific expression formula is shown in equation (9). There are many devices in the virtual power plant represented by discrete data, and the output of each device together constitutes the action space of the micro-energy network. The specific action space of this model is composed of three different types of controllable loads and energy storage systems.

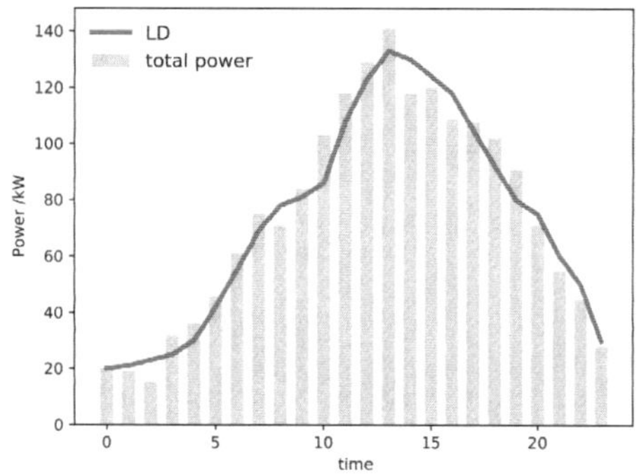

Figure 3. Final scheduling optimization results of virtual power plants.

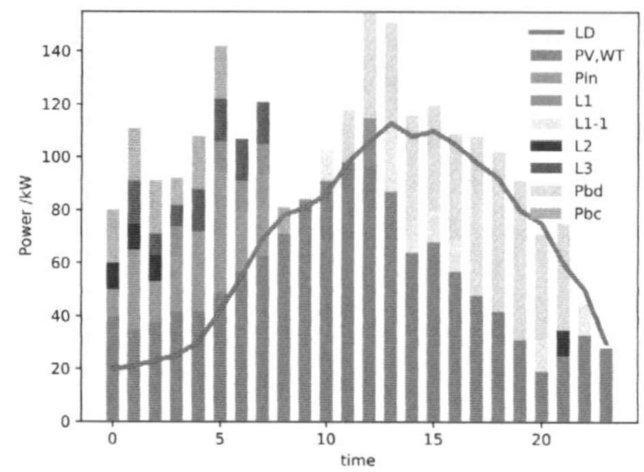

Figure 4. Energy management final result allocation.

It can be seen from Figure 2 that all energy and load are balanced, and the expected scheduling effect is achieved. It can be seen from Figure 4 that the load scheduling results of the 12 and 13 types will be selected in the time period from 0:00 am to 8:00 am the next morning, while the 11 temperature control type load will cut the load during the peak period of electricity consumption around noon, making full use of the electricity price Load shifting during underestimation plays the role of peak shaving and valley filling. The battery charging is concentrated in the middle of the night, and the discharge is concentrated in the noon load peak period. Since most of the new energy is consumed at night, purchasing low-cost electricity from outside is necessary for controllable load transfer operations. Finally, the value change of the reward function after the deep reinforcement learning training of the algorithm model in this paper is from Figure 5. The above results realize the complete response of the demand side, verify the effectiveness of the method, and demonstrate the feasibility of the virtual power plant to aggregate and package countless scattered small resources to form a complete power plant scheduling.

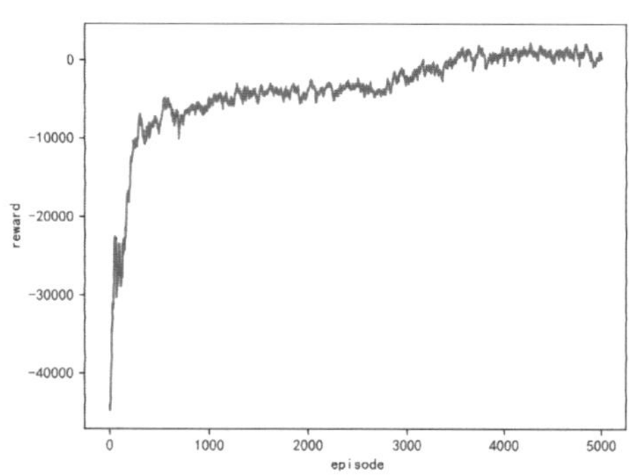

Figure 5. The value change of the reward function after the deep reinforcement learning training of the algorithm model in this paper.

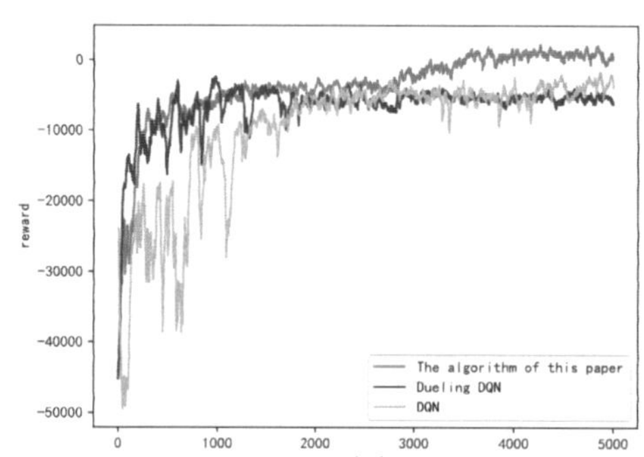

Figure 6. Comparison of training effects using different methods.

It can be seen in Figure 6 that this method has the smallest penalty value; that is, compared with the scheduling strategies of other methods, the energy optimization strategy of this method can achieve good economy.

5. Conclusion

The virtual power plant has good resource integration ability, which can aggregate some scattered small resources to form a unified whole and participate in the optimal scheduling of the power grid. In this paper, the demand-side response of virtual power plants is combined with the energy storage system to deal with the uncertainty of wind and solar output and enhance the economy of power generation in the distribution network. Experiments show that if the controllable load can be well utilized and combined with other backup energy sources, it will bring redundant space for distribution network scheduling, which is green and environmentally friendly. The virtual power plant can combine the above and find a good strategy under the deep reinforcement learning algorithm model training to solve the energy management optimization problem of the aggregated resources. The above experiments provide a useful reference for the energy management optimization of virtual power plants.

Acknowledgments

Jilin Province Young and Middle-aged Science and Technology Innovation Excellent Team Project 20220508149RC.

Reference

[1] Yu, S., Fang, F., Liu, Y. and Liu, J. 2019. Uncertainties of virtual power plant: Problems and Countermeasures. *Applied Energy*, 239:454-470.

[2] Yavuz, Levent, et al. 2019 Transformation of microgrid to the virtual power plant–a comprehensive review. *IET generation, transmission & distribution* 13.11: 1994-2005.

[3] Yan, Q., Zhang, M., Lin, H. and Li, W. 2022 Two-stage adjustable robust optimal dispatching model for multi-energy virtual power plant considering multiple uncertainties and carbon trading. *Journal of Cleaner Production*, 336:130400.

[4] Pasetti, Marco, Stefano Rinaldi, and Daniele Manerba.2018 A virtual power plant architecture for the demand-side management of smart prosumers. *Applied Sciences* 8.3: 432.

[5] Baringo, Ana, and Luis Baringo. 2016 A stochastic adaptive robust optimization approach for the offering strategy of a virtual power plant. *IEEE Transactions on Power Systems* 32.5: 3492-3504.

[6] Kasaei, Mohammad Javad, Majid Gandomkar, and Javad Nikoukar. 2017 Optimal management of renewable energy sources by the virtual power plant. *Renewable Energy* 114: 1180-1188.

[7] Qiu J, Meng K, Zheng Y, et al. 2017 Optimal scheduling of distributed energy resources as a virtual power plant in a transactive energy framework. *IET Generation, Transmission & Distribution*, 11(13): 3417-3427.

[8] Bahadur M D, Sakamoto K, Sogabe T. 2020 Multi-agent based energy balancing management algorithm for smart grid system. *Bulletin of Networking, Computing, Systems, and Software*, 9(1): 58-62.

[9] Li Y, Wang R, Yang Z. 2021 Optimal scheduling of isolated microgrids using automated reinforcement learning-based multi-period forecasting. *IEEE Transactions on Sustainable Energy*, 13(1): 159-169.

[10] Ya L, Deliang Z, Xuanyuan W. 2019 A peak regulation ancillary service optimal dispatch method of virtual power plant based on reinforcement learning.*2019 IEEE Innovative Smart Grid Technologies-Asia (ISGT Asia)*. IEEE, 2019: 4356-4361.

[11] Salehizadeh M R, Soltaniyan S. 2016 Application of fuzzy Q-learning for electricity market modeling by considering renewable power penetration. *Renewable and Sustainable Energy Reviews*, 56: 1172-1181.

Journal of Physics: Conference Series **2384** (2022) 012042 doi:10.1088/1742-6596/2384/1/012042

Application in Dynamic Stabilization System of Matrix Eigenvalue Estimation Based on Iterative Algorithm

Zhiying Hu

College of Information Engineering, Xi'an FanYi University, Xi'an, Shaanxi, China

huzhiying@xafy.edu.cn

Abstract. The role of matrix eigenvalue in engineering technology and dynamic process and the necessity of eigenvalue estimation are described. Jury criterion and iterative recursion method are used to estimate the magnitude of matrix eigenvalue modulus, at last extending the application of eigenvalue to the discrete dynamic system.

1. Introduction

In the 1850s, the concept of "Matrix Matrix" was first defined by Kelley, and then the theoretical system of Matrix was developed and perfected by Freinius et al. The "feature" of Matrix was first proposed and used by Hilbert, and many famous mathematicians (Cauchy, etc.) successively studied the feature problem of Matrix. Matrix characteristic problems can not only solve all kinds of mathematical calculation problems directly, but also play an important role in engineering technology and physics[1]. Characteristics of the matrix are an important direction of matrix calculation. At present, many scholars at home and abroad are trying to estimate the smallest range of characteristic values, the matrix eigenvalue estimates for a large amount of literature, such as the use of easy to score of the bounded set to roughly estimate the eigenvalues of a matrix, the most representative is the disc theorem and its extension[2], etc. Another example is the expression of eigenvalue estimation of symmetric matrices based on the variational principle[3]; that is, the matrix eigenvalue problem is transformed into the problem of finding functional extremum (or standing value). Rayleigh-ritz theorem and Courant-Fischer theorem are based on the variational principle[4]. It is seldom possible to get accurate analytical solutions in practical applications, so approximate calculation methods are mostly used. In conclusion, matrix characteristics are an important part of numerical calculation. Some problems in many science and engineering often boil down to the characteristics of a square matrix. Matrix characteristics caused the scholar's study interest. The study of the characteristic value of matrix estimation has become a hot research topic in the current rapid development of computer science in numerical algebra. Under such a background, it is necessary to study the estimation of matrix eigenvalues deeply.

2. Eigenvalues of the Matrix

2.1 Eigenvalues of Matrix

For the square matrix A of order n, if there is a number λ and a non-zero vector $\vec{\xi}$, making $A\vec{\xi} = \lambda\vec{\xi}$, λ is the eigenvalue of square matrix A and the vector $\vec{\xi}$ is the eigenvector of the square matrix A corresponding to the eigenvalue λ[5].

Set $A=\begin{pmatrix}1 & -1\\ 2 & 4\end{pmatrix}$ and factor $\vec{\xi}=\begin{pmatrix}x\\ y\end{pmatrix}$, to solve the characteristic equation $|\lambda E-A|=0$, get the eigenvalues of $\lambda_1=2$, $\lambda_2=3$, then solve the homogeneous linear equations, respectively $(\lambda_i E-A)x=0$. The basic solution $\vec{\xi}_1=\begin{pmatrix}1\\ -1\end{pmatrix}$ and $\vec{\xi}_2=\begin{pmatrix}1\\ -2\end{pmatrix}$ are the feature vector of matrix A respectively corresponding to $\lambda_1=2$ and $\lambda_2=3$.

2.2 *The function of Eigenvalues in Linear Transformation basic Solution is Deduced*

For $A\xi=\lambda\xi$, n-order square matrix A can be regarded as A linear transformation on R^n, $A_{n\times n}$ changing points to points and lines to lines, and the eigenvalue gives the growth rate in one direction after the transformation once. The "maximum" eigenvalue is often used in practical applications, because it estimates the maximum growth rate[6]. In other words, if you only care about is the feature vector is deduced $\vec{\xi}$ stretch or compression, so we need to examine whether $|\lambda|$ greater than 1.

2.3 *Geometric meaning of matrix eigenvalues*

Set $A=\begin{pmatrix}1 & 1\\ 0 & 2\end{pmatrix}$, $\vec{\xi_1}=\begin{pmatrix}1\\ 0\end{pmatrix}$, $\vec{\xi_2}=\begin{pmatrix}1\\ 1\end{pmatrix}$, then the geometric meaning of $A\vec{\xi_i}=\lambda\vec{\xi_i}(i=1,2)$ versus $\vec{\xi_1}$ and $\vec{\xi_2}$ is shown in Figure 1.

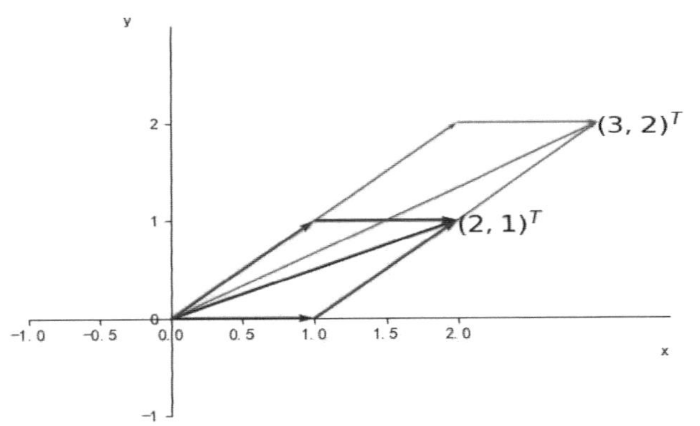

Figure 1. The Geometric Meaning of Vectors

The matrix A applied to the linear space V is the same thing as expanding by A factor in the direction of the vector $\begin{pmatrix}1\\ 0\end{pmatrix}$ and stretching by A factor of 2 in the direction of the vector $\begin{pmatrix}1\\ 1\end{pmatrix}$. And this transformation transforms $\begin{pmatrix}2\\ 1\end{pmatrix}$ to $\begin{pmatrix}3\\ 2\end{pmatrix}$.

3. The Estimation of Matrix Eigenvalues

To solve an order square matrix eigenvalues of A, we need to solve the characteristic equation $|\lambda E-A|=0$, namely $|\lambda E-A|=\lambda^n+a_1\lambda^{n-1}+\cdots+a_{n-1}\lambda+a_n=0$. If n is large and requires a high-order equation, we can get some characteristic values of numerical calculation when matrix elements are known. What is the maximum magnitude of the eigenvalues? You can figure that out by comparison. But when some elements of the matrix are not known constants, the numerical calculation to obtain eigenvalues is difficult or even impossible. Suppose we only need to distinguish whether the modulus of the matrix eigenvalue is greater than 1. In that case, we can use iterative recursion to see whether the vector increases or decreases after iterative recursion.

3.1 When n = 2

To solve characteristic equation $|\lambda E-A|=\lambda^2+a_1\lambda+a_2=0$, the extract roots formula

$$\lambda_1 = \frac{1}{2}\left(-a_1 + \sqrt{a_1^2 - 4a_2}\right); \lambda_2 = \frac{1}{2}\left(-a_1 - \sqrt{a_1^2 - 4a_2}\right). \quad (1)$$

It is not difficult to calculate if the length of $\lambda_1$or λ_2 is less than 1:

$$1+a_1+a_2 > 0,\ 1 - a_1 + a_2 > 0,\ 1 - a_2 > 0. \quad (2)$$

3.2 When n > 2

Jury criterion[7]: All the roots of $|\lambda E - A| = \lambda^n + a_1\lambda^{n-1} + \cdots + a_{n-1}\lambda + a_n = 0$ are in the unit disk. The following Sufficient and necessary conditions must be met

$$P(1)>0, (-1)^n P(-1) > 0, |a_n| < 0, |b_0| > |b_{n-1}|,$$
$$|c_0| > |c_{n-2}|, \ldots, |s_0| > |s_2| \quad (3)$$

Among them

$$b_0 = \begin{vmatrix} a_0 & a_n \\ a_n & a_0 \end{vmatrix}, b_1 = \begin{vmatrix} a_0 & a_{n-1} \\ a_n & a_1 \end{vmatrix}, \ldots, b_{n-1} = \begin{vmatrix} a_0 & a_1 \\ a_n & a_{n-1} \end{vmatrix},\ a_0 = 1,$$
$$c_0 = \begin{vmatrix} b_0 & b_{n-1} \\ b_{n-1} & b_0 \end{vmatrix}, c_1 = \begin{vmatrix} b_0 & b_{n-2} \\ b_{n-1} & b_1 \end{vmatrix}, \ldots, c_{n-2} = \begin{vmatrix} b_0 & b_1 \\ b_{n-1} & b_{n-2} \end{vmatrix}, \ldots \quad (4)$$

Until there are only three elements in the same row.

P(1)>0 is the stability criterion directly applied in the domain. Juli criterion directly determines whether the root of the discrete system is located in the unit circle on the plane according to the coefficient of the closed-loop characteristic equation, so as to judge whether the system is stable.

4. Application of Matrix Eigenvalues in Discrete Dynamic Systems

To understand and predict the long-term behavior or evolution of the dynamic system described by the difference equation$x_{n+1} = Ax_n$, the key is to master the eigenvalues and eigenvectors of the matrix[8], such as:

In a certain city, there are 150,000 people with a bachelor's degree or above, and 15,000 of them are teachers. According to the survey, 10% of the people turn from teachers to other professions every year and another 1% turn from other professions to teachers. Try to predict how many of the 150,000 people are still engaged in the teaching profession ten years later.

Resolution:x_n is the number of teachers and other professions after the n year, then $x_0 = \begin{pmatrix} 1.5 \\ 13.5 \end{pmatrix}$, matrix $A = (a_{ij}) = \begin{pmatrix} 0.90 & 0.01 \\ 0.10 & 0.99 \end{pmatrix}$ represents the transfer of teaching and other professions, where $a_{11} = 0.90$ represents 90% of the people who used to be teachers and are still teachers every year; $a_{21} = 0.10$ indicates that 10% of people switch from teaching to other professions each year. Clearly

$$x_1 = Ax_0 = \begin{pmatrix} 0.90 & 0.01 \\ 0.10 & 0.99 \end{pmatrix}\begin{pmatrix} 1.5 \\ 13.5 \end{pmatrix} = \begin{pmatrix} 1.485 \\ 13.515 \end{pmatrix}. \quad (5)$$

A year later, there were 14,485 teachers and 135,515 others. And because

$$x_2 = Ax_1 = A^2x_0, \cdots, x_n = Ax_{n-1} = A^n x_0. \quad (6)$$

So $x_{10} = A^{10}x_0$, and the direct calculation A^{10} requires a lot of calculation. In order to calculate A^{10}, A can be diagonalized first; that is, the eigenvalues and eigenvectors of A can be solved.

By

$$|\lambda E - A| = \begin{vmatrix} \lambda - 0.9 & -0.01 \\ -0.1 & \lambda - 0.99 \end{vmatrix} = (\lambda - 0.9)(\lambda - 0.99) - 0.00$$
$$= \lambda^2 - 1.89\lambda + 0.891 - 0.001 = \lambda^2 - 1.89\lambda + 0.890 = 0 \quad (7)$$

We get $\lambda_1 = 1$, λ_2=0.89, $\lambda_1 \neq \lambda_2$. Obviously, A can be diagonalized.

Substitute $\lambda_1 = 1$ into $(\lambda E - A)x = 0$, then get its corresponding eigenvector

$$\vec{\xi}_1 = \begin{pmatrix} 1 \\ 10 \end{pmatrix}. \tag{8}$$

Substitute $\lambda_2 = 0.89$ into $(\lambda E - A)x = 0$, then get its corresponding eigenvector

$$\vec{\xi}_2 = \begin{pmatrix} 1 \\ -1 \end{pmatrix}. \tag{9}$$

Let

$$P = (\vec{\xi}_1 \quad \vec{\xi}_2) = \begin{pmatrix} 1 & 1 \\ 10 & -1 \end{pmatrix} \tag{10}$$

Then

$$P^{-1}AP = \Lambda = \begin{pmatrix} 1 & 0 \\ 0 & 0.89 \end{pmatrix}, A = P\Lambda P^{-1}, \tag{11}$$

$$\mathrm{A}^{10} = \mathrm{P}\Lambda^{10}\mathrm{P}^{-1} \tag{12}$$

$$P^{-1} = -\frac{1}{11}\begin{pmatrix} -1 & -1 \\ -10 & 1 \end{pmatrix} = \frac{1}{11}\begin{pmatrix} 1 & 1 \\ 10 & -1 \end{pmatrix},$$

$$x_{10} = P\Lambda^{10}P^{-1}x_0 = \frac{1}{11}\begin{pmatrix} 1 & 1 \\ 10 & -1 \end{pmatrix}\begin{pmatrix} 1^{10} & 0 \\ 0 & 0.89^{10} \end{pmatrix}\begin{pmatrix} 1 & 1 \\ 10 & -1 \end{pmatrix}\begin{pmatrix} 1.5 \\ 13.5 \end{pmatrix} = \frac{1}{11}\begin{pmatrix} 1 & 1 \\ 10 & -1 \end{pmatrix}\begin{pmatrix} 1 & 0 \\ 0 & 0.311817 \end{pmatrix}\begin{pmatrix} 1 & 1 \\ 10 & -1 \end{pmatrix}\begin{pmatrix} 1.5 \\ 13.5 \end{pmatrix} = \begin{pmatrix} 1.4062 \\ 13.5938 \end{pmatrix}.$$

The bar chart simulated by excel software is as Figure 2

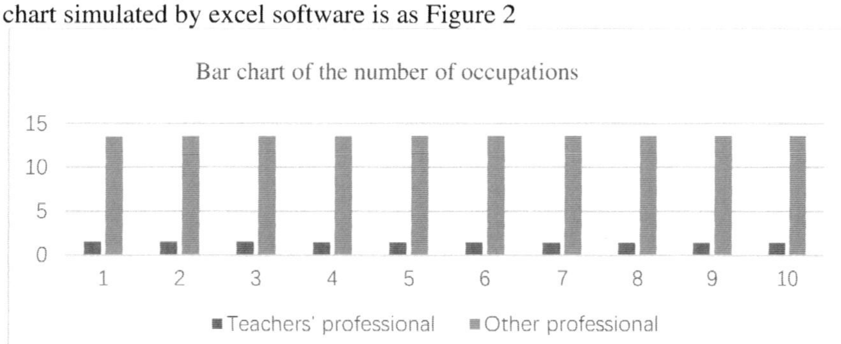

Figure 2. Bar chart of the occupations number after 1 to 10 years later

Ten years later, about 14,100 of the 150,000 were still teachers, and about 135,900 were in other professions.

5. Conclusions

Matrix characteristic problem is an important part of numerical computation and an active research topic in computer science and numerical algebra. When estimating the eigenvalues, when the order n of the matrix is less than or equal to 3, it can be directly computed by iterative recursion, while when n is greater than or equal to 4, the direct calculation is heavy, and Jury Criterion is used. The eigenvalues of matrices can not only directly solve various mathematical problems, but also be applied to the analysis of statically stable systems. The determination of some critical values in engineering design also boils down to the eigenvalue problem of the matrix. In short, eigenvalue problems of matrices are widely used.

Journal of Physics: Conference Series **2384** (2022) 012042 doi:10.1088/1742-6596/2384/1/012042

Acknowledgements
This work was supported by the 2021 university-level regular scientific research project of Xi 'an Fanyi University: Estimation of eigenvalues of matrices with unknown elements and its application (21A04).

References

[1] Xue J 2009 *J.*Estimation of matrix eigenvalues and its application. Journal of Shandong University (Natural Science). 44(12):48-51.
[2] Zhang X F 2022 *J. The properties of Gale circles and their applications. Journal of Jiamusi University (Natural Science Edition).* 40(01): 157-160.
[3] Cheng K 1 2019 *J.*Estimation of eigenvalues of matrices. Journal of Anyang Institute of Technology.18(02):87-88.
[4] Wu X Y 2015 *J.* Estimation of eigenvalues of real symmetric matrices. Science & Technology Vision.(36):70-71.
[5] Deng L Z 2014 *J.* Discussion on the estimation of matrix eigenvalues. Journal of Jilin Engineering and Technology Normal University.30(02):88-90.
[6] Wu G C 2017 *M. Linear algebra. China Renmin University Press. Beijing.*
[7] Liu X P 1991 *M.*Journal of Applied Mathematics and Computational Mathematics. Xi 'an Jiaotong University Press. Xi 'an.
[8] Wang R 2018 *J.* Eigenvalues and eigenvectors and their application cases. Modernization of Education. 5(27): 258-261.

EEG Emotion Classification and Correlation Research Based on Deep Learning

Zhengping Li, Hongsheng Xia and Lijun Wang

College of Information, North China University of Technology, No.5 Jinyuanzhuang Road, Shijingshan District, Beijing, China

Corresponding author's e-mail: ljwang@outlook.com

Abstract. In the classification of EEG emotions, the improvement of the generalization performance of deep learning networks is particularly important for the information extraction of EEG raw signals. In this paper, we use label smoothing (LS) as a regularization method to improve the generalization performance of the network by alleviating the overfitting of the network. The method is evaluated on the DEAP (Database for Emotion Analysis using Physiological signals) dataset, and the differences in emotion classification in different brain regions are compared. The experimental results show that our method has improved the generalization performance of the network, increasing the mean accuracy by 1.34% in the arousal dimension and 2.28% in the valence dimension.

1. Introduction

Emotion plays an important role in human cognition, which affects human decision-making, perception, communication, and learning. In the field of human-computer interaction, affective computing simulates the emotional interaction between humans and computers by combining technology with emotion. With the improvement of computer data processing performance, deep learning technology has gradually emerged in electroencephalogram (EEG) emotion recognition because it can learn good feature representations from raw data.

Schirrmeister et al. [1] used convolutional neural networks (CNN) for deep learning, changing the conventional way of learning in computer vision through end-to-end learning. Ding et al. [2] proposed an end-to-end multi-scale convolutional neural network TSception for learning temporal dynamics and spatial asymmetry from emotional EEG signals. Regularization methods can help the network obtain better generalization performance. Lawhern et al. [3] used dropout to prevent overfitting when training on small-scale samples and added a penalty item to encourage weight sparsity. Yang et al. [4] improved the generalization ability of the model by adopting the dropout operation as a form of regularization and adding an L2 regularization term to the cost function.

Better generalization performance can improve the accuracy of the model on classification problems. Therefore, to improve the generalization performance of the TSception multi-scale convolutional neural network model proposed in [2], we adopt label smoothing (LS) and test on the DEAP dataset to improve the accuracy of EEG emotion classification in the two dimensions of arousal and valence. Then, we explore the correlation between the classification accuracy with different brain regions from two emotional dimensions of arousal and valence.

2. Multi-scale convolutional neural network

Multi-scale convolutional neural network (TSception) is shown in Figure 1 [2]. The f_s is the EEG signal sampling rate, C is the number of channels, BN is batch normalization, AvgPool is the average pooling operation, and GAP is the global average pooling operation. TSception is an end-to-end multi-scale convolutional neural network with four main parts: dynamic temporal layer, asymmetric spatial layer, high-level fusion layer, and classifier. The dynamic temporal layer uses a multi-scale 1D convolution operation to extract time and frequency representations of multiple EEG data from channels. Then, the asymmetric spatial layer uses a multi-scale 1D convolution operation to learn emotional asymmetry features from both hemispheres and the global space. The high-level fusion layer is used to fuse the information from the hemispheric and global representations and finally uses softmax as the activation function to pass the fused feature representation to the fully connected layer.

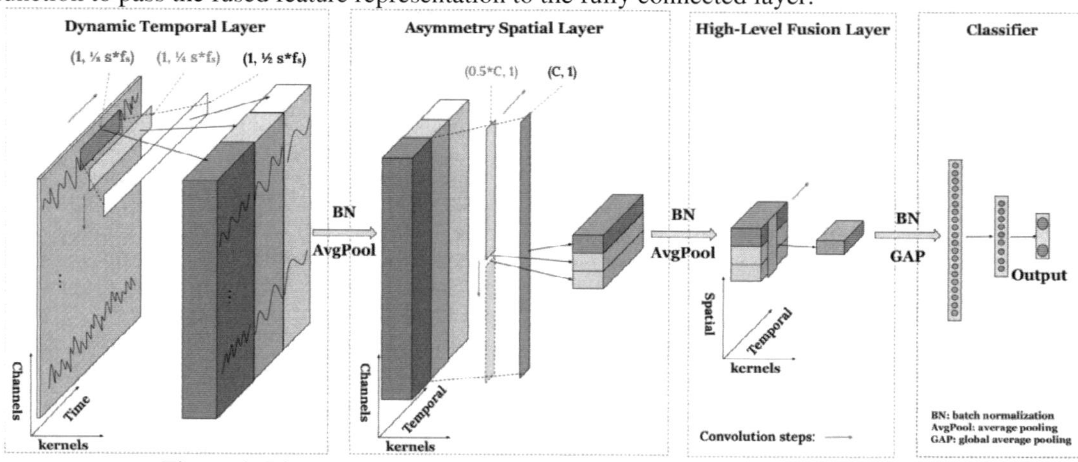

Figure 1. The structure of the multi-scale convolutional neural network.

3. Label regularization

Label smoothing is a label regularization technique proposed by Szegedy et al. [5]. Its central idea is to perform regularization by adding noise to the real data label distribution to suppress the overfitting problem. Let the target no longer be a one-hot label, but become a form such as equation (1):

$$q(k)=\begin{cases} 1\text{-}\varepsilon\,, & \text{if } k=y \\ \varepsilon/K\,, & \text{otherwise} \end{cases} \tag{1}$$

where K is the total number of categories, ε is the smoothing index.

General data labels are all in the form of one-hot encoding; that is, the correct category is marked as 1, and the wrong category is marked as 0. In the classification task, the neural network will output the confidence corresponding to each category of the current data, normalize it through softmax, and finally get the probability that the current data belongs to each category, that is

$$p(k|x)=\frac{e^{z_k}}{\sum_{i=1}^{K} e^{z_i}} \tag{2}$$

where $k \in \{1\ldots K\}$ denotes a certain class among multiple classes, and z_i denotes the log probability.

Then calculate its cross-entropy loss function, as shown in equation (3):

$$loss=-\sum_{k=1}^{K} log(p(k))q(k) \tag{3}$$

Its label distribution is a Dirac function, as shown in equation (4):

$$q(k)=\begin{cases} 1\,, & \text{if } \;k=y \\ 0, & \text{otherwise} \end{cases} \tag{4}$$

Minimizing loss is equivalent to maximizing the log-likelihood of the correct label, where the label is chosen according to its true distribution $q(k)$. The model score output z_k is processed by softmax to obtain the prediction distribution $p(k)$, and the way to minimize loss is to make $p(k)$ approach 1, but due to z_k is bounded, it cannot be achieved, so only let the correct label score much higher than the others, that is, z_y>> z_k. There will be two problems: 1. It may lead to overfitting; 2. It reduces the generalization ability of the model.

To solve the above problems, a new method is proposed in [5], that is, considering the real distribution $u(k)$ and smoothing index ε to replace the label distribution $q(k)$ of equation (4) to obtain a new label probability distribution, as shown in equation (5):

$$q'(k)=(1-\varepsilon)q(k)+\varepsilon u(k) \tag{5}$$

For $u(k)$, it is generally the mean distribution, that is, $u(k)=1/K$, then we get

$$q'(k)=(1-\varepsilon)q(k)+\frac{\varepsilon}{K} \tag{6}$$

This ensures that the true probability of the incorrect label is not 0, and the predicted probability of the correct label is balanced with the predicted probability of other labels. It avoids the phenomenon that when $q(k)$ approaches 1, and other terms approach 0, the loss function will be a large value.

Finally, the cross entropy is expressed as equation (7):

$$H(q',p)=-\sum_{k=1}^{K} logp(k)q'(k)=(1-\varepsilon)H(q,p)+\varepsilon H(u,p) \tag{7}$$

4. Experiments

4.1. Dataset

The dataset used in this paper is the DEAP (Database for Emotion Analysis using Physiological signals) dataset [6]. The DEAP dataset is a multi-modal emotional dataset, including physiological signals such as EEG, electrocardiogram (ECG), and galvanic skin response (GSR). The researchers asked 32 subjects to watch music video clips while recording their EEG, facial expressions, and galvanic skin. Each subject participated in a total of 40 trials. The duration of each trial is 63 seconds, and the baseline time before the trial is 3 seconds. After each trial, subjects will evaluate their emotional states in terms of arousal, valence, dominance, and liking on a questionnaire with nine discrete levels for each dimension. The EEG signal acquisition electrode is 32 channels, and the sampling frequency is 512Hz.

4.2. Data preprocessing

The 3-second baseline time for each trial was first removed. The data were then down-sampled to 128 Hz, and blind source separation technique was used to remove the electrooculogram (EOG) according to [6], and a band-pass filter was used to filter the signal at 4.0 Hz-45.0 Hz. Finally, the data were averaged to the common reference. The label level of each dimension is 1-9, selected 5 as the threshold, and mapped these 9 discrete values into the low and high classes of each dimension [2]. The research conducted in this paper focuses on two dimensions: arousal and valence. For deep learning networks that require large amounts of data, data augmentation was used to divide each trial into non-overlapping 4-second segments for network training.

4.3. Research details

In the generalization performance research using the TSception network, we use label smoothing, a label regularization technique, to optimize the training process. The initial learning rate is 1e-3, dropout is 0.5, and cross-entropy is selected as the loss function to guide the training process, using 10-fold cross-validation to prevent potential data leakage problems. The 10-20 electrode system method is used to place EEG collecting electrodes. Input EEG electrodes include Fp1, AF3, F3, F7, FC5, FC1, C3, T7, CP5, CP1, P3, P7, PO3, O1, Oz, Pz, Fp2, AF4, Fz, F4, F8, FC6, FC2, Cz, C4, T8, CP6, CP2, P4, P8, PO4, O2 a total of 32.

In the research on the correlation between EEG signals and emotions, we removed the middle four electrodes, Oz, Pz, Fz, Cz, and the remaining 28 symmetrical electrodes were paired in pairs to obtain a total of 14 groups of left and right brain symmetrical electrodes. These EEG signal data are used as inputs in turn to study the correlation between the EEG emotion classification accuracy and different brain regions from two emotional dimensions of arousal and valence.

5. Results and analysis

5.1. Generalization performance

After using label smoothing, the accuracy results of emotion classification in the two dimensions of arousal and valence are shown in Figure 2 and compared with the experimental results without label smoothing, as shown in Table 1 and Table 2.

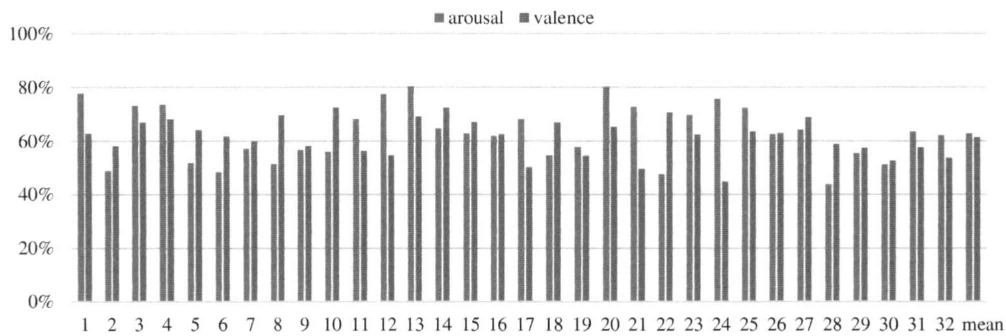

Figure 2. Mean accuracy of each subject for arousal and valence.

Table 1. Accuracy of arousal.

Subjects	Not using LS	Using LS	Comparison	Subjects	Not using LS	Using LS	Comparison
1	77.50%	77.83%	0.33%	17	66.50%	68.17%	1.67%
2	47.67%	48.67%	1.00%	18	52.83%	54.67%	1.84%
3	73.00%	73.33%	0.33%	19	57.50%	57.83%	0.33%
4	73.50%	73.67%	0.17%	20	80.50%	80.17%	-0.33%
5	49.50%	51.83%	2.33%	21	72.33%	72.83%	0.50%
6	47.83%	48.33%	0.50%	22	42.83%	47.67%	4.84%
7	56.33%	57.17%	0.84%	23	69.67%	69.83%	0.16%
8	50.83%	51.33%	0.50%	24	71.33%	75.67%	4.34%
9	54.50%	56.67%	2.17%	25	72.17%	72.33%	0.16%
10	56.00%	56.17%	0.17%	26	62.50%	62.67%	0.17%
11	68.83%	68.33%	-0.50%	27	64.17%	64.33%	0.16%
12	76.50%	77.50%	1.00%	28	42.00%	43.67%	1.67%
13	78.50%	80.50%	2.00%	29	47.00%	55.50%	8.50%
14	62.00%	64.83%	2.83%	30	50.67%	51.17%	0.50%
15	62.50%	62.83%	0.33%	31	61.83%	63.50%	1.67%
16	60.00%	62.00%	2.00%	32	61.50%	62.17%	0.67%
				mean	61.57%	62.91%	1.34%

Table 2. Accuracy of valence.

Subjects	Not using LS	Using LS	Comparison	Subjects	Not using LS	Using LS	Comparison
1	60.17%	62.83%	2.66%	17	46.67%	50.17%	3.50%
2	55.33%	58.33%	3.00%	18	65.17%	67.00%	1.83%
3	65.33%	67.00%	1.67%	19	51.83%	54.33%	2.50%
4	68.00%	68.17%	0.17%	20	64.67%	65.33%	0.66%

5	60.83%	64.17%	3.34%	21	45.67%	49.67%	4.00%
6	57.83%	61.67%	3.84%	22	68.00%	70.67%	2.67%
7	55.50%	60.00%	4.50%	23	58.00%	62.50%	4.50%
8	68.33%	69.67%	1.34%	24	43.00%	44.83%	1.83%
9	55.17%	58.17%	3.00%	25	67.00%	63.67%	-3.33%
10	68.33%	72.50%	4.17%	26	58.50%	63.00%	4.50%
11	54.50%	56.33%	1.83%	27	66.83%	68.83%	2.00%
12	54.00%	54.67%	0.67%	28	55.17%	58.83%	3.66%
13	68.83%	69.17%	0.34%	29	53.33%	57.50%	4.17%
14	71.33%	72.50%	1.17%	30	49.50%	52.67%	3.17%
15	65.67%	67.17%	1.50%	31	57.00%	57.67%	0.67%
16	61.17%	62.67%	1.50%	32	51.83%	53.67%	1.84%
				mean	59.14%	61.42%	2.28%

The experimental comparison shows that the method used in this paper improves the mean accuracy of classification by 1.34% in the arousal dimension and 2.28% in the valence dimension. Except that the experimental results of subjects 11, and 20 in the arousal dimension and subject 25 in the valence dimension were not improved, the other experimental results were improved to varying degrees. In the arousal dimension, the experimental results of subjects 22, 24, and 29 improved the most, with an increase of 4.84%, 4.34%, and 8.50%, respectively. In the valence dimension, the experimental results of subjects 7, 23, and 26 improved the most, with an increase of 4.50%.

Studies in [7] show that label smoothing forces the classification results to be grouped more closely and encourages more equidistant spacing between each category; label smoothing makes the clustering tighter and the distance between categories larger. In this paper, because the DEAP dataset is scored by subjects after watching music videos, the emotional classification is not a specific discrete emotion, so the final emotion classification will be different. For the three experiments that have not been improved, the reason may be that the positive and negative samples of the collected data are not balanced, resulting in a poor clustering effect. But overall, the method used in this paper has improved the accuracy of EEG emotion classification in the two dimensions of arousal and valence and increased the generalization performance of the network.

5.2. Correlation research

In order to better study which areas of the input contain classification-related information, we use the EEG signal data of 14 groups of left and right brain symmetrical electrodes as input in turn, and their classification accuracy in arousal and valence dimensions are shown in table 3.

Table 3. Mean accuracy of brain symmetrical electrode.

Symmetrical electrode	**Arousal**	**Valence**	**Symmetrical electrode**	**Arousal**	**Valence**
FP1, FP2	61.59%	59.09%	T7, T8	59.74%	57.59%
AF3, AF4	60.99%	59.23%	CP5, CP6	60.45%	58.92%
F3, F4	59.63%	58.92%	CP1, CP2	59.82%	58.95%
F7, F8	59.84%	58.69%	P3, P4	60.28%	59.73%
FC5, FC6	59.77%	58.62%	P7, P8	60.78%	57.47%
FC1, FC2	60.57%	58.47%	PO3, PO4	60.73%	57.62%
C3, C4	59.80%	57.66%	O1, O2	60.00%	59.14%

From table 3, for the arousal dimension, the classification accuracy of symmetric electrodes FP1, FP2, and AF3, AF4 (frontal lobe electrodes) is the highest, 61.59% and 60.99%, respectively, followed by symmetric electrodes P7 and P8 (parietal lobe electrodes), the classification accuracy is 60.78%, then symmetrical electrodes PO3 and PO4 (occipital lobe electrodes), the classification accuracy is 60.73%. This indicates that the classification of frontal, parietal, and occipital regions of the brain in the arousal dimension provides more information. For the valence dimension, symmetric electrodes P3 and P4 (parietal lobe electrodes) have the highest classification accuracy of 59.73%, followed by symmetric electrodes AF3 and AF4 (frontal lobe electrodes), the classification accuracy is 59.23%, then symmetric electrodes O1and O2 (occipital lobe electrodes), the classification accuracy is 59.14%. This indicates that the parietal, frontal, and occipital regions of the brain are more informative in the valence dimension. These are consistent with [2].

In general, for arousal and valence, the frontal, parietal and occipital lobes provide different degrees of information on the classification effect, the information provided by the frontal lobe has the best classification effect on arousal, and the information provided by the parietal lobe has the best classification effect on valence.

6. Conclusion

EEG signals are complex signals, and different brain regions contain a lot of different information, which requires more advanced signal processing methods and training models to be interpreted correctly. The emergence of deep learning has enabled the learning of good feature representations from raw EEG signals. The generalization performance of its network determines the effect of the network in multi-task learning. We use label smoothing and study the contribution of different brain regions to emotion classification. Experiments in two dimensions of arousal and valence have achieved performance improvements.

Acknowledgments

This paper is supported by the key national research and development program (NO. 2020YFC0811004), a high-level talent scientific research start-up fund (NO. 110051360022XN1).

References

[1] Tibor S, Tobias S, Josef F, Martin G, Katharina E, Michael T, Frank H, Wolfram B and Tonio B. (2017) Deep learning with convolutional neural networks for EEG decoding and visualization. *Hum Brain Mapp*, 38(11): 5391-5420.

[2] Ding Y, Robinson N, Zhang S, Zeng Q and Guan C. (2022) TSception: capturing temporal dynamics and spatial asymmetry from EEG for emotion recognition. *IEEE T Affect Comput.*

[3] Lawhern V, Solon A, Waytowich N, Gordon S, Hung C and Lance B. (2018) EEGNet: a compact convolutional network for EEG-based brain-computer interfaces. *J Neural Eng*, 15(5): 056013.1-056013.17.

[4] Yang Y, Wu Q, Qiu M, Wang Y and Chen X., 2018. Emotion recognition from multi-channel EEG through parallel convolutional recurrent neural network. In: *2018 Int. Joint Conf. on Neural Networks (IJCNN).* Rio de Janeiro, Brazil. pp 1-7.

[5] Szegedy C, Vanhoucke V, Ioffe S, Shlens J and Wojna Z., (2016). Rethinking the inception architecture for computer vision. In: *2016 IEEE Conf. on Computer Vision and Pattern Recognition (CVPR).* Las Vegas, USA. pp 2818-2826.

[6] Koelstra S, Muhl C. Soleymani M, Lee J, Yazdani A, Ebrahimi T, Pun T, Nijholt A and Patras I. (2012) DEAP: a database for emotion analysis using physiological signals. *IEEE T Affect Comput*, vol. 3, no. 1, pp 18-31.

[7] Müller R, Kornblith S and Hinton G., 2019. When does label smoothing help? In: *NeurIPS 2019.* Vancouver, Canada. pp 4694–4703.

Journal of Physics: Conference Series **2384** (2022) 012044 doi:10.1088/1742-6596/2384/1/012044

Anomaly Detection of Distribution Network Based on Adversarial Dual Autoencoder

Yu Chen, Wang Luo, Yunhe Hao, Houming Jiang

NARI Group Corporation (State Grid Electric Power Research Institute), China

213161422@seu.edu.cn

Abstract. Anomaly detection on attributed graphs aims to identify individuals or groups whose patterns deviate significantly from most samples, which is very important for internet security, power system, and other applications. Especially in the power system, the energy flow pattern of the distribution network is very similar to the graph topological constraints. Taking each equipment and bus as nodes and the relevant electrical measurement data as attributes, the distribution network fault location problem is mapped to the anomaly detection task on the attributed graph. The traditional GNN-based anomaly detection algorithm for distribution networks has some problems, such as over-smoothing and inefficient embedding learning. To solve the problem of anomaly detection in the distribution network, an anomaly detection algorithm based on adversarial dual autoencoder (DGAE_GAN) is proposed, which is composed of a structural encoder and attribute encoder based on residual connection GAT. Through joint learning, the attributed graph is mapped to the low-dimensional space, and the decoders reconstruct the attributed graph. Then, a discriminator is invited to learn the sample similarity, and the game strategy can effectively alleviate the over-smoothing problem. Finally, the anomaly node detection task is realized by calculating the reconstruction errors from both the structure and attribute levels. We conducted multiple experiments on three real-world datasets, and the IEEE standard example shows that our proposed model is superior to the existing methods.

1. Introduction

Aside from observing the interaction information between nodes, the attributed network also adds an extra set of abundant features for each node. It has become a powerful tool to deal with the heterogeneous data that we often encounter and model vast complex systems, such as social networks, power system networks, and transportation networks. Some social science research has shown that the correlation of data can well reflect the interaction between individual attributes [1]. Recently, various novel paradigms of deep neural networks (DNN) in the graph domain, such as GCN [2] and GAT [3], have been continuously updated to deal with complex graphs and show excellent performance.

The graph-oriented data analysis technology, especially graph anomaly detection technology, has been widely concerned by academia and industry because of its universality and efficiency. In the DOMIANT model proposed by Ding et al. [4], node embeddings are generated by graph autoencoder, and then the anomaly score for each node is calculated based on the reconstruction errors of graph structure and attributes. More recently, a generative adversarial Network (GAN) has been proposed as an effective tool to approximate the data distribution, and achieves superior performance in anomaly detection of image and sequence data [5]. Similar to the generator of GAN, the GANAN [6] uses the encoded node embeddings to generate the reconstructed attributed graph through the decoder, and the discriminator learns to distinguish whether the input graph is from the real data or the reconstructed network. Similarly, GANAN measures the reconstruction error to realize the anomaly detection task.

In the distribution network, different power plants, lines, and other power facilities constitute graphs and are affiliated with other measurement information (e.g., voltage and phase angle). Anomaly phenomena in the power system are mainly manifested in damaged equipment or faults, which usually cause great damage to the real system. However, the traditional anomaly detection methods on the power grid cannot consider the topological information comprehensively [7]. As a powerful tool for processing graph-structure data, GCN can effectively process such power grid data with topological information and abundant attributes. Khorasgani et al. [8] retained the spatial information of the buses by the GCN model, and aggregated features of multiple measurement devices to achieve the task of fault bus classification. Zhang et al. [9] used the detrended cross-correlation analysis (DXA) technology to build a multivariate coupling network based on the distributed control system of nuclear power plants and input it into the variational graph autoencoder. At the same time, the reconstruction error of the coupling network will be used to evaluate the abnormality of nodes.

In order to solve the problem of anomaly detection on the attributed graph, especially anomaly location on the distribution grid, this paper uses the dual autoencoder (DGAE) [10] with residual connection to effectively capture the complex interaction between node attributes and topology. At the same time, through parameter sharing and joint training, the DGAE can be combined with GAN to achieve sample similarity measurement. Moreover, both the residual connection and adversarial network can effectively alleviate the over-smoothing problem of graph neural networks. After the DGAE_GAN model training, the evaluation measurement of node level anomaly detection is realized by calculating the reconstruction from both the structure and attribute perspectives.

The remaining paper is organized as follows. Section 2 introduces the details of the proposed DGAE_GAN. In Section 3, multiple experiments on real-world datasets and IEEE standard examples are presented. And conclusions are drawn in Section 4.

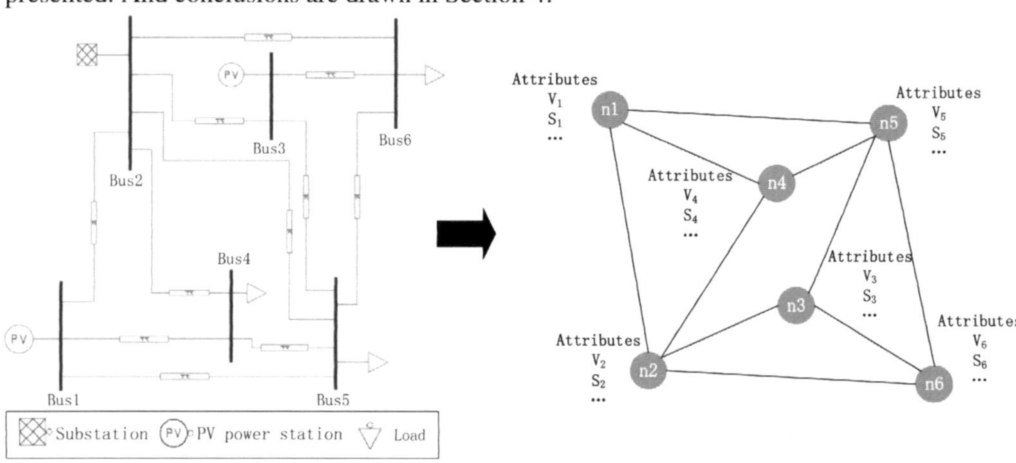

Figure 1. The attributed graph construction based on the distribution network.

2. DGAE_GAN Model Design

Considering that the information transfer between nodes constrained by topological relationship is similar to the energy flow in the power system physical model. As shown in Figure 1, we convert the distribution network into an attribute graph, and the buses and branches in the system are modeled as the nodes and edges of the graph. The electrical measurements on each bus are taken as the node attribute. Therefore, the attributed graph can be defined as an undirected graph $\mathbf{G} = (\mathbf{V}, \mathbf{E}, \boldsymbol{X})$, in which $\mathbf{V} = \{v_1, v_2, \cdots, v_n\}$ represents the set of all nodes, $\mathbf{E}$ denotes the set of edges between nodes, and the d-dimensional attribute matrix $\boldsymbol{X} \in \mathbb{R}^{n \times d}$ is affiliated with the nodes $\mathbf{V}$. In addition, the topological information of the attributed graph $\mathbf{G}$ can be effectively denoted by constructing an adjacency matrix $\boldsymbol{A} \in \mathbb{R}^{n \times n}$, so the attributed graph can also be denoted as $\mathbf{G} = (\boldsymbol{A}, \boldsymbol{X})$. On this basis, the objective of anomaly detection on attributed graphs is to find anomaly nodes whose attributes and structure patterns significantly deviate from most nodes.

The proposed DGAE_GAN is mainly composed of dual-autoencoder (DGAE) and discriminator based on GAN. What's more, DGAE can be divided into a structure encoder for processing topological information and an attribute encoder for aggregating node attribute features. As shown in Figure 2, when inputting the given graph, the structure hidden representation $\boldsymbol{Z}^s$ and the attribute latent representation $\boldsymbol{Z}^a$ can be learned through two types of encoders, respectively. Then, the structure decoder and the attribute decoder can jointly capture the interactive information between node attributes and topology by taking $\boldsymbol{Z}^s$ and $\boldsymbol{Z}^a$ as inputs to reconstruct the network $\hat{\boldsymbol{G}}=\left(\hat{\boldsymbol{A}},\hat{\boldsymbol{X}}\right)$. The role of the discriminator is to distinguish whether the input samples come from the real graph or the reconstructed one. In order to obtain more anomaly-aware features of various node anomalies in the potential space, we consider adding a re-encoding network to reconstruct the feature information extracted by the first encoder [11].

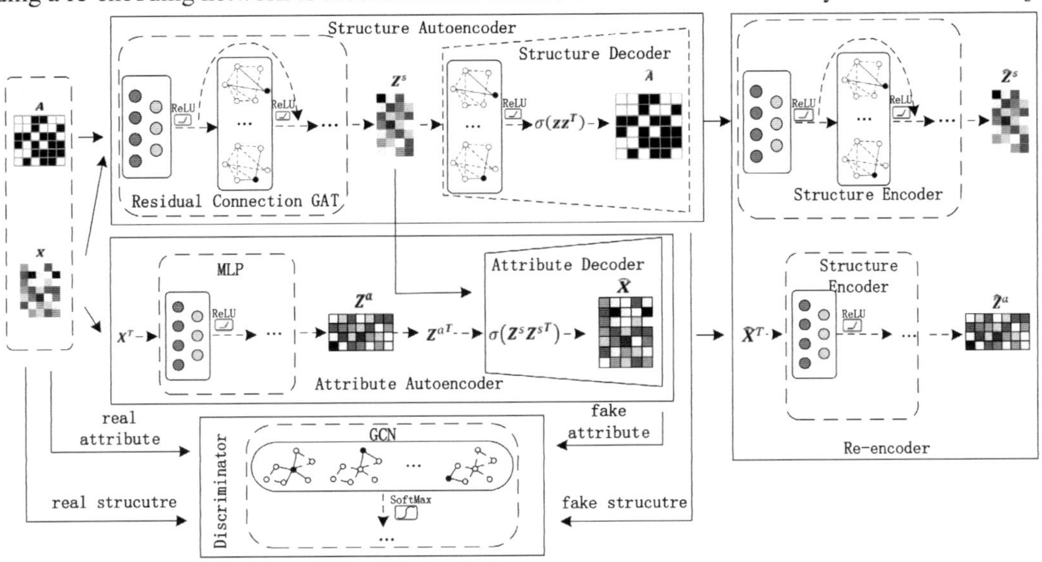

Figure 2. The framework of DGAE_GAN.

2.1. Structure Autoencoder

Given an input attributed graph $\mathbf{G}$, the structure encoder (SE) firstly maps the node attribute matrix X into the low-dimensional latent representation $\tilde{\boldsymbol{Z}}^s$, which can obtain sufficiently representative features for the following attention calculation:

$$\tilde{\boldsymbol{Z}}^s=\sigma\left(\boldsymbol{W}_s^1 X+\boldsymbol{b}_s^1\right) \tag{1}$$

Where, $\boldsymbol{W}_s^1$ is the learnable weight parameter matrix, $\boldsymbol{b}_s^1$ is the corresponding offset vector, and σ denotes the activation function $ReLU(\cdot)$.

Based on the transformed node embedding $\tilde{\boldsymbol{Z}}^s$, the attribute and adjacency matrixes are encoded by aggregating the information from neighbor nodes through three-layer graph attention (GAT) with residual connection for the over-smoothing problem. In the residual connection GAT layer, the self-attention mechanism based on weight sharing is applied to each node to calculate the corresponding attention coefficient, so as to encourage the model to learn structural information more efficiently. The attention coefficient can be formulated as follow,

$$e_{ij}=attn\left(\tilde{\boldsymbol{Z}}_i^s,\tilde{\boldsymbol{Z}}_j^s\right)=\sigma\left(\boldsymbol{a}^T\left[\mathbf{W}_s^2\tilde{\boldsymbol{Z}}_i^s\|\mathbf{W}_s^2\tilde{\boldsymbol{Z}}_j^s\right]\right) \tag{2}$$

Where e_{ij} represents the importance weight between the nodes v_i and v_j. $\boldsymbol{a}$ and $\boldsymbol{W}_s^2$ are the learnable parameters shared by all nodes, and denotes the concatenate operation.

Then, to facilitate the learning and comparison of attention coefficients between nodes, the $SoftMax(\cdot)$ function is introduced to normalize the attention coefficients of the node v_i and all its neighbor nodes. So far, the normalized attention coefficients between different nodes can be utilized to predict the final output embedding of each node by weighted sum, which can be expressed as:

$$\alpha_{ij} = exp\left(\frac{e_{ij}}{\sum_{v_k \in \mathrm{N}_i} e_{ik}}\right), \boldsymbol{Z}_i^s = \sum_{v_k \in \mathrm{N}_i} \left(\alpha_{ik}\tilde{\boldsymbol{Z}}_k^s + \tilde{\boldsymbol{Z}}_k^s\right) \tag{3}$$

Where N_i denotes the neighbors of the node v_i. Taking the learned hidden representation $\boldsymbol{Z}_i^s$ as the input of the structure decoder, the graph structure can be reconstructed by dot product, and the reconstructed adjacency matrix $\hat{\boldsymbol{A}}$ can be expressed as:

$$\hat{\boldsymbol{A}} = sigmoid\left(zz^T\right),\ z = GAT\left(\boldsymbol{Z}^s\right) \tag{4}$$

The deviation between original samples and estimated one, reconstruction error, is a powerful indicator for evaluating anomaly data [12]. The node structure information cannot be accurately reconstructed, which means that it does not match the relation pattern of the majority of normal nodes. Therefore, the topology reconstruction error, $R_s = \boldsymbol{A} - \hat{\boldsymbol{A}}$, can be used to detect anomaly structures on the graph. Simply put, the larger $R_s(i,:)$ indicates a higher probability of the ith node being identified as an anomaly from the topology aspect.

2.2. Attribute Autoencoder

In the attribute encoder, take the node attributes as input and employ two-layer fully connected layers to generate the attribute embeddings through linear transformation and nonlinear mapping, as in:

$$\boldsymbol{Z}^a = \sigma\left(\boldsymbol{X}^T\boldsymbol{W}_a^1 + \boldsymbol{b}_a^1\right)\boldsymbol{W}_a^2 + \boldsymbol{b}_a^2 \tag{5}$$

Where, $\boldsymbol{W}_a^1$, $\boldsymbol{W}_a^2$, $\boldsymbol{b}_a^1$ and $\boldsymbol{b}_a^2$ are the learnable parameters of the fully connected layer.

Finally, the attribute decoder captures the interaction between the network topology and the node attributes by receiving $\boldsymbol{Z}^s$ and $\boldsymbol{Z}^a$, and implements the inner product between two node embeddings. The reconstruction task of node attributes is formulated as in:

$$\hat{\boldsymbol{X}} = \boldsymbol{Z}^s\boldsymbol{Z}^{aT} \tag{6}$$

Similarly, by calculating the reconstruction error, $R_a = \boldsymbol{X} - \hat{\boldsymbol{X}}$, of the node attribute matrix, we can detect the anomaly information of the attributed graph at the node attribute level.

2.3. Discriminator

In the above DGAE, the encoders map the attributed graph to the low-dimensional embeddings, and the decoders hope to restore the graph data based on the low-dimensional embedding. Obviously, the function of the decoders is similar to that of the generator in GAN $\left(\hat{\boldsymbol{G}} = \mathrm{G}\left(\boldsymbol{Z}^s, \boldsymbol{Z}^a\right)\right)$. However, GAN has obvious defects: collapse, irreversible, and unable to provide density estimation, which is crucial for anomaly detection [13]. Unlike the GAN network learning to generate samples indirectly, the autoencoder (AE) directly models the correlation between the samples and the embeddings [14]. Moreover, AE has the advantages of reversibility and probability of generation pattern, so the ideas of GAN and AE are complementary and can be combined by sharing parameters and jointly training [15].

According to the demand of attributed graph processing, we utilize three-layer GCN to construct the graph-oriented adversarial network D. Benefiting from the outstanding ability of the discriminator in learning the similarity of samples, and it can distinguish whether the input graph pair $<\boldsymbol{G},\hat{\boldsymbol{G}}>$ are from the real attributed graph or the generated one.

2.4. Optimization and Anomaly Scoring Function

The encoder compresses the input attributed graph into low-dimensional representations $\boldsymbol{Z}^s$ and $\boldsymbol{Z}^a$, and then employs $\boldsymbol{Z}^s$ and $\boldsymbol{Z}^a$ to reconstruct the attribute matrix $\hat{\boldsymbol{X}}$ and adjacency matrix $\hat{\boldsymbol{A}}$ through the decoders (also viewed as generators). The DGAE learns the anomaly-aware representations of nodes, and expects to map normal and anomaly nodes to different regions in the low-dimensional space. Therefore, the training objective of DGAE is to minimize the deviation of graph topology and node attributes, and the corresponding loss function is expressed as,

$$\mathrm{L}_{rec} = \alpha\left\|\left(\boldsymbol{A}-\hat{\boldsymbol{A}}\right)\odot\lambda\right\|_2 + (1-\alpha)\left\|\left(\boldsymbol{X}-\hat{\boldsymbol{X}}\right)\odot\rho\right\|_2 \tag{7}$$

Of which α controls the importance of topology reconstruction error and node attribute reconstruction error to the loss function L_{rec}, and $\odot$ denotes the Hadamard product. What's more, we introduce the penalty factors $\lambda>1$, $\rho>1$ and the reconstruction error caused by edge or attribute deviation will be punished more severely when non-zero elements actually exist in the adjacency matrix and node attribute matrix. λ, ρ can be defined as,

$$\lambda_{ij} = \begin{cases} 1, & \boldsymbol{A}_{ij}=0 \\ \lambda, & \boldsymbol{A}_{ij}\neq 0 \end{cases}, \quad \rho_{ij} = \begin{cases} 1, & \boldsymbol{X}_{ij}=0 \\ \rho, & \boldsymbol{X}_{ij}\neq 0 \end{cases} \tag{8}$$

Based on the learned anomaly-aware representations, the second stage aims to train a discriminator D that can accurately estimate the data similarity. During training, discriminator D attempts to distinguish whether the node pair is from the real or fake data distribution. Therefore, here we only optimize the cross-entropy loss of the binary classifier for discriminator D, which can be expressed as,

$$\mathrm{L}_{adv} = \left\|D(\boldsymbol{X},\boldsymbol{A}) - D\left(G\left(\boldsymbol{Z}^s,\boldsymbol{Z}^a\right)\right)\right\|_2 \tag{9}$$

At the same time, the re-encoder loss $\mathrm{L}_{\mathrm{enc}}$ is also introduced to minimize the gap between the embeddings of the real graph and the reconstructed one during the secondary encoder training, as in:

$$\mathrm{L}_{enc} = \beta\left\|\boldsymbol{Z}^s - \hat{\boldsymbol{Z}}^s\right\|_2 + (1-\beta)\left\|\boldsymbol{Z}^a - \hat{\boldsymbol{Z}}^a\right\|_2 \tag{10}$$

Where, $\hat{\boldsymbol{Z}}^s$ and $\hat{\boldsymbol{Z}}^a$ denotes the hidden representations of the reconstructed graph, and the weight β is used to balance the importance between the node embedding error and attribute embedding error.

Therefore, by applying the weighted summation method to the above loss functions, the joint training of GAN and DGAE is realized. The loss function of the whole DGAE_GAN can be expressed as:

$$\mathrm{L} = \omega_{rec}\mathrm{L}_{rec} + \omega_{adv}\mathrm{L}_{adv} + \omega_{enc}\mathrm{L}_{enc} \tag{11}$$

Inspired by the gap between anomaly nodes and the majority of normal nodes in structure or attribute pattern, after model training, we can estimate the anomaly score S_v of each node based on the reconstruction error L_{rec}, which is shown as follow,

$$S_v = \mathrm{L}_{rec} = \alpha\left\|\left(\boldsymbol{A}-\hat{\boldsymbol{A}}\right)\odot\lambda\right\|_2 + (1-\alpha)\left\|\left(\boldsymbol{X}-\hat{\boldsymbol{X}}\right)\odot\rho\right\|_2 \tag{12}$$

Therefore, the higher the anomaly score S_v of a node, the greater the probability of being an anomaly.

3. Experiments

3.1. Datasets
In order to verify the effectiveness of the proposed DGAE_GAN, the evaluations were performed on three real-world datasets, including BlogCatalog, Flickr, and ACM. BlogCatalog and Flickr are social networks composed of interactive behaviors such as following and sharing among users. Each user is attached with tags and interests as node attributes. ACM is an academic citation network. The citation relationship between different literature forms the network, and the abstracts are used as node attributes. At the same time, to verify the effectiveness of the anomaly detection task on the distribution network, the IEEE 2868 standard example was used in the experiments, in which the topological structure of the distribution network is formed based on the connection relationship between the buses, and the electrical quantity of the equipment (or loads) is taken as the node attributes.

There is no ground truth about anomaly nodes in these datasets, so we randomly generate exceptions following the method proposed by [16]. By randomly modifying the connection relationship and node attribute features, a group of anomaly samples is injected into each dataset. The detailed statistics of the four data sets are shown in Table 1.

Table 1. Statistics of the used real-world datasets.

Dataset	BlogCatalog	Flickr	ACM	IEEE 2868
# nodes	5,196	7,575	16,484	2,868
# edges	171,743	239,738	71,980	6,942
# attributes	8,189	12,047	8,337	6,949
# anomalies	300	450	600	90

3.2. Baseline
At the same time, we compare the proposed DGAE_GAN with the existing anomaly detection models:

(1) **LOF** [17]: LOF only focuses on the attribute features of nodes and determines whether the nodes are anomalies by calculating the deviation degree between each node and its neighbor nodes.

(2) **DOMINANT**: DOMINANT uses GCN to generate node embedding based on the topology and attributes. Anomaly nodes are detected by calculating the reconstruction errors of topology and attributes.

(3) **AnalogyDAE**: AnalogyDAE uses dual autoencoder and attention mechanisms to capture the interaction between node attributes and topology for hidden representations learning. It also uses reconstruction errors to detect anomaly nodes.

(4) **EfficientGAN**: EfficientGAN only focuses on the attribute features of nodes, carries out anomaly detection by a jointly training encoder, generator, and discriminator, and combines attributed graph reconstruction loss and discriminator loss.

(5) **GANAN**: GANAN uses two encoders to transfer the real graph and reconstructed graph into a low-dimensional space. The decoder uses these structures and attributes embeddings to estimate the reconstruction attributed graph. The reconstruction error is also used to detect anomaly nodes.

3.3. Experiment Setup
By optimizing the loss function with the Adam optimization algorithm, the experiment performs 100 iterations on the DGAE_GAN, and the learning rate is set to 0.001. Two structure encoders adopt a single fully connected layer (128-neuron) followed by three-layer graph attention (128-neuron, 128-neuron, and 64-neuron, respectively). Two attribute coders adopt the architecture of a two-layer fully connected layer, and the dimensions are 256 and 64, respectively. The discriminator comprises a three-layer graph convolution (256-neuron, 128-neuron, and 2-neuron, respectively). Finally, the anomaly detection task is implemented on the four datasets, and the objective is to spot the anomaly nodes from the datasets. And the parameters $(\alpha, \beta, \lambda, \rho, \omega_{rec}, \omega_{adv}, \omega_{enc})$ are set as $(0.4,\ 0.4,\ 10,\ 15,\ 1,\ 1,\ 0.5)$,

$(0.4, 0.4, 12, 12\ 1, 1, 0.3)$, $(0.5, 0.5, 20, 20, 1, 1, 0.5)$ and $(0.2, 0.2, 25, 5, 1, 1, 0.5)$ for BlogCatalog, Flickr, ACM, and IEEE 2868, respectively.

We performed each experiment on an AMD Ryzen 7 5800H CPU and an NVIDIA GeForce RTX 3070 8GB. The main programming environments used are Python 3.9.13, Pytorch 1.11.0 + cu113, and sklearn 1.1.1. During the experiment, AUC is used as the evaluation metric to measure the model performance.

3.4. Results Analysis

Table 2 lists the AUC scores of the above anomaly detection methods on the four datasets, and the bold values indicate the best anomaly detection performance. The results show that the classical anomaly detection algorithms, LOF based on local neighborhood deviation measurement, only focus on node attributes or graph topology, so their performance is the poorest. Because the node attributes and topology can be captured jointly based on AE, the performance of DOMINANT and AnomalyDAE is superior to the former model, which evaluates the effectiveness of the anomaly detection algorithm based on reconstruction loss. EfficientGAN only considers the attribute feature of nodes and adopts the framework based on GAN to achieve similar results on BlogCatalog and Flickr. GANAN improves it by capturing the interactive information between node attributes and topology through GAN combined with AE, and the results indicate the potential of the GAN framework. Compared with the most advanced method GANAN, the AUC scores of DGAE_GAN proposed in this paper on BlogCatalog, Flickr, and ACM datasets increased by 2.1%, 0.7%, and 3.3%, respectively. DGAE_GAN uses residual connection dual autoencoders to obtain effective node embedding and attribute embedding jointly and fully capture the complex modality interaction between topology and node attributes. Therefore, in all the mentioned models, DGAE_GAN achieves the most superior on three real-world datasets.

Table 2. AUC scores of all methods on three datasets.

Dataset	BlogCatalog	Flickr	ACM	IEEE 2868
LOF	0.665	0.674	0.723	0.678
DOMINANT	0.722	0.739	0.751	0.721
AnomalyDAE	0.727	0.721	0.778	0.766
EfficientGAN	0.719	0.708	0.640	0.688
GANAN	0.797	0.788	0.805	0.817
DGAE_GAN	0.818	0.795	0.838	0.941

The distribution network simulation dataset based on IEEE 2868 standard example is sparser than the previous real-world datasets about social networks and literature citation networks. The existing anomaly detection algorithms based on AE or GAN are more sensitive to the topological sparsity of the network. In the proposed DGAE_GAN, the dual-autoencoder framework for topological information and attribute features was constructed to better adapt to highly sparse attributed graphs. Moreover, the IEEE 2868 standard example has fewer nodes, and the scale is smaller. Therefore, the detection performance of DGAE_GAN on the IEEE 2868 simulation dataset was improved significantly.

4. Conclusion

This paper proposes a joint embedding learning framework based on autoencoder and GAN to study the node anomaly detection task for the attributed graph. In this model, a dual-autoencoder is constructed for learning node and attribute embedding jointly, GAN is introduced into the dual-autoencoder framework, and a discriminator is used to estimate the node distribution. The experiment results show that the proposed DGAE_GAN achieved state-of-the-art performance, and the detection performance of anomaly nodes on the distribution network dataset was also improved significantly. Considering the anomaly events of various complex systems (including power systems) in the real world may have a long incubation period, and the anomaly detection task on the dynamic graph will be more complex and widely encountered, future work will investigate the anomaly detection task on the dynamic graph based on autoencoder and GAN.

Reference

[1] Shalizi C R and Thomas A C 2011 Homophily and contagion are generically confounded in observational social network studies *Sociological Methods & Research* **40** no 2 pp 211–239
[2] Welling M and Kipf T N 2016 Semi-supervised classification with graph convolutional networks *J. International Conference on Learning Representations (ICLR 2017)*
[3] Velickovic P, Cucurull G, Casanova A, Romero A, Lio P, and Bengio Y 2017 Graph attention networks *Stat* **1050** p 20
[4] Ding K, Li J, Bhanushali R, and Liu H 2019 Deep anomaly detection on attributed networks *Proceedings of the 2019 SIAM International Conference on Data Mining (Calgary)* pp 594–602
[5] Zenati H, Foo C S, Lecouat B, Manek G, and Chandrasekhar V R 2018 Efficient gan-based anomaly detection *arXiv preprint* arXiv:1802.06222
[6] Ouyang B, Tan M, Ding L, and Kang Y 2021 Ganan: Gan-based for anomaly detection on attributed networks *2021 3rd International Academic Exchange Conference on Science and Technology Innovation (Guangzhou)* pp 458–462
[7] Hosseini Z S, Mahoor M and Khodaei A 2018 Ami-enabled distribution network line outage identification via multi-label SVM *IEEE Transactions on Smart Grid* **9** no 5 pp 5470–72
[8] Khorasgani H, Hasanzadeh A, Farahat A, and Gupta C 2019 Fault detection and isolation in industrial networks using graph convolutional neural networks *2021 IEEE International Conference on Prognostics and Health Management (San Francisco)* pp 1–7
[9] Zhang L, Cheng W, Liu X, Chen X, Chang F and Li X 2021 System-level anomaly detection for nuclear power plants using variational graph auto-encoders *2021 IEEE International Conference on Sensing, Diagnostics, Prognostics, and Control (Weihai)* pp 180–5
[10] Fan H, Zhang F and Li Z 2020 Anomalydae: Dual autoencoder for anomaly detection on attributed networks *ICASSP 2020-2020 IEEE International Conference on Acoustics, Speech and Signal Processing (Virtual Barcelona)* pp 5685–89
[11] Liu J, Song K, Feng M, Yan Y, Tu Z and Zhu L 2021 Semi-supervised anomaly detection with dual prototypes autoencoder for industrial surface inspection *Optics and Lasers in Engineering* **136** p 106324
[12] Zhou C and Paffenroth R C 2017 Anomaly detection with robust deep autoencoders *Proceedings of the 23rd ACM SIGKDD International Conference on Knowledge Discovery and Data Mining (Halifax)* pp 665–74.
[13] Salimans T, Goodfellow I, Zaremba W, Cheung V, Radford A, and Chen X 2016 Improved techniques for training gans *Advances in Neural Information Processing Systems* **29**
[14] Makhzani A, Shlens J, Jaitly N, Goodfellow I and Frey B 2015 Adversarial autoencoders *arXiv preprint arXiv:1511.05644*
[15] Bao J, Chen D, Wen F, Li H, and Hua G 2017 Cvae-gan: fine-grained image generation through asymmetric training *Proceedings of the IEEE International Conference on Computer Vision (Venice)* pp 2745– 54
[16] Ding K, Li J and Liu H 2019 Interactive anomaly detection on attributed networks *Proceedings of the Twelfth ACM International Conference on Web Search and Data Mining (Melbourne)* pp 357–65
[17] Breunig M M, Kriegel H P, Ng R T and Sander J 2000 Lof: identifying density-based local outliers *Proceedings of the 2000 ACM SIGMOD International Conference on Management of Data (Dallas)* pp 93–104.

Journal of Physics: Conference Series **2384** (2022) 012045 doi:10.1088/1742-6596/2384/1/012045

Inverse Design of Metamaterials via Deep Learning for Electromagnetically Induced Transparency

Fushan Lu[1], Jianwei Wang[1,2], Handong Li[1], Chanchan Qing[1], Tao Lei[1], Tianyu Xiang[1], and Shengbo Hu[1]

[1] School of Big Data and Computer Science, Guizhou Normal University, Guiyang 550000, China

[2] School of Information Science and Engineering, Southeast University, Nanjing 210000, China

Author to whom correspondence should be addressed: jianweiwang@gznu.edu.cn

Abstract. The physical limitations of metamaterial structures cannot be solved under the conditions of high time cost and complex algorithms in metamaterial inverse engineering in the past. This paper proposes limiting the value range of metamaterial structural parameters through a single structural parameter acquisition method (SSPAM) for the first time, which will meet the expected values of our predictions and obtain high-quality and effective data in a relatively short time. This is the first attempt to use this method to solve the problem of physical limitations in the inverse design of metamaterials effectively. Furthermore, it is a further improvement of inverse design, enhances the reliability of metamaterial inverse design once again, and realizes the idea of on-demand design. The mean squared error of our best deep learning model is 0.00075 and 0.00026 in the training set and validation set, respectively, and 3.0×10^{-5} in the test set. We input three specific points of the EIT spectrum into our optimal model to predict the corresponding EIT structural parameters inversely, verified by numerical simulation calculation, and obtained satisfactory results. This work can provide new ideas and methods for the inverse design of metamaterials for other models.

1. Introduction

Electromagnetically induced transparency (EIT), a quantum interference effect that eliminates light absorption in opaque media [1-2], has associated non-optical properties that allow light to pass through the medium in a narrow spectral range [3]. The high transparency and strong dispersion of EIT make it potentially promising for applications in slow-light devices [4-5]. For example, EIT has a wide range of applications in sensing [6], slow light generation [7], frequency conversion [8], optical quantum memory [9], and enhanced light-medium interactions and nonlinear effects [10]. Metamaterials [11], as an artificial electromagnetic medium with a negative refractive index at the subwavelength scale, have become a paradigm for engineering electromagnetic space and controlling wave propagation. However, experimental conditions of extremely low temperature and high laser intensity are usually required to realize the EIT effect in atomic systems [12]. Thus, metamaterials, as an extremely effective method that can generate EIT phenomena at room temperature [13], not only avoid the strict experimental conditions, but also can cause more pronounced EIT phenomena. Metamaterials are greatly necessary to get further study. However, research on metamaterials usually uses the finite difference time domain method (FDTD) or finite element method (FEM) [14] and requires the calculation of the electromagnetic response of all structural parameters by using numerical simulations, which often need a lot of time and numerous searches due to the lack of empirical relationships between structural parameters and the corresponding electromagnetic response. Nevertheless, deep learning that is based on neural networks has been mathematically shown to be able to approximate any arbitrarily complex function, such as nonlinear functions. A sufficiently large neural network can theoretically be modeled for anything [15-16]. In recent years, deep learning has been widely used for the structural and functional design of metamaterials that promote the research of related systems and devices in photonics [17-24].

Deep learning methods for the inverse design of metamaterials generally start with the simulation of metamaterials using conventional numerical analysis tools to obtain a large set of data parameters, which are then used for training neural networks. A well-trained neural network model is sure of an exact mapping relationship between the training parameters, which can accurately and quickly predict the metamaterial structure and the corresponding electromagnetic parameters [25-27]. Zhen Zhang et al. [28] proposed an adaptive inverse design strategy to realize the intelligent design of metamaterial structures with the EIT phenomenon. However, the generated spectra of the proposed GAN model do not match well with the predefined expected spectra for structures that are beyond the physical limits resulting in the absence of electromagnetic response. In the same year, Wei Huang et al. [29] proposed a deep neural networks (DNN) model for the inverse design of EIT metamaterials. Their method did not consider how to exceed the physical limitations and failed to avoid those unknown structures that cannot realize the EIT spectrum during the inverse prediction, so the correct predicted values were not obtained. In inverse design, this was called a physical limitation problem. The physical limitation here is that the metamaterial structural parameters predicted in the inverse design do not form the expected spectrum when the electromagnetic control reaction is performed. In terms of the effects produced by light in metamaterials, the EIT metamaterial structure can be physically limited due to the complex coupling mechanism within the metamaterial unit cell, thus making the inverse design on metamaterial not yet highly reliable for practical use, and hence, people are highly concerned about this urgent problem that needs to be solved. However, the deep learning-based inverse design proposed above cannot make good predictions when the physical limitations of the metamaterial structure exist. Until now, for this challenge, it is still impossible to show the relationship between the range of values of the structural parameters of EIT metamaterials and the physical limitations by using the available mathematical means as well as complex physical theories. Therefore, we consider finite-size effects and the radiation that goes beyond the limits of the conventional effective medium theory. For this reason, we propose SSPAM for the first time. This method is also the first attempt to solve the problem of physical limitations in the inverse design of metamaterials, by compressing the metamaterial design space and reducing the degrees of freedom of the metamaterial structure to obtain a realistic and valid range of structural parameters. In this paper, the operation of SSPAM is to collect data only within the range of values limited by a single structural parameter at a time, and the data obtained can all achieve EIT responses normally.

The structural parameters that cannot form the EIT spectrum by numerical simulation are called dirty data in this paper. Here, after extensive iterations of data acquisition experiments, we learn that changing multiple structural variable parameters simultaneously during data collection not only makes it easy to obtain dirty data, but also the time cost is higher compared to SSPAM, all of which are not in line with the principle of simplicity, high efficiency, and accuracy of the inverse design. Accordingly, the experiments in this paper used SSPAM for the first time; that is to say, the approximate range of values of the desired structure and the corresponding spectra data were predefined by means of the parameter scanning method [30]. Within this interval, it is ensured that the acquired structure data can form the EIT spectrum in the 3D-FDTD simulation, and thus all the structure parameters predicted in the inverse design can generate the expected spectra data. There are three roles in the method of SSPAM: Firstly, in the process of running the program to automatically collect data (manual data collection will not be discussed in this article), it can avoid the generation of dirty data due to the change of multiple structural parameter variables during data collection. Secondly, it can solve the problem encountered by previous research of the same spectral spectrum generated by different group structural parameters, called the spectrum of metamaterials are not unique. Thirdly, it is possible to inverse design on demand simply and conveniently on the defined physical structure. Therefore, we only need to input a certain part of the data set in the interval into the built deep learning model for training within the limited structure range. Subsequently, the optimal model obtained after model analysis is taken out for prediction, and the structural parameters predicted by the model output are all valid and can form our expected EIT spectrum. In this way, the physical limitations of EIT metamaterials can be solved simply and efficiently.

2. Data generation and description

In this paper, the transmission spectral data collection is performed by the co-simulation method (PYTHON-VB programming interface provided by CST2020 version). Three specific points of the spectrum (two minimal and one maximal value) are obtained by data processing methods to avoid the complexity of computation because these three specific points can determine whether an EIT spectrum is formed. We use SSPAM to collect data corresponding to one structural parameter at each time, three sets of data (frequency and amplitude of three specific points of the EIT spectrum and the relevant structural parameters) corresponding to W, L1, and d of the EIT structure (see Figure 1) that has two straight metal structures as the bright-bright mode are collected, where the length of L1 ranges from 4.5 to 6.5 mm; the distance d between the L1 and L2 ranges from 0.3 to 3 mm; and the width W ranges from 0.2 to 2 mm; 5000 samples are collected separately for each structure, with a total of 15,000 samples of data. All data are grouped in a 16:4:5 ratio into the training set, validation set, and test set.

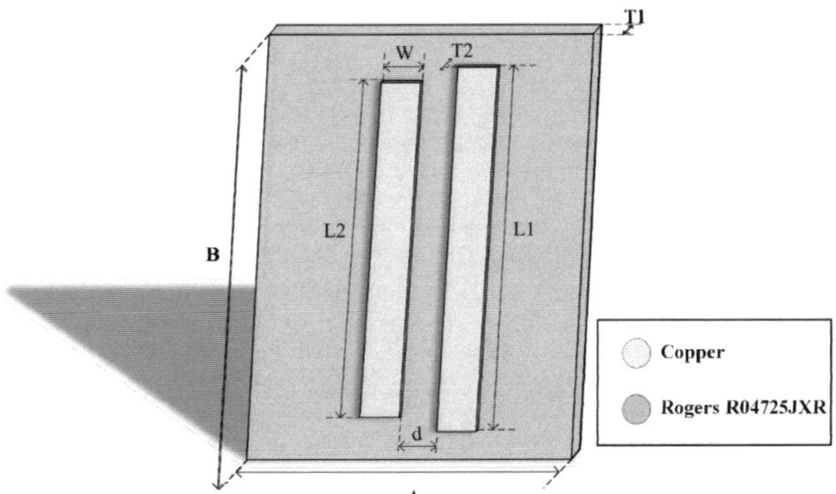

Figure 1. Schematic figure of the structure of the EIT metamaterial, including the surface bright-bright mode of two straight metal structures that lengths L1 and L2, width W, height T2, and distance d, and its substrate with length B, width A, as well as thickness T1.

3. Deep learning model

To implement our proposed inverse design approach, a neural network structure (see Figure 2) was designed in this paper. Our network consists of fully connected layers with one input layer, one output layer, and six hidden layers. The six inputs to the deep learning model are three specific points of the EIT spectrum, each including two physical values (frequency and amplitude). The output is a structural parameter. The activation function in our model is the Parametric Rectifier Linear Unit (PReLU), which can adaptively learn the rectifier parameters and improves accuracy at a negligible additional computational cost. We used a gradient descent-based Adam algorithm in the network [31]. Since the large-batch methods tend to poorer generalization [32], we choose a suitable small batch to train our model. The learning rate directly affects the convergence of the model. If it is too high, the model will not converge. On the contrary, the model will converge slowly or fail to learn. We set the learning rate to 1e-4 after several trials and trained our deep learning model with 300 iterations on the training set.

Journal of Physics: Conference Series **2384** (2022) 012045 doi:10.1088/1742-6596/2384/1/012045

The loss function used in our deep learning model is the mean squared error (MSE), which is the expected value of the square of the difference between the parameter estimated value and the true value, and is expressed as follows

$$MSE = \frac{1}{n}\sum_{i=1}^{n}(y_i - \bar{y}_i)^2 \tag{1}$$

Where y_i denotes the i-th true value, $\bar{y}_i$ for the i-th predicted value, n is the sample size.

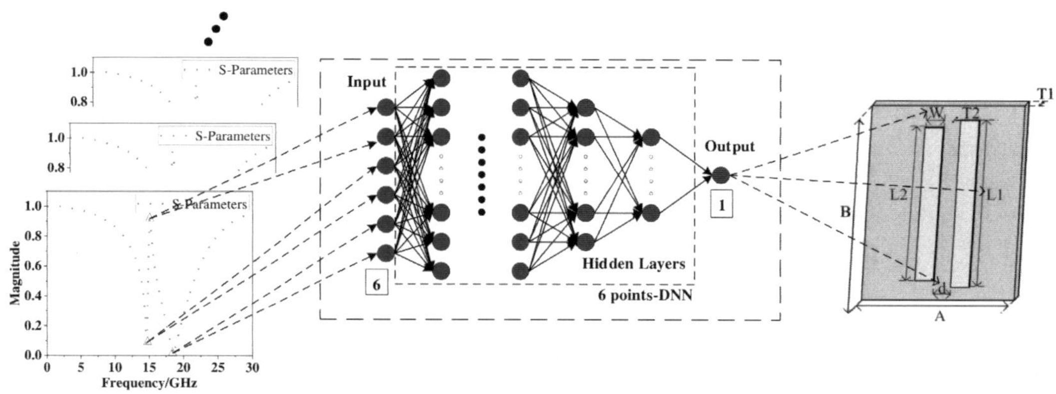

Figure 2. Schematic representation of the proposed six points-DNN. Three special points of the EIT spectrum (frequency and amplitude) are used as inputs for deep learning. The numbers of nodes in each hidden layer were set to 16, 32, 32, 32, 32, and 32.

4. Identifying the best deep learning models

Aiming to find the optimal deep learning model, by showing the MSE of the training set and test set to analyze. We first train the model by training neurons at different numbers of layers. From the theory of deep learning models, more neurons can solve more complex nonlinear problems; however, there are overfitting problems and greater computational resource consumption. It can be seen from Figure 3(a) that the optimal number of layers is 5. As the number of layers increases, the influence of gradient dispersion will be enlarged, making it difficult for the model to converge, so the MSE tends to rise. Furthermore, the number of training samples is also a trade-off between our prediction accuracy and computational resources. Therefore, we divide the data set into different sample numbers for training. As can be seen from Figure 3(b), the MSE of the training set and the test set decreases with the increase in the number of samples in the training set. It is because the high-quality training set data obtained through SSPAM always has a positive effect on the performance of the model. The amount of data in this paper is small, so the waste of computing resources is not considered. All in all, we select the model with five layers and 5000 data as the optimal model for analysis and discussion in this paper.

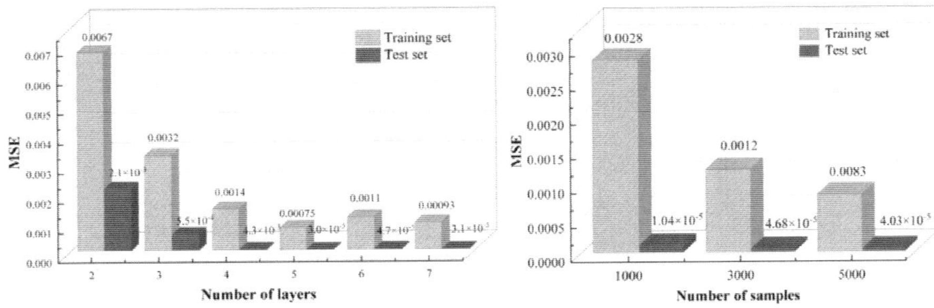

Figure 3. (a) The MSE of the training set and the test set varies with changes in the number of neurons in different layers. (b) The MSE of the training set and the test set varies with the number of samples in the data set.

5. Results and discussion

The optimal model we select is trained with the W data set, and the loss function is shown in Figure 4. The MSE of the training set and validation set is 0.00075 and 0.00026, and the MSE of the test set is 3.0×10^{-5}, which indicates that the accuracy of prediction is good. We directly use this optimal model to train the L1 and d data sets, respectively, record the maximum and minimum MSE of the test set in the L1 and d data sets on the trained model, and count the corresponding structural parameters and predicted values of structural parameters, it is the same operation on W data set, then get the statistics as in Table 1, it is clear from the data that the predicted value of our deep learning model is almost equal to the value of the direct numerical computation simulation.

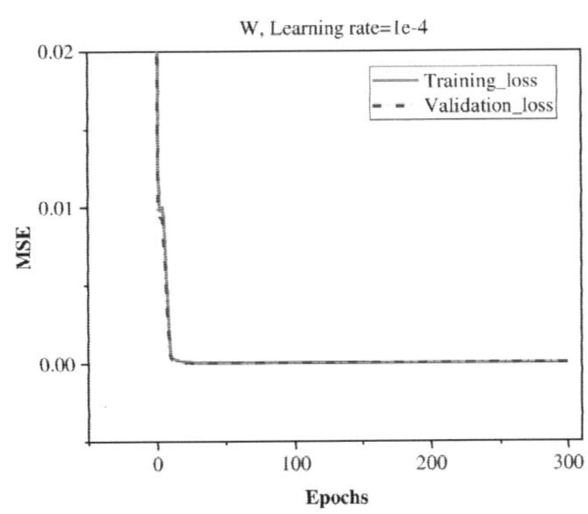

Figure 4. The schematic shows the MSE of the optimal model as a function of the number of iterations.

To visualize the predicted results, the title of the simulation in Table 1 lists the structural parameters, and the corresponding EIT spectrum is obtained through numerical calculation, as shown by the solid red line in Figure 5. Subsequently, input the obtained EIT spectral data into our deep learning model. The data in the prediction row of the title in Table 1 is the structural prediction value of our deep learning model. The predicted structural parameters are also numerically calculated to obtain the EIT spectrum, as shown by the blue dashed line in Figure 5. It can be seen from Figure 5 that our deep learning model can successfully predict the geometrical parameters of the EIT metamaterial structure from the given EIT spectrum. If our model is used to learn structural parameters and spectral data of other metamaterials with three specific spectral points, we believe that the desired results can also be successfully predicted.

Journal of Physics: Conference Series **2384** (2022) 012045 doi:10.1088/1742-6596/2384/1/012045

Table 1. In the optimal model, Figures 5(a), (b), and (c) correspond to the simulated and predicted values of the maximum MSE and minimum MSE of the test set on the W, L1, and d data sets, respectively.

	Figure 5(a)		Figure 5(b)		Figure 5(c)	
	MSE_Max 5.2×10^{-4}	MSE_Min 6×10^{-13}	MSE_Max 4.0×10^{-3}	MSE_Min 7.9×10^{-10}	MSE_Max 1.3×10^{-1}	MSE_Min 7.3×10^{-10}
Simulation	0.3767953	0.5651130	4.7195377	5.4698453	2.9562511	2.3767154
Prediction	0.3540023	0.5651122	4.6561375	5. 4698170	2.5925279	2.3766885

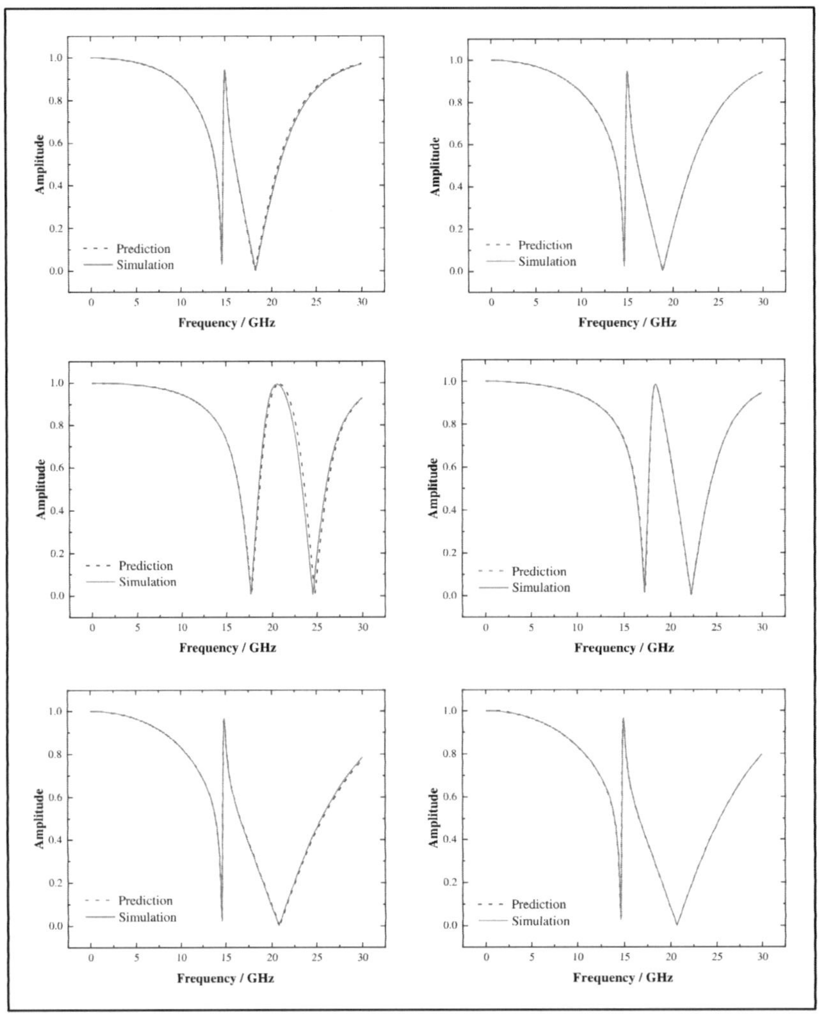

Figure 5. The left and right plots in (a), (b), and (c), represent the simulation results corresponding to the predicted and simulation values of the maximum and minimum values of MSE of the test set in datasets W, L1, and d, respectively. It can be seen from the figure that the spectral curves of the predicted and simulation values in the three datasets are consistent, attaining satisfactory accuracy.

The reason for not discussing the prediction of structural parameters beyond the data set is that when we collect data, we pass the simulation test of numerical calculation in advance to obtain the approximate value range of the spectral data that the EIT structure can produce the EIT effect. However, structural parameters predicted from spectral data outside this range are difficult to form distinct EIT spectra. Further, the SSPAM limits the EIT metamaterial structure to a range of values that can be mapped to the spectral data we need, enabling on-demand design. In our work, we only need to input the spectral data to obtain the required structural parameters by inverse prediction, so the work in this paper is effective.

6. Conclusions

We propose for the first time to use SSPAM to effectively solve the physical limitations in the field of metamaterials in this paper, and we expect a better approach. The validity of the experimental data in this paper benefits from SSPAM, which requires less data to train the model and also realizes the purpose of on-demand design. Our deep learning model is composed of fully connected layers. In the optimal model, three specific points on the transmission spectrum are used as input, and the predicted structural parameters of the EIT metamaterial can be output, avoiding the complex computational problems of traditional numerical simulations. Compared with previous work on inverse engineering using image processing techniques, our work shows the simplicity of algorithm design implementation and reduces the time cost of training neural network models. Our model is highly accurate, and we believe that the work in this paper will provide new ideas for designers who suffer from physical limitations in the study of metamaterial inverse design.

Acknowledgments

This study acknowledges funding from the Natural Science Research Foundation of the Education Department of Guizhou Province (Grant No.: [2021]305); the Science and Technology Foundation of Guizhou Province of China (Grant No.: ZK [2021] 306); the Department of Education, Guizhou Province (Grant No.: KY [2020] 007).

References

[1] Ning R X, Xiao Z Q, Chen Z H, and Huang W 2021 *J. Electron. Mater* 50 7
[2] Harris S E, Field J E and Imamoğlu A 1990 *J. Phys. Rev. Lett* 64 10
[3] Liu C, Dutton Z, Behroozi C H and Hau L V 2001 *J. Nature* 409 6819
[4] Marangos J P 1998 *J. Mod. Opt* 45 3
[5] Tassin P, Zhang L, Koschny T, Economou E N and Soukoulis C M 2018 *J. Phys.Rev.Lett* 102 5
[6] Zhong J T, Xu X C and Lin Y S 2021 *J. Nanomater* 11 9
[7] Wang J Q, Zhang J, Fan C Z, Mu K J, Liang E J and Ding P 2017 *J. Opt. Commun* 383
[8] Fleischhauer M, Imamoglu A and Marangos J P 2005 *J. Rev. Mod. Phys* 77 2
[9] Chen L R, Xu Z X, Zeng W Q, Wen Y F, Li S J and Wang H 2016 *J. Sci. Rep* 6 1
[10] Wang C Q, Jiang X F, Zhao G M, Zhang M Z, Hsu C W, Peng B, Stone A D, Jiang L, andYang L 2020 *J. Nat. Phys* 16 3
[11] Zheludev N I and Kivshar Y S 2012 *J. Nat. Mater* 11 3431
[12] Xia H, Sharpe S J, Merriam A J and Harris S E 1997 *J. Phys. Rev. A* 56 5
[13] Hu S, Liu D, Lin H, Chen J, Yi Y Y and Yang H L 2017 *J. J. Appl. Phys* 121 12
[14] Zhang T, Liu Q, Dan Y H, Yu S, Han X, Dai J and Xu K 2020 *J. Opt. Express* 28 13
[15] Zhang S, Li Z W and Xing F 2020 *J. Int. J. Mol. Sci* 21 5
[16] Cen C L, Zhang Y B, Chen X F, Yang H, Yi Z, Yao W T, Tang Y J, Yi Y G, Wang J Q and Wu P H 2020 *J. Phys. E* 117
[17] Ma W, Cheng F and Liu Y M 2018 *J. ACS Nano* 12 6326
[18] Ashalley E, Acheampong K, Besteiro L V, Yu P, Neogi A, Gand A O and Wang Z M 2020 *J. Photonics Res* 8 7
[19] Yan R Q, Wang T, Jiang X Y, Zhong Q F, Huang X, Wang Lu and Yue X Z 2020 *J. Nat. Mater* 31 37
[20] Zheng B, Yang J, Liang B and J C Cheng 2020 *J. Appl. Phys* 128 134902
[21] Inampudi S and Mosallaei H 2018 *J. Appl. Phys. Lett* 112 24

[22] Itzik M, Michael M, Achiya N, Uri A, Lior W and Haim S 2018 *J. Light Sci. Appl* 7 60
[23] Wiecha P R, Arbouet A, Girard C and Muskens O L 2021 *J. Photonics Res* 9 5
[24] Wen F F, Jiang J Q and Fan J A 2020 *J. ACS Photonics* 7 2098
[25] Ammari H, Li B and Zou J 2020 *J. Multiscale Model. Simul* 18 2
[26] Soltani I, Soltani T and Aguili T *J. INT J RF MICROW C E* 30 9
[27] Zhang T, Wang J, Liu Q, Zhou J Z, Dai J, Han X, Zhou Y and Xu K 2019 *J. Photon. Res* 7
[28] Zhang Z, Han D, Zhang L Y, Wang X Q and Chen X F 2021 *J. J. Appl. Phys* 130 3
[29] Huang W, Wei Z M, Tan B Y, Yin S and Zhang W T 2021 *J. J. Phys. D: Appl. Phys* 54 13
[30] Li Y B, Wan X, Cai B G, Cheng Q and Cui T J 2014 *J. Appl. Phys* 4 6921
[31] Kingma D P and Ba J L 2015 *C. ICLR*
[32] Keskar N S, Mudigere D, Nocedal J, Smelyanskiy M and Tang P T P 2017 *C. ICLR*

Journal of Physics: Conference Series **2384** (2022) 012046 doi:10.1088/1742-6596/2384/1/012046

Forest fire detection algorithm based on Improved YOLOv5

Shuailong Yu[1], Chunxia Sun[1*], Xiaopeng Wang[1] and Baomin Li[1]

[1]School of electronic and information engineering, Lanzhou Jiaotong University, Lanzhou, 730070, China

[*]Corresponding author's e-mail: come.4.so.hyun@gmail.com

Abstract. Aiming at the problem that the existing target detection algorithms are difficult to detect sudden fire with high precision in real-time in the complex forest environment, an improved forest fire detection algorithm based on YOLOv5 is proposed by us. Firstly, the k-means clustering algorithm based on $1-\mathrm{IoU}$ distance is used to preprocess the forest fire data set. Secondly, the Transformer attention mechanism is added to the model backbone network to improve the detection accuracy, and the regression box loss function GIoU-NMS is changed to EIoU-NMS. Finally, the network structure is changed, and the deep separable convolution is added to reduce the parameters. Comparing the algorithm we propose with the original YOLOv5 algorithm and the mainstream target algorithm, the experimental results show that the algorithm can have more efficient and high-precision detection results in fewer samples and complex environments and prove its feasibility and effectiveness.

1. Introduction

Forest resources are an important part of the ecological environment and play an important role in air purification, natural epidemic prevention, oxygen production, and climate regulation. However, the sudden occurrence of forest fire will cause huge damage to the ecological environment of the forest and bring a burden to economic development. Forest fire have the characteristics of sudden occurrence, rapid spread, widespread, and difficult to deal with quickly. Therefore, it's particularly important to use a detection method with high real-time performance and relatively low-cost detection of fire and smoke, which is of great significance to the construction of forest ecology and protection.

At present, some scholars have proposed forest fire target detection algorithms based on deep learning. Kim[1] used the fire detection method combining R-CNN and LSTM neural networks, the R-CNN network to determine non-fire areas and suspected fire areas, LSTM[2] network to further discriminate and screen the R-CNN judgment results. Kaabi[3] proposed a DBN-based smoke detection method to detect the smoke of forest fire. Benjamin[4] used RGB, HSV, and YCbCr color space to segment the flame area, but the non-flame light source is easy to cause false detection by this method. Some above algorithms have many parameters, and the detection speed is difficult to meet the requirements of real-time applications. Some can detect quickly, but the missed detection rate is large, and the detection accuracy is not enough.

We propose an improved algorithm based on YOLOv5[5]. The calculation of anchors of the dataset is changed to a method based on 1-IOU distance[6] to perform k-means[7] clustering on the training dataset. The early forest fire were small; the Transformer[8] attention mechanism is added to the algorithm's model backbone network to improve the detection accuracy of the overall network for small targets in order to detect the early stage of the fire. The boundary loss function of box regression is changed to EIoU[9], which improves the regression accuracy of anchor boxes. And depth-separable convolution is added to reduce the number of parameters of the network model, quantify parameters, reduce the memory occupied by each parameter and make training the algorithm's model gets compressed for the purpose. Experiments show that the algorithm has a good detection effect on the forest fire dataset with few samples and has achieved a certain degree of improvement in the detection accuracy compared with the original YOLOv5 algorithm while meeting the real-time application.

2. Improved YOLOv5 algorithm

2.1 Change the calculation of anchors

In original YOLOv5, the distance between samples is defined by Euclidean distance, and the anchors of the samples are obtained from the k-means algorithm by clustering. However, when the difference between the width and height of the samples is relatively large, clustering based on Euclidean distance will lead to a large error in the results, as shown in formula(1), where x_i and y_i are the coordinates of points. We use IoU distance clustering instead, where w, h are the width and height of the intersection of the two boxes, w_1, w_2, h_1, h_2 are the width and height of the two boxes, respectively. The range of the IoU value is [0, 1], and it's not as large as the calculation of the Euclidean distance.

$$d(x,y)=\sqrt{\sum_{i=1}^{n}(x_i-y_i)^2} \tag{1}$$

$$S_j = w*h \tag{2}$$

$$S_b = w_1*h_1 + w_2*h_2 - S_j \tag{3}$$

$$1-IoU = 1-\frac{S_j}{S_b}\in[0,1] \tag{4}$$

When the two frames have no intersection, it takes 0, and takes 1 when they completely overlap. It can be seen that the IoU value is negatively correlated with the distance between the two frames. Therefore, we use 1-IoU to define the sample distance with the k-means algorithm to cluster samples' anchors.

2.2 Add Transformer attention mechanism

The transformer module is constructed by a multi-head, self-attention module and forward propagation algorithm's module network. Each module is connected by a residual structure. The application layer is normalized after the addition calculation of the residual connection. For each position corresponding to the input sequence, the Transformer encoder will output a dimensional vector Z, which is calculated by Formula(5). Q (Query), K (Key), V (Value) is the parametric feature matrix and d_k is the number of columns of the matrices Q and K. The correlation size of each word to the current position word is obtained by the softmax function, which is divided by $d_k^{1/2}$ can make the training process have a more stable gradient, and then the obtained correlation size is multiplied by the matrix V to get the final vector Z.

$$Z=\mathrm{soft\,max}\left(\frac{QK^T}{d_k^{1/2}}\right)V \tag{5}$$

Self-attention learning is about the relationship between current tokens and other tokens in Transformer. This relationship is calculated by the formula above and gets the new features of the current token. If the tokens are taken into account in a sequence, then the relationship between them will be calculated even if they are far away. The addition of the Transformer attention mechanism can strengthen the attention of small targets and reduce the problem of insufficient detection accuracy caused by complex environments and noise.

2.3 NMS non-maximum suppression improvement

In the original YOLOv5 algorithm, GIoU is used as the loss function of the bounding box, and the GIoU loss function is defined as shown in formula(6), where C is the smallest rectangular anchor box containing anchor boxes A and B at the same time. But when the prediction box and the real box are included, or they are coincided in a certain dimension, the effect of GIoU is nothing improved compared with the original IoU.

$$GIoU = IoU - \frac{|C-(A\cup B)|}{|C|} \tag{6}$$

We use EIou instead of GIou, and the definition is shown in formula(7). Different from GIoU, EIoU adds the loss value of the detection frame scale and the loss value of length and width. Calculates the difference value of width and height, respectively, so that the predicted frame is more in line with the real frame.

$$EIoU = 1 - IoU + \frac{\rho^2(b,b^{gt})}{c^2} + \frac{\rho^2(w,w^{gt})}{C_w^2} + \frac{\rho^2(h,h^{gt})}{C_h^2} \tag{7}$$

The EIoU loss function calculation includes overlap loss, center distance loss, and loss of width and height of boxes. Where b represents the center point of the prediction frame. b^{gt} , ρ , c , C_w and C_h represent the center point of the target frame, Euclidean distance, diagonal distance, width, and height of the smallest rectangle formed between the intersecting prediction frame and the target frame, respectively. The addition of the loss value calculation of width and height further reduces the gap between the predicted anchor box and the real anchor box, making the algorithm's network model converge faster, and more accurately in the positioning of the regression box.

2.4 Change the convolutional layer to be a depth-separable convolution

Considering the real-time requirements of forest fire detection, we modify some of the convolutional layers in the algorithm's network model to be depth-separable convolutions to reduce the number of parameters, reducing the algorithm's network model and speeding up the detection speed to meet the needs of real-time detection. The FLOPs(Float Point Operations) of original convolution as shown in formula(8). The size of the input feature map is(H , W , C_{in}), the output feature map is (H_{out} , W_{out} , C_{out}), the kernel is(K , C_{out}), steps are S , and padding is P .

The depth-separable convolution is divided into two steps: depth-convolution and pointwise convolution. A single convolution kernel is applied to each input channel in the depth-convolution, its kernel is(K , C_{in}) and the feature map output by the depth-convolution is combined with a multi-layered convolution of 1×1 size in the pointwise convolution, the layers perform convolution and output the final feature map. The change of the convolution method reduces the number of network parameters when the input image has the same number of channels. The reduction of calculation is shown in formula(9).

$$FLOPs = \left((H+2*P-K+1)/S*(W+2*P-K+1)/S*C_{in}*C_{out}\right)*K^2 \tag{8}$$

$$\Delta FLOPs = \left(C_{in}*C_{out}*K^2-\left(C_{in}*K^2+C_{in}*C_{out}\right)\right)*\left((H+2*P-K+1)/S*(W+2*P-K+1)/S\right) \tag{9}$$

The improved YOLOv5 algorithm structure by us is shown in Fig 1.

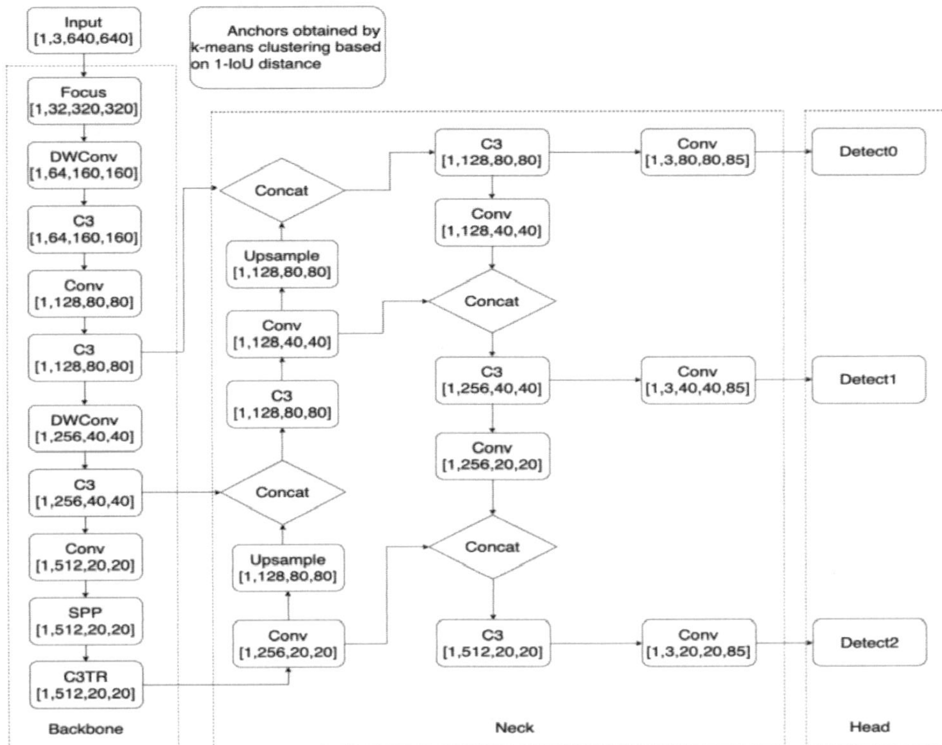

Fig 1. Improved YOLOv5 algorithm structure

3. Experiments of detection and comparative analysis of results

3.1 Dataset and experimental environment

The dataset that we use comes from forest fire pictures downloaded and collected by Kaggle and China Forest and Grassland Fire Fighting Network. The experimental environment of this paper is Windows 10 operating system, 32GB memory, NVIDIA TeslaV100 graphics card, Inter(R) Xeon(R) Platinum 8260 CPU @ 2.40GHz processor, and the construction and training of experimental projects are realized under Pytorch 1.8.0 and CUDA10.0. The initial anchor fires' parameters for dataset training are set as [33,14,43,85,91,97], [271,250,395,237,601,182], [328,341,407,394,585,308]. The image resolution size is 640×640, the initial learning rate is 0.01, the training batch is set to 64, and the number of model training iterations is 300.

3.2 Ablation experiment

In order to verify the effectiveness of the improved algorithm we propose, four groups of control experiments with different changed parts were designed to compare. Each group of experiments was running under the same training environment and the same training parameters. A, B, C, and D stands for changing the calculation of anchors, adding Transformer attention mechanism, non-maximum suppression improvement, and changing the convolutional layer to depth-separable convolution, respectively. "√" indicates that this part of the change has been added, mAP is the average detection accuracy of the algorithm, and FPS is the detection frame rate of the algorithm.

Table 1. Ablation experiment.

Algorithm	A	B	C	D	mAP	FPS(frames/s)
YOLOv5					0.887	66.67
YOLOv5+A	√				0.899	64.51
YOLOv5+A+B	√	√			0.909	64.10
YOLOv5+A+B+C	√	√	√		0.923	62.50
Ours	√	√	√	√	**0.929**	**76.92**

3.3 Comparison with the mainstream target detection algorithm

To verify the feasibility and effectiveness of the improved algorithm in this paper, our algorithm is also compared with the current mainstream target detection algorithm.

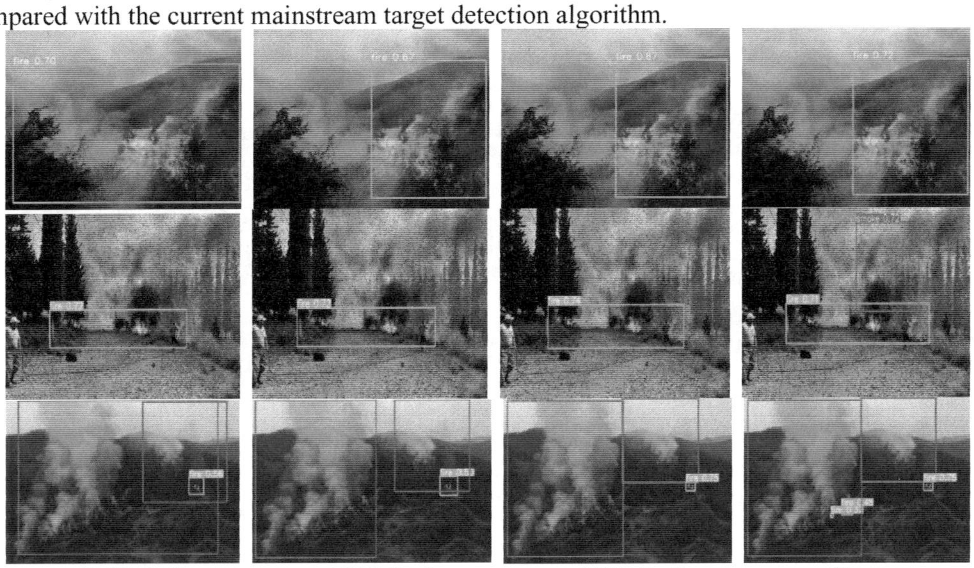

YOLOv3 YOLOv4 YOLOv5 **Ours**

Fig 2. Comparison of results before and after improvement

In our work, we compare the improved YOLOv5 algorithm with the mainstream algorithms of YOLOv3, YOLOv4, and YOLOv5. It can be seen from Fig 2 that compared with other mainstream algorithms, our improved algorithm has higher confidence in fire and smoke detection, and the range of anchors is more accurate. As shown in Table 2, the mAP value of the improved YOLOv5 algorithm in this paper reaches 0.929, which is 4.2% higher than the baseline YOLOv5 algorithm and has different degrees of improvement compared to other mainstream algorithms in the table.

Table 2. Comparison of the mainstream object detection algorithm

Algorithm	Map (fire)	mAP (smoke)	mAP	FPS (frame/s)
YOLOv3	0.813	0.875	0.844	74.55
YOLOv4	0.844	0.906	0.875	75.68
YOLOv5	0.853	0.921	0.887	62.50
Ours	**0.904**	**0.953**	**0.929**	**76.92**

4. Conclusion

Our improved algorithm ensures the accuracy of detection, and the detection speed meets the real-time and fast detection requirements simultaneously. We changed the calculation of IoU, added the Transformer attention mechanism into the algorithm, used EIou instead of GIoU, and changed some convolution layers to be depth-separable convolutions. The improvement in detection accuracy and speed compared with the mainstream models verifies the feasibility and effectiveness of our improved algorithm. The next step will be to improve the detection accuracy in the night environment and the forest background with poor exposure to achieve the robustness of the algorithm model in various environments and reduce missed detection in complex environments.

Acknowledgments
The author of this paper wishes to acknowledge the support and assistance provided by Associate Professor Sun Chunxia in the work of my thesis, as well as the encouragement of my work by a colleague at Li Baomin. This study was funded by the National Natural Science Foundation of China (grant number 61761027), and the Science and Technology Program of Gansu Province (grant number 20YF8GA036).

References

[1] Kim B and Lee J. (2019) A video-based fire detection using deep learning models. *J. Appl Sci.***9** 2862

[2] Yu Y, Si X, Hu C, et al. (2019) A review of recurrent neural networks: LSTM cells and network architectures. *J. Neural comput.***31** 1235-70

[3] Kaabi R, Sayadi M and Bouchouicha M, et al. (2018) Early smoke detection of forest wildfire video using deep belief network. In Proceedings of the 2018 4th International Conference on Advanced Technologies for Signal and Image Processing (ATSIP), Sousse, Tunisia, 21–24 March.1–6

[4] Benjamin S G, Radhakrishnan B, and Nidhin T G, et al. (2016) Extraction of fire region from forest fire images using color rules and texture analysis. In Proceedings of the 2016 International Conference on Emerging Technological Trends (ICETT). Kollam, India, 21–22 October.1-7

[5] Pham M T, Courtrai L, Friguet C, et al. (2020) YOLO-Fine:one-stage detector of small objects under various backgrounds in remote sensing images. *J. Remote Sens-Basel.* **12** 2501

[6] Yan J, Wang H, Yan M, et al. (2019) IoU-adaptive deformable R-CNN: Make full use of IoU for multi-class object detection in remote sensing imagery. *J. Remote Sens-Basel.* **11** 286

[7] Likas A, Vlassis N, Verbeek J J. (2003) The global k-means clustering algorithm. *J. Pattern recogn.***36** 451-61

[8] Hao Y, Dong L, Wei F, et al. (2021) Self-attention attribution: Interpreting information interactions inside Transformer. In Proceedings of the AAAI Conference on Artificial Intelligence. 2021 May.**35** 12963-71

[9] Peng H, Yu S. (2021) A systematic IoU-related method: Beyond simplified regression for better localization. *J. IEEE T.I.P.***30** 5032-44

Optimization of Interface layout of electrical equipment using Heuristic Algorithm

Jianqin Yu, Guoying Lu, Ting Zhang[1] and Tianyu Cheng

Shanghai Dianji University, Shanghai China

[1] zhangting@sdju.edu.cn

Abstract. Facility Layout Problem(FLP) is a major matter in the process of human-machine interface(HMI) design, which has a direct impact on how well operators perform their duties. Electrical equipment, which controls the power system, has a direct impact on how safely the system operates. The current optimization techniques prioritize improving the equipment's performance characteristics at the expense of user aspects. This work aims to apply heuristic algorithms that introduce cognitive psychology and adopt the layout principle as a constraint condition to solve this problem. The improved analytic hierarchy process (AHP) is used to conduct the optimal design practice of the HMI of electrical switching equipment, develop the optimization mathematical model, and quantify the subjective and objective layout principles. Comparing the optimization results reveals that the genetic algorithm's optimization impact is superior and the ant colony algorithm's iterative pace is faster. The HMI's optimized design based on a heuristic algorithm can improve the usability of interface operation, and has a certain reference for the optimization design of HMI of electrical equipment.

1. Introduction

Electricity[1] is connected to the national financial system and the style in which individuals live because it is a clean, sustainable energy source[2]. Therefore, electricity is very significant to individuals today. The current center of the study is on the proper operation of electrical equipment because it is a crucial component of the nation's industrial production and serves as control equipment for the entire power system. Therefore, the world and China have a huge possibility to develop electrical equipment speedily.

The HMI acts as a conduit and type of interaction for information transfer among humans and machines. How logically and scientifically the interface is created will directly affect how quickly and accurately the equipment user operates the device[3]. In recent years, research on HMI has received more and more attention. Deng Li et al. [4] used the factor analysis method to extract aesthetic factors to evaluate the interface comprehensively from the perspective of aesthetic cognition. Şenol [5] developed a relative layout design (RLD) model for the interface design of aircraft cockpit display panels, which took into account engineering needs, human factors, ergonomics, and pilot preferences. Sheng [6] et al. optimized the design of the console interface based on the human eye physiological comfort criterion and evaluated the optimization scheme. Xu et al. [7] optimized the layout design of the high-speed rail operating interface by studying the attention distribution of the high-speed rail operating interface and summarized design suggestions. The optimization scheme is constructed based on different physiological criteria, and the optimization design is completed based on different indicators, but the quantitative relationship between the construction of cognitive laws and the HMI is not involved. In addition, subjective data and objective data are not considered together.

In summary, at present, China has begun to pay attention to the practical application of interface layout in various fields, while researches focus more on technological innovation and ignore the optimization of HMI. In this study, the layout principle of HMI is summarized from cognitive psychology and applied to the interface optimization of electrical equipment. The optimization mathematical model is built using this strategy, and an algorithm is utilized to generate the ultimate layout design.

2. Method

2.1. Principles of HMI layout design[8]

- Correspondence between cognition and objectivity
- Operation sequence
- Matching cognitive strategy
- Information organization
- The reachable area
- Importance and frequency of the operating elements
- Spacing

2.2. The optimization objective function of the HMI layout

The process is shown in Figure 1:

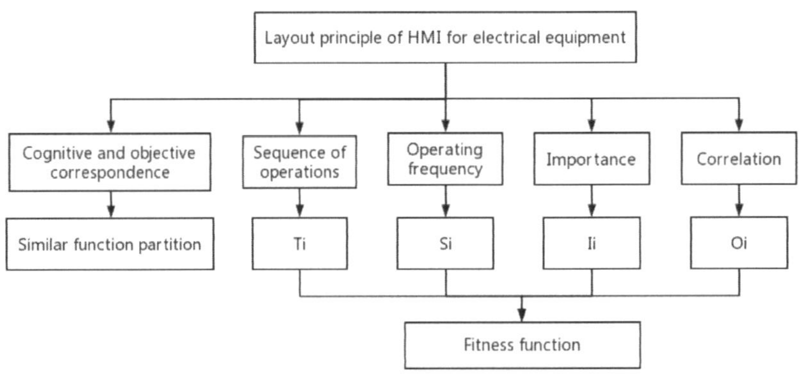

Figure 1. Establishment process of the objective function

The final objective function is defined as:

$$fit(i)_{best} = \max\left[\sum_{i=1}^{n}(\delta_{1i}\cdot(T_i + S_i + I_i) + O_i)\right] \tag{1}$$

$$\delta_{1i} = 1 - i/n \tag{2}$$

Where T_i represents the weight of the sequence of operations; S_i represents the weight of operating frequency; I_i represents the weight of importance; O_i represents the weight of correlation; δ_{1i} represents the control coefficient.

2.3. The layout design model of HMI based on heuristic algorithms

The layout of the HMI will be optimized in this research using the genetic algorithm (GA) and the ant colony algorithm (ACA), and the effectiveness of the two algorithms will be compared based on real-world examples.

(1) Genetic algorithm

- Coding: Generate initialization population

The sequence encoding is as follows:

$$C_m = \left[C_1^m, C_2^m, C_3^m ..., C_n^m \right] \quad (3)$$

Where m represents the population size, n represents the number of layout objects.

- Construct the fitness function

By calculating the fitness value of each randomly generated permutation and combination, which will be treated as an individual, the ideal solution to the scheme will be found. The following is the fitness function:

$$fitness(i) = \sum_{i=1}^{n} (\delta_{1i} \cdot (T_i + S_i + I_i) + O_i) \quad (4)$$

- Selection

The probability formula for being selected is as follows:

$$P(C_i) = \frac{fitness(C_i)}{\sum_{i=1}^{n} fitness(C_i)} \quad (5)$$

- Crossover
- Mutation
- Termination conditions

(2) Ant Colony Algorithm

The specific process is as follows:

- Initialization
- Constructing the solution space

The state transition rule of the ant transferred from the operating element i to the operating element j at time t is as follows:

$$p_{ij}^k(t) = \begin{cases} \dfrac{[\tau_{ij}(t)]^\alpha [\eta_{ij}]^\beta}{\sum_{l \in U} [\tau_{il}(t)]^\alpha [\eta_{il}]^\beta}, & j \in U \\ 0, & j \notin U \end{cases} \quad (6)$$

$$\eta_{ij} = \frac{1}{D_{ij}} \quad (7)$$

$$D_{ij} = |T_i - T_j| + |S_i - S_j| + |O_i - O_j| + |I_i - I_j| \quad (8)$$

Where $\tau_{ij}(t)$represents the pheromone between operating element i and operating element j at t moment; $\eta_{ij}(t)$represents the heuristic information between operating element i and operating element j at t moment; D_{ij} represents the difference between the comprehensive weights of operating element i and operating element j; α represents the pheromone factor; β represents the heuristic function factor; U represents the set of feasible points.

- Update pheromone
- Termination conditions

3. Application Verification

The HMI system of electrical equipment is a rather complicated piece of technology that includes numerous layout components. This research uses, as an example, the HMI of an armored extraction AC metal closed switchgear to confirm the effectiveness of the design methodology. The layout and arrangement of the 16 operational elements on the operation interface are shown in Figure 2.

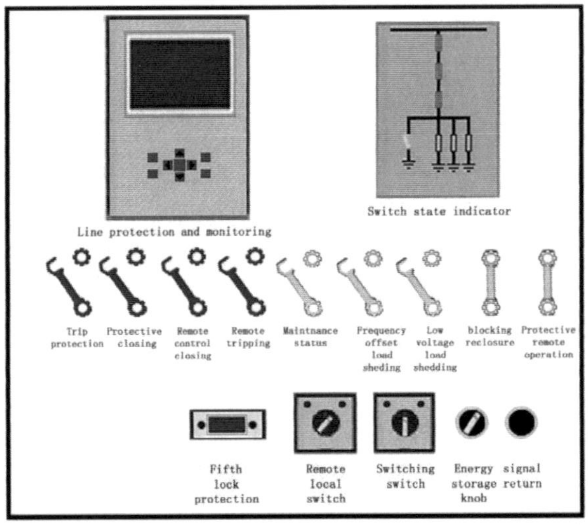

Figure 2. The layout of operation interface for armored exhaust AC metal closed switchgear (before optimization)

First of all, we need to code these 16 operating elements and arrange the operating elements with similar functions in the same area to reduce the operator's search time. As shown in the Table 1, the 16 operating elements are divided into four groups of codes, namely {h1, h2}∈{1, 2}, {h3, h4, h5, h6}∈{3, 4, 5, 6}, {h7, h8, h9, h10, h11}∈{7, 8, 9, 10, 11}, {h12, h13, h14, h15, h16}∈{12, 13, 14, 15, 16}.

Secondly, simplify the layout area (HMI of electrical equipment) and the objects to be distributed (operating elements), and build a mathematical model, which is conducive to the mathematical description of design variables. After simplification, the layout of the operating elements can be treated as an ordering problem. Use heuristic algorithms to determine the optimal ordering, then decide on the actual layout plan in light of the findings.

In the process of operating the electrical switchgear, there are usually the following four working conditions: hot standby to cold standby, hot standby to line maintenance, line maintenance to a hot standby, and cold standby to hot standby. Ti of each operating element under different working conditions according to the "Operation Ticket for Reversing Gate" is shown in Table1. Si and Ii through questionnaires, on-site interviews, and the improved AHP are shown in Table1. Oi is shown in Table 2.

Table 1. Operation sequence, frequency, and importance weights Ti, Si, Ii.

	1	2	3	4	5	6	7	8	9	10	11	12	13	14	15	16
T_i	0.1817	0.0649	0.0378	0.0189	0.0378	0	0.0413	0.0162	0.0142	0.0284	0	0.4542	0.0790	0	0.0255	0
S_i	0.2546	0.1032	0.0364	0.0124	0.0096	0.0055	0.0236	0.0113	0.0111	0.0092	0.0036	0.1956	0.1725	0.0192	0.1142	0.0180
I_i	0.4374	0.0552	0.0382	0.0119	0.0065	0.0250	0.0076	0.0028	0.0030	0.0212	0.0108	0.1756	0.1047	0.0495	0.0377	0.0129

Table 2. Correlation weight Oi.

O_{ij}	1	2	3	4	5	6	7	8	9	10	11	12	13	14	15	16
1	1	0.672	0	0	0	0	0	0	0	0	0	0	0	0	0	0
2	0.672	1	0	0	0	0	0	0	0	0	0	0	0	0	0	0
3	0	0	1	0.318	0.192	0	0	0	0	0	0	0	0	0	0	0
4	0	0	0.318	1	0.018	0	0	0	0	0	0	0	0	0	0	0
5	0	0	0.192	0.018	1	0.036	0	0	0	0	0	0	0	0	0	0
6	0	0	0	0	0.036	1	0	0	0	0	0	0	0	0	0	0
7	0	0	0	0	0	0	1	0.446	0.428	0.5	0.09	0	0	0	0	0
8	0	0	0	0	0	0	0.446	1	0.82	0.234	0	0	0	0	0	0
9	0	0	0	0	0	0	0.428	0.82	1	0.784	0	0	0	0	0	0
10	0	0	0	0	0	0	0.5	0.234	0.784	1	0	0	0	0	0	0
11	0	0	0	0	0	0	0.09	0	0	0	1	0	0	0	0	0
12	0	0	0	0	0	0	0	0	0	0	0	1	0.964	0.036	0.8	0.018
13	0	0	0	0	0	0	0	0	0	0	0	0.964	1	0.036	0.674	0.018
14	0	0	0	0	0	0	0	0	0	0	0	0.036	0.036	1	0.018	0
15	0	0	0	0	0	0	0	0	0	0	0	0.8	0.674	0.018	1	0
16	0	0	0	0	0	0	0	0	0	0	0	0.018	0.018	0	0	1

4. Results and Discussion

Ten tests are run on each algorithm, and Table 3 displays the test results. After 10 runs, the layout scheme selects the outcome with the lowest objective function value, and the final result is the average objective function value and the number of iterations. Two optimization schemes are shown in Figure 3. It can be seen from the test results that the value given by GA is 4.368, while the objective function value of ACA is 4.438. While ACA's solution speed beats that of GA's, GA is superior to ACA. We have also verified and concluded that GA's initial optimization speed is slower, but the global search ability is stronger, while ACA's initial optimization speed is faster, but it is simple to slip into the regional optimum.

Table 3. Comparison of GA and ACA.

Algorithm	Operating element layout scheme	Objective function value
GA	1, 2, 4, 3, 5, 6, 7, 10, 9, 8, 11, 16, 14, 13, 12, 15	4.368
ACA	1, 2, 6, 3, 4, 5, 10, 8, 7, 9, 11, 16, 15, 14, 13, 12	4.438

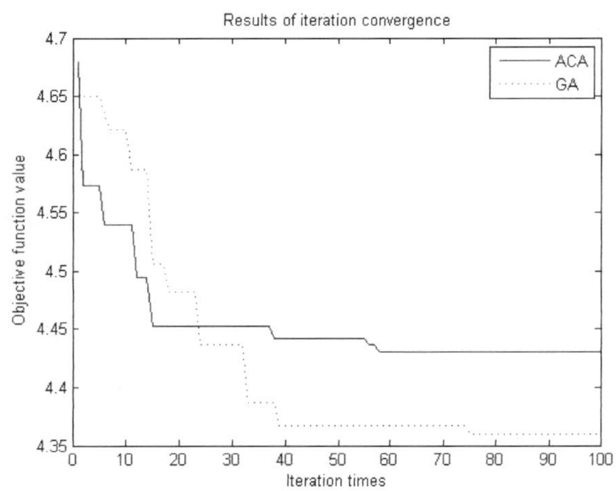

Figure 3. Comparison of running curves of GA and ACA.

The layout principle that is appropriate for electrical equipment is summarized, quantized, and used in this paper's analysis of the information processing process of operators. Heuristic algorithms are used to handle the optimization scheme. The acquisition of the layout principle combines objective rules and subjective evaluation data. The layout principle has to be further refined because some cognitive activities are challenging to quantify due to the complexity of the human cognitive behavior process.

5. Conclusion

This paper proposes an interface layout optimization method based on cognitive psychology. This method summarizes the layout principles of HMI based on human information organization rules, visual search rules, and memory characteristics, and then constructs optimization mathematics. The solution to the interface layout optimization strategy uses heuristic GA and ACA algorithms. The layout optimization of armored extraction AC metal-enclosed switchgear is proved by an example. The shortcomings of each strategy are illustrated by contrasting the optimization results of two heuristic algorithms. It is also shown that the layout optimization challenge for HMIs can be attached to the form using heuristic techniques. It can assist designers and engineers in creating HMI and increase the effectiveness of layout design as a layout ergonomic design method.

Acknowledgments

This work was supported by the Ministry of Education Humanities and Social Sciences Youth Fund Project (18YJCZH114)

References

[1] Sepasi R R, Abbas A, Fariba B, et al. (2016) Nurses' Perceptions of the Concept of Power in Nursing: A Qualitative Research. JOURNAL OF CLINICAL AND DIAGNOSTIC RESEARCH, 10(12): 10-15.

[2] Goldemberg J. (2006) The promise of clean energy. Energy policy, 34(15): 2185-2190.

[3] Ma C M, Zhao D H and Xin H. (2020) Human-machine interaction interface design of intelligent equipment based on user experience. Computer Integrated Manufacturing Systems, 26(10):2650-2660.

[4] Deng L, Wang G.(2020) Quantitative Evaluation of Visual Aesthetics of Human-Machine Interaction Interface Layout. Computational Intelligence and Neuroscience, 2020:1-14.

[5] Şenol M B. (2020) A new optimization model for the design of traditional cockpit interfaces. Aircraft Engineering and Aerospace Technology, 92(3):404-417.

[6] Sheng S, Sen G, Yiqian Z, et al. (2020) Human-Ocular-Physiological-Characteristics-Based AdaptiveConsole Design. IEEE Access, PP(99).

[7] Xu Y S, Li L L, Wu Y D, et al. (2022) Optimization Design of Interface Layout of High-speed Railway Control Console Based on Attention Distribution. Journal of Southwest Jiaotong University,57(02):401-409.

[8] WANG B C, HUANG S H, YI B, et al. (2020) State-of-art of Human Factors/Ergonomics in Intelligent Manufacturing. Journal of Mechanical Engineering, 56(16):240-253.

Journal of Physics: Conference Series **2384** (2022) 012048 doi:10.1088/1742-6596/2384/1/012048

Design and Implementation of Microservice Secure Communication Framework based on National Secret Algorithm and Dynamic Key

Hongyu Ding[1,*], Qingqiang Meng[1], Lihua Sun[1], Ming Liu[1], Xu Chen[1], Chang Liu[1], Duanjie Leng[1]

[1] Nanjing NARI Information and Communication Technology Co., Ltd, Nanjing, 210096, China

*Correspondence: dinghongyu@sgepri.sgcc.com.cn

Abstract. With the acceleration of the pace of the national electricity reform and the wide application of the micro-service architecture, various network threats in the field of information security are increasing, and the power industry has become the "hardest hit area" for cyber attacks. In order to improve the safe implementation of the National Power Grid's electronic projects and effectively resist the "malicious exploration" of various network threats, this paper proposes a microservice secure communication framework based on the combination of dynamic keys and national secret algorithms for a high concurrency web system based on microservices. The framework uses the national secret SM2 algorithm to generate session-level paired keys and then generates a digital signature through the SM3 algorithm, which effectively guarantees the integrity of the communication. Finally, it extracts the encryption key and encryption offset from the paired keys and uses the SM4 algorithm to encrypt the original request text payload. Timestamps and random numbers are added to each request to prevent tampering and replay attacks effectively. The system framework has been practically applied on the State Grid's digital capability open platform, which ensures the security of micro-service communication and the safety and efficient transmission of data.

1. Introduction

In recent years, microservice architecture[1] has been widely used in complex web systems. It is based on a completely "uncomplicated" concept that encapsulates each major function of the system as a small, semi-independent web service. Microservices provide application programming interfaces (APIs) for various business requirements and describe exchangeable data formats, commands, and tasks. They are smaller, easier to maintain, and adapt to changing business needs. Technically, microservices can be stateless or stateful, and the language used to develop microservices is not fixed. Combining all vulnerabilities of language and third-party plugins, microservice systems are put in a wider variety of attacks.

1.1. Microservice Secure Communication

1.1.1. OAuth2 unified authentication. OAuth2[2] implements unified authentication of microservices and provides token access security. A federated identity mechanism was added in 2014. The OAuth2 relies on the authorization server, can use the access token to communicate to any other internal service, and is easy to manage. But if one service's token is compromised, all services are at risk.

1.1.2. Digital digests are tamper-proof. Digital digest, also known as message digest[3], is a method to prevent messages from being tampered with during the communication process. Generally, it is generated by the Hash function, which encrypts the communication message. The digest value and the message are communicated and transmitted together. The receiver needs to recalculate the message and compare it with the received message digest to determine whether the communication process has tampered with. But for tamperers, the message is still plaintext, and the digital digest cannot prevent the message from being exposed.

1.1.3. Transport Layer Key Technology. The server and the client are based on a common public key. The client processes the public key and data[4] to obtain encrypted data and transmits it to the server; the server restores the original communication data in reverse. This symmetric encryption system uses a single key for encryption and decryption. Hackers easily crack the key through brute force, causing security problems.

1.2. Dynamic key technology

The dynamic key technology[5] combines traditional and symmetric encryption systems and adds a digital signature. The signature is non-repudiation, which increases information security to a certain extent. When the sender sends data, firstly, it must obtain the receiver's public key, then encrypts the data to generate ciphertext and signs it. After receiving the data, the receiver decrypts the data with its own private key and can see the signature of the other party and perform a consistency check on the signature. Even if the hacker intercepts the public key, the attack is abandoned by the private key cannot be cracked. At present, dynamic key technology is used in various fields.

1.3. National secret algorithm

The domestic cryptographic algorithms are referred to as national cryptographic algorithms, with independent intellectual property rights recognized by the State Cryptography Administration. Currently, SM2, SM3, and SM4 are mostly used[6].

The SM2 algorithm[7] is a public key cryptographic algorithm with a strength of 256 bits based on the ECC elliptic curve cryptographic mechanism, which can realize digital signature and encryption. Compared with the currently most used RSA asymmetric encryption algorithm, it has higher computational complexity and security, requires fewer public key digits under the same security performance, has faster encryption and decryption, is faster than the key generation speed, and is about a hundred times faster than RSA.

The SM3 algorithm fills and iteratively compresses the message data and then generates a hash value. Usually used in the field of digital signature and message authentication code technology, it can meet the integrity and security requirements of various communication password application scenarios.

The SM4 algorithm is a symmetric encryption algorithm based on a blocked cipher that has 128 bits. Real-time encryption and decryption of data can be achieved to ensure data confidentiality. The plaintext message is cut into fixed blocks and then iteratively encrypted. Compared with the AES symmetric encryption algorithm, the security and usage of the two algorithms are the same. The difference is that the AES algorithm uses the key at the end of each round, while the SM4 algorithm uses the key at the beginning of each round. In terms of the complexity of the key scheduling algorithm, SM4 is simpler[8].

The secure communication framework designed in this paper performs dynamic key encryption on the request message during the interaction and uses the national secret algorithm with higher security and faster speed, which is more in line with the national independent and controllable security development needs.

2. System Architecture Design and Implementation

2.1. General idea

The architecture adopts distributed microservices and front-end and back-end separation technology, which can effectively improve the robustness and scalability of the platform. The back-end microservice is built based on the SpringCloud framework[9] and provides RESTful API interfaces, which are convenient for each end to access. The microservice security communication architecture model is shown in Figure 1.

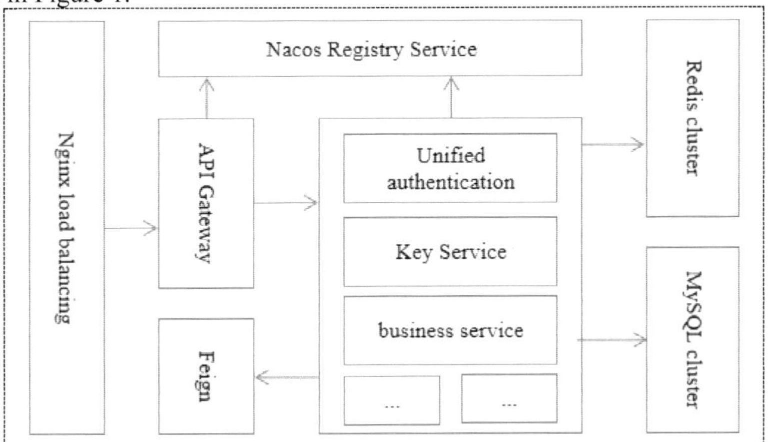

Figure 1. Overall architecture.

Front-load balancing Nginx is a lightweight, high-performance HTTP and reverses proxy web server. It also provides IMAP/POP3/SMTP services, occupies less memory, and has strong concurrency capabilities.

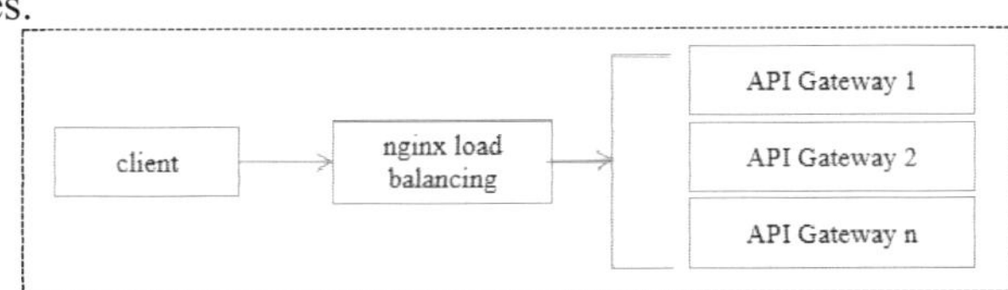

Figure 2. Nginx load balancing.

API Gateway is a more intelligent microservice application developed based on Spring Cloud Gateway. All external requests must go through the API Gateway first, and then the API Gateway implements functions such as request routing, load balancing, and permission verification, and finally, each access request is processed and distributed to the corresponding microservice service. Usually, the API gateway is deployed in a cluster, and the front-end Nginx load balancing cluster performs routing and distribution requests to better solve the traffic pressure problem.

The Key Service provides complete lifecycle management based on the generation, storage, and destruction of user session-level dynamic keys and provides key query services.

2.2. Dynamic session keys

Based on the dynamic key technology, the microservice secure communication key in this paper automatically generates a user session paired key string $(Pub, \mathrm{Pr}v)$ when the user logins the system authentication and is automatically destroyed after the user logout. The specific process is:

2.2.1. Key Generate. After the user successfully logins the system, the key query service is accessed to obtain the current user's session key. The key management microservice uses the SM2 algorithm to generate a session-level paired key string $(Pub, \mathrm{Pr}v)$ for the current user.

2.2.2. Key Storage. After the session-level paired key string $(Pub, \mathrm{Pr}v)$ is generated, the current user ID is used as the unique identifier and $(Pub, \mathrm{Pr}v)$ is used as the value to form the key-value pair of the user session key string $M(\mathrm{k}, v)$, which is stored in the distributed cache in Redis.

$$M(k,v) = (uid, M) \tag{1}$$

2.2.3. Key Destroy. When the user logs out of the system, the key management microservice performs the deleting operation $M(\mathrm{k}, v)$ according to the current user ID. Considering that the user's behavior is changeable and indefinable, the key management microservice will automatically initiate a task to clear the timed-out sessions periodically and then clear all timed-out keys.

2.2.4. Dynamic Key Model. The key management microservice uses the SM2 algorithm to generate a session-level dynamic paired key for the current user. The public key $S2Pub(i)$ is a hexadecimal string of length 130, and the private key $S2\mathrm{Prv}(i)$ is a hexadecimal string of length 64. The data structure of $S2Pub(i)$ is shown in Table 1:

Table 1. Session-level SM2 public key data model.

Bits 1 to 16	used as the message encryption key
Bits 17~32	used as the packet encryption offset
Bits 33~130	other information

2.3. Secure Communication Process

In this paper, a combination of national secret algorithms is used to ensure the security and efficiency of the communication process.

Table 2. Application scenarios of the national secret algorithm.

Scenes	Apply Algorithms	Implemented functionality
Pairing key	SM2	generates pairing public key and private key
Digital Signature	SM3	asymmetric Encryption, Integrity
Message encryption and decryption	SM4	requests message encryption and decryption

In terms of digital signature, the asymmetric encryption mechanism combining the SM2 algorithm and SM3 algorithm is used to enhance digital signature; in terms of request message security, the SM4 algorithm and session-level dynamic key are used for symmetric encryption; in terms of integrity, SM3 algorithm is used for integrity encryption guarantee.

Based on the above ideas, the complete microservice security communication process is:

- Step 1: The user $U(i)$ logins the system, and the front end send a request for its session key. The key microservice generates a paired dynamic public key $S2Pub(i)$ and private key

$S2Prv(i)$, and stores them in the distributed cache Redis. The front end page stores $S2Pub(i)$ in its local session.

- Step 2: The front-end page obtains the encryption key string $M(i)$ and the offset vector $IV(i)$ from $S2Pub(i)$ through the interception rule before sending business requests.
- Step 3: According to the current request resource address, request message $T(i)$, and the encryption key $M(i)$, a tamper-proof digital signature S(i) is generated using SM3:

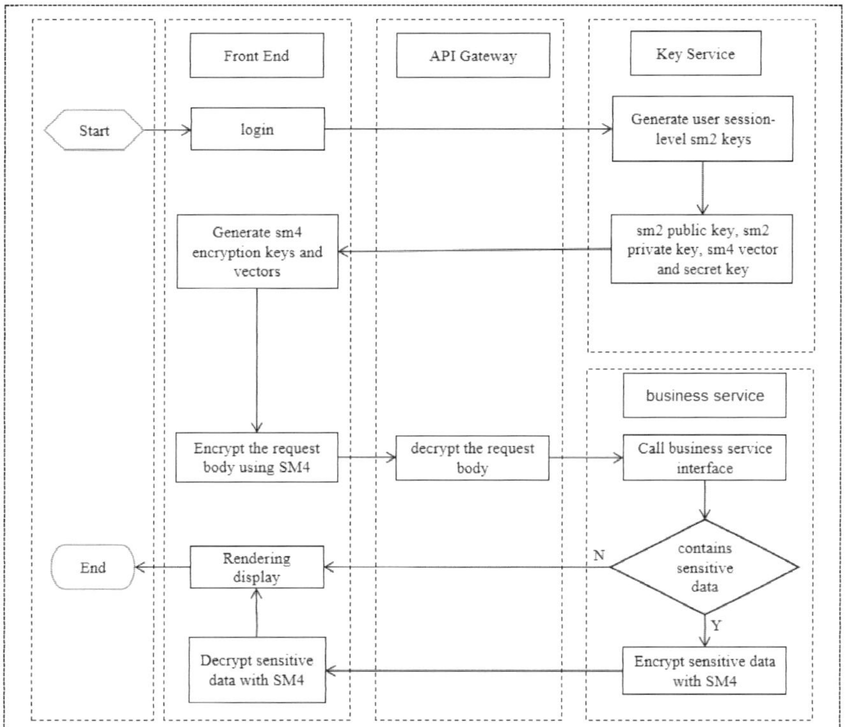

Figure 3. Secure communication mechanism.

$$S(i) = SM3(uri, T(i), M(i)) \tag{2}$$

- Step 4: Encrypt the request message to generate the ciphertext $E(i)$:

$$E(i) = SM4(T(i), M(i), IV(i)) \tag{3}$$

- Step 5: The digital signature $S(i)$ is put into the request header before the front end sends the request, then sends the ciphertext $E(i)$ to the background.
- Step 6: The API Gateway intercepts the request, judges whether the current request is legal or not, whether there is tampering, replay, unauthorized behavior, etc., and then obtains the public key $S2Pub(i)$ and private key $S2Prv(i)$ of the current user, and repeats the Step 2, to decrypt the ciphertext $E(i)$ to get the original request message $T(i)$, and continue to forward it to the business microservice.
- Step 7: After the business microservice processes the request, desensitize sensitive fields such as password, ID card, mobile phone number, email, etc., and repeat Step 4.
- Step 8: After the front end obtains the return value, repeat Step 6 to decrypt the ciphertext.

The SM4 algorithm encrypts the request and response message, and the encryption relies on the dynamic encryption key $M(i)$ and the offset vector $IV(i)$. In this scheme, $M(i)$ and $IV(i)$ correspond to the head 16 bits and the 17th to 32nd bits of the public key $S2Pub(i)$, respectively, and are transcoded by BASE64.

2.4. Communication encryption
Before sending a request, the digital signature $S(i)$ must be generated firstly using the public key $S2Pub(i)$ through SM3. Then, in combination with the encryption key $M(i)$ and the offset vector $IV(i)$, the request message $T(i)$ is encrypted to generate the ciphertext $E(i)$. Finally, the digital signature $S(i)$ is put into the request header, and the payload is replaced by the ciphertext $E(i)$.

2.5. Communication decryption
When the API Gateway intercepts the communication request, the digital signature $S(i)$ is checked first. Then, the request ciphertext $E(i)$ is decrypted to the real message $T(i)$ and replaced with the request payload; finally, the request is transferred to the actual business microservice.

2.6. Anti-replay attack
The secure communication framework in this paper generates an instant timestamp Tc and a random number Rc for each request to improve the anti-replay attack capability.

The timestamp Tc ensures that it is in the allowed safe threshold time range. If it exceeds the threshold range, the gateway immediately discards the request and records the current client address as an illegal one, refusing to continue communication.

The nonce Rc is another signing mechanism for the request. The Gateway will verify the non-retransmission random number for each request to prevent replay attacks. Therefore, a random number generation algorithm with good performance needs to be selected. According to the quality evaluation criteria of random number generators given by the German Federal Office for Information Security: stationarity, irrelevance, and degree of fit. This paper selects the Mersenne Twister algorithm because of its low memory occupation, high speed, long period, and uniform distribution.

3. Application and Testing

3.1. Test environment
The microservice secure communication architecture in this paper is based on cloud deployment. The cloud server includes three main service nodes and five container nodes, and the operating systems are all CentOS 7.6. The container node is used to deploy microservices, and the hardware configuration is a high-efficiency cloud disk of 150G, with a memory of 64 GB and 32 cores CPU. The database server is MySQL 5.7, and the hard disk capacity is 500GB, with 64GB memory and 32 cores CPU. The distributed cache component is Redis4.0.

3.2. Dynamic key generation efficiency
The performance of the dynamic key, in addition to considering the generation efficiency, also needs to verify the correctness of the dynamic key generation. The experimental verification analysis is carried out under the condition that the request volume continues to increase from 100,000 to 1 million. Figure 4 shows the average key generation time in milliseconds. Under the condition that the request volume keeps increasing, the key algorithm generation efficiency maintains good stability.

Journal of Physics: Conference Series **2384** (2022) 012048 doi:10.1088/1742-6596/2384/1/012048

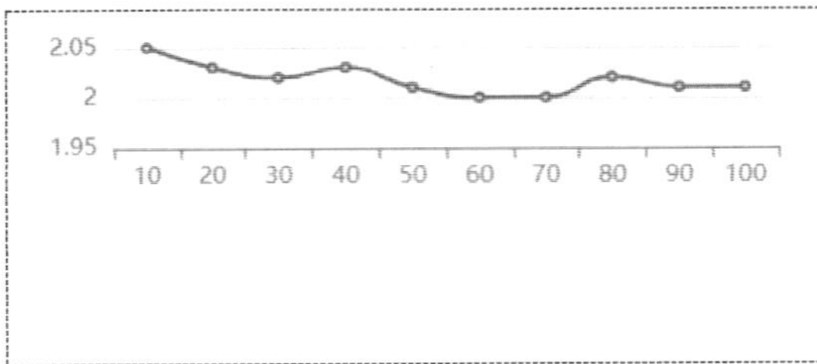

Figure 4. Generation time efficiency.

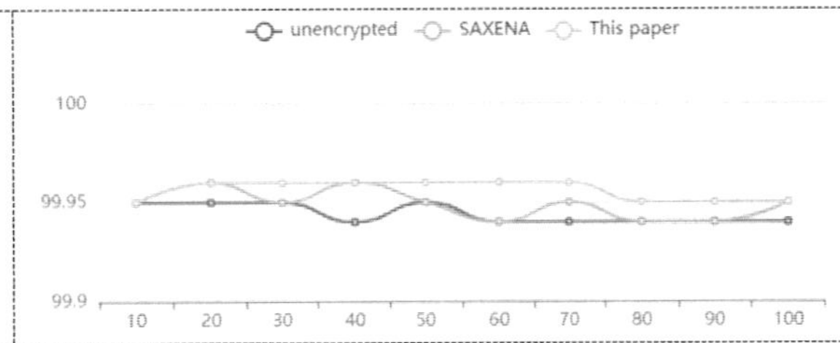

Figure 5. Correctness verification.

In terms of key correctness, under different request volume pressures, the average correctness rate of unencrypted requests and the dynamic keys proposed by SAXENA[10] and others were tested, respectively, and the change curve of the correctness rate was finally obtained, as shown in Figure 5.

Under the continuous pressure of high concurrency, it is difficult for unencrypted requests to have good correctness performance, and the dynamic key performance of this paper is more stable than that proposed by SAXENA et al.

3.3. Time-consuming analysis of encryption and decryption

Encryption and decryption operations will increase each request for a certain amount of time. To ensure efficient access to requests, it is necessary to ensure that the time-consuming is as little as possible. The key length is 64 bits of AES, while RSA's key length is 1024 bits. Our dynamic key length, while based on national secrets in this paper, is 130 bits. This paper compares the time-consuming encryption and decryption of classical encryption algorithms. The experiment performs encryption and decryption operations on data of different lengths (1B~1024B). Each length performs 1 million operations under different concurrency and takes the average value in milliseconds. Figure 6 and Figure 7 show the final statistics of the time overhead of the three encryption and decryption algorithms.

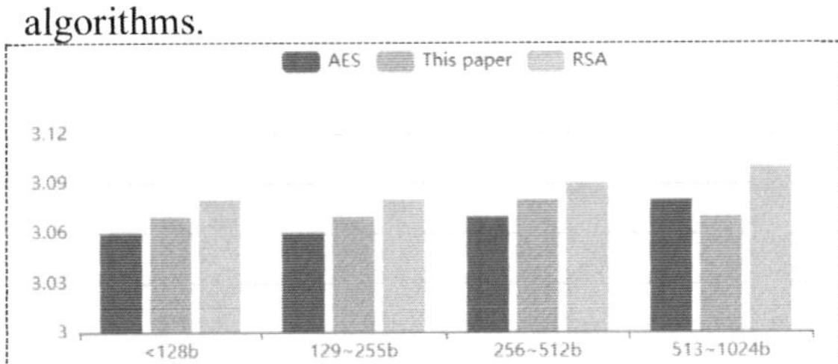

Figure 6. Encryption time comparison.

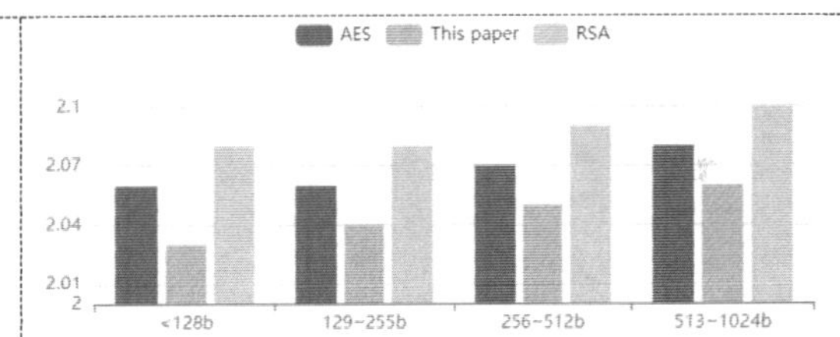

Figure 7. Decryption time comparison.

The dynamic encryption and decryption mechanism based on the national secret algorithm performs the middle range of AES and RSA in terms of encryption time, and the decryption time is lower than that of AES and RSA. Relative stability can be maintained as the length of the message body increases.

3.4. Security Analysis

The security of secure communication depends on the security of the key during its life cycle and the ability to resist tampering and replay during the communication process.

3.4.1. Key Security. The key is generated in real-time and is paired. A tamperer needs to decipher both keys simultaneously within a limited session time. And the public and private keys are stored in the distributed cache in the form of hexadecimal strings. For attackers, the primary thing is to break the outer security capabilities of the system. The entire system is currently deployed in the cloud service environment, and network security is absolutely reliable.

3.4.2. Anti-tampering and anti-replay. Based on the triple communication guarantee of real-time digital signature mechanism, timestamp Tc , and random number Rc, it can realize anti-tampering and anti-replay attacks in real-time and ensure communication security.

4. Conclusion

This paper generates session-level paired keys based on the national secret SM2 algorithm and then generates a digital signature through the SM3 algorithm, which effectively guarantees the integrity of the signature. Finally, the encryption key and encryption offset are extracted based on the paired key, and the original request text is encrypted and decrypted using the SM4 algorithm. Timestamps and random numbers are added to each request to prevent tampering and replay attacks effectively. Based on the microservice secure communication framework designed in this paper, it has been fully verified on the national Power Grid Digital Capability Open Platform Project, which not only effectively guarantees the system function, but also effectively guarantees the communication security. Since the microservice architecture integrates many third-party tools, each tool may have its security problems. It is necessary to pay attention to the security vulnerabilities of each tool and fix them in time. Further research is needed to improve the timeliness and stability of encryption and decryption algorithms.

References

[1] Villamizar M, Garces O, Castro H, Verano M, Salamanca L, Casallas R and Gil S., 2015. Evaluating the monolithic and the microservice architecture pattern to deploy web applications in the cloud. In: *Computing Colombian Conference*. Colombia, 583-590.

[2] Knoche H and Hasselbring W. (2018) Using microservices for legacy software modernization. *IEEE Software*, **35**(3): 44-49.

[3] WANG H X, GU C X and ZHENG Y H. (2014) Formal security analysis of oauth2. 0 authorization protocol. *Journal of Information Engineering*, **15**(2):141-147.

[4] Zhang S Y. (2003) Network security principles and applications. *Science Press*, **5**:134-141.

[5] LI M H, QI X C and WU Q Q. (2019)Smart grid wireless communication data encryption transmission scheme based on dynamic key. *Netinfo Security*, **19**(12): 10-21.

[6] Xie Y Q.(2016) Application of cryptographic algorithms in my country. *Journal of Information Security Research*, **2**(11):969—971.

[7] Huang J.(2016) Security analysis and improvement of RSA public key encryption system. *Computer and Network*, **42**(1): 70-73.

[8] Yang J. (2017) Application research of three symmetric encryption algorithms in wireless blocking center. *Automation and Instrumentation*, **4**:135-137.

[9] Huang Q W and Zeng D. (2019) Design of distributed microservice architecture based on spring cloud and docker. *Microcomputer Application*, **35**(6): 98-101.

[10] Saxena N and Grijalva S. (2017) Dynamic secrets and secret keys based scheme for securing last mile smart grid wireless communication. *IEEE Transactions on Industrial Informatics*, **13**(3): 1482-1491.

Journal of Physics: Conference Series **2384** (2022) 012049 doi:10.1088/1742-6596/2384/1/012049

SERs of Different Modulation Systems in 1G-4G

Zhenxiang Chen[1, a, *, †], Yining Ding[2, b, *, †], Luying Ruan[3, c, *, †]

[1] School of Communication & Information Engineering, Shanghai University, Shanghai, 200444, China

[2] School of Communication & Information Engineering, Shanghai University, Shanghai, 200444, China

[3] School of Communication & Information Engineering, North China Electric Power University, Hebei, 100096, China
[*]Corresponding author email: [a] zxchen@shu.edu.cn,
[†]These authors contribute equally.

Abstract. The Digital sub-band system is an indispensable technology for the development of modern society, which has a wide range of applications. This work is particularly concerned with developing applications for modulation techniques used in mobile communication. Based on MATLAB, this paper compared the modulation and demodulation of 2ASK, 2FSK, 2PSK, and QPSK horizontally and analysed the practical and theoretical error rates. With the constellation maps of 16QAM, 32QAM, 64QAM, and 256QAM, the relationship between the error rate and the signal-to-noise ratio was studied. In binary modulation systems, 2PSK has the best noise resistance and requires 6dB less SNR than 2ASK when employing the same demodulation approach. In addition, this paper outlined the technologies and modulation standards employed in the development of 1G-4G. To make the comparison clearer, the advantage, disadvantages, and modulation comparisons among 1G-5G are displayed. The analysis and research of digital sub-band modulation technology is the basis for developing new technology in the future.

1. Introduction

Communication technology has permeated every part of people's lives, providing a wide range of services and evolving into a vital production component. The types and quality of supported services are regularly expanded and upgraded when generations of updates and upgrades from 1G to 5G [1] are released. Each iteration of mobile development employs a different set of approaches and relative criteria.

The core technology in 1G is FDMA, which divides the communication system's entire frequency band into evenly spaced frequency channels. The AMPS standard is used in 1G, and analog modulation is the primary modulation [2]. The most significant advancement in 2G is digital modulation rather than analog modulation, which allows for multiplexing with TDMA technology. The GSM standard is used by 2G, and the primary modulation is GMSK [3]. 3G networks can transfer voice and data, with mobility determining the maximum speed. 3G communication employed CDMA as a standard, with QPSK being the most often used modulation type [4]. 4G, also called by Long Term Evolution (LTE), can be utilized in speech, data, multimedia, and other applications due to low latency, high speed, and IP packet switching. As the name of 4G shows, LTE is also used as criterion in this progress. LTE employs a maximum of 64QAM as the main modulation technique [5]. Cellular mobile communication is a fifth-generation communication technology with performance targets such as rapid data throughput, low latency, energy-efficient cost reduction, and large-scale device connectivity [6-9].

This paper focuses on the fundamentals of digital signal processing by merging the basic modulation process with the advancement of cellular technology. This study is extended based on the modulation process and bit error rate results to evaluate why 3G and 4G use QPSK and QAM from a bit error rate perspective, demonstrating the integration of the modulation process and cellular evolution. This review is divided into three sections: 1) Integrating the modulation and demodulation simulations of 2ASK, 2FSK, 2PSK, and QPSK with the constellation plot and bit error rate of 16QAM, 64QAM, and 256QAM. The results will be evaluated for validity and reliability. 2) Mapping the modulation result based on various parameters into the relevant generation (1G-4G) and comparing each generation's modulation performance. 3) Briefly overviewing and forecasting of future developments in 5G. In the future, 5G, services such as audio, video, and images for large-scale users will grow rapidly, and the explosive growth of network traffic will greatly affect the service quality of users accessing the Internet. How to effectively distribute large-traffic service content and reduce users' delay in obtaining information has become an urgent problem to be solved [10].

2. Research of shift keying modulation

2.1. Background

Virtual communication channels typically have band-pass characteristics and cannot directly transmit digital baseband signals with rich low-frequency components. The high-frequency carrier must be modulated with the baseband signal to generate the frequency band signal for transmission. Shift keying is a method that uses the discrete value features of digital signals to adjust the carrier parameter. It has a quick modulation transformation rate and an easy modulation test. The shift keying of the carrier's amplitude, frequency, and phase correlates to Amplitude Shift Keying (ASK), Frequency Shift Keying (FSK), and Phase Shift Keying (PSK).

2.2. 2ASK

Binary Amplitude Shift Keying (2ASK) is a digital modulation whose amplitude of a sine carrier varies with the digital baseband signal [11]. The binary symbol sequence to be sent $s(t)$ can be expressed as:

$$s(t) = \sum_n a_n g(t - nT_s) \tag{1}$$

where a_n is the value of the binary sequence, T_s is the period of the sequence, and $n = 0, 1, 2, ...$ In amplitude keying, the carrier amplitude varies with the baseband signal, which means if there is a carrier, it represents one, and if there is no carrier, it represents zero. And the time-domain mathematical expression of the 2ASK signal is:

$$S_{2ASK}(t) = a_n \cdot A \cos \omega_C t \tag{2}$$

In the formula, A is the unadjusted carrier amplitude and the carrier angle frequency. And if A is defined to be 1, the general time-domain expression of the 2ASK signal is:

$$S_{2ASK}(t) = [\sum_n a_n g(t - nT_s)] \cos \omega_C t = S(t) \cos \omega_C t \tag{3}$$

where, $S(t)$ is a random monopolar pulse sequence representing the binary information.

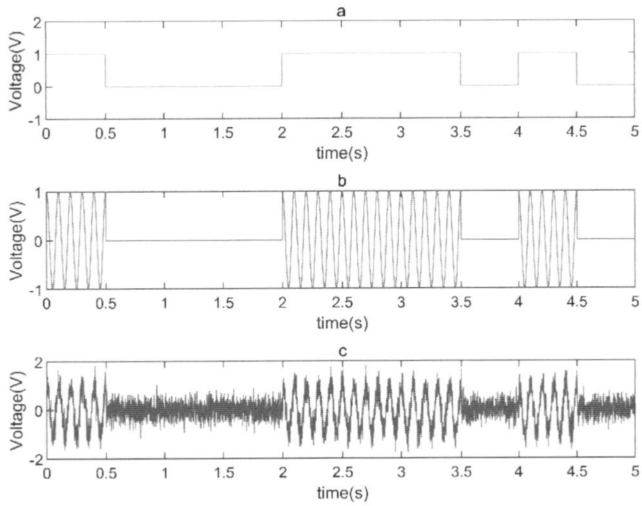

Figure 1 The simulation of 2ASK modulation: (a) baseband st.
(b) 2ASK signal. (c) 2ASK signal with noise.

As shown in Figure 1, after a random binary sequence st is generated in MATLAB, the carrier defined previously will be multiplied by the sequence value to be sent, and thus, the modulated 2ASK signal is created, which equals 0 when $st = 0$, and $cos\, 20\pi\, t$ when $st = 1$. Also, Additive White Gaussian Noise is added to the modulated signal to simulate the signal passing through an AWGN channel as depicted.

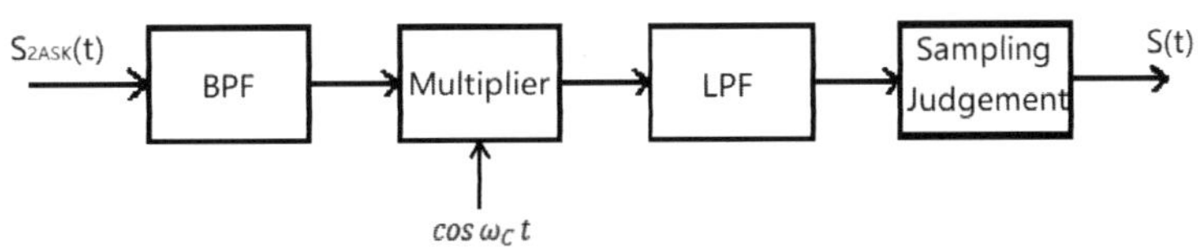

Figure 2 The corresponding demodulation system structure.

Figure 2 shows the corresponding demodulation system structure of the 2ASK signal, and Figure 3 shows the simulation of 2ASK demodulation. Correlation demodulation, also known as the synchronous detection method, is used in this project to demodulate the 2ASK signal. In correlation demodulation, a signal same as the carrier is multiplied by the 2ASK signal. Thus, the signal turns out to be:

$$s(t) \cdot cos\, \omega_C\, t \cdot cos\, \omega_C\, t = \frac{1}{2} s(t)[1 + cos\, 2\omega_C\, t] \tag{4}$$

And by using a low-pass filter to filter the high-frequency components, the output turns to be $s(t)/2$, and the system can recover the modulated signal $s(t)$ without distortion by sampling judgment.

Since the binary sequence is to be transmitted st is $[1, 0, 0, 0, 1, 1, 1, 0, 1, 0]$ as Figure 1 shows, the 2ASK signal multiplied with the carrier is $\frac{1}{2}[1 + cos\, 40\pi\, t]$ when $st = 1$, and 0 when $st = 0$ as Figure 2 depicted. Also, the filtered signal equals 0.5 when st = 1, and 0 when st = 0. Some fluctuations are existed because of the noise added during transmission. The simulation value is correspondent with the theoretical value. And the binary sequence after sampling judgment is [1, 0, 0, 0, 1, 1, 1, 0, 1, 0], which is exactly the same as the baseband, which means the signal is demodulated perfectly.

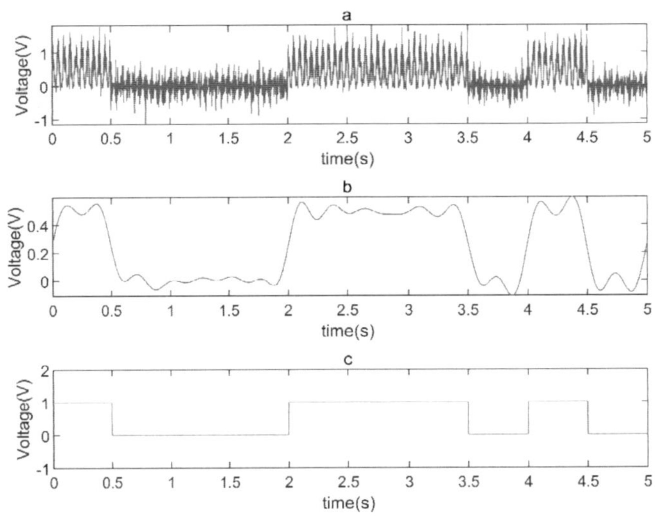

Figure 3 The simulation of 2ASK demodulation: (a) 2ASK signal multiplied with the carrier. (b) multiplied signal after low-pass filtering. (c) sampling judgment.

2.3. 2FSK

Frequency shift keying (FSK) transmits a digital message with the carrier frequency change, which means the carrier's frequency is controlled by the transmitted digital message. In binary digital modulation, the 2FSK signal is generated if the carrier frequency switches between 2 different frequencies f_1 and f_2 according to the value of the baseband signal[12]. The time-domain mathematical expression of 2FSK is as follows:

$$S_{2FSK}(t) = [\sum_n a_n g(t - nT_s)] \cos \omega_1 t + [\sum_n \overline{a_n} g(t - nT_s)] \cos \omega_2 t \tag{5}$$

where ω corresponds to the angular frequency, respectively, $\omega_1 = 2\pi f_1, \omega_2 = 2\pi f_2, f_1 \neq f_2$.

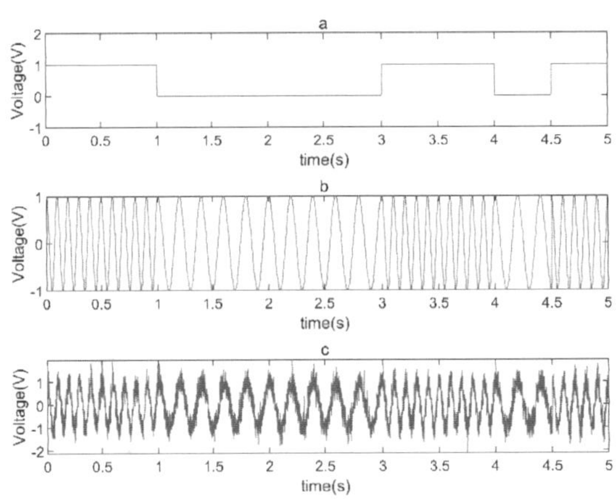

Figure 4 The simulation of 2FSK modulation: (a) baseband st. (b) 2FSK signal. (c) 2FSK signal with noise.

In MATLAB, the inversed code $st2$ of the generated random sequence $st1$ is calculated and $st1$ is multiplied by the carrier $s1 = \cos \omega_1 t$ while $st2$ is multiplied by $s2 = \cos \omega_2 t$. Correspondingly, the

2FSK signal is generated. Figure 4 presents the binary data and the 2FSK signal, modulating 0/1 levels to $f1/f2$ frequencies and the frequency transition between two frequencies. The 2FSK signal equals $cos\ 20\pi\ t$ when $st = 1$, and $cos\ 10\pi\ t$ when $st = 0$. And, the signal fluctuates during transmission since noise are added.

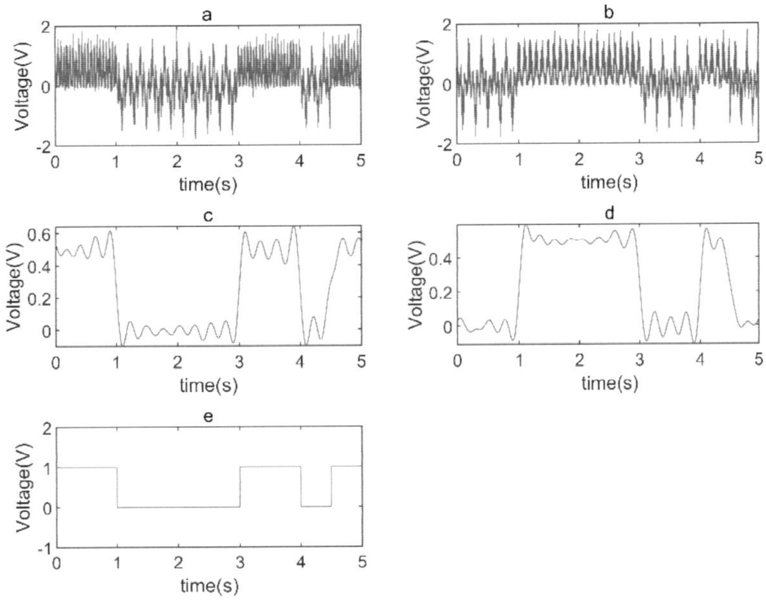

Figure 5 The simulation of 2FSK demodulation: (a) 2FSK signal multiplied with carrier s1. (b) 2FSK signal multiplied with carrier s2. (c) multiplied signal fs1 after low-pass filtering. (d) multiplied signal fs2 after low-pass filtering. (e) sampling judgment.

2FSK demodulation, which is demonstrated in Figure 5, is identical to 2ASK demodulation. The modulated signal is multiplied to $s1$ and $s2$ independently, and the multiplier outputs are then passed through the low pass filters. As viewed in Figure 5, by comparing the filtered signal $fs1$ and filtered signal $fs2$, the original signal can be recovered. If $fs1$ is bigger than $fs2$, it represents "1". Otherwise, it represents "0".

2.4. 2PSK

Phase shift keying, also known as PSK, means modulating the phase of the carrier wave and leaving the amplitude without any changes [13]. 2PSK is a form of phase-shift keying, which uses two initial carriers separated by 180 to transmit binary information. Assume the binary monopolar code is a_n, and its corresponding bipolar binary code is b_n, then the general time-domain mathematical expression of the 2PSK signal is:

$$S_{2PSK}(t) = [\sum_n b_n g(t - nT_s)] \cos \omega_c t \tag{6}$$

which can also be written as:

$$S_{2PSK}\ (t) = \begin{cases} [\sum_n g(t - nT_s)] \cos(\omega_c t + \pi) & a_n = 0 \\ [\sum_n g(t - nT_s)] \cos(\omega_c t + 0) & a_n = 1 \end{cases} \tag{7}$$

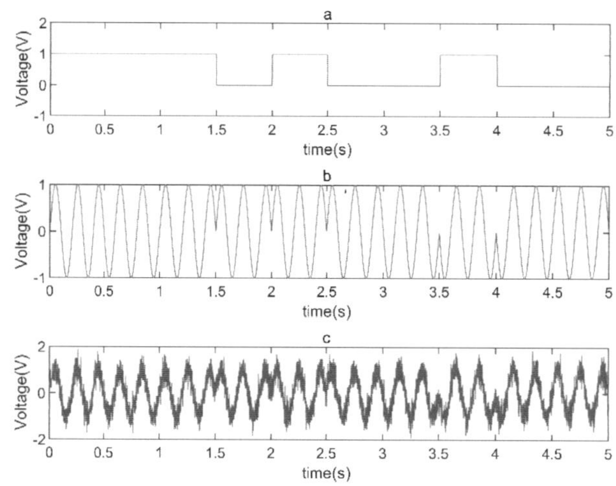

Figure 6 The simulation of 2PSK modulation: (a) baseband $st1$.
(b) 2PSK signal. (c) 2PSK signal with noise.

In MATLAB coding, the bipolar form of a random sequence $st1$ can be obtained by adding itself with its base minus complement. Afterward, the 2PSK signal can be engendered by multiplying the bipolar sequence with the carrier. The whole modulation process is depicted in Figure 6. When $st1 = 0$, the 2PSK signal is equal to $\cos 10\pi\, t$, and the 2PSK signal equals $-\cos 10\pi\, t$ when st1 = 1. There're changes in the trends when time = 1.5, 2, 2.5, 3.5, 4, which is when the value of the baseband changes, and the changes indicate the variation of phases. And also, the signal fluctuates during transmission since noise are added.

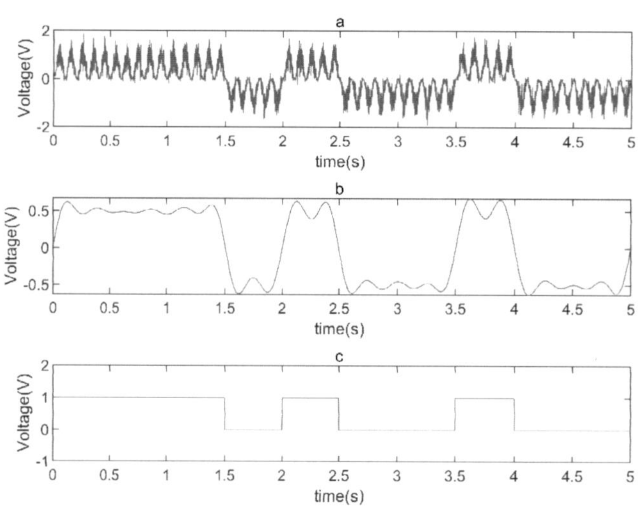

Figure 7 The simulation of 2PSK demodulation:
(a) 2PSK signal multiplied with the carrier.
(b) multiplied signal after low-pass filtering.
(c) sampling judgment.

This work uses correlation demodulation, similar to 2ASK, to demodulate the 2PSK signal. The original signal can be precisely recovered by multiplying the 2PSK signal by carrier s1 with a multiplier, then passing through a low-pass filter and sampling judgment. And the results are plotted in Figure 7. After multiplying the 2PSK signal with the carrier, which equals $cos\ 10\pi\, t$, the signal turns into $\frac{1}{2}[1 + cos\ 20\pi\, t]$ when 2PSK signal equals $cos\ 10\pi\, t$, and $-\frac{1}{2}[1 + cos\ 20\pi\, t]$ when 2PSK signal equals $-\,cos\ 10\pi\, t$, which is demonstrated in Figure 7 (a). By low-pass filtering, the signal becomes 0.5 when the multiplied signal is $\frac{1}{2}[1 + cos\ 20\pi\, t]$, and -0.5 when multiplied signal equals $-\frac{1}{2}[1 + cos\ 20\pi\, t]$. Finally, by comparing with 0, we can recover the original signal. If filtered signal is bigger than 0, $st = 1$, otherwise $st = 0$.

2.5. QPSK

Quadrature Phase Shift Keying(QPSK) modulation is also known as 4PSK, which uses 4 different phases of the carrier to express digital information. The time-domain mathematical expression of QPSK is as follows:

$$S_i(t) = cos(\omega_C\, t + \theta_i) \quad (i = 1,2) \tag{8}$$

In QPSK, each carrier phase carries two binary symbols, and the corresponding relationship is shown in Table 1.

Table 1. The corresponding relationship between symbols and phrases.

Symbols		Phases
I	Q	
0	0	225°
1	0	315°
1	1	45°
0	1	135°

The orthogonal phase method of QPSK, illustrated in Figure 8, is used in this project, where the QPSK signal can be considered to synthesize two orthogonal 2PSK signals. The string-to-parallel converter divides the input binary sequences into two parallel bipolar sequences I and Q, with rate halving, which means the period of each symbol was twice as long as the input symbol. After two orthogonal binary phase syntheses, a QPSK signal is formed.

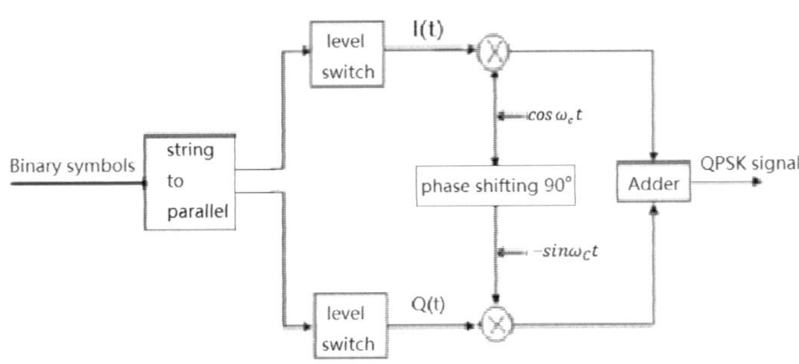

Figure 8 The structure of the Orthogonal phase method of QPSK.

In MATLAB, transforming the original binary sequence to be bipolar NRZ sequence after generating, then the odd number of symbols are moved to I, and those even ones to Q. Value of sequence I are multiplied by $cos\ \omega_C\, t$ while Q are multiplied by $cos(\omega_C\, t + \pi/2)$. A QPSK signal which contains 4 phases, is produced after adding I to Q. Figure 9 depicts the whole process of QPSK modulation. The baseband is $[1, 1, 0, 0, 0, 0, 1, 0, 1, 0, 1, 0, 1, 1, 0, 1, 1, 0, 1, 1]$. As a result, the I stream equals $[1, 0, 0, 1, 1, 1, 1, 0, 1, 1]$, and Q stream equals $[1, 0, 0, 0, 0, 0, 1, 1, 0, 1]$. By multiplying I with carrier, I stream PSK signal becomes $cos\ 4\pi\, t$ when I stream equals 1 and $-cos\ 4\pi\, t$, when I stream, equals 0, as depicted in Figure 9 (d). Similarly, Q stream PSK signal equals $cos(4\pi t + \pi/2)$ when Q stream equals 1 while Q stream PSK signal equals $-cos(4\pi t + \pi/2)$ when Q stream equals 0, as Figure 9 (e) shows. The 4 phases QPSK signal is created by adding I stream PSK signal to Q stream PSK signal as Figure 9 (f) shows.

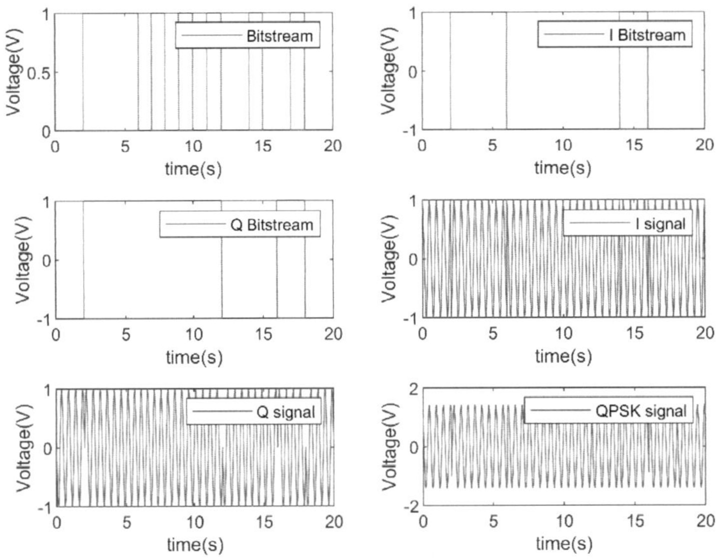

Figure 9 The simulation of QPSK modulation: (a) baseband.
(b) I Bitstream. (c) Q Bitstream. (d) I stream PSK signal.
(e) Q stream PSK signal. (f) QPSK signal.

The demodulation of QPSK can be decomposed into the demodulation of two 2PSK. As illustrated in Figure 10, the same-phase branch and the orthogonal branch are demodulated with coherent demodulation, and the two parallel codes a and b are restored into a serial binary sequence.

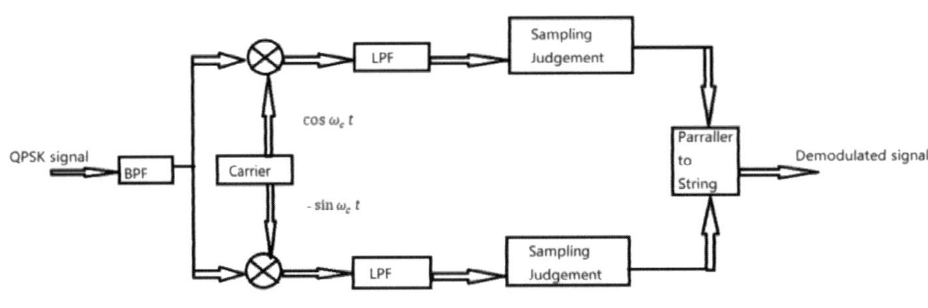

Figure 10 The structure of coherent demodulation of QPSK.

For a deeper understanding of the differences in several modulations, the Symbol Error Rate (SER) of 2ASK, 2FSK, 2PSK and QPSK are calculated by generating a binary sequence that contains 10000 symbols, then modulating and demodulating it with different methods. The SER of each modulation can be plotted by comparing the recovered signal with the original signal while changing the SNR simultaneously, as shown in Figure 11(a-d). The blue line is the simulation value from this system, and the pink one is the theoretical value. The simulation values accord with the theoretical ones, indicating the proposed system's correctness.

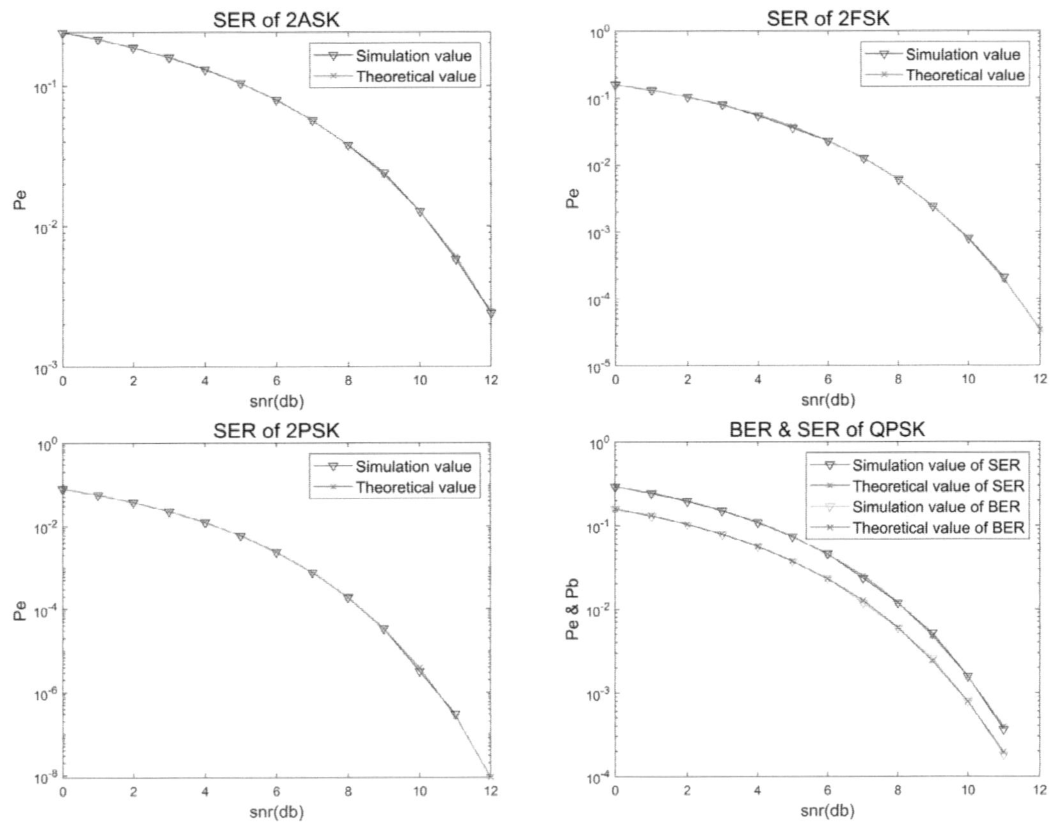

Figure 11 (a) SER(BER) of 2ASK system in AWGN channel.
(b) SER(BER) of 2FSK system in AWGN channel.
(c) SER(BER) of 2PSK system in AWGN channel.
(d) SER & BER of QPSK system in AWGN channel.

Figure 12 illustrates the SER of different systems in the AWGN channel. Observably, 2PSK has the smallest SER for binary modulation systems under the same SNR, and 2ASK has the biggest when using the same demodulation method. This implies that 2PSK has the best noise-resistance performance while 2ASK has the worst. The SNR requirement between 2PSK and 2ASK are 6dB. What's more, 2PSK has an SNR requirement of 4dB smaller than QPSK, which indicates that for PSK, the binary system has a better noise-resistance performance than the M radix system since the SNR requirements will increase 4dB when M is doubled if using the same demodulation method.

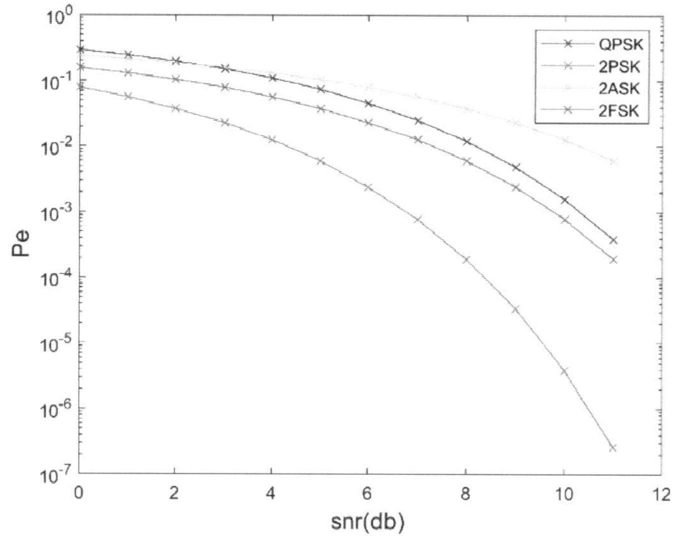

Figure 12 SER of different systems in AWGN channel.

3. Research of QAM

3.1. Background

ASK, FSK, and PSK are restricted due to transmission and bit error rates. Now QAM technology is more used [14].

QAM (or MQAM) is a multiband amplitude and phase joint keying modulation, which also can be combined with DSP [15]. QAM modulation is a kind of modulation applied to the digital signal. The input signal must be a sequence of 0 and 1. If the input signal modulated is analog, we must first convert the analog signal to a digital signal through digital-to-analog conversion technology.

3.2. Modulation

M stands for the order of QAM or PSK, which M represents in the following articles. When $M > 4$, the Euclidean distance between adjacent points of the QAM signal will be greater than that of the Mask, MPSK, and other modulation methods, so it has a more vital interference ability. One symbol of the QAM signal can be expressed as

$$e_k(t) = A_k \cos(\omega_c t + \theta_k),\ kT_B < t \le (k+1)T_B \tag{9}$$

It also can be expanded to:

$$e_k(t) = X_k \cos \omega_c t + Y_k \sin \omega_c t \tag{10}$$

where X_k、A_k、θ_k、Y_k can take multiple discrete values, respectively. Multiply sin and cos to get orthogonal sequences. And the number of values depends on the constellation structure of the QAM signal.

Take 16QAM as an example, the input binary sequence passes through two parallel sequences firstly, whose output rate is halved by the serial-to-parallel converter. Then, the input sequence goes through the transformation mapping from 2 to 4 levels[16] to form a 4-level baseband signal X_k、Y_k, which are multiplied by two mutually orthogonal carriers to form two mutually orthogonal 4ASK signals. Finally, the 16QAM modulated signal is completed.

The distinction between 16QAM, 64QAM, 256QAM and MQAM is that the first 2 levels need to be converted to different binary levels with higher M of QAM. However, the above method cannot be used if the graphics on a constellation map are not square, such as 32QAM or 128QAM. Instead, the

MQAM constellation is utilized to specify the amplitude of the symbol[17] and punctuation is used directly on the constellation map.

3.3. Demodulation

Since the 16 signal points of the 16QAM signal project 4-levels on the horizontal and vertical axes, corresponding to the 4-level baseband signal output by the low-pass filter, the 4-level judger should have 3 decision levels: -1, 0, 1. The 4-level judge judges and detects the 4-level baseband signal, then converts it to 2-level by 4-level. Finally outputs the original signal by the parallel-to-serial converter.

3.4. Simulation results

In this paper, 16QAM, 32QAM, 64QAM, and 256QAM are simulated using Matlab, and their constellation and theoretical and actual error rate curves are observed.

Figure 13(a), (b), (c), (d)are the constellation maps got from Matlab. This paper adds white Gaussian noise between 0-1 to all the QAM modulation processes. The regional distribution of point concentration can still be seen in the constellation of 16QAM and 32qam after adding noise, while the boundary between each region is blurred in 64QAM and 256qam.

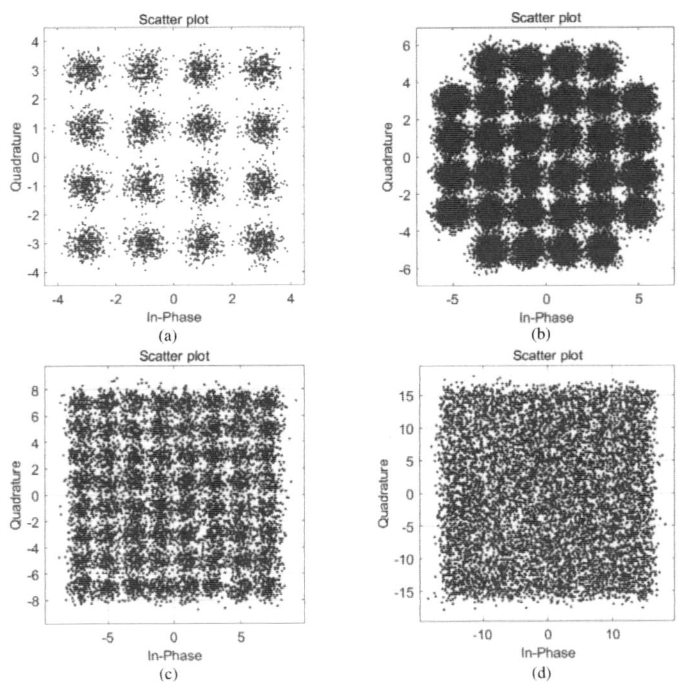

Figure 13 (a)16QAM constellation map (b) 32QAM constellation map
(c)64QAM constellation map (d)256QAM constellation map

Journal of Physics: Conference Series **2384** (2022) 012049 doi:10.1088/1742-6596/2384/1/012049

3.5. Error rate analysis

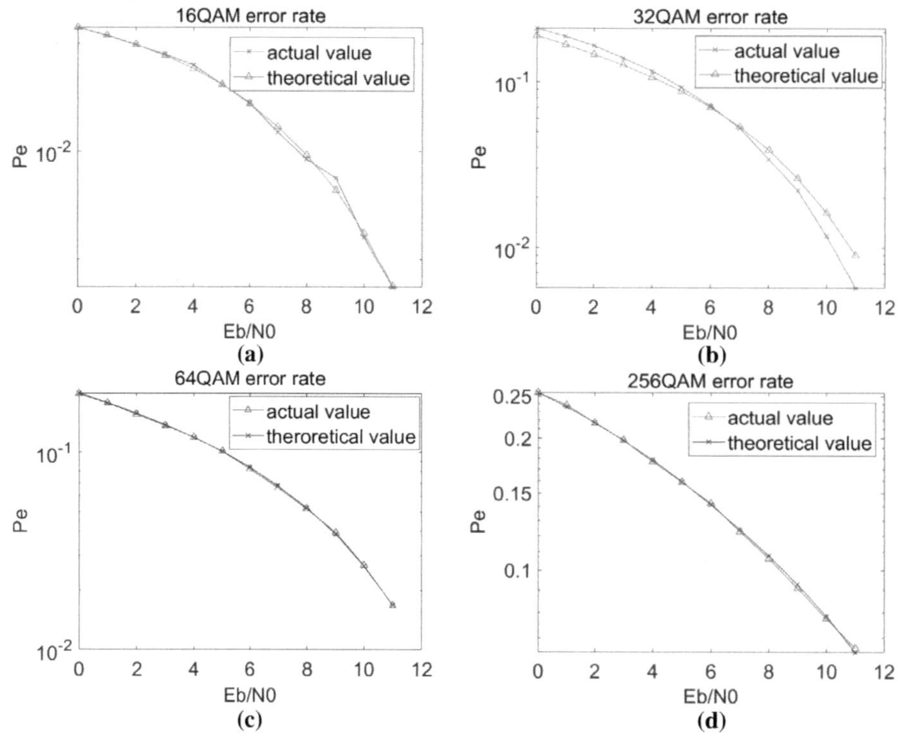

Figure 14 (a)16QAM signal error rate (b)32QAM signal error rate
(c)64QAM signal error rate(d)256QAM signal error rate

In Figure 14, the error rate of QAM with varying M under different SNR is recorded, and the theoretical and experimental error rates are compared. In this paper, the error rate of QAM with different M under different SNR is recorded, and the theoretical error rate and the actual error rate are compared. The results are shown in figure 20. The error rate study is essential to determine the quality of a modulation mode, which corresponds to the reliability of one of the two main indicators of the communication system. When the SNR increases from 1 to 12, each increase of 1 SNR can significantly reduce the error rate. That means that the acquisition of a high transmission rate is accompanied by the sacrifice of error rate, which corresponds to the validity and reliability of the communication system.

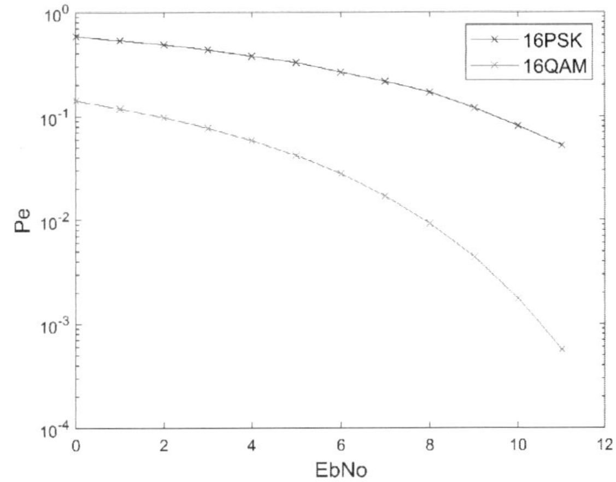

Figure 15 16QAM and 16PSK error rate

By comparing the error rate curves in Figure 15, When the SNR is 0, the error rate of 16QAM is 0.5 lower than that of 16PSK. With the increase of the SNR, the error rate difference between 16QAM and 16PSK can reach two orders of magnitude, proving that 16QAM is more reliable than 16PSK.

4. Development of cellular technology

Table 2. The progression of mobile communication over generations

	General Techniques	Criteria	Main Modulation Technique
1G	FDMA	AMPS	Analog Modulation
2G	TDMA	GSM	GMSK
3G	CDMA	CDMA2000	QPSK
4G	OFDM	LTE	Maximum 64QAM

The progression of mobile communication over generations is shown in Table 2. 1G is a 'narrowband' technology that is mainly used for voice services. However, the 1G process introduces a delay issue and poor sound quality. The general technique used in 1G is FDMA, which may divide the communication system's full frequency band into many equally spaced frequency channels.

The most major enhancement in 2G mobile communication technology over 1G is digital modulation rather than analog modulation. Multiplexing is accomplished through the use of TDMA technologies. Time can be divided into periodic frames via TDMA, and each frame can be partitioned into numerous time slots.

Simultaneous transmission of speech and data information is possible with 3G. The maximum speed is primarily determined by mobility. CDMA is the method employed. Separate users' signals for transmitting information are identified not by different frequencies or time periods but by different coding sequences. The same bit-error probability can be derived by comparing the bit-error probability of QPSK and BPSK. However, QPSK band utilization is higher than BPSK because, while QPSK has the same shape as BPSK, its bandwidth is half that of BPSK. The narrower the band, the greater the effectiveness. As a result, QPSK outperforms BPSK in terms of effectiveness.

In the 4G, based on the characteristics of low latency, high speed, and IP packet switching, it can be used in voice, data, multimedia, etc., and the highest rate can reach 100Mbps [18]. OFDM is mainly utilized. The characteristics of OFDM include parallel transmission, sub-carrier orthogonal, FFT implementation, and dynamic sub-channel allocation and bit allocation. In the 4G modulation, 64QAM is used, which is high-order modulation. High-order modulation is adapted because it can realize high-speed data transmission under limited bandwidth and improve spectral efficiency when $M >= 8$. The 64 points are evenly and clearly distributed in the constellation map. Also, the constellation diagram is approximately without noise, making the signal quality is in good condition.

5G is the latest generation of cellular mobile communication technology. Performance goals of 5G are high data rates, reduced latency, energy savings, lower costs, increased system capacity, and massive device connectivity.5G can be applied in the internet of vehicles and, autonomous driving surgery, smart grid.

To make the comparison clearer, the advantage, disadvantages, and modulation comparisons among 1G-5G are displayed in Table 3.

Table 3. Contrast of cellular development

	Advantage	Disadvantage	Modulation Comparison
1G	Mainly for voice services	Delay problem	Analog modulation
2G	Digital modulation	Information cannot be delivered directly	More concentrated main lobe energy
3G	Transmit voice and data information at the same time	Inherits the circuit switching of the 2G system	Higher utilization of band of QPSK, exist maximum phase change
4G	Low latency, high speed	Can't meet the needs of ubiquitous network	Larger minimal distance of 64QAM, better reliability
5G	High data rates, reduced latency, lower costs, increased system capacity	High cost, RF and antenna are too large	256 QAM, $\pi/2$-BPSK

5. Conclusion

This work is particularly concerned with developing applications for modulation techniques used in mobile communication. Furthermore, SER results compare development from 1G to 4G in terms of dependability and efficacy. Additionally, this paper shows why and how to develop new modulation strategies for the next generation. When employing the same demodulation approach, 2PSK has the best noise-resistance performance for binary modulation systems and has a 6dB lower SNR need than 2ASK. Moreover, when operating at the same radix, QAM outperforms PSK in terms of noise resistance and has a lower bit error rate by approximately two orders of magnitude than PSK when the signal-to-noise ratio is high. Meanwhile, when the radix M increases, the noise-resistance performance of the identical modulation and demodulation system degrades. With higher band effectiveness, QPSK is employed in 3G modulation rather than BPSK. With superior reliability than 64PSK, 64QAM is utilized in 4G modulation. Applied with 256 QAM, 5G is widely used with its High data rates, reduced latency, lower costs, increased system capacity.

In conclusion, with the SERs of different modulation systems and the development of 1G to 5G, the applications in the development of terrestrial mobile communication and satellite communication will be highly potential.

References

[1] Goyal, M., & Lather, Y. (2015). Advancement of communication technology from 1G to 5G. *International Journal of Advanced Research in IT and Engineering*, *4*(5), 1-17.

[2] Nambi, E. T., Sakthivel, P., & Krishna, T. V. Survey of Evolution Analytical Report on Upgradation Technology from 1G to 5G. *International Journal of Computer Applications*, *975*, 8887.
[3] Sood, R., & Garg, A. (2014). Digital society from 1G to 5G: a comparative study. *International Journal of Application or Innovation in Engineering & Management (IJAIEM)*, *3*(2), 186-193.
[4] Dahlman, E., Parkvall, S., Skold, J., & Beming, P. (2010). *3G evolution: HSPA and LTE for mobile broadband*. Academic press.
[5] Dahlman, E., Parkvall, S., & Skold, J. (2013). *4G: LTE/LTE-advanced for mobile broadband*. Academic press.
[6] Wu, Q. (2021,June). 4G Communication Technology Wireless Network Secure Communication. In *2021 International Wireless Communications and Mobile Computing (IWCMC)* (pp. 915-918). IEEE.
[7] An, D., Woo, H., Yoon, H., & Yeom, I. (2013). Enhanced cooperative communication MAC for mobile wireless networks. *Computer Networks*, *57*(1), 99-116.
[8] del Peral-Rosado, J. A., Raulefs, R., López-Salcedo, J. A., & Seco-Granados, G. (2017). Survey of cellular mobile radio localization methods: From 1G to 5G. *IEEE Communications Surveys & Tutorials*, *20*(2), 1124-1148.
[9] Morinaga, N., Yokoyama, M., & Sampei, S. (1996). Intelligent radio communication techniques for advanced wireless communications systems. *IEICE transactions on communications*, *79*(3), 214-221.
[10] Toskala, A., & Ranta-aho, K. (2015). Further Outlook for LTE Evolution and 5G. *LTE Small Cell Optimization: 3GPP Evolution to Release 13*, 423-432.
[11] Li, Q., Zhang, H., Liu, Z., & Fu, S. (2021, October). SPM-based Data Format Conversion from one 2ASK/4PSK to two 4PSK Signals with Wavelength Preservation. In *2021 Asia Communications and Photonics Conference (ACP)* (pp. 1-3). IEEE.
[12] Delshadpour, S., & Geng, M. (2020, February). A PLL based FSK demodulator with auxiliary path. In *2020 IEEE 11th Latin American Symposium on Circuits & Systems (LASCAS)* (pp. 1-4). IEEE.
[13] Suhartomo, A., & Vincent, V. (2020, December). Comparison of BER Performance for M-ary QAM and PSK on DWT-based OFDM System with PTS Technique Through AWGN Channel. In *2020 IEEE International Conference on Sustainable Engineering and Creative Computing (ICSECC)* (pp. 5-10). IEEE.
[14] Panem, C., Vetrekar, N., & Gad, R. (2022). Data reduction and recovery in wireless communication system: An extensive experimental evaluation using PSK and QAM modulations. *International Journal of Communication Systems*, e5198.
[15] Sivakumar, P., Boopathi, C. S., Sumithra, M. G., Singh, M., Malhotra, J., & Grover, A. (2020). Ultra-high capacity long-haul PDM-16-QAM-based WDM-FSO transmission system using coherent detection and digital signal processing. *Optical and Quantum Electronics*, *52*(11), 1-18.
[16] Wang X., Yu Y., Gu H., Song Y., Pang F., Li Y., Zhang Q., Zhuang L., Yang S., He X., Yang Y. 2022 120 km low-loss propagating OAM beams enabled by OAM-mode group multiplexing and a ring-core fiber *Optics Communications*, 516, art. no. 128264
[17] Abdulzahrah A.S., Hmood J.K. 2022 Proposal of PPM-RZ-mQAM scheme for suppressing nonlinear phase noise in high spectral ultra-DWDM system *Optical and Quantum Electronics*, 54 (1), art. no. 45
[18] Feher, K. (2021). *U.S. Patent No. 11,146,431*. Washington, DC: U.S. Patent and Trademark Office.

Conformal Antenna Design for Smart Swimming Goggles Surface

Qiheng Wu[1,2], Wencong Chen[1], Xingyan Li[1,2], Dawei Xin[1,2], Shuai Wang[1,*] and Baojin Zhai[3,*]

[1] Department of Mechanical Engineering, Dongguan University of Technology, 523808 Dongguan, China

[2] Department of Mechanical Engineering, Yanshan University, 066004 Qinhuangdao, China

[3] BiosMed Technology Co. Ltd., 518000 Shenzhen, China

[1,*]wangs@dgut.edu.cn, [3,*]zhaibaojin@sina.com

Abstract. Based on the irregular undevelopable surface of the smart swimming goggles, an electric dipole conformal antenna array is proposed in this article. Through the simulation of the conformal antenna, the beam scanning area of the conformal antenna array is studied to achieve a wider radiation range. In addition, the signal strength loss of the conformal antenna in the underwater environment is discussed, and the monitoring method based on the signal strength is proposed, which is helpful for the positioning monitoring of the swimming goggles to achieve risk prevention and control of children drowning.

1. Introduction

A conformal antenna is an antenna that is consistent with the object in shape [1], and it has the characteristics of low profile, easy bending, and bearing. Based on the traditional microstrip antenna, the conformal antenna can fully fit with the carrier and further reduce the additional antenna structure. Therefore, the conformal antenna is widely used in aerospace [2,3], wearable equipment [4,5], and medical devices [6,7].

In the current research on the conformal antenna, Chen X et al. proposed a new broadband conformal omnidirectional antenna with vertical polarization [8]. The omnidirectional radiation is realized using symmetrical radiation and feeding structure, and the radiation efficiency is improved by using a groove structure. Bai T M et al. designed an 8×8 conformal antenna array on the cylindrical surface based on the microstrip conformal antenna [9]. The simulation results show that the gain of the cylindrical conformal array is 1.2 dB lower than that of the planar array. The deflection of the beam is 4°, verifying the feasibility of the design based on the conformal antenna array through a physical object. Le Goff D et al. studied the RF substrate based on silicone rubber [10], and fabricated a conformal antenna working at 2.45 GHz, which can maintain stable electromagnetic radiation characteristics when the conformal material undergoes obvious deformation. These studies not only broaden the application scenarios of conformal antennas, but also provide a view of the variation and influence of electromagnetic radiation characteristics of conformal antennas. However, conformal characteristics and electromagnetic radiation characteristics of the current conformal antenna study have not been well combined in some fields.

Based on the current research on the conformal antenna on the surface, a new conformal antenna array on the smart swimming goggles is designed. The working frequency of the antenna is 2.45 GHz, and the input impedance is 50 Ω. The antenna gain of 4.3 dBi is achieved by adjusting the position distribution of the conformal antenna element. The electromagnetic radiation characteristic of the conformal antenna in water is verified by simulation. When the swimmer enters the water, goggles cannot communicate effectively. The drowning prevention and control warning will be executed after a certain time period, so as to achieve the goal of real-time monitoring and protection of swimmers.

2. Analysis and Design of Conformal Antenna Structure

The antenna unit is in the form of a dipole patch. The dielectric material of the selected swimming goggles is modified epoxy resin. The relative dielectric constant ε_r is set to be 4.2, and the loss angle tangent is 0.0009. The approximate length L_0 of the dipole antenna oscillator can be obtained by the relative constant of the working frequency f_0 and the selected dielectric material, and then the final size L can be obtained by the simulation and optimization of the length.

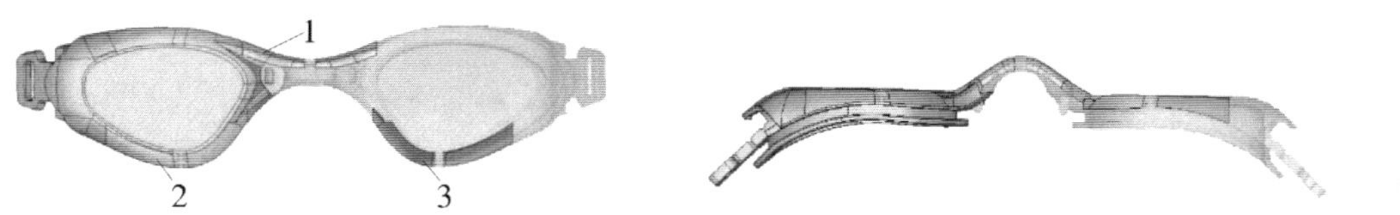

Figure 1. Structure of conformal antenna array

In terms of the electromagnetic radiation characteristics of the antenna, in order to achieve the isotropy of its directional pattern in space and meet the special surface structure of the swimming goggles and the limited conformal area of the antenna, the superposition of the directional pattern is considered, and the length of the conformal dipole antenna is repeatedly adjusted to optimize its S_{11} and bandwidth at the working frequency. The three-pair dipole conformal antenna is reasonably distributed on the lower goggles frame and the nose beam frame of the swimming goggles. The three-pair dipole conformal antenna is symmetrically distributed to achieve the effect of radiation electromagnetic wave beam superposition and increase the antenna gain. The dipole conformal antenna structure is shown in Fig. 1.

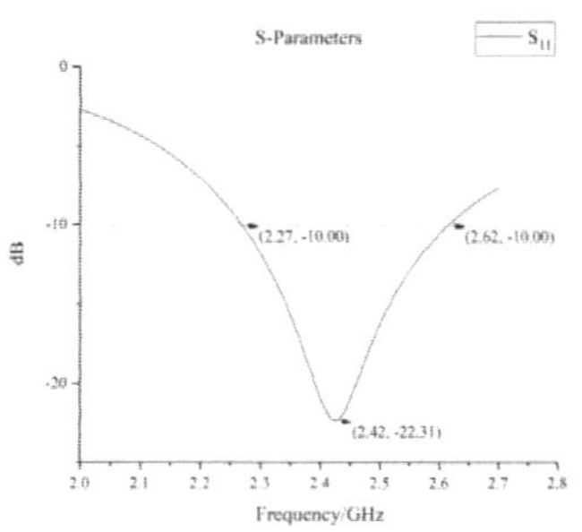

Figure 2. Working S_{11} diagram of conformal antenna element 1

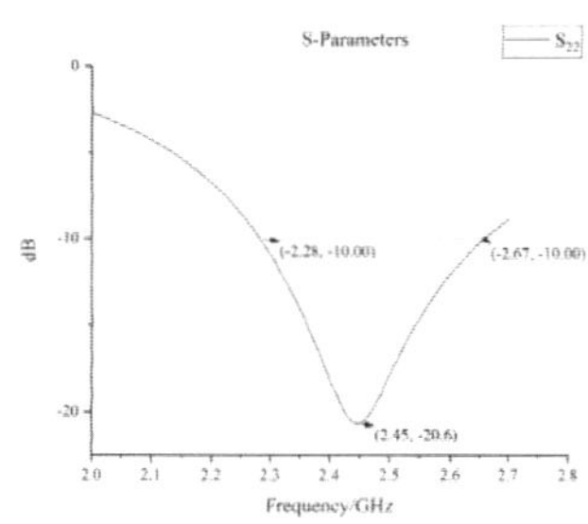

Figure 3. Working S_{11} diagram of conformal antenna element 2

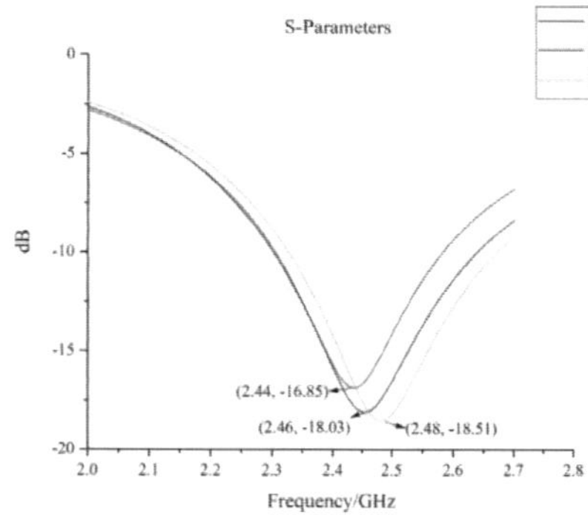

Figure 4. Working S_{11} diagram of conformal antenna array

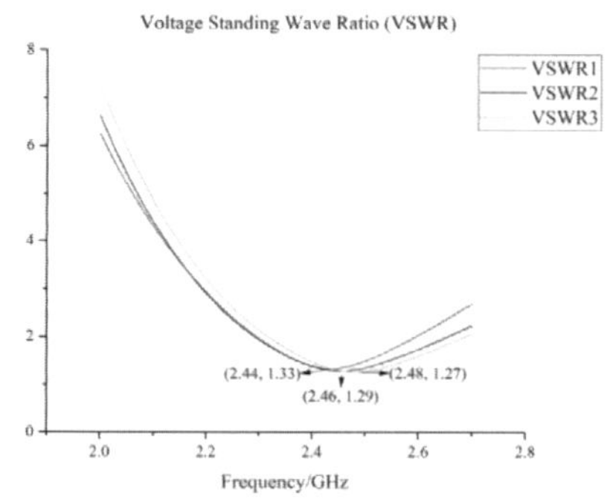

Figure 5. Working VSWR diagram of conformal antenna array

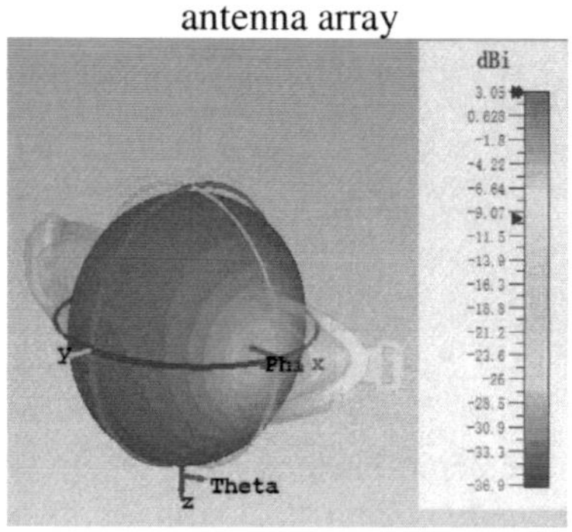

Figure 6. Working gain diagram of antenna element 1

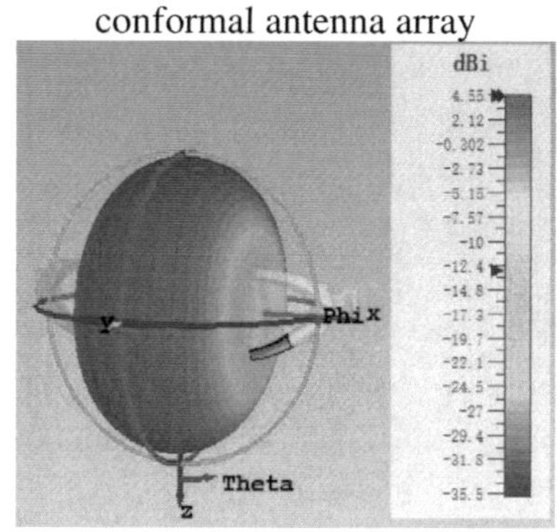

Figure 7. Antenna array 2 working gain diagram

The bright part in the figure is the designed dipole conformal antenna, and CST is used to simulate and analyze the antenna element and antenna array. While conformal antenna element 1 is working, conformal antenna elements 2 and 3 are not. The S_{11} is shown in figure 2. While the conformal antenna array element 2 is working and the conformal antenna array elements 1 and 3 are not, the S_{11} is shown in figure 3. When the conformal array element works simultaneously, the S_{11} is shown in Figure 4, and the voltage standing wave ratio is shown in figure 5.

The working frequency of conformal antenna array element 1 is 2.42 GHz. In the ISM Bluetooth band, the working bandwidth at -10 dB is 350 MHz, accounting for 14.6%. The working frequency of conformal antenna array element 2 is 2.45 GHz, and the operating bandwidth at -10 dB is 390 MHz, accounting for 16.3 %. It can be found that electromagnetic coupling occurs when antenna elements work together. Subsequently, the energy consumed between antennas will be absorbed by other conformal antenna elements, decreasing radiation energy and increasing return loss. However, the voltage standing wave ratio remains below 1.5 at 2.45 GHz. The comparison between Fig. 5 and Fig. 6 shows that the far-field electromagnetic waves of each antenna element generate beam superposition. The gain is from 2.9 dBi to 4.5 dBi.

3. Discussion

When a single dipole conformal antenna array element works or a conformal antenna array works, the pattern is similar to that of the dipole antenna, and the beam angle of radiation is fixed. In order to get a larger electromagnetic radiation angle, the antenna array element can work according to a certain rule to achieve a certain conformal antenna array beam scanning.

When the conformal antenna element 1 is controlled to work, and the conformal antenna elements 2 and 3 do not work, the gain is shown in Fig. 6. When the conformal antenna elements 1 and 2 are controlled to work, and the conformal antenna element 3 does not work, the gain is shown in Fig. 8.

When the conformal antennas 1 and 3 are controlled to work, and the conformal antenna element 2 does not work, the gain is shown in Fig. 9. It can be seen that the beam control of electromagnetic wave can be realized by switching the antenna because the problem of weak signal caused by angle has been improved.

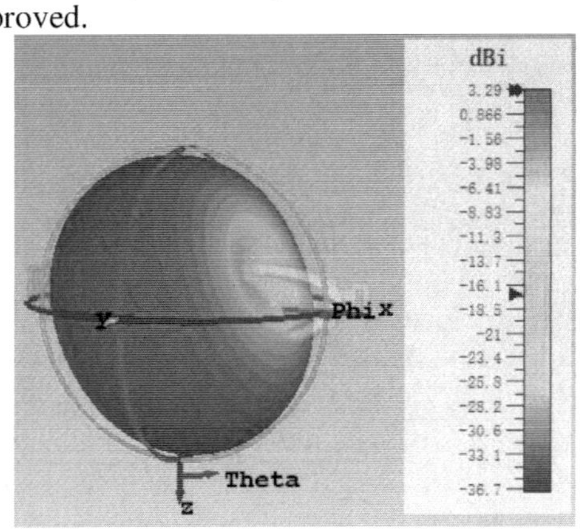

Figure 8. Working gain of conformal antenna element 1,2

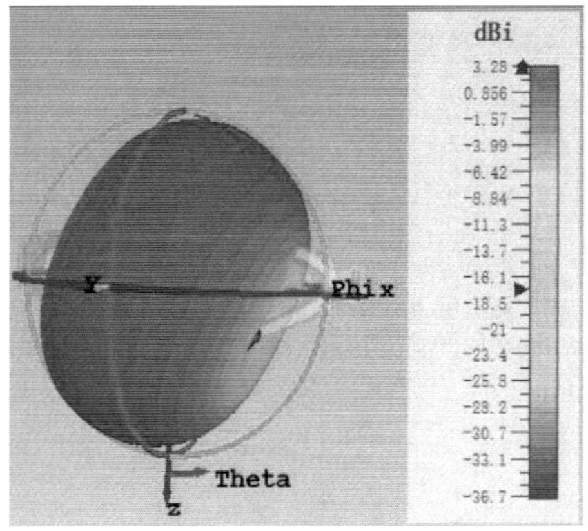

Figure 9. Working gain of conformal antenna elements 1 and 3

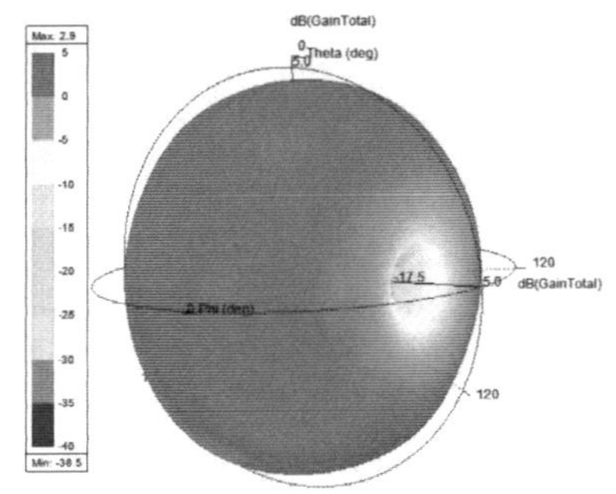

Figure 10. 2.45 GHz electric dipole antenna gain

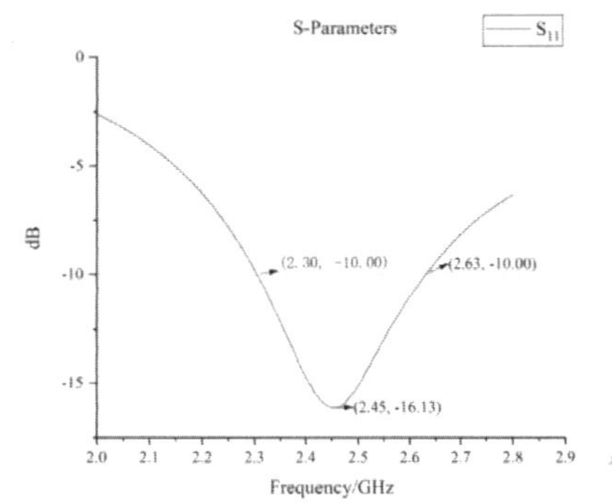

Figure 11. 2.45 GHz electric dipole antenna S_{11}

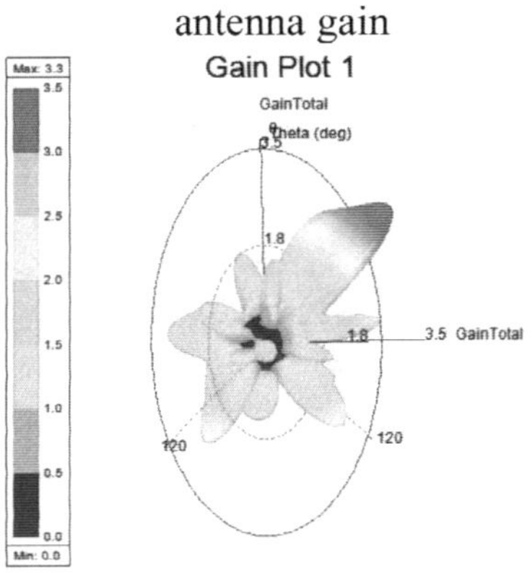

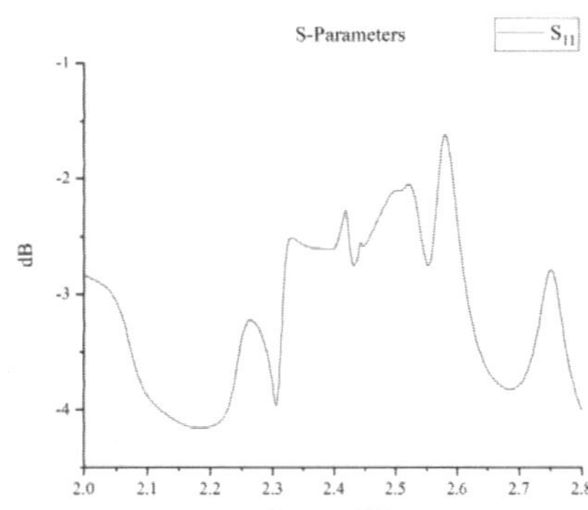

Figure 12. 2.45 GHz electric dipole in underwater environment Antenna gain

Figure 13. 2.45 GHz electric dipole in underwater environment Antenna S_{11}

Since the conformal antenna on the surface of the smart swimming goggles can be affected by the water when it works underwater, in which the gain changes tremendously, leading to greatly weakened signal strength. Figs. 10 and 11 show the simulation of the S11 figure and gain figure of a 2.45 GHz electric dipole antenna working in a normal environment. When it is set in water with a depth of 10 cm, the S11 figure and gain figure are shown in Figs. 12 and 13, the S11 constant, gain deflection angle. Based on this change, taking a certain signal strength index as the reference threshold and time, an effective drowning warning system can be established, realizing the drowning prevention and control of intelligent swimming goggles wearers.

4. Conclusion

A dipole conformal antenna array based on a complex undevelopable surface was presented in this article, which is conformal to the swimming goggles frame. The conformal antenna array realizes the regular change of radiation pattern by controlling the actions of the array element, which is feasible for the simulation results.

In future research, the influence of the human face model on the conformal antenna on the surface of intelligent swimming goggles will be taken into further consideration to obtain more accurate electromagnetic radiation characteristics and then achieve more accurate real-time monitoring of drowning.

Acknowledgments

Project supported by Dongguan Science and Technology Specialist Project (No. 20211800500332); Shenzhen Science and Technology Plan Project (No. KCXFZ20201221173202007); KEY Laboratory of Robotics and Intelligent Equipment of Guangdong Regular Institutions of Higher Education (Grant No.2017KSYS009) and The DGUT Innovation Centre of Robotics and Intelligent Equipment (No. KCYCXPT2017006).

References

[1] L Joseffsson and P Persson 2006 *Conformal Array Antenna Theory and Design* (New York–Wiley-IEEE Press)

[2] Fan X H and Li S N *et al.* 2022 *Conformal multiple input multiple output antenna array for wireless avionics intra-communications system in the aircraft cabin.* International Journal of Rf and Microwave Computer-Aided Engineering

[3] Feng L and Wei L *et al.* 2019 *Airborne Aperture Synthesis Radiometers with Conformal Antenna Arrays.* IEEE International Geoscience and Remote Sensing Symposium (Yokohama) pp 9125-9128

[4] Yang H C and Liu X Y *et al.* 2022 *Design of Broadband Circularly Polarized All-Textile Antenna and Its Conformal Array for Wearable Devices.* IEEE Transactions on Antennas and Propagation Vol. 70 pp209-220

[5] Johnson A D and Nichols M W *et al.* 2020 *Reconfigurable log-periodic dipole array on textile.* Iet Microwaves Antennas & Propagation vol. 14 pp 1797-1794

[6] A Ale 2021 *Compact wideband antenna for wireless capsule endoscopy system.* Applied Physics A 127(4): p 271

[7] Bao Z D and Guo Y X *et al.* 2018 *Conformal Capsule Antenna With Reconfigurable Radiation Pattern for Robust Communications.* IEEE Transactions on Antennas and Propagation vol. 66 pp 3354-3365

[8] Chen X and Wu H *et al.* 2022 *Wideband VHF/UHF Wing Conformal Vertical Polarized Omnidirectional Antenna.* IEEE Antennas and Wireless Propagation Letters vol. 21 pp 953-957

[9] Bai T M and Jiang D *et al.* 2021 *Design of Cylindrical Conformal Array Antenna based on Microstrip Patch Unit.* Applied Computational Electromagnetics Society Journal vol. 36 pp 1008-1014

[10] Le Goff D and Song Y *et al.* 2019 *Low-Cost and Highly Flexible Antenna for 2.4 GHz IoT applications.* USNC-URSI Radio Science Meeting/IEEE International Symposium on Antennas and Propagation (Atlanta, GA) pp 871-872

Heterogeneous communication network planning based on ant colony algorithm in smart grid

Jiangang Lu [1], **Wenjie Zheng** [1], **Ruifeng Zhao** [1], **Kaiwen Zeng** [1], **Shuwen Qiu** [2] **and Ting Yang** [2, 3]

[1] Guangdong Power Grid Limited Liability Company, China.
[2] School of Electrical Automation and Information Engineering, Tianjin University, China.
[3] yangting@tju.edu.cn.

Abstract. With the development of smart grids, highly automated power systems increasingly rely on communication networks, and research on the reliability, safety, and economy of heterogeneous communication networks is particularly important. Traditional network planning methods do not take into account network load problems, resulting in serious network congestion and long delays. Aiming at the above problems, this paper proposes a heterogeneous communication network planning method for a smart grid based on an improved ant colony algorithm. This method improves the original ant colony algorithm by updating the pheromone concentration of the network node in real time and realizes the planning of heterogeneous communication networks in the smart grid. Simulation results show that the proposed method has the advantages of avoiding local network congestion, reducing the average delay, and improving network stability.

1. Introduction

With the continuous advancement of the intelligent construction of the power grid, not only the composition of the access power supply in the power grid has changed, but also the load composition in the power grid is changing [1]. The traditional power grid is unable to adapt to the needs of a large number of distributed source load access. A heterogeneous network provides safe and reliable communication services through a variety of access technologies [2]. The so-called heterogeneity means that two or more communication systems under the same network use different access technologies, or although the same access technology is used, they belong to different network operators. It is of great practical significance to study an effective heterogeneous communication network planning method to achieve resource integration and unified services and accelerate the transmission efficiency of heterogeneous communication data.

2. Related work

Smart grid communication networks have considerable requirements for the real-time and reliability of communication networks [3]. The literature [4] studies a distributed multiservice resource planning algorithm in heterogeneous wireless communication, however, the proposed model is only suitable for solving the problem of maximizing self-utility in heterogeneous communication networks. A multiservice resource planning algorithm based on Markov's decision-making process is studied in the literature [5] that dynamically allocates computing resources to a set of business requirements. However, the problem of wireless resource allocation in wireless networks is not considered. A multiservice resource planning algorithm based on Markov's decision-making process has been studied in the literature [6], but the wireless resource allocation problem in wireless networks is not considered. In order to avoid the problems of the above methods, a heterogeneous communication network planning method based on an improved ant colony algorithm is proposed.

3. Smart grid heterogeneous communication network planning method

3.1. Smart grid heterogeneous communication planning model

Assuming that the smart grid heterogeneous communication planning model includes n terminals and m heterogeneous communication data, it can be described in the following quadruple form:

$$M = (U, V, F, \theta) \tag{1}$$

$$V = \left\{v_1, v_2, \cdots, v_m\right\} \tag{2}$$

where U represents the set of terminals composed of n terminals in the model; V represents the server set composed of m heterogeneous communication data distribution terminal servers in the model; F stands for the objective function of smart grid heterogeneous communication planning; θ represents the planning approach adopted. The specific process is as follows:

1) Refine the number of cores, memory size, and disk space of each terminal server in collection V.

$$v_i = \left\{\lambda_i, \mu_i, \varphi_i\right\}, i = 1, 2, \cdots m \tag{3}$$

where i represents the terminal server number; λ_i, μ_i, and φ_i represent the number of terminal server cores, memory size, and disk space respectively.

2) Assuming that T is a collection of heterogeneous communication data of a smart grid, representing n independent multi-source heterogeneous communication data planning tasks of a terminal, the calculation formula is as follows:

$$T = \left\{t_1, t_2, t_n\right\} \tag{4}$$

3) It is supposed that s_{ij} represents heterogeneous communication planning task j time to perform planning on Terminal Server i; $S_{n\times m} = (s_{ij})$ represents the smart grid heterogeneous communication planning task time moment, j = 1,2,... n。

4) Smart grid heterogeneous communication planning matrix $E_n = (e_i)$ represents the terminal server e_i performing heterogeneous communication intelligent planning tasks j.

5) The smart grid heterogeneous communication usage matrix $X_{n\times m} = (x_{je_i})$ is generated according to E_n, representing that during the execution of the planning task, the terminal server is used by the heterogeneous communication planning task j, x_{je_i} =1.

The above-mentioned heterogeneous communication planning model of the smart grid is established to meet the following assumptions [7]:

Condition 1: The performance of the terminal server can meet any heterogeneous communication planning requirements;

Condition 2: A heterogeneous communication planning task can only be assigned to one terminal server to perform;

According to the above calculations, the time required to complete the planning of heterogeneous communication data on the terminal server v_i can be obtained, and the calculation formula is as follows:

$$T_j = \max_{1 \leqslant i \leqslant n}\left\{ s_{ij} \cdot x_{je_i} \right\} \tag{5}$$

m heterogeneous communication data is available to allocate the total time required for the terminal server to complete the planning task:

$$makespan = \sum_{j=1}^{m} \max_{1 \leqslant i \leqslant n}\left\{ s_{ij} \cdot x_{je_i} \right\} \tag{6}$$

3.2. Heterogeneous communication planning of smart grid based on ant colony algorithm

1) Ant colony algorithm

It is supposed N represents the number of nodes contained in the communication network; i represents the number of ants placed in the network; τ(s,u) represents the pheromone concentration value between any two adjacent nodes in the network, s and u; η_{su} represents the heterogeneous communication data planning heuristic, in the process of ant foraging, k is set to be currently located in nodes, and $Z_k(s)$ represents the set of nodes in the network that have not been accessed by ant k. It is needed to calculate the probability that ant k next hop chooses to access node u according to the following equation (7) [8].

$$P_k(s,u) = \begin{cases} \arg \max\limits_{(s,v)\in E, v\in Z_k(s)} \left\{ [\tau(s,v)]^a \left[\eta_{su}(s,v)\right]^b \right\}, & q \leqslant q_0 \\ J, & \text{else} \end{cases} \tag{7}$$

$$J = \begin{cases} \dfrac{[\tau(s,u)]^a \left[\eta_{su}(s,u)\right]^b}{\sum\limits_{(s,v)\in E, v\in Z_k(s)} [\tau(s,v)]^a \left[\eta_{su}(s,v)\right]^b}, u \in Z_k \\ 0, else \end{cases} \tag{8}$$

where q∈[0, 1] is a random number; q_0 represents the pre-set threshold; a and b represent the pheromone concentration weights and heuristic factor weights of intelligent programming of heterogeneous data, respectively; (s,υ) ∈ E represent heterogeneous communication links; Z_k represents the set of nodes in the network that will be accessed by ant k.

Balancing the energy consumption of network nodes by calculating the remaining energy of the nodes, the calculation formula is as follows

$$\eta_{su} = \frac{e_u}{d_{su}} \tag{9}$$

where e_u represents the residual energy information of node u; d_{su} represents the communication distance between nodes s and u.

Journal of Physics: Conference Series **2384** (2022) 012051 doi:10.1088/1742-6596/2384/1/012051

The pheromone concentration $\tau(s,u)_{new}$ of the network node is updated in real time using the following equation (10).

$$\tau(s,u)_{new} \leftarrow (1-\delta)\cdot\tau(s,u)+\delta \left(\Delta\tau(s,u)+\omega\cdot \max_{(s,v)\in E, v\in Z_k(s)} \tau(s,u)\right) \tag{10}$$

where δ represents the nodal pheromone concentration volatilization coefficient; $\Delta\tau(s,u)$ indicates the concentration of pheromones released by ant k after visiting the communication link (s, u), which is calculated as follows：

$$\Delta\tau(s,u)=\begin{cases} Q/L_k & (s,u)\in T \\ 0 & (s,u)\notin T \end{cases} \tag{11}$$

where Q is constant and represents the total pheromone concentration released by the ant after iterating through all links; L_k represents the number of communication links; T represents the search tree of all communication links.

2) The spread of interest in smart grid heterogeneous communication planning

Based on the above model, the data planning target node periodically broadcasts the smart grid heterogeneous communication planning interest message to the whole network, the specific calculation formula is as follows:

$$interest=(type, interval, rect, timestamp, expiresat, energy) \tag{12}$$

When the target node n_j receives a smart grid heterogeneous communication planning interest message from neighbor node n_i, node n_i is added to its own gradient table, and the remaining energy information of node n_i is obtained, the network node pheromone concentration is updated according to the above equation (10), and the interest message is broadcast to each neighbor node until the interest message floods to the planning task execution source node n_s.

4. The simulation test and result analysis

The simulation network used in this section is a power distribution communication network. It can be seen from the experimental results of Figure 1 that when performing the same number of heterogeneous communication data planning tasks, the smart grid heterogeneous communication planning method based on the improved ant colony algorithm has better load sharing energy than the traditional method, which greatly reduces the maximum bandwidth of the communication link in the planning process. Figure 2 shows three different ways to plan the latency of heterogeneous communication packets separately. The proposed method has the shortest average delay in the process of performing the heterogeneous communication planning of the smart grid.

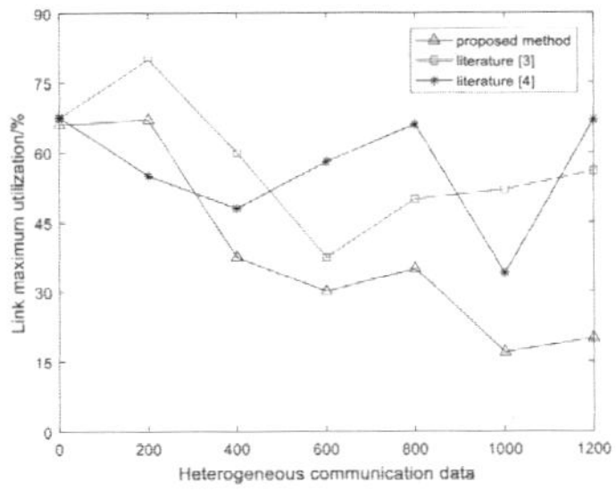

Figure 1 Comparison of maximum link utilization

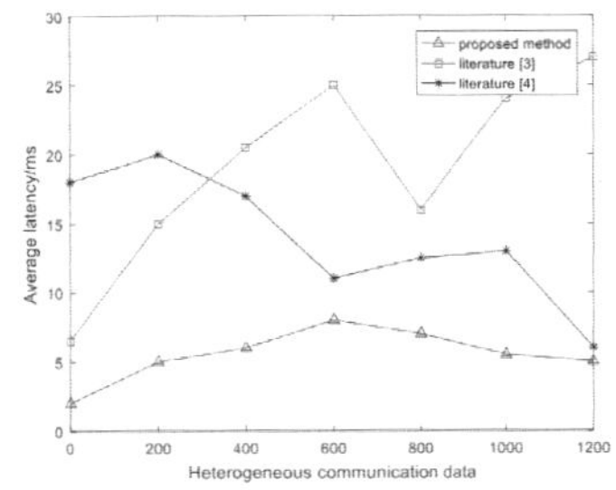

Figure 2 Comparison of average delays

5. Conclusion

The heterogeneous communication network planning method based on the ant colony algorithm in the smart grid proposed in the study makes full use of the adaptability and dynamic optimization advantages of the ant colony algorithm on the basis of the directional diffusion routing protocol. This algorithm better balances the energy consumption of nodes, which avoids premature node death, realizes the intelligent scheduling of heterogeneous communication networks and provides an effective reference for future network resource sharing and resource allocation.

Acknowledgments

The work is funded by China Southern Power Grid Corporation under Grant (GDKJXM20210063).

References

[1] Xu T H, Lu Z X, Qiao Y and An J 2019 *J.* High-proportion renewable energy power supply planning for multi-type flexible resource coordination of source and load storage. *Global Energy Internet* **(1)** p 8.

[2] Merwaday A and Guvenc I 2015 *C.* UAV assisted heterogeneous networks for public safety communications. *Wireless Communications & Networking Conference Workshops. IEEE.*

[3] Wang D P and Zhang C 2017 *J.* Reliability analysis of relay protection of intelligent substation.*Modern Manufacturing*. 30 pp 8-9.

[4] Ismail M and Zhuang W 2012 *J.* A distributed multi-service resource allocation algorithm in heterogeneous wireless access medium. *IEEE Journal on Selected Areas in Communications.* 30(2) pp 425-432.

[5] Oddi G, Panfili M, Pietrabissa A, Zuccaro L and Suraci V 2013 *C.* A Resource Allocation Algorithm of Multi-cloud Resources Based on Markov Decision Process. *Proceedings of the 2013 IEEE International Conference on Cloud Computing Technology and Science*.

[6] Si P, Zhang Q, Yu F R and Zhang Y 2014 *J.* Qos-aware dynamic resource management in heterogeneous mobile cloud computing networks. *Communications* **11(5)** pp 144-159.

[7] Li H H, Fu Y W, Zhan Z H and Li J J 2015 *J.* Renumber strategy enhanced particle swarm optimization for cloud computing resource scheduling. *Evolutionary Computation*.

[8] Zhang J, Gao Q and Xinhe X U 2000 *J.*A self-adaptive ant colony algorithm. *Control Theory & Applications* **17(1)**.

Distributed power supply and energy storage configuration method in the active power distribution network

Lu Zhao, Zhaozhao Peng *, Wenbin Ni

Power China Hubei Electric Engineering Co. Ltd, Wuhan Hubei, 430040, China

*Corresponding author's e-mail: pengzhaozhaosj@powerchina-hb.com (Zhaozhao Peng)

Abstract. Distributed energy access and energy storage configuration are important links in the design of an active distribution network, and research on their design methods is essential to support the scientific development of an active distribution network. Therefore, on the basis of full research and demonstration, the content of the scheme design of the active distribution network and the technical principle of the overall design of the active distribution network are first determined, so that they can play the role of macro instruction in the scheme design. In addition, this paper deeply studies the capacity allocation of distributed generation and energy storage systems in the active distribution network, puts forward a general method for optimal allocation of distributed generation access and energy storage capacity, and provides design information and technical support for the active distribution network method.

1. Introduction

As the main means of distributed energy utilization, the distribution network with distributed power is also called an active distribution network, which plays an important role in alleviating energy crises, optimizing energy structure, promoting energy conservation and emission reduction, and peak shaving. It is very important to eliminate the system load difference and improve the electrical quality. As an effective method of using distributed power to access the power system, a microgrid can realize the local consumption and plug-and-play of large-scale and multi-type new energy, so it is gradually becoming an important part of an active distribution network [1]. When a large number of microgrids exist in the distribution network and develop into a multi-microgrids system (microgrids group), the connection scheme between microgrids, that is, an active distribution network based on microgrids, can be reasonably constructed to construct a new type of an active intelligent distribution network. With the concept of an active distribution network and the extensive development of empirical projects, the optimal configuration and adjustment of an active distribution network system have become the research focus of experts and scholars at home and abroad. At present, domestic and foreign scholars have done a lot of research on modeling, simulation, and optimal control of active distribution network systems [2]. However, the research on system coordinated development and overall scheme design for engineering needs to be improved. In order to further promote the local consumption of clean energy and the healthy development of active distribution networks, it is necessary to study and establish the scheme of an active distribution network system according to the industrial development trend. Therefore, this paper will make full use of the existing research results, refer to the relevant standard design, summarize and improve the general rules of active distribution network design, and provide technical support for the active distribution network scheme.

2. Capacity optimization of active distribution network

2.1 Active distribution network capacity optimization allocation method

It can be seen from the content of the active distribution network scheme design that the type and capacity of each distributed generation and energy storage system in the active distribution network must be determined when designing the scheme, which is also the basis for improving safe and

 2384 (2022) 012052 doi:10.1088/1742-6596/2384/1/012052

reliable economic operation and clean power efficiency of distribution network and large-scale power grid.

(1) Processing the data of scenery resources. Wind speed, light intensity, and air temperature are important for the output of wind turbines and solar power generation. The capacity optimal allocation method introduced in this paper is based on the hourly data of wind and solar energy resources. However, in the project planning and design stage, it is difficult to obtain historical data on wind and solar energy resources at the installation site. (2) Synthesis of wind speed data. Many studies show that the wind speed data is closest to the Weibull distribution of two parameters. This research is also synthesized on this basis, and the corresponding hourly wind speed data are generated by using the inverse function transformation method and random number generation method. (3) Synthesis of light intensity data. Influenced by the astronomical relationship between the sun and the earth, the light intensity changes periodically by year. Based on this, in this study, the monthly average sunshine coefficient is calculated according to the monthly average luminosity value, and the annual hourly sunshine coefficient related to the monthly average sunshine coefficient is generated by the empirical formula.

2.2 Optimization objectives and constraints

(1) Economics. For active distribution network operators, the economic problem is the first issue. This paper comprehensively studies the initial investment, operation, maintenance cost, and equipment renewal cost of each decentralized power supply in the whole operation cycle. At the same time, studying the existing distributed generation operation policies, including government subsidies, is also the input condition of economic goals [3]. The present value of the total cost of an active distribution network is shown in Formula 1:

$$C_{total} = (\sum_{j=1}^{K} N_{DGj} \times P_{DGj} + C_g - C_s)/[F_{RF}(i, N)] \tag{1}$$

1: NDGj is the capacity of the jth distributed power supply; PDGj is the present value of the cost in the whole life cycle of the jth distributed generation; Cg is the cost of purchasing electricity from the distribution network; Cs subsidizes the power generation of active distribution network; K is the number of distributed power sources included in the active distribution network; FRF(i,N) is the capital recovery coefficient, where i is the annual interest rate and n is the engineering life.

(2) Reliability and utilization rate of renewable energy. In order to realize the advantages of an active distribution network, it is very important to ensure the high-power supply reliability of users. However, the annual load loss probability PALL of the active distribution system is used to explain the power supply reliability of the active distribution network, as shown in Formula 2:

$$P_{ALL} = [\sum_{t=1}^{T} P_{loss}(t) \times \Delta t]/[\sum_{t=1}^{T} P_L(t) \times \Delta t] \tag{2}$$

2: Ploss(t) is the cut-off load power, and PL(t) is the online load power.

The utilization rate of renewable energy is an index to reflect whether the distributed capacity design of the active distribution network is reasonable, and it is also the object of optimization in this survey. The annual renewable energy utilization rate of the active distribution network is shown in Formula 3:

$$P_{DGu} = 1 - \frac{\sum_{t=1}^{T} P_{waste}(t) \times \Delta t]}{\sum_{t=1}^{T} [P_{pv}(t) + P_{wt}(t)] \times \Delta t} \tag{3}$$

Where: Pwaste(t) reduces the output of photovoltaic or wind turbines to meet the safe operation of an active distribution network; Ppv(t) is the available output of PV; Pwt(t) is the available output of wind power.

(3) Distributed generation capacity and other constraints. As wind turbines, solar modules, and batteries are often affected by the limitations of installation sites and users' own needs during installation, this paper takes the upper and lower limits of distributed generation capacity as input conditions to participate in the optimization process. In order to ensure the safe and reliable operation

of the active distribution network, this paper studies the charging and discharging power limit, charging state limit, and cycle number limit of the energy storage system in operation [4].

3. Scheme design

3.1 Content of scheme design

On the basis of the existing empirical project research of active distribution networks and referring to the relevant design specifications of power systems, this paper summarizes the relevant contents of active distribution network design, and draws the following differences from the planning and design of traditional distribution network:

(1) Systematic once. Determination of the type, installed capacity, long-term planned installed capacity, load type, long-term predicted capacity, energy storage device type, installed capacity, and long-term planned capacity of distributed generation and inverter in the active distribution network. (2) Protection and safety of automatic device of system relay. According to the design scheme of the primary active distribution network of the system, the composition principle scheme of the system relay protection and safety automatic device is put forward. (3) Automation of system scheduling. According to the needs of users, the necessity of establishing an energy management system for an active distribution network is analyzed.

According to various access modes of distributed power, the active distribution network can be divided into four modes, including low-voltage user microgrid, distribution branch microgrid, feeder microgrid, and substation microgrid [5]. (1) Low-voltage customer microgrid mode and distribution branch microgrid mode. In the low-voltage user microgrid mode, the distributed power is directly connected to the 0.4kV side of the distribution transformer and is partially consumed by the 0.4kV low-voltage load. After the system is connected, the grid-connected circuit breaker and reverse power protection are installed at the grid-connected points. This mode is suitable for 0.4kV low voltage users with appropriate scale and high-power reliability requirements. The system consists of a fan, solar power generation, energy storage, micro turbine, fuel cell, and household cluster load. The capacity is as follows, usually less than 2MW. Low-voltage user microgrid mode is shown in Figure 1:

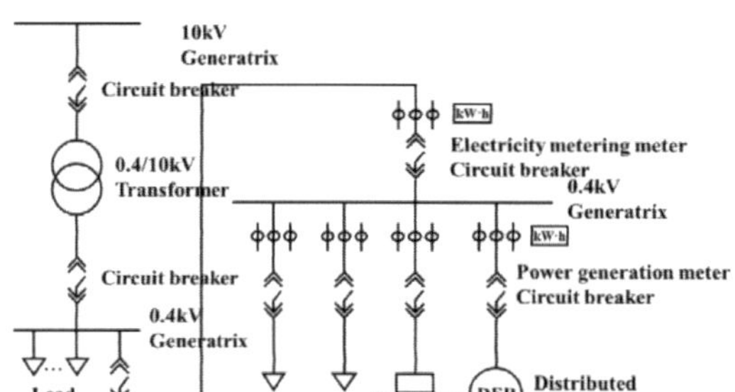

Figure 1 Low-voltage user microgrid mode

(1) Distribution microgrid mode. Branch microgrid mode is a mode of connecting distributed power to the distribution transformer or 0.4kV side of the branch line, which will be consumed locally under the regional load of the branch line and finally connected to the distribution trunk line. It consists of 10-110kV distribution branch lines and 0.4kV users, and the capacity is usually 2-5MW. The distribution microgrid mode is shown in Figure 2:

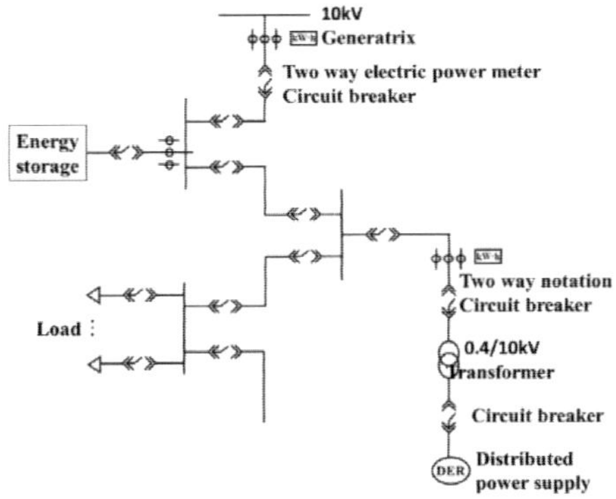

Figure 2 Distribution branch microgrid mode

3.2 Basic technical principles of scheme design

According to relevant standards, the basic technical principles of the scheme design are summarized as follows: (1) Voltage level. The selection of voltage level of the active distribution network should be based on the distributed power and load capacity of the active distribution network and the current carrying capacity of grid-connected lines, and follow the principles of safety, flexibility, and economy. (2) Cross section of grid-connected tie wire. According to the required exchange capacity, grid-connected voltage, and the influence of the active distribution network operation control strategy, the conductor cross-sectional area of the grid-connected voltage of the active distribution network is selected. (3) Switchgear. The grid-connected switch is located at the point of common coupling between the active distribution grid and the large-scale grid. Static switches must be able to automatically switch the active distribution network to island operation when large-scale faults, power quality, and other events occur. (4) Relay protection and automatic devices. The configuration of relay protection and safety automatic devices in an active distribution network should meet the requirements of reliability, selectivity, sensitivity, and quickness, and its technical conditions should conform to the current national standards GB/T 14285—2006 Technical Specification for Relay Protection and Safety Automatic Devices, DL/T 584—2017 Operation Setting Specification for 3 kV ~ 110 kV Power Network Relay Protection Devices and GB 50054—2011 Low Voltage. The protection device of distributed power grid connection is shown in Figure 3:

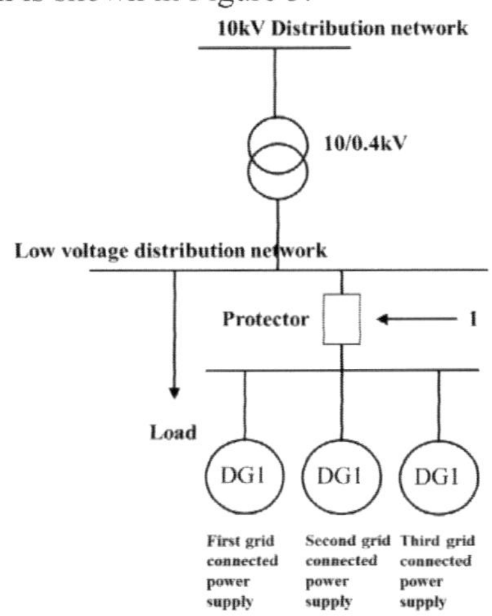

Figure 3 Protection device of distributed power grid connection

(5) Operation control of active distribution network and switching from grid to grid. The active distribution network system needs to be equipped with frequency and voltage control devices so that

the power generation system can be controlled when the voltage and frequency of conductors in isolated islands are abnormal. (6) Online monitoring of power quality. If the currently active distribution network system needs to be connected to the distribution network, the online power quality monitoring device must be installed at the public access point, and the relevant data should be sent to the high-level operation management department.

3.3 Comparison of microgrid mode and access mode between feeder level and substation level

The feeder microgrid consists of 10-110kV distribution trunk lines and their branch lines and 0.4kV users and is finally connected to the distribution bus of the substation through circuit breakers, usually with a capacity of 5-10MW. The substation microgrid consists of several feeders provided by the substation, and its capacity usually exceeds 10MW.

Access to distributed power has a great influence on the distribution network. If different locations and capacities have different effects, when connecting distributed power, the principle of "safety, reliability, resource-saving, and efficient operation" should be followed. Secondly, it is necessary to comprehensively study factors such as short-term and long-term installed capacity to scale, distribution network status, and planning, and finally discuss their impacts. Access modes can be divided into centralized access and distributed access. The advantages and disadvantages of centralized access and decentralized access are shown in Table 1:

Table 1 Advantages and disadvantages of centralized access and decentralized access

Project	Centralized access	Decentralized access
electric energy	Influence of large capacity on power quality Larger; Single point centralized access to voltage Support is weaker than distributed multipoint access.	Influence of small capacity on power quality Smaller; Then, the closer the entry point is to the end node, the more obvious the voltage support effect is.
economy	Centralized collection of dispersed photovoltaics in First, the cable investment is large; Network loss Larger; Relating to distribution automation Coordinate, operate, and manage costs. Higher; To achieve the same harmonic Level, higher requirements for the inverter.	Local access and cable investment are small; The network loss is small; the operating management cost Low; To achieve the same harmonic Level, with less requirement on the inverter, The investment of power producers is small.
run	High requirements for relay protection; need Coordinated control with distribution automation	The relay protection is simple; Does not affect Operation of distribution automation system
manage	Facilitate the management of power producers; non-existent Access caused by property rights problems Difficult problem	Inconvenient for the management of power producers; with regard to Low voltage distribution transformer and distributed power assets Different owners, may It is difficult to access low voltage.

4. Capacity configuration of energy storage system

4.1 Constant volume principle of energy storage system

Considering typical application schemes, such as peak shaving and valley filling, tracking planned output, emergency support power supply, demand response, etc., the capacity allocation of the energy storage system in an active distribution network must be configured according to the requirements of comprehensive application programs. Engineering is usually used in the planning and design of software for distributed generation development, such as HOMER and Hybrid2 software of the Renewable Energy Research Institute of the U.S. Department of Energy.

4.2 Energy storage capacity allocation based on wind power, solar power generation and load forecasting

For an active distribution network that can access both wind and sunlight, in the grid-connected system, wind turbines, solar generators, local loads, gas turbines, energy storage systems, and power compensation equipment (system connection) are directly or indirectly connected to the low-voltage bus B0 of the step-up transformer T (system connection) after startup, and then connected through the high-voltage bus B1 integrated into the system. Therefore, the influence of three factors on energy storage configuration must be considered at the same time. The typical structure of an active power distribution system connected with wind power and photovoltaic is shown in Figure 4:

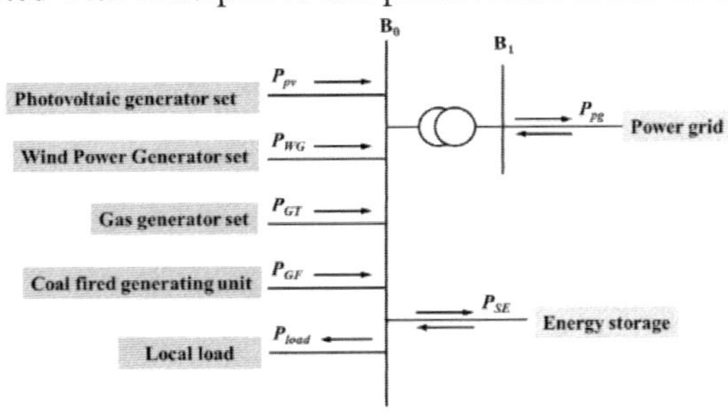

Figure 4 Typical structure of active power distribution system connected with wind power and photovoltaic

Firstly, the quick response start-up time tr of the active distribution network system is counted, and the safety margin coefficient η of the calculated capacity can be calculated as the unbalanced power change ΔP (Δt≤tr) of the minimum energy storage capacity composition W of the microgrid system. The minimum capacity of energy storage configuration of an active distribution network including wind power stations and solar power stations can be shown in Formula 4:

$$w = \max \sum_{i,j=1}^{t_r} [(P_{wti} - P_{wtj})(t_j - t_i) + (P_{pvi} - P_{pvj})(t_j - t_i) \times \eta \qquad (4)$$

In the formula 4: TJ-ti ≤ δ t, j > I; Pwti and Pwtj are the random output of wind power generation at the time I and J in δ T period, respectively; Ppvi and Ppvj are the random output of photovoltaic power generation at the time I and J in δ T period respectively.

5. Conclusion

With the development and utilization of distributed new energy, the benign coordination and interaction of "source network and load storage" makes the distribution network bound to be efficiently integrated with distributed energy, energy storage, and adjustable load. From the technical level, the principles and general methods of active distribution network design are put forward, and the contents of active distribution network design, the design principles of distributed generation, and the technical principles of the overall design of active distribution network are determined.

Common methods of distributed power access and optimal allocation of energy storage capacity provide technical support for the design of an active distribution network. From the viewpoint of economy, reliability, utilization of renewable energy, capacity limitation of distributed power and reliable utilization of energy, the optimal composition of distributed power and energy storage capacity is the foundation of an active distribution network.

Compared with the traditional distribution network, the active distribution network has the advantages of a strong network structure, strong expansion flexibility, and easy realization of electricity marketization. Therefore, the research on new topology and related key technologies of the active distribution network is the future research direction.

Acknowledgments
The authors would like to show sincere thanks to those techniques who have contributed to this research.

References
[1] Martin-Ramos, J. A., Pernía, A. M., Díaz, J., Nuño, F., & Martínez, J. A. (2008). Power supply for a high-voltage application. *IEEE Transactions on power electronics*, 23(4), 1608-1619..
[2] Heydt, G. T., Liu, C. C., Phadke, A. G., & Vittal, V. (2001). Solution for the crisis in electric power supply. *IEEE Computer Applications in Power,* 14(3), 22-30.
[3] Gutnik, V., & Chandrakasan, A. P. (1997). Embedded power supply for low-power DSP. *IEEE Transactions on Very Large Scale Integration (VLSI) Systems*, 5(4), 425-435.
[4] Akhmatov, V., Knudsen, H., & Nielsen, A. H. (2000). Advanced simulation of windmills in the electric power supply. *International Journal of Electrical Power & Energy Systems,* 22(6), 421-434.
[5] Li, C., Zhang, S., Zhang, J., Qi, J., Li, J., Guo, Q., & You, H. (2018). Method for the energy storage configuration of wind power plants with energy storage systems used for black-start. *Energies,* 11(12), 3394.

Distribution Network-Oriented Distributed Energy Storage Optimization Planning and Economic Quantitative Evaluation

Qing Gong, Lu Zhao[*]**, Li Fan**

Power China Hubei Electric Engineering Co. Ltd, Wuhan Hubei, 430040, China

Corresponding author's e-mail: zhaolusj@powerchina-hb.com (Lu Zhao)

Abstract. Distributed energy storage can provide node support for energy consumption, but there are still some shortcomings in the formulation of technical standards and overall development planning. Due to the current cost of electric energy storage, it is difficult to give full play to the advantages of energy storage application under the current electricity price mechanism and favorable policy conditions. From the perspective of external support and technical methods for electric energy storage, this paper studies the typical application mode and comprehensive quantitative evaluation of distributed energy storage in the distribution network and takes the actual operation scheme as a verification case.

1. Introduction

There are still some shortcomings in the formulation of technical standards and overall development planning of distributed energy storage. Due to the current cost of energy storage, it is difficult to balance the advantages of energy storage applications under the current electricity price mechanism and supporting policies.

However, in areas where distributed power sources are connected to the power system on a large scale, the utilization rate of new energy generation is greatly improved by setting distributed energy storage devices, and the carrying capacity of the power grid side is also greatly improved, which is convenient for better control [1].

The country has gradually realized the importance of intermittent fluctuating power in distributed generation and began to pay attention to the development plan of power storage technology in the 13th Five-Year Plan. In 2015, power storage technology was officially listed in the top 100 projects of the plan. In the direction of scientific and technological research and development, how to rationally distribute distributed energy storage through the best plan to achieve better social and economic benefits is an urgent problem to be studied and solved in the power industry.

2. China's energy storage development status and typical scenario operation mode

2.1 Definition and development status of energy storage in China

According to the two aspects of transmission and consumption in the power system, energy storage can also be divided into three categories: source, user, and grid. What is different is the investment owner, including the location where the energy storage is accessed, the power supply owner, the power grid company, and various power customers. From the perspective of technology distribution, pumping energy storage accounts for more than 95% of China's energy storage scale. Therefore, it is still the main energy storage mode in China, while electrochemical energy storage accounts for only about 3.4% [2]. From the perspective of development speed, the scale of electrochemical energy storage in China is gradually increasing, from 40.7 MW in 2011 to 1072.7 MW in 2018. The types of electrochemical energy storage are mainly dominated by lithium-ion batteries, which account for more than 70% of capacity. From China's perspective of energy storage application solutions, almost half of energy storage is suitable for the current market. At present, it is scattered in more than 30 provinces and cities, among which the cumulative installed capacity in Jiangsu is 327.8MW, accounting for 30.1%.

2.2 Typical application mode based on the power grid

(1) Provide auxiliary service modes such as peak shaving and valley filling directly. In recent years, in many places, the power grids show different levels of peak pressure in summer peak hours, and energy storage is one of the few peaks that can flexibly adjust the charging and discharging state and cut off the power. (2) Improve the reliability mode of the power supply. By using distributed energy storage system to select the location and time of charging and discharging, many problems related to distribution lines can be reduced. On the other hand, energy storage can be used as a backup power supply for quick start-up in case of emergency to ensure the power supply of the regional power grid. (3) Delay the upgrade and expansion mode of distribution equipment. Due to the constraints of land use and the environment, the land resources of transmission line corridors and substations are increasingly insufficient, which makes it difficult for local areas to meet the demand of load growth in time. With the reduction of the cost of energy storage components, energy storage equipment can gradually replace or delay the investment in power grid upgrading. (4) Support the grid-connected mode of high-permeability distributed generation. In many scenarios, grid-side energy storage guarantees transmission and distribution services. The problems caused by large-scale centralized grid connection of distributed photovoltaics can be effectively solved by installing energy storage on the grid side and combining it with measures such as grid transformation and power restriction [3].

2.3 based on the typical application mode of users

By applying distributed energy storage in response to the demand of electricity price, small-capacity electricity storage and large-scale electricity generation can be realized, and weak electricity can be stored in the cold, and the electricity can be transmitted and stored at the peak, thus realizing cost reduction. It is arranged at the load end to realize voltage support corresponding to the load demand. In industrial and commercial situations, reasonable capacity storage can reduce the power loss of important users when power is cut off due to sudden system failure or overhaul and maintenance [4]. The "operation-failure" model is shown in Figure 1:

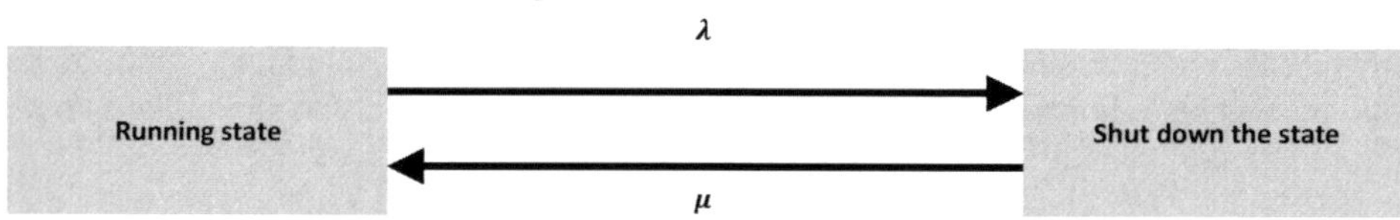

Figure 1 "Operation-failure" model

As an emergency power supply, it is necessary to ensure the power quality of products required by high-precision processes when they are produced, aiming at the problems of voltage imbalance and voltage drop in the production process. Through the demand analysis of electricity price, distributed energy storage can be applied to realize small capacity power storage and large power generation, and power transmission and power storage can be carried out at peak times, thus realizing cost reduction [5].

3. Comprehensive evaluation method of distributed energy storage life cycle

3.1 Cost composition of the whole life cycle of electric energy storage

(1) the initial investment cost of the system. The cost of investment in the first stage of energy storage includes: equipment cost, project construction cost, grid connection cost, initial cost, etc., which can be expressed by the product of energy storage capacity and unit cost, as shown in Formula 1:

$$C_g = \sum_{i=1}^{n} C_{kw_i} P_i \tag{1}$$

Where Cg is the initial investment cost of energy storage, Pi is the capacity of energy storage I, the unit installed capacity cost of distributed photovoltaic or energy storage I, and n is the total amount of energy storage.C_{kw_i}

(2) Operation and maintenance management costs. Operation management fee mainly refers to the equipment maintenance fee, labor fee, and other expenses in the project operation, usually calculated according to a certain proportion of the initial investment amount, as shown in Formula 2:

$$C_{om} = \sum_{i=1}^{n} C_{kw_i} P_i K_{om_i} \tag{2}$$

Where: Com is the maintenance cost; Is the ratio coefficient of annual maintenance cost of energy storage I to investment cost.K_{om_i}

(3) Energy storage charging cost. No matter whether you buy electricity from distributed solar power generation or a large-scale power grid, charging requires a certain cost, as shown in Formula 3:

$$C_b = \sum_{j=1}^{k} EI_j \times Prb_j \tag{3}$$

In formula 3: Cb is the charging cost, EIj is the charging amount of the accumulator J, k≤n, and Prbj is the charging unit price of the accumulator J.

3.2 Income composition of the whole life cycle of electric energy storage

(1) Discharge subsidy income. At present, the potential subsidies available for energy storage include electricity subsidies and investment subsidies. The product of available installed capacity and unit capacity subsidy standard for investment in subsidy income is expressed, and the product of available power generation and unit capacity subsidy standard for electricity in subsidy income is expressed, as shown in Formula 4:

$$B_t = \sum_{t=1}^{8760} \sum_{i=1}^{n} EG_{i,t} \cdot PG_{i,t} + \sum_{i=1}^{n} CaG_i \cdot CaPG_i \tag{4}$$

Where: Bt is the subsidy income of power generation; EGi,t is the actual power generation of energy storage I at time T; PGi,t is the electricity subsidy standard; CaGi is the installed capacity of energy storage I; CaPGi is the capacity subsidy standard of energy storage I; For those who have annual power generation data directly, they can be used directly, instead of accumulating in 8760 hours.

(2) Income from energy storage and release. Electrochemical energy storage and discharge can obtain some benefits, but there is charge and discharge efficiency, which needs to be taken into account, as shown in Formula 5:

$$B_s = \sum_{j=1}^{k} EI_j \times Eff_j \times Prb_j \tag{5}$$

Where: Bs is the energy storage discharge income, Effj is the charging and discharging efficiency of energy storage J, that is, the total discharge amount/total charge amount, and Prbj is the discharge price of energy storage J. The storage and release of electrochemical energy have some advantages, but charging and discharging efficiency should be considered. The discharge curves in many typical periods can't be connected to an 8760-hour curve. We can now consider the scene reduction technology to find out the key curve and calculate the annual discharge amount. The theory of scene reduction technology is as follows: a set of scenes that can show all the features of scenes is the basis of scene reduction. However, in power grid planning, the data often obtained are historical data in the actual operation scene, so the scene set at this time can be composed of all historical scenes on a certain time scale.

The mathematical meaning of scene reduction can be expressed as follows: assuming that the probability of the scene and the scene are sum, respectively, and m < n, the functional of scene reduction is shown in Formula 6:$\{\omega_1, \omega_2, K, \omega_n\}\{\omega_1, \omega_2, K, \omega_m\}\{p_1, p_2, K, p_n\}\{q_1, q_2, K, q_m\}$

$$f = \min\{\sum_{i=1}^{n} \sum_{j=1}^{m} c(\omega_i, \omega_j)\eta_{ij} : \eta_{ij} \geq 0, \sum_{j=1}^{m} \eta_{ij} = q_j, \sum_{j=1}^{m} \eta_{ij} = p_j, \forall i, j\} \tag{6}$$

At present, the synchronous substitution elimination method is widely used. Before introducing the synchronous substitution elimination method in detail, several concepts included in the algorithm are introduced first:

(1) Nr represents the data dimension contained in the scene;

(2) Ns represents the total number of original scenes;

(3) Scenario ωi is defined as a sequence, as shown in Formula 7;

$$\omega_i = \{\lambda_i^1, K, \lambda_i^k, K, \lambda_i^{n_r}\}, i = 1, \mathrm{L}, \mathrm{n_s};\ \ k = 1, \mathrm{L}, \mathrm{n_r} \tag{7}$$

Where: is the numerical value of the k-th dimension data of the ith scene.λ_i^k

(4) Pi is the probability of occurrence of the scene, and it satisfies Formula 8:ω_i

$$\sum_{i=1}^{n_s} p_i = 1 \tag{8}$$

(5) The distance between scene Ω i and scene Ω j is shown in Formula 9:

$$\mathrm{d}(\omega_i, \omega_j) = \left(\sum_{k=1}^{n_r} (\lambda_i^k - \lambda_j^k)^2\right)^{\frac{1}{2}} \tag{9}$$

4. Method and planning of distributed generation and energy storage configuration

4.1 Constraints that affect the access to distributed generation

Large-scale access to distributed generation will lead to constant changes in the peak value of power generation and power consumption of the distribution network, which will lead to the difficulty of voltage control or overload of transmission lines. The tendency of distributed power generation is reversed from medium and low voltage distribution networks to medium and high voltage distribution networks. However, the power output and power factor of distributed photovoltaic power supply are difficult to predict accurately due to weather, which makes it more difficult to control the line voltage quality of the distribution network.

(1) Exceeding the limit of voltage problem. The distribution voltage exceeds the limit mainly due to the voltage rise of the parallel point caused by the power flow from the power supply side. For example, in the daytime, when the sunshine conditions are the best, and the solar power output is the largest, and when the industrial park is shut down and resting, it is difficult for the peak value of the power generation curve to be positively correlated with the power load curve, especially on weekends and holidays. Moreover, the centralized system connection of several solar power projects scattered in the same area also brings difficulties to the operation and management of the power grid. On the other hand, the poverty alleviation projects of solar power generation are mostly located in remote areas. When there is enough solar power, when the number of system connection points increases or exceeds 20% of the reference value by introducing solar power into the Internet, the solar power inverter will be forced to stop. On the other hand, the randomness of the solar power generation curve will reduce the power quality near the system connection point.

(2) The problem of current overload means that the countercurrent generated by distributed generation exceeds the operating limit of power transmission and distribution equipment. With the increase in the scale of distributed power generation equipment, reverse power generation will be more frequent.

4.2 Demand and scheme of large-scale distributed generation connected to the distribution network for energy storage

Since the 12th Five-Year Plan, the installed capacity of distributed solar power generation in China has shown a rapid growth trend, especially in large industrial parks and high-quality seminars. However, the concentrated solar power poverty alleviation projects in remote rural areas are connected to the system too intensively, which will lead to problems such as voltage deviation and instrument overload. Sensitivity analysis of energy storage project yield and kilowatt-hour income in the metal processing zone is shown in Figure 2:

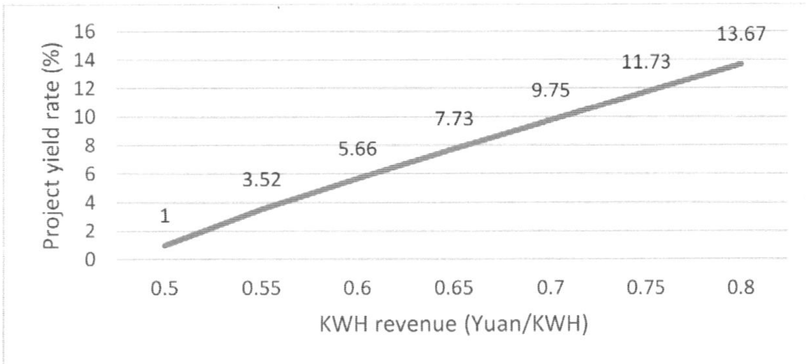

Figure. 2 Sensitivity analysis of the rate of return of energy storage projects in metal processing zones and the income of electricity generation

The space load forecast results of metal resource recycling industrial base are shown in Figure 3:

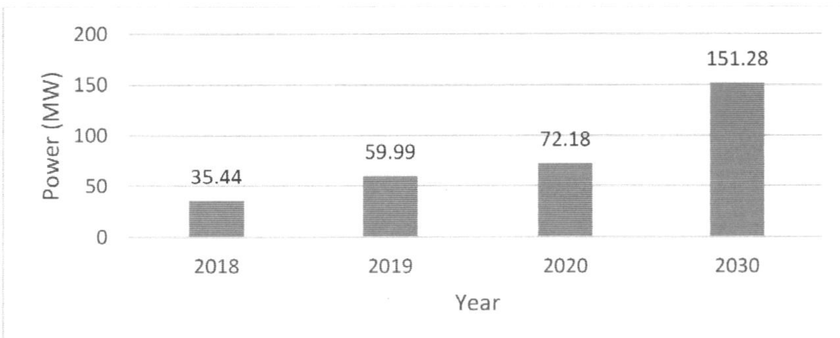

Figure 3 Forecast results of space load of metal resource recycling the industrial base

The summary of space load forecast results of metal resource recycling industrial base is shown in Table 1:

Table 1 Summary of space load forecast results of metal resource recycling the industrial base

Year	2018	2019	2020	2030
Utility load (MW)	22.96	31.32	36.28	67.48
Juke industry load	12.48	28.67	35.9	83.8
add up to	35.44	59 99	72.18	151.28

4.3 Technical ideas and measures to support large-scale distributed power grid connection

Generally speaking, the model aims at building the whole network and minimizing the operating cost and consists of two parts: one is to minimize the cost of feeder construction and reconstruction, substation construction and additional cost, transformer construction cost, new energy construction cost, and power storage module construction cost, and then reduce the power grid generation cost, maintenance cost, and load limitation cost. The changes in photovoltaic benchmark price/guide price (RMB /kW) in China are shown in Table 2:

Table 2 Changes in photovoltaic benchmark price/guide price (RMB /kW) in China

Resource area	Changes in benchmark electricity price (RMB /kWh)							
	2011	2012-2013	2013-2015	2016	2017	2018.1-5	2018.5-2019.7	2019.7
Class I			0.90	0.80	0.65	0.55	0.50	0.40
Class II	1.15	1.00	0.95	0.88	0.75	0.65	0.60	0.45
Class III			1.00	0.98	0.85	0.75	0.70	0.55

The combination of various technical means can be used in various applications, but the costs of various technical means are also different. For example, a survey conducted by the European Photovoltaic Industry Association used an example of solar power generation connected to a low-voltage distribution network below 0.4kV to study the cost of adopting strategies such as grid upgrading, conversion, and power adaptive grid. Technical measures to improve the photovoltaic acceptance capacity of the distribution network are shown in Table 3:

Table 3 Technical measures to improve photovoltaic acceptance capacity of distribution network

Type	Technical measures
Distribution network side	Power grid transformation On-load voltage regulation of medium/low voltage transformer Advanced voltage control of high/medium voltage transformer Static reactive power compensation Grid side energy storage Power network reconfiguration Advanced closed loop operation
Sender side	User side energy storage Electricity price incentives for self-use. Feeder power limit Active power control of inverter P(U) Inverter reactive power control Q(U), Q(P)
Interaction measures with power grid users	Demand side response based on a local price signal Demand side response based on a market price signal SCADA+ direct load control SCADA+ photovoltaic inverter control

5. Conclusion

Based on the analysis of the current development of energy storage, the general overview of current national and regional energy storage policies is given, and the typical application schemes of distributed energy storage in power sources, transmission networks, and users are given.

Therefore, this paper will sort out the energy storage strategy and operation mode, start with the requirements of practical work, and study the best allocation method of distributed energy storage with high quality. To sum up, this paper mainly studies the whole cycle cost and benefits composition of electric energy storage and analyzes the cycle cost and benefits composition of distributed energy storage connected to the distribution network.

Acknowledgments

The authors would like to show sincere thanks to those techniques which have contributed to this research.

References

[1] Picci, L. (2006). The quantitative evaluation of the economic impact of e-government: A structural modeling approach. *Information Economics and Policy*, 18(1), 107-123.

[2] Dai, S., Zhang, W., Zong, J., Wang, Y., & Wang, G. (2021). How effective is the green development policy of China's Yangtze River Economic Belt? A quantitative evaluation based on the PMC-index model. *International Journal of Environmental Research and Public Health*, 18(14), 7676.

[3] Kaplinski, O., & Peldschus, F. (2011). The problems of quantitative evaluation of socio-economic systems' development. *Engineering Economics,* 22(4), 345-355.

[4] Ginevičius, R. (2009). Some problems of quantitative evaluation of the state of social-economic systems. *Business: theory and practice*, 10(2), 69-83

[5] Townsend, R. M., & Ueda, K. (2006). Financial deepening, inequality, and growth: a model-based quantitative evaluation. *The Review of Economic Studies*, 73(1), 251-293.

Improvement on the DV-hop localization algorithm based on average hop distance optimization

Jia Xu, Yue Tian, Xianling Wang and Chen Zhu

Fujian Key Laboratory of Communication Network and Information Processing, Xiamen University of Technology, Xiamen 361024, China

Corresponding author: Xianling Wang (e-mail: xianling.wang@ieee.org).

Abstract. Wireless sensor networks, which have numerous uses in numerous industries, have attracted more attention over the past few years. Node localization is one of the gordian technology to locate in it. This research sought to address the low precision of the classic DV-hop algorithm in the context of wireless sensor networks. This novel DV-hop localization algorithm is given some practical solutions in this paper. The results of the simulations demonstrate that the new method can improve accuracy and increase communication range. First of all, the least squares method is applied to correct the average length of each hop between beacon nodes. In addition, an error distance matrix is defined as an intermediate variable to reckon the average length between the unknown node and the beacon node for each jump and select the minor error distance matrix. Finally, this improved algorithm chooses the minimum error distance matrix from the previous steps and modifies it. Simulation consequences reveal that this modified algorithm can better enhance the accuracy and extend the communication range.

1. Introduction

Great importance has been attached to the progress of wireless sensor networks (WSN) in the past few years. Because of its characteristics of small compact size, lightweight, and low energy consumption. WSN is considered a good way to obtain information. This network consists of a number of discrete devices that detect physical phenomena through sensors. It is a complex network made up of a lot of nodes. Beacon node general proportion is very small, usually by manual configuration or equipped with a GPS receiver to obtain its own location information. With the use of node localization technology, the measurement of positioning accuracy becomes very important. For a wireless sensor network, a reliable and efficient positioning scheme should meet the following requirements: 1) high positioning accuracy 2) Low communication energy consumption. For the sake of raising the accuracy of the conventional DV-hop algorithm, the least square method is used to modify the mean distance per hop.

Due to the disadvantages above, many scholars applied the swarm intelligence algorithm to the node location algorithm. In literature [1], a research strategy combined Improved Whale Optimization Algorithm (IWOA) with a localization model. In the depth of the whale population initialization algorithm, using a neural network to improve the initial position diversity, surrounded by the feeding behavior of the parameters with a nonlinear and adaptive strategy to avoid premature algorithm trapped in local optimum, in the process of iterative search used the quadratic interpolation method to alleviate the whale position diversity attenuation problem, ultimately this method achieved better location search result; In literature[2], the author used sparrow search algorithm to optimize DV-hop localization algorithm. Gao et al. [3] applied a particle swarm optimization algorithm to modify node position estimation, effectively raising the accuracy of locating. Furthermore, the author also suggested a fresh Hop distance Corrected DV-hop (HDCD) localization algorithm on the strength of reference [4]. Consider how the uneven distribution of anchor nodes may affect positioning accuracy. The hop distance is adjusted using global anchor node information based on the distribution features of anchor nodes. In literature [5], the author combines the jump distance correction with the sparrow search algorithm, through the communication radius of refining accurate hop count, and adds the weighted correction factor to reduce the error of average jump from weighted correction jump distance, reusing a sparrow algorithm instead of least square method. The localization issue of a wireless sensor network is turned into a function optimization problem, and the position of the node which needs to locate is then determined. In view of average hop distance optimization, Hu Yulan et al. [6] came up with an original DV-hop algorithm (AHDD). By optimizing the quantity of hops, the precision of the estimation of mean hop distance is improved. L Deng [7] optimized the quantity of hops in the DV-hop algorithm. To improve the ranging accuracy, the minimum mean square error criterion is used to estimate the mean hop distance. The experimental results made clearly that this method was a very effective way to improve the ranging accuracy. Based on optimizing hop distance and correcting hop numbers, Zhang Yuan et al. suggested the DV-hop localization algorithm (NHDL) in reference [8]. The signal strength value is used to amend the hop number and the information on which the known position is between anchor nodes. At the same time, the minimum and maximum method is applied to estimate the node position. Sana Messous et al. [9] suggested a new recursive DV-hop positioning method on account of the homogeneous multi-hop rangeless positioning algorithm. They applied the optimization formula to determine the average quantity of jumps of the anchor node to reduce distance estimation error. In reference [10], opportunistic transmission NOMA was applied to multi-point cooperative systems, which effectively improved the efficiency of successive-interference-cancellation (SIC) decoding. In collaborative networks, a complementary NOMA was proposed in reference [11]. This scheme solved the problem of edge user coverage and reduced the computational complexity.

2. DV-hop localization algorithm

The classic DV-hop algorithm is the same as the conventional distance vector routing mechanism. There are three primary stages in the classical DV-hop localization algorithm, which are the following:

Stage 1: Calculation of the shortest hops between unknown nodes and beacon nodes.

A beacon node notifies its neighbors of its location by broadcasting a packet containing that information. This package has a hop count. Its initial value is zero. The receiving node keeps a record of the minimum quantity of jumps from each beacon node, which automatically discards any beacon nodes with a higher hop count. Secondly, the jump value plus one and transferred to the following neighbor node. Therefore, the minimum hop count of every beacon node can be noted by all nodes in the network through multiple transmissions.

Stage 2: Reckon the average hop length among nodes.

Through the broadcast of the packet from the previous stage, each beacon node receives the packet with the shortest hop distance from other anchor nodes. Using the known position information for the anchor node and the minimum hop length discovered in the first stage, we can apply formula (1) to reckon up the average hop distance that is based on anchor node i. In addition, hop_{ij} is the minimum number of hops between i and j, which are both anchor nodes. Because each beacon node broadcasts a different average hop length, the unknown node just obtains the data information about the mean hop distance that arrives here first.

$$HopSize_i = \frac{\sum_{i \neq j} \sqrt{(x_i - y_i)^2 + (x_j - y_j)^2}}{\sum_{i \neq j} hop_{ij}} \qquad (1)$$

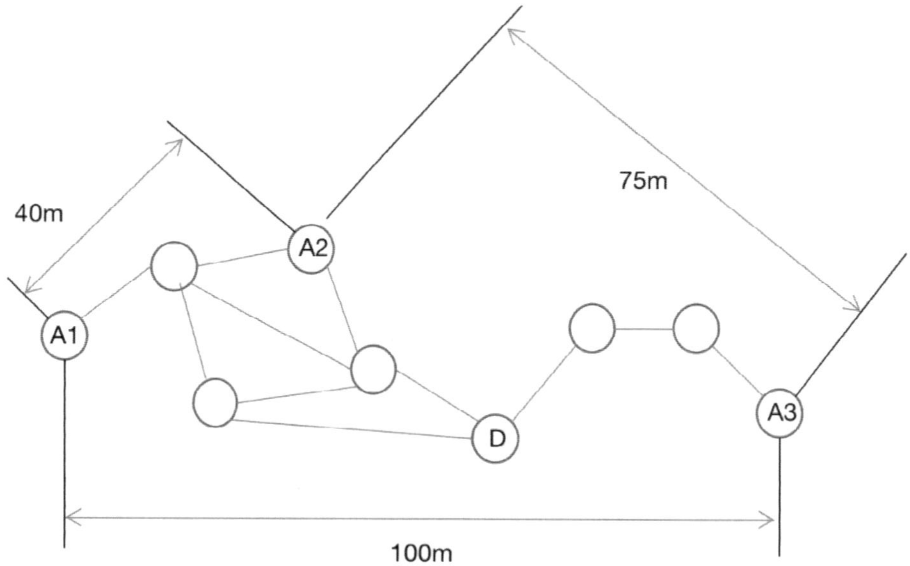

Figure 1. Schematic diagram of the second phase of DV-hop

In Figure1, the beacon nodes A1, A2, and A3 are shown, whereas D is the unknown node that has to be located. There are two and five hops from A2 to A1 and A3, and the lengths are 40 and 75 meters, respectively. Consequently, based on anchor node A, the typical hop distance is:

$$HopSize_i = \frac{75+40}{2+5} = 16.42m \tag{2}$$

The average hop distance calculated using A1 which is 17.5 meters, whereas the average hop distance calculated using A3 which is 15.91 meters. Because it is the fewest number of hops away from A2, the mean hop distance of A2 will be utilized as the unknown node D's mean hop distance.

Stage 3: Localization using the localization estimation method.

According to the previous steps, we can clearly estimate the hop count and average hop distance of the unknown node. It is easy to reckon the distance from the uncharted node to the anchor node. When we obtain the distance and more than three anchors, the trilateral measurement method can be used to locate the node.

Error analysis:

(1) Precision of the minimum hop count. The acquisition of the minimal hop number of unknown nodes contains errors because of the disorganization, random distribution of sensor nodes, and error in the broadcasting of packets. When the estimation error of the quantity of jumps and the average jump distance is larger, the distance between the uncharted node and the anchor point is larger, which will lead to a larger number of jumps and deviation.

(2) Applicability of average distance per jump. The classic DV-hop algorithm estimates the mean distance per hop on account of the distances between all anchor nodes other than the node and the unknown node, which yields an average distance per jump for the universal network range rather than the network density distribution within the area around the anchor node. Therefore, the average path length per hop obtained by this method has little influence in isotropic networks with uniform density, but it will not be applicable in anisotropic networks with uneven density, resulting in large errors.

(3) Because of the mistakes in the minimum hop and the deviation of mean length each hop in the previous two steps, the more beacons jump from unknown nodes, the greater the error in estimating the jump distance between them. Hence, using the distance information of distant beacon nodes to handle position estimation which reduces the precision of positioning.

3. Proposed algorithm: improved DV-hop localization algorithm

In the DV-hop algorithm, after each beacon node receives the broadcast message of the others, the average $HopSize_i$ value of each hop distance is reckoned on the basis of the unbiased estimation criterion, that is, the average $HopSize_i$ estimation error obtained by equation (1) is zero. In this paper, the least squares criterion is applied to amend the average length between beacon nodes for each hop.

In the first section of the DV-hop algorithm, assuming that beacon node *i* receives the location information of other beacon nodes, the average $HopSize_i$ reckoned by beacon node *i* shall meet the following criteria:

$$d_{ij} = \sqrt{(x_i - y_i)^2 + (x_j - y_j)^2} = HopSize_i hops_{ij} + \varepsilon_{ij} \tag{3}$$

Where *j* represents other beacon nodes in the data table, $hops_{ij}$ denotes the quantity of jumps between beacon node *i* and beacon node *j*, d_{ij} indicates the real length between beacon nodes *i* and *j*, and ε_{ij} stands for the error caused by replacing the actual length with the product of the mean distance per jump and the quantity of jumps. A reasonable $HopSize_i$ should minimize the total error. According to the principle of the least square method, $\sum_{i\neq j} \varepsilon_{ij}^2$ is used as the total error, then formula (4)

$$\begin{aligned} \sum_{i\neq j} \varepsilon_{ij}^2 &= \sum_{i\neq j}\left(d_{ij} - HopSize_i hops_{ij}\right)^2 \\ &= \sum_{i\neq j}\left(d_{ij}^2 - 2HopSize_i hops_{ij} d_{ij} + HopSize_i^2 hops_{ij}^2\right)^2 \end{aligned} \tag{4}$$

When $\dfrac{\partial \sum_{i\neq j} \varepsilon_{ij}^2}{\partial HopSize_i} = 0$, then Formula (5) means the average length of each jump which is calculated by beacon node *i*.

$$HopSize_i = \frac{\sum_{i\neq j}\left(hops_{ij} d_{ij}\right)}{\sum_{i\neq j} hops_{ij}^2} \tag{5}$$

4. Simulation and performance comparison

Before the simulation of the localization algorithm, the network structure and communication model based on the algorithm are established. Consider a two-dimensional wireless sensor network with *N* sensor nodes that are freely dispersed around a square area with *L* side lengths. A small part of them is defined in advance by manual deployment or GPS positioning method. These nodes are called anchor nodes (that is beacon nodes, the exact location of the point is known)).

Communication model: Assume that the distance of communication radius is R meters and that the communication model of a node is a circle with the node at its center. The uncharted node and anchor node have the same communication ability, and the transceiver ability of nodes is symmetric (that means the communication ability of nodes is equal at symmetric positions).

Experiment 1: In order to validate the usefulness and practicability of the ameliorated algorithm, the conventional DV-hop algorithm and the ameliorated algorithm were implemented in a Matlab R2022a emulation environment and compared. In the experimental simulation scenario, 100 points are randomly generated within the two-dimensional plane range of 100m×100m as nodes of the sensor network, among which beacon nodes and unknown nodes are generated randomly. The communication scope of the node is 50 meters. There are ten anchor nodes in total.

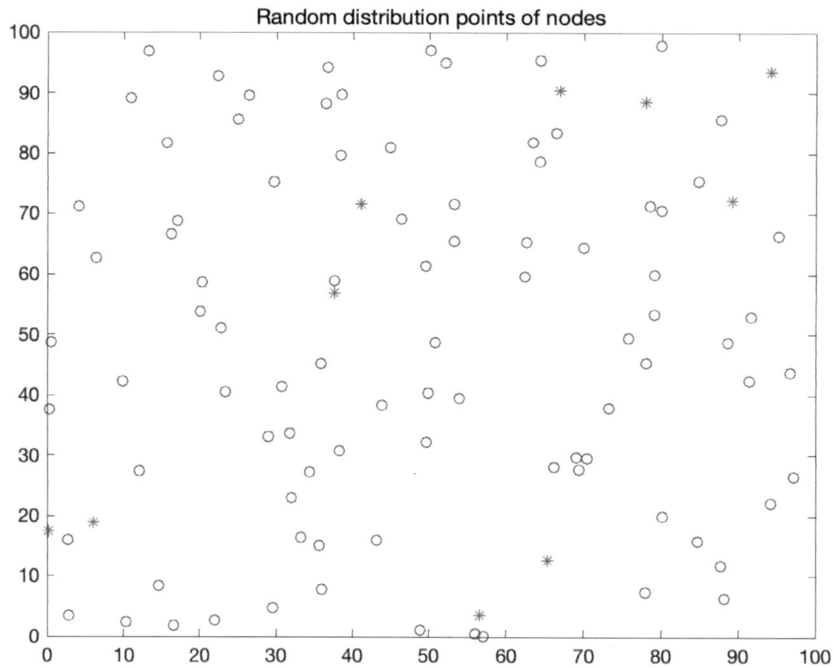

Figure 2. Distribution of random deployment nodes

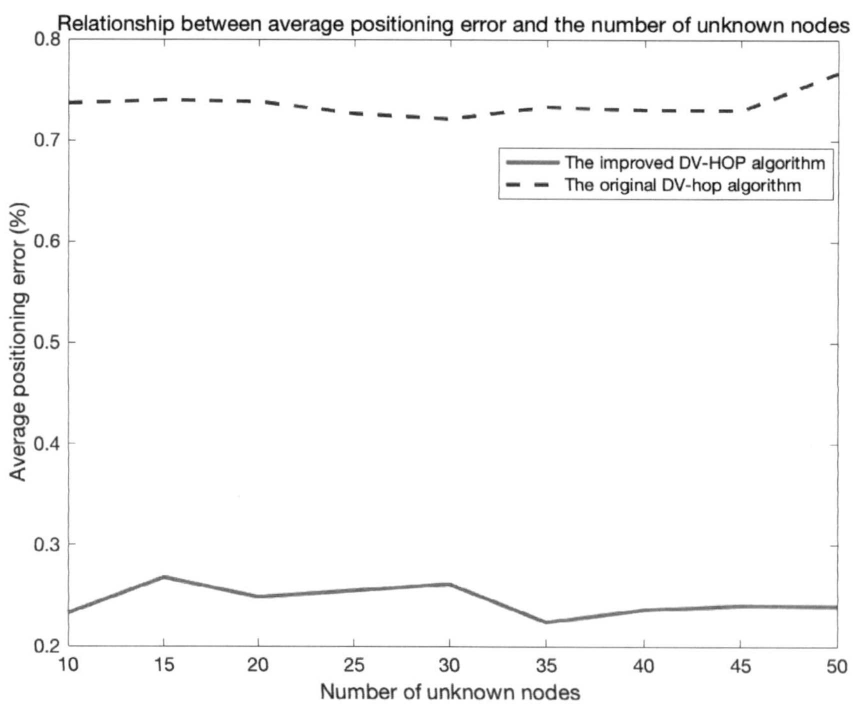

Figure 3. Comparison between the average positioning error and the quantity of unknown nodes before and after improvement

Figure 3 clearly demonstrates that the ameliorated algorithm can reduce the average location error with effect. With the quantity of unknown nodes growing, the average positioning deviation of the modified algorithm varies at 0.25, which is dramatically lower than that of the previous algorithm. The advantage of the performance is clear. It is well proved the effectiveness of the revised approach and its high practical value.

Experiment 2: Investigate the effect on positioning accuracy as the quantity of anchor nodes and communication radius vary

Finally, the effect positioning accuracy is analyzed in terms of the quantity of anchor nodes and the communication radius R. The parameters of this emulation are shown in the form below:

Table 1. The parameters of this experiment.

Parameters	Number/range
unknown nodes(n)	100
anchor nodes	10~30
communication radius	15~50

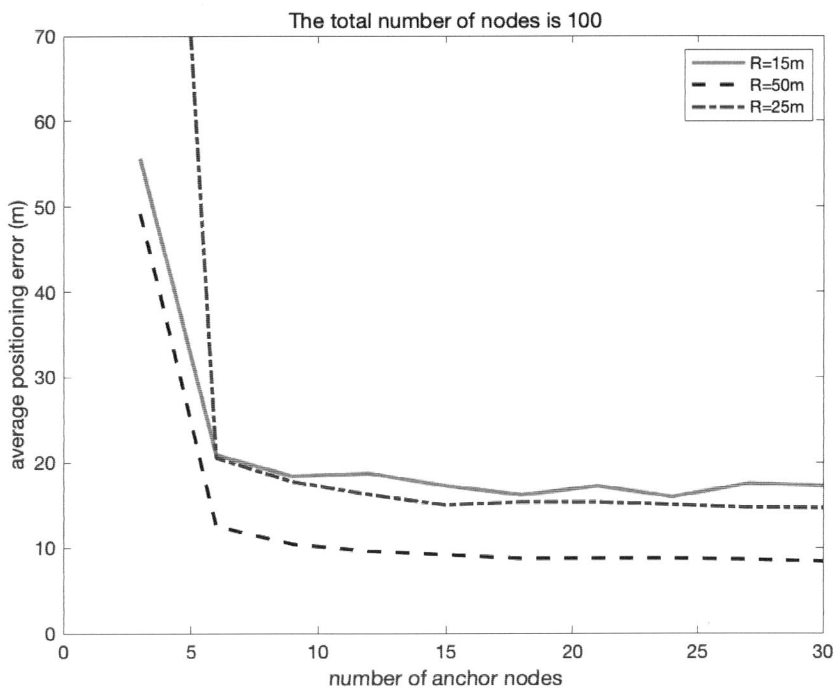

Figure 4. Effects on positioning accuracy as the quantity of anchor nodes and communication radius vary

The figure above shows the variation of the average positioning deviation of the improved DV-hop algorithm with the communication radius R. The chart shows that as communication radius and anchor nodes increase, the average positioning error reduces. In addition, it can be clearly seen that when the anchor node is larger than 5, the average location deviation decreases rapidly. The cause of it is that the larger the communication radius, the more anchor nodes, the larger the communication range of nodes, and the more positioning information can be obtained, which is conducive to improving positioning accuracy.

5. Conclusion

Traditional DV-hop algorithms do not have a high level of positional precision. On account of optimizing hop count and correcting hop distance, a better DV-hop placement algorithm is suggested.

The least squares criterion is applied to correct the mean length per hop between beacon nodes. The matrix with the minimum inaccuracy is chosen to determine the mean distance per hop from the updated uncharted node to the beacon node. The performance of the algorithm is simulated.

The modified algorithm can modify the mean distance per jump and reduce the positioning deviation effectively. Additionally, the impact of the communication radius and the quantity of anchor nodes on positioning precision is analyzed. The results of the simulations demonstrate clearly that the algorithm this paper put forward can successfully minimize the inaccuracies of location and that the average positioning error lowers as the quantity of anchor nodes and communication radius increases.

Acknowledgments

This work is supported by the Youth Innovation Foundation of Xiamen (3502Z20206067) and the Natural Science Foundation of Fujian Province, China (2022J011276, 2021J011219). This work is supported in part by the Natural Science Foundation of China under Grant 62201482.

References

[1] X Chen, J Gu, K J Mao and J Y Wu, Research on localization optimization of wireless sensor networks based on improved whale algorithm [J]. *Journal of Sensor Technology*,2021,34(10):1366-1373.

[2] L Yin, D Gu and F Liu, Optimization of dv-hop localization algorithm based on improved sparrow search algorithm [J]. *Chinese journal of sensor technology*,2021,34(05):670-675.

[3] M Gao and F Li, Genetic PSO improved dv-hop localization algorithm[J]. *Chinese journal of sensors and actuators*, 017, 30 (7): 1083-1088.

[4] Y F Li, dv-hop localization algorithm based on anchor node distribution characteristics [J]. *Chinese journal of sensor technology*,2021,34(10):1379-1384.

[5] H C Yan, Z W Zeng and G Wang. Fusion jump from the revision and the improvement of the sparrow search dv-hop algorithm [J]. *Journal of electronic measurement technology*, 2021, 44 (21): 133-138.

[6] Y L Hu, X Yu and Q S Liu, An improved dv-hop localization algorithm [J]. *Journal of Taiyuan Normal University (Natural Science Edition)*, 2019, 18(3): 40-44, 49.

[7] L Deng. Research and improvement of dv-hop localization algorithm in wireless sensor networks [D]. Ganzhou: Jiangxi University of Science and Technology, 2019.

[8] Y Zhang, M Wang, X L Miao and H Ding. A dv-hop localization algorithm based on Hop count correction and Hop distance optimization [J]. *Global positioning system*,2021,46(04):22-26.

[9] Messous S, Liouane H and Liouane N. Improvement of dv-hop localization algorithm for randomly deployed wireless sensor networks[J]. *Telecommunication Systems*, 2020, 73(1): 7 5-86.

[10] Y Tian, A R Nix and M Beach, "On the Performance of Opportunistic NOMA in Downlink CoMP Networks," in *IEEE Communications Letters*, vol. 20, no. 5, pp. 998-1001, May 2016.

[11] Y Tian, X Wang, Z Wang and Y H Kho, "On the Performance of Mutual-Aid NOMA Strategy in Cooperative Networks," in *IEEE Communications Letters*, vol. 24, no. 2, pp. 282-286, Feb. 2020.

Journal of Physics: Conference Series **2384** (2022) 012055 doi:10.1088/1742-6596/2384/1/012055

Construction of Safety Evaluation Model for Distribution Network with Distributed Generation

Xiaojuan Zhang, Liangyan Yu, Yue Liu*, Tong Du, Xuan Li

State Grid Dingxi Electric Power Company, Dingxi Gansu 743000, China

*ly76148978@163.com

Abstract. With the continuous change in urban power demand, many problems in the urban power system are exposed, and distribution network planning is one of the more prominent ones. In the operation of the distribution network, the access of distributed generation will have a certain degree of impact on power quality, relay protection, current protection of the distribution network, and energy consumption loss. Therefore, in order to improve the evaluation accuracy of the distribution network security evaluation model, obtain a more realistic situation evaluation value, and expand the overall evaluation scope, this paper constructs the distribution network operation security evaluation model under distributed generation access. Firstly, the subjective weight of evaluation is determined, and the evaluation index system of the power grid operation security situation is designed under the background of power access. Secondly, the multi-source extension fuzzy evaluation matrix is constructed, and the model is established by comprehensive weighting processing. Finally, the test experiment is carried out. The test results show that compared with the traditional power grid operation security evaluation model, the power grid situation evaluation value obtained by the distribution network operation security evaluation model constructed in this paper is relatively high. This shows that the model has a good evaluation effect on the operation security of the distribution network, with high evaluation authenticity and small error, and has a high practical application value.

1. Introduction

Distributed Generation (DG) refers to a small power generation system with a power generation capacity ranging from several kilowatts to tens of megawatts, which is miniaturized, modularized, decentralized, and arranged near the user to supply power to the user[1]. According to whether Distributed Energy Resource (DER) or Distributed Generator (DG) is renewable or not, distributed generation can be divided into two categories: one is renewable energy, including solar energy, wind energy, geothermal energy, ocean energy, and other forms of power generation; the other is non-renewable energy, including internal combustion engines, cogeneration, micro-gas turbines, fuel cells and other forms of power generation.

In the modern power system, distributed generation is a very important part. At this stage, the research and application of distributed generation have become more and more extensive. There are two main operation modes of distributed generation, namely island operation mode and grid-connected operation mode [2]. In the island mode, the distributed generation and part of the load will form a self-sufficient island, and the distributed generation will supply power to the load independently. Island operation mode may cause some security problems, but it is still considered one of the most important operation modes of distributed generation.

The distribution system includes distribution stations of different voltage levels, distribution transformers, distribution lines, and other electrical facilities connecting different users [3]. China's distribution network has a large coverage, strong variability, and is vulnerable to external factors in the operation process. To avoid the occurrence of large-scale related power accidents and gradually create a more stable and safe power dispatching environment, an operation safety evaluation model is constructed in combination with different power conditions [4]. As the basis of ensuring the safe operation of the power grid system, the evaluation model of distribution network operation security has the characteristics of flexibility and changeability, which can also ensure the normal operation of the distribution network in complex power environments, but in recent years, with the improvement of people's power demand and standards, some distribution networks can not bear the heavy burden, and there are more or fewer problems in the daily operation process[5].

In the context of distributed generation access, the efficient application of the distribution network can not be guaranteed [6]. To improve the overall operational efficiency of the power grid and enhance the safety of power consumption, the safety evaluation model of distribution network operation under distributed generation access is designed and analyzed. Considering the stability and reliability of the final test results, a bidirectional security evaluation structure is established. Through the analysis of the different power indicators, the corresponding security evaluation problems are extracted, and the description of power grid security and benefit characteristics is realized. According to the access status of distributed generation, the evaluation model is adjusted slightly to ensure the authenticity of the final evaluation results and provide a reference for the subsequent distribution network operation and technological innovation [7].

2. Construction of evaluation model

In order to successfully realize the design of the evaluation model for the safe operation of the distribution network, the specific design process is shown in Figure 1.

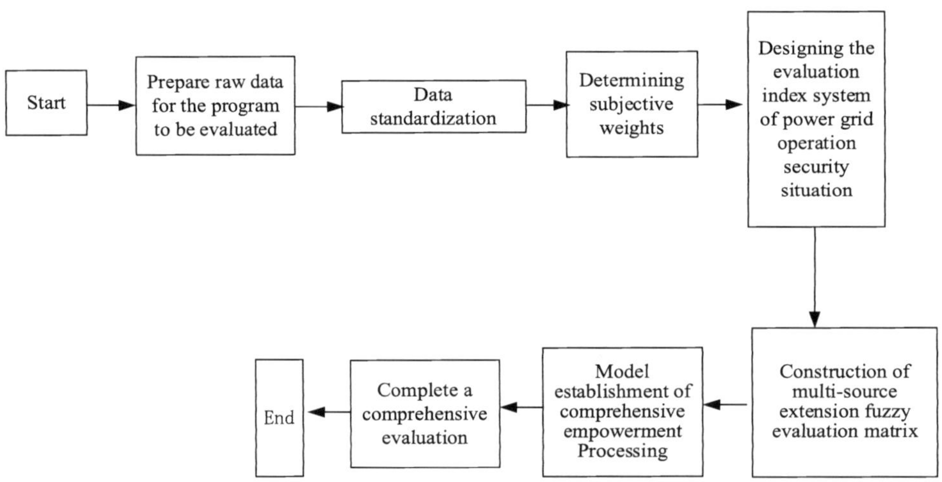

Figure 1 Construction process of distribution network safety operation evaluation model

2.1. Determination of subjective weight

Under the environment of distributed generation access, the evaluation weight of basic distribution network operation is determined. This part can be realized by subjective weighting and objective weighting. Usually, a certain number of data nodes can be laid out in the operation program of the distribution network, after the access of the distributed generation, the data and information are classified by means of combined measurement. At the same time, combined with the characteristics of the data and the distribution network area to be evaluated, the basic index weight is quantified, and the dynamic evaluation level is set. The evaluation subjective weight is calculated as shown in the following formula (1):

$$G = (\beta_1 + \sqrt{\beta_1 \beta_2 \times v})^2 - \sum_{i=1} \beta_2 \varpi + vi \quad (1)$$

In Formula (1), G represents the evaluation subjective weight, β_1 represents the evaluation preset range, β_2 represents the evaluation measured range, v represents the directional evaluation difference, ϖ represents the conversion constant, and i represents the number of evaluations. Through the above calculation, the actual evaluation subjective weight can be finally obtained, and the distribution structure can be adjusted in combination with the obtained value to lay a foundation for the subsequent operation safety evaluation [8].

2.2. Evaluation index system

After the determination of the subjective weight of the evaluation, the evaluation index system of the power grid operation security situation is designed under the background of distributed generation access. According to the change of subjective weight, a fuzzy overlapping evaluation matrix can be constructed within a reasonable range by combining the fuzzy analytic hierarchy process [9]. In this part, it should be noted that for the nodes at different locations in the distribution network, in addition to obtaining data and information, the actual operating frequency of the power grid dispatching can also be analyzed according to the fluctuating state of its operation. An evaluation module [10] is set in the matrix to evaluate factors such as operating frequency, power dispatching, user feedback, and power distribution. A calibrated operation safety evaluation structure is formed, as shown in Figure 1 below:

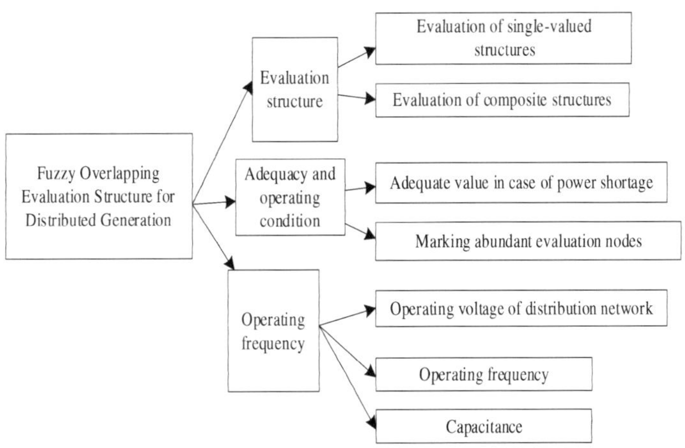

Figure 2 Diagram of fuzzy overlapping evaluation structure under distributed power

According to Figure 2, the design and adjustment of the fuzzy overlapping evaluation structure under the distributed generation can be completed. Then, in the scope of the calibrated distribution network, combined with the evaluation index difference calculated by the matrix, the basic evaluation structure is formed. Design the multi-dimensional safety evaluation situation [11]. This part can adjust the actual security situation evaluation standard according to the operation fluctuation state of the distribution network, and integrate its own characteristics and information security management requirements, which can not only strengthen the daily monitoring of the evaluation distribution network but also expand the coverage of the evaluation to ensure the authenticity and reliability of the final evaluation results.

2.3. Construction of multi-source extension fuzzy evaluation matrix

After completing the design of the evaluation index system, it is necessary to construct the design of the evaluation index system for the security situation of power grid operation under power access. Under the background of the distributed generation access, the operation status and mode of the distribution network will also change accordingly.

At this time, a standard extension fuzzy evaluation sequence can be constructed according to the power data and information obtained by the node, and the absolute value of the corresponding component evaluation of the safety operation can be calculated in combination with the grey correlation model, as shown in the following formula (2):

$$h = \int \eta \times \frac{\chi\eta + \sum_{e=1} \varsigma\eta \times e}{\Re^2 - (0.5\chi \times \frac{1}{2})} + \chi^2 \tag{2}$$

In formula (2), h represents the evaluation absolute value, η represents the conversion difference, $\Re$ represents the directional evaluation range, χ represents the extension constant, ς represents the number of sequence nodes, and e represents the one-way operation frequency of the power grid. Through the above calculation, the actual evaluation absolute value can be finally obtained. Combined with the obtained evaluation absolute value, a corresponding dynamic operation evaluation standard is constructed; meanwhile, before and after the distributed power supply is accessed, the designed security situation evaluation index system is added to the matrix, and the evaluation results before and after are measured, and are compared and verified; and within the calibrated extension fuzzy evaluation range, the fluctuation of the operation frequency after the power supply is accessed is eliminated, which is to improve the application ability of the multi-source extension fuzzy evaluation matrix.

2.4. Establishment of integrated weighting processing implementation model

After completing the construction of the multi-source extension fuzzy evaluation matrix, next, this paper adopts the comprehensive weighting method combining the subjective improvement AHP method and the objective entropy weight method. According to the comprehensive weighting method, the design of the evaluation model for the safe operation of the distribution network is realized.

2.4.1 Calculation of objective weight by entropy weight method

In the first decision-making scheme, the information entropy H_j of the j decision index is shown in formula (3)

$$\omega_j = \frac{1 - H_J}{n - \sum_{j=1}^{n} H_J} \tag{3}$$

From this, the information entropy weight ω_j of the j index can be calculated.

2.4.2 Calculating Subjective Weights

According to the operation state of the power grid, it is needed to adjust the corresponding coverage evaluation mode, partially evaluate the equivalent conversion, measure the correlation between subjective weights, objective weights, and combined weights, and calculate the correlation evaluation difference in the process, as shown in the following Formula (4).

$$K = (\Im + n)^2 - \sqrt{\omega_1} \times n\omega_2 \tag{4}$$

In Formula (4), K represents the correlation evaluation difference, $\Im$ represents the grid conversion standard value, n represents the equivalent mean value, ω_1 represents the evaluation result before conversion, and ω_2 represents the evaluation result after conversion. Then, combine the obtained correlation evaluation difference value, delimit a corresponding evaluation range in the evaluation matrix, combine the access fluctuation of the distributed generation, set an evaluation standard, and complete the acquisition of an operational safety evaluation result.

3. Model testing

3.1. Test preparation

This paper selects the distribution network of a city as the evaluation object to verify the evaluation system. The indicators of the index layer are the proportion of heavy overload of 10kv line, the proportion of light load of 10kv line, the proportion of heavy overload of distribution transformer, the proportion of self-provided power supply, and the proportion of dual power supply automatic switching user capacity.

3.2. Test process and result analysis

In order to improve the application ability of the model, the test is analyzed in the form of comparison, and the measurement evaluation is carried out according to the test environment built above and combined with the actual evaluation needs. Set 4 operation evaluation cycles, each cycle is 3 hours, according to the specific classification, the corresponding power grid area is classified and evaluated, with the changing state of the weight, the security and stability of the power grid operation, so as to map and analyze the power grid. The changing status of the situation evaluation value, combined with the actual evaluation needs, is measured and evaluated, as shown in Table 1 and Table 2

Table 1 Comparative analysis table of performance weights of different methods

Function	Traditional extension power grid operation safety evaluation model	Traditional comprehensive power grid operation safety evaluation model	Fuzzy Hierarchical Power Grid Operation Safety Evaluation Model
Stability	1.00	1.20	2.00
Safety	1.00	1.11	3.00
Economy	0.45	0.89	1.96
Reliability	0.96	1.54	2.48

Table 2 Comparative analysis table of situation evaluation values of different methods

Test cycle	Power grid situation evaluation value of traditional extension power grid operation security evaluation model test group	Power grid situation evaluation value of traditional comprehensive weighting pow grid operation security evaluation model test group	Power grid situation evaluation value of fuzzy hierarchy power grid operation security evaluation model test group
Period 1	2.05	3.69	4.66
Period 2	2.91	2.67	4.25
Period 3	3.67	3.54	4.07
Period 4	2.88	3.57	4.81

According to Table 1 and Table 2, the analysis of the performance of different methods and the evaluation value of the power grid situation can be completed: The final evaluation value of the power grid situation obtained by the fuzzy-level power grid operation safety evaluation model test group designed in this paper is relatively high, and its safety The performance, economy, reliability, and stability of the model are all high, indicating that the model has a good effect on the evaluation of operation safety during the operation of the distribution network, and has practical application value.

4. Conclusion

This paper analyzes and studies the practical application effect of the distribution network operation security evaluation model under distributed generation access. Compared with the basic operation evaluation model, the specific power dispatching environment is set. Combined with the access status of the distributed generation, the actual needs in the operation process are gradually extracted, and the evaluation index system of safe operation is established. In addition, for the power dispatching and operation data and information collected by each monitoring node, combined with the combination weighting method, the weight of the evaluation index is calculated. A detailed analysis of the operation status of the distribution network before and after the power access is made to realize the multi-dimensional evaluation verification, increase the authenticity and effectiveness of the evaluation results, and promote China's distribution network to a new level of development.

References

[1] Li Xiao, X. J., Li, M. (2022) A Security Evaluation Method for Cyber-Physical Distribution System Considering Influence of Information Failure. Electric Power, 55(02): 73-81.
[2] He, Z., Xia, T., Chen, Z. (2021) Research on quantified evaluation model of personal safety risk of power grid enterprises based on principal component neural network. ZHONG GUO GAO XIN KE JI, (02): 21-22+33.
[3] Zhang, L., Xu, C., Yu, X. (2020) Evaluation Model of Power Grid Operation Safety Capability Based on Comprehensive Weighting Grey Correlation Model. Journal of Data Acquisition & Processing, 35(06): 1147-1153.
[4] Lin, J., Fan, S., Xu, Z. (2020) A Situation Awareness Model for Power Grid Operation Based on Fuzzy Analytic Hierarchy Process and LSTM-Attention Mechanism. Electric Power Information and Communication Technology, 18(04): 58-66.
[5] Qian, S., Yang, T. (2020) Research on security risk assessment of smart microgrid based on fuzzy set theory. Energy Conservation, 39(02): 6-9.
[6] Tan, L., Li, J., Ren, B. (2020) Health evaluation model of smart grid dispatch and control system based on RB-XGBoost algorithm. Electric Power Automation Equipment, 40(02): 189-198.
[7] Huang, W., Kang, M. (2022) Research on intrinsic safety evaluation method of power grid based on improved cloud model. Computer Applications and Software, 39(04): 133-137+153.
[8] Du, Y., Yang, X., Guo, L. (2021) Coupling Evaluation of Green Development and Safety and Efficiency of Distribution Network in Context of Double Carbon. Science Technology and Engineering, 21(30): 12973-12981.
[9] Ai, X., Qin, J., Hu, H. (2021) Comprehensive Evaluation of Power Grid Security and Benefit Based on BWM Entropy Weight TOPSIS Method. Modern Electric Power, 38(01): 60-68.
[10] Yu, Q., Li, H., Qu, Y. (2022) Research on security situation awareness of large power grid based on deep neural network and internal and external factors. Electrical Measurement & Instrumentation, 59(02): 16-23+67.
[11] Liu, Y. (2014) Research on Reliability Evaluation Model of Distribution Network with Distributed Generation. Science and Technology and Innovation, No.09, 23-28.

Health state prediction of lithium-ion batteries based on XGBoost algorithm

Ju-qi BAI[1], Yan-hua XIAN*[1,2]

[1]Guangxi Normal University School of Electronic Engineering Guilin, 541000, Guangxi

[2] Key Laboratory of Integrated Circuits and Microsystems (Guangxi Normal University), Education Department of Guangxi Zhuang Autonomous Region

[*] Corresponding author's: xyhua@mailbox.gxnu.edu.cn

Abstract. The state of health of lithium-ion batteries is an extremely important indicator for the safe operation of electrical equipment. In order to solve the problem that the capacity of Li-ion batteries is difficult to measure online, this paper proposes an online method based on the XGBoost algorithm to estimate the state of health (SOH). First, health characteristics are extracted from the discharge voltage profile to characterise the capacity of the Li-ion battery. XGBoost is then used to predict SOH based on capacity drop. results show that the proposed method is able to estimate SOH with a root mean square error of less than 1.2% respectively. Furthermore, the proposed method was used for model training and prediction in less than 1 second.

1. Introduction

Lithium-ion batteries are becoming an critical energy storage device because of its high output voltage, long cycle life, and high energy density. The long-term stable operation of the battery is ensured by a Battery Management System (BMS) that monitors the status of the battery in real time. One of the key parts of the BMS is the accurate prediction of the state of healthy (SOH) of the battery [1]. The SOH is usually defined as the ratio of the current available capacity to the initial capacity. An effective SOH estimation method is particularly important because of the complex ageing mechanism within lithium cells [2]. A large number of SOH estimation methods are available and can be divided into three categories [3]. The direct measurement method [4] is to directly measure certain parameters of the battery, such as current, voltage, ohmic internal resistance, etc. The model-based approach generally requires the construction of an electrochemical model or equivalent circuit model, which identifies the model parameters and then calculates the SOH [5]. Such methods include joint estimation, collaborative estimation, and fusion estimation. However, there are various drawbacks to these methods. In recent years, data-driven methods have become a hot research topic because of their flexibility and versatility [6]. The data-driven approach is to learn the battery aging behavior from a large amount of statistical information, find the HF that reflects the degradation of the lithium battery, and predict SOH using deep learning. Currently, researchers focus on non-probabilistic methods such as support vector regression (SVR), neural networks and random forest regression, and XGBoost.

A new method based on the XGBoost algorithm for estimating SOH is proposed. Firstly, the health features are extracted. Then an estimation model is built using the XGBoost to prediction the SOH.

The comparison experiments with SVR show that the proposed method has high estimation accuracy and short computation time for SOH.

2. XGBOOST ALGORITHM

The XGBoost algorithm is one of the most prediction algorithms due to its high accuracy, fast calculation speed, and difficulty in over-fitting.

XGBoost is an integrated model. The integrated model is a set of classification and regression trees. Set the data, (x_1, y_1), (x_2, y_2), ..., (x_i, y_i), $i = 1, 2..., n$, a tree is used as the initial value. The gradient descent method is used as follows and then repeated. First, setting

$$\hat{y}_i^{(t)} = \sum_{k=1}^{t} f_k(x_i) = \hat{y}_i^{(t-1)} + f_t(x_i) \tag{1}$$

Eq. (1) is the predicted value of the model in round t.

The objective function of XGBoost consists of a loss function and regularization. The loss function is used to measure the error in the predicted values. Regularization is used to prevent over-fitting. Expressions are as follows:

$$\text{obj}(w) = L(w) + \Omega(w) = \sum_{i=1}^{n} l\left(y_i, \hat{y}_i^{(t-1)} + f_i(x_i)\right) \tag{2}$$

where w is the estimated parameter, $L(w)$ is the loss function $\Omega(w)$, and is the regularization term. Eq. (3) is expanded using the second-order Taylor formula to obtain:

$$\text{obj}^{(t)} \approx \sum_{i=1}^{n} \left[l\left(y_i, \hat{y}_i^{(t-1)}\right) + g_i f_t(x_i) + \frac{1}{2} h_i f_t^2(x_i) \right] + \Omega(f_t) + c \tag{3}$$

Where defined $g_i = \partial_{\hat{y}_i t-1)} l\left(y_i, \hat{y}_i^{(t-1)}\right), h_i = \partial^2_{y \leqslant t-1)} l\left(y_i, \hat{y}_i^{(t-1)}\right)$.

$$\text{obj}^{(t)} = \sum_{i=1}^{n} \left[g_i f_t(x_i) + \frac{1}{2} h_i f_t^2(x_i) \right] + \Omega(f_t) \tag{4}$$

Define the tree $f(x)$ as:

$$f_t(x_i) = \omega_{q(x)} \quad q: \mathbb{R} \to T, \omega \in \mathbb{R}^T \tag{5}$$

where q is the structure of each tree and T is the number of leaf nodes in the tree.
Define the complexity $\Omega(f)$ of the tree as:

$$\Omega(f) = \gamma T + \lambda \frac{1}{2} \| \omega \|^2 \tag{6}$$

The objective function is updated as:

$$\text{Obj}^{(t)} = \sum_{j=1}^{T} \left[\left(\sum_{i \in I_j} g_i \right) w_j + \frac{1}{2} \left(\sum_{i \in I_j} h_i + \lambda \right) w_j^2 \right] + \gamma T \tag{7}$$

$$= \sum_{j=1}^{T} \left[G_i \omega_j + \frac{1}{2} (H_i + \lambda) \omega_j^2 \right] + \gamma T \tag{8}$$

Among them, $I_j = \{ i \mid q(x_i) = j \}$ $\quad G_i = \sum_{i \in I_j} g_i \quad H_i = \sum_{i \in I_j} h_i$.

Then the optimal weight W of node A with the best objective function solution O as:

$$\omega_j^* = -\frac{G_i}{H_i + \lambda} \tag{9}$$

$$\text{obj}^{(*)} = -\frac{1}{2} \sum_{j=1}^{T} \frac{G_i^2}{H_i + \lambda} + \gamma T \tag{10}$$

$\text{obj}^{(*)}$ is a scoring function.

Calculate the corresponding gains：

$$\text{Gain} = \frac{1}{2}\left[\frac{G_L^2}{H_L+\lambda}+\frac{G_R^2}{H_R+\lambda}-\frac{(G_L+G_R)^2}{H_L+H_R+\lambda}\right]-\gamma \tag{11}$$

γ denotes the complexity of the introduction. The above equation allows the optimal decision tree structure to be obtained, followed by a prediction.

3. DATA PROCESSING

3.1 The experimental data analysis

To verify the validity of the proposed method, four cells from the CS series were selected from the CALCE Li-ion battery dataset [5]. This battery was tested at 1 °C, with a rated capacity of 1.10 Ah, using the Arbin BT2000 lithium battery test system from the USA. Figure 1 shows the charge and discharge processes in the cycle. For the experiments in this paper the first 700 charge/discharge cycles were chosen, using voltage versus time data.

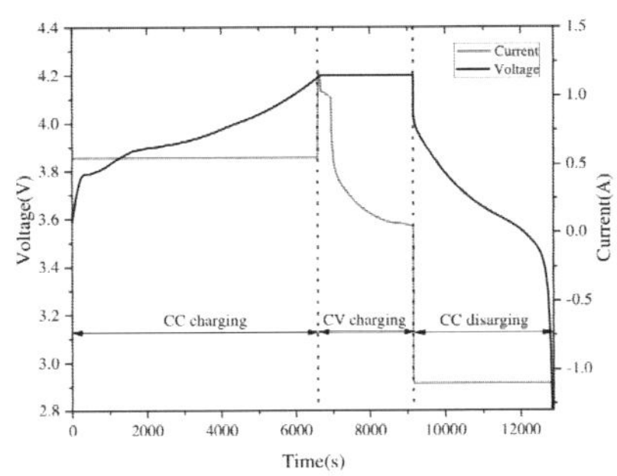

Figure 1. Charge and discharge process in the cycle of battery.

3.2 Health feature extraction

Figure 2 shows that as the battery ages, the time required for the voltage to drop to the cut-off voltage decreases. Therefore, the voltage curve for the interval from the start of discharge (4.1V) to the cut-off voltage (2.7V) was extracted and the integral of voltage variation over time (IVT) [7] was used to predict the health characteristics of SOH. By integrating the voltage over time, as shown in Figure 2, it can be used to describe the energy released by the battery.

$$S_i = \int_{t_0}^{t_{\text{cutoff}}} u_i dt \tag{12}$$

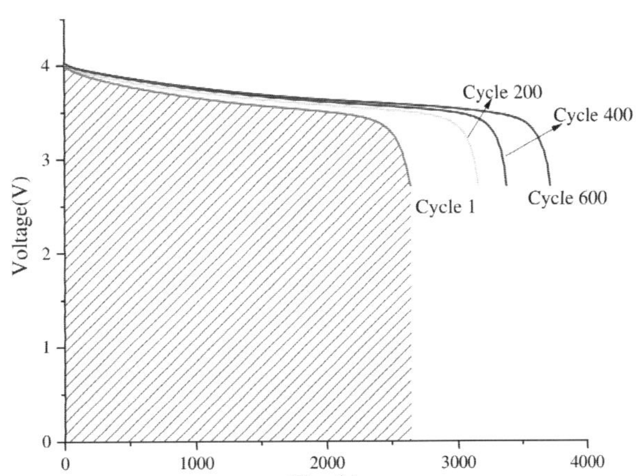

Figure 2. Area decreases as the number of cycles

The steps of extracting health features (IVT) are divided into three steps:

Step 1: Extraction of voltage and time data for each cycle during the constant current discharge phase

Step 2: Polynomial fit of the voltage.

Step 3: Calculate the integral based on Equation 12.

3.3 Pearson's Correlation Analysis

The Peel correlation coefficient is used to measure whether the two matrices are on the same straight line, and to measure the linear relationship between the distance variables. The closer the absolute value of the obtained value is to 1, the stronger the correlation. Pearson's correlation coefficient analysis was performed on the proposed health characteristics according to Equation 13.

$$r = \frac{\sum_{i=1}^{n}\left(U_i - \bar{U}\right)\left(V_i - \bar{V}\right)}{\sqrt{\sum_{i=1}^{n}\left(U_i - \bar{U}\right)^2}\sqrt{\sum_{i=1}^{n}\left(V_i - \bar{V}\right)^2}} \tag{13}$$

Table 1. Pearson correlation coefficients

CS	Pearson's Correlation Coefficient
35	0.991
36	0.995
37	0.991
38	0.988

Table 1 shows that the Pearson correlation coefficients between the battery health SOH and IVT are all above 0.98. This indicates a strong correlation between the eigenvalues taken and SOH. It can be used as a prediction model input for SOH prediction.

4. Algorithm validation

4.1 The verification method

The SOH estimation model based on the XGBoost algorithm is shown in Figure 3 and consists of three parts: feature selection, XGBoost model building and XGBoost estimation. Firstly, features were extracted from the Maryland Li-ion battery dataset; then the dataset was divided into a training set and a test set, with the training set used for XGBoost model building and training, the test set for validating the XGBoost model by estimating the SOH. Where in the phase of dividing the training and test sets, four CS series of batteries are divided using the classical leave-one-out method in this paper which means that three batteries are used as the training set and one battery is used as the test set..

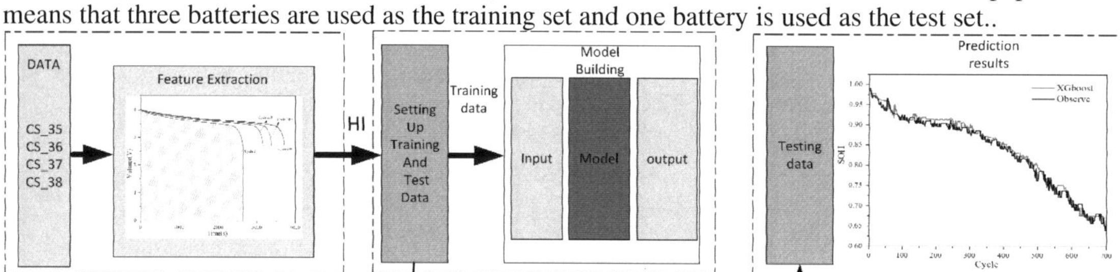

Figure 3. SOH estimation model based on XGBoost algorithm

4.2 Prediction Results

In this section, the comparison test is completed by validating the data from each of the four cells by the XGBoost model and the SVR model.

The SOH prediction results of CS_35 and CS_36 by XGBoost were shown in Figure 4 and 5. In Figure 4, there is a large deviation in 100 to 150 cycles, but the deviation is quickly corrected to close to the true value. And the prediction ability is relatively strong. From Figure 5, it can be seen that XGBoost in has good prediction results.

Figure 6 and Figure 7 show the comparison of prediction errors of XGBoost and SVR in CS_37 and CS_38. The results in Fig.6 show that a large deviation arose in the first 300 cycles and the maximum error is close to 0.08 by SVR, while the error of XGBoost can still keep within 0.03.

Therefore, compared with the SVR algorithm, the XGBoost algorithm can guarantee the stability of SOH prediction.

Table 2 and Table 3 show the prediction error of SOH on the four-battery data set.

According to Table 2, the root means square error is less than 1.2 percent. The mean absolute error is less than 1 percent. The mean maximum absolute error is controlled by six percent. The confidence level of the model is above 0.96. Therefore, XGBoost has well estimation accuracy on four batteries.

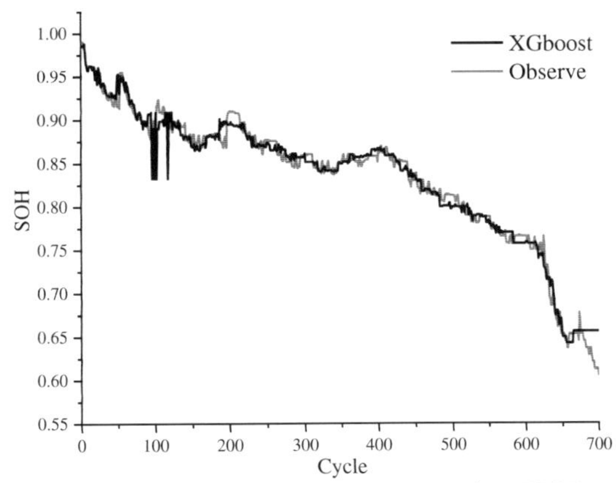

Figure 4. Prediction results of CS_35 using XGBoost

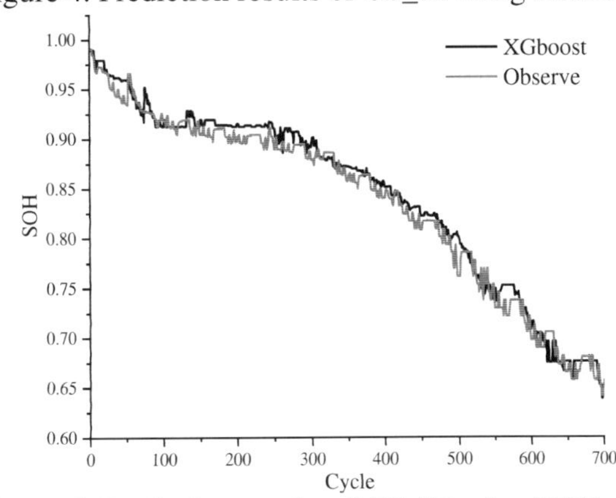

Figure 5. Prediction results of CS_36 using XGBoost

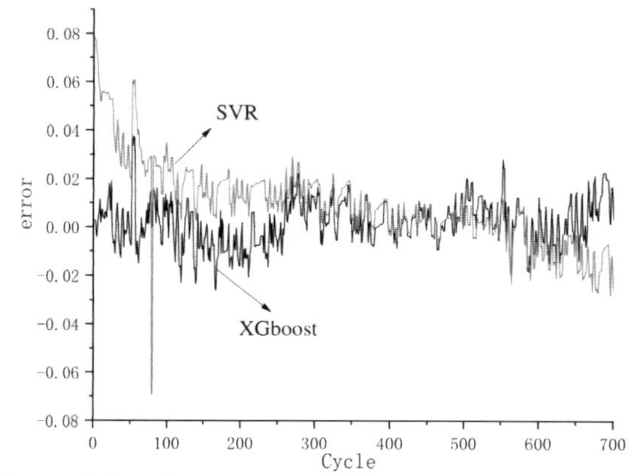

Figure 6. Prediction results of CS_37 using XGBoost

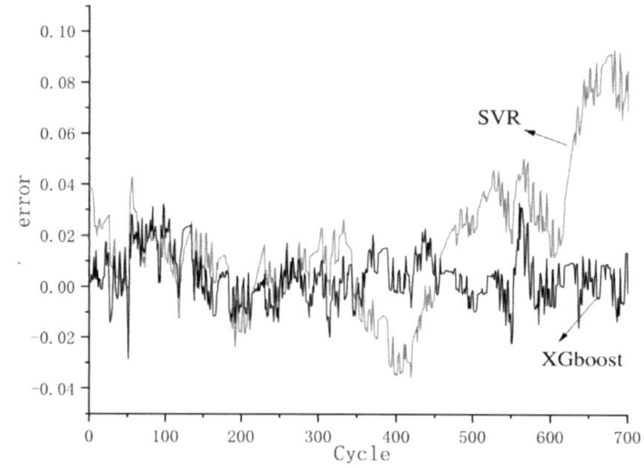

Figure 7. Prediction results of CS_38 using XGBoost

Table 2： XGBoost prediction results

CS	RMSE	MAE	MAXAE	R^2	Time
35	1.14%	0.78%	7.57%	0.97	0.1650s
36	1.20%	0.97%	4.25%	0.98	0.1520s
37	1.04%	0.75%	6.94%	0.97	0.1570s
38	1.04%	0.81%	3.07%	0.96	0.1470s

It can be seen from table 2 and table3 that the values of XGBoost in RMSE MAE and MAXAE are lower than the SVR algorithm on four batteries. The root mean square errors of XGBoost predictions were 1.14%, 1.20%, 1.04% and 1.04% respectively, all less than 1.5 percent. The MAE of XGBoost predictions were 0.78%, 0.97%, 0.75% and 0.81% respectively, all less than 1percent. The MAXAE of XGBoost predictions were 7.57%, 4.25%, 6.94% and 3.07% respectively, all less than 8 percent. Also, the confidence level of XGboost is higher than SVR. SVR should consume more time in train and testing than XGBoost.

Table 3： SVR prediction results

CS	RMSE	MAE	MAXAE	R^2	Time
35	2.34%	1.22%	15.54%	0.91	42s
36	1.70%	1.21%	5.94%	0.96	19s
37	1.98%	1.47%	7.80%	0.90	44s
38	3.22%	2.39%	9.32%	0.69	23s

In the test, the proposed method accomplished highly accurate online estimation of SOH at most test sites. The root mean square errors of SVR predictions were 2.34%, 1.70%, 1.98% and 3.22% respectively, all less than 4 percent. Comparison with the SVR-based SOH estimation model shows that the proposed SOH estimation method has higher accuracy and less computation time.

5. Conclusions

In order to solve the problem that the capacity of Li-ion batteries is difficult to measure, this paper proposes an XGBoost-based algorithm to estimate the health status.The integral of the voltage over time during the constant current discharge phase was extracted and used to predict by the XGBoost model. The following conclusions are drawn.

(1) Health features extracted in this paper are easily available without filtering and with high correlation.

(2) The XGBoost model was built and compared with the SVR model. XGBoost showed excellent prediction capability, which can control the RMSE within 1.5% and the training testing time within 1S, verifying the accuracy and real-time performance of the algorithm.

(3) The method proposed in this paper is real-time, highly accurate and easy to operate, and can meet the SOH prediction requirements.

Acknowledgments

This work was supported by the National Science Fund subsidized project (61976063).

References

[1] Fan T, Liu M X, Wang W X. 2022, Optimisation of electric vehicle functionality by BMS working offline [J]. Automotive Industry Research, 3: 38-40.

[2] Zhang Z G, Zhang T, Tang A H, et al. 2022, A review of fast charging methods for automotive lithium batteries in healthy conditions [J]. Journal of Southwestern University, 44(02):194-206.

[3] Li C, Wang C H, Wang G, et al. 2021, A data-driven technique for estimating the health status of lithium-ion batteries [J]. China Electricity, 55(08):73-86 +95.

[4] Fan Y X, Xiao F, Xu J, et al.2021, SOH estimation for Li-ion batteries based on charging voltage fragmentation and kernel ridge regression [J]. Chinese Journal of Electrical Engineering, 41(16):5661-5670.

[5] Pecht M. Battery data set. Maryland, MD, USA: CALCE battery research group;2017 [Online] Available: https://web.calce.umd.edu/batteries/data.htm.

[6] Fei C, Zhao L, Wang Y K. 2022, Health state estimation of lithium-ion batteries based on XGBoost algorithm [J]. Zhejiang Electricity, 41(05):14-21.

[7] Ji C J, Yuan X Q, Zhang Y. 2022, A comparative study of RUL prediction for lithium batteries based on improved PF [J]. Computer Simulation, 39(08):86-91.

Research on unmanned driving path planning algorithm based on Robot Operating System

Xinwang Zheng[1], Chen Liang[1], Yixin Huang[1], Bing Li[2] and Guangsong Yang[2*]

[1] Chengyi University College, Jimei University, Xiamen, 361021, China

[2] School of Ocean Information Engineering, Jimei University, Xiamen, 361021, China

Corresponding author: gsyang@jmu.edu.cn

Abstract. With the rapid development of communication and computer technology In recent years, automatic driving has attracted much attention. By using path planning technology, the AV can automatically find an optimal path to the destination while avoiding obstacles, so as to reduce transportation costs, improve driving efficiency, improve customer satisfaction and reduce traffic accidents, which is of important meaning to spurring the further development of intelligent transportation. Two algorithms of driverless path planning are studied, one is and the A* algorithm is for global path planning, another is the timed elastic band (TEB) algorithm for local path planning. Firstly, the direction of the exploration domain of the traditional A* algorithm is improved by adding an exploration domain to smooth the planned path.t. Secondly, some parameters of the TEB algorithm are adjusted to make the car stay away from obstacles and run more safely. Finally, on the mobile robot platform equipped with ROS operating system, the hybrid path planning is realized by combining these two algorithms, and the simulation experiment under a real environment is carried out. Experiment results show that the proposed method can autonomously plan path and navigation smoothly.

1. Introduction

At present, unmanned driving technology has been researched and developed on a large scale worldwide. Automated Vehicles (AVs) rely on various environmental sensing sensors to obtain external information and analyze the surrounding environment. Therefore, Automated Vehicles can independently plan routes according to the information of the surrounding environment and realize functions such as autonomous navigation and obstacle avoidance [1][2].

Automatic driving has attracted much attention. There are three kinds of core technologies of autonomous vehicle systems, i.e., perception, control, and planning[3]. Perception is the ability of autonomous systems to collect information and get relevant environment knowledge. Control is the ability of a robot to perform planned actions generated by higher-level processes. Planning is the process of making decisions to close to the goals. It is usually a heuristic algorithm that brings the vehicle from the start to the end while avoiding obstacles.

Path planning in unmanned driving technology determines whether the unmanned driving platform can take the shortest and safest path to the destination according to the information of the surrounding environment, It is an indispensable part of unmanned driving technology to predict the traffic situation and deal with the unexpected situation in the traffic process [4].

Path planning mainly includes the following two path planning strategies. One is the path planning that relies on the external sensor equipped by the autonomous vehicle to obtain the surrounding environment information in an unknown environment. It does not need to give the environmental information in advance and has strong adaptability. However, due to the lack of global information, it can only realize the locally optimal path planning [5]. The other is path planning based on environmental map information, which obtains the optimal path through offline calculation. This method can achieve global optimization but requires detailed environmental information [6].

In this paper, we combined the two strategies by modifying the A* and TEB to get better path planning. The remainder content of the paper is as follows. In section 2, the improvement of the A* algorithm is discussed. Section 3 gives the improved TEB local path planning algorithm. In Section 4, we evaluated the performance of the proposed algorithm by experiments based on Robot Operating System (ROS). Finally, we presented conclusions and future work in Section 5.

2. Improvement of A* global path planning algorithm

2.1. Traditional A algorithm*

A* algorithm is a kind of path planning algorithm, which combines the Dijkstra and Breath First Search (BFS) algorithm. It is a heuristic artificial intelligence algorithm that can be applied to a variety of situations. It has strong robustness [7].

Equation (1) is the evaluation function f(n) :

$$f(n)=g(n)+h(n) \tag{1}$$

g (n)——the estimated cost of an initial node in a grid and one of its grid nodes
h(n)——estimated cost during the path from the current node to the destination node

In order to make path planning more diverse, the prediction function of h(n) can be obtained by the Euclidean distance method [8]. The current coordinate grid is (xn, yn) and the target coordinate grid is (xT, yT), h(n) is shown in Equation (2) :

$$h(n)=\sqrt{(x_T-x_n)^2+(y_T-y_n)^2} \tag{2}$$

When planning the path to node n, f (n) is shown in Equation (3):

$$f(n)=g(n)+\sqrt{(x_T-x_n)^2+(y_T-y_n)^2} \tag{3}$$

For path planning, it needs to explore the information of the eight nodes around the current parent node when using the A* algorithm. This information is sorted, compared, classified and used to determine the next parent for the next exploration. The exploration process is repeated until the current node becomes the target node [9]. Figure1 is the exploration domain of the conventional A* algorithm.

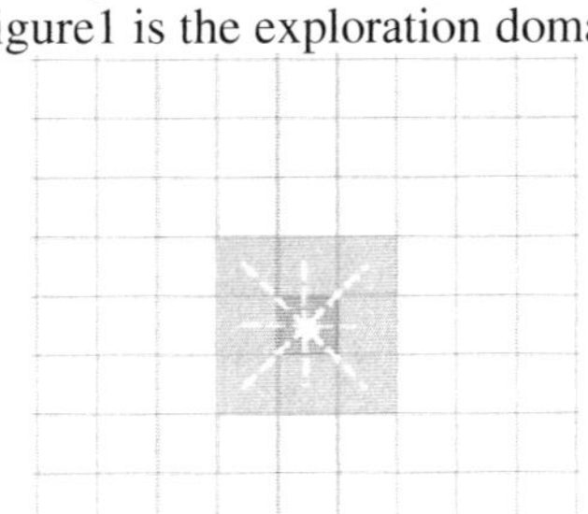

Figure 1. Exploration domain of the conventional A* algorithms

2.2. Improved A algorithm for adding exploration domains*

A* algorithm has some shortcomings, such as too many broken lines and uneven paths. It can be seen that the exploration domain of A* algorithm is the surrounding 8 grid nodes. To improve the exploration efficiency of the A* algorithm, the size of its exploration domain should be increased.The improved A* algorithm will take the surrounding 16 nodes as the exploration domain to explore the 16 nodes around the current parent node as shown in figure 2.

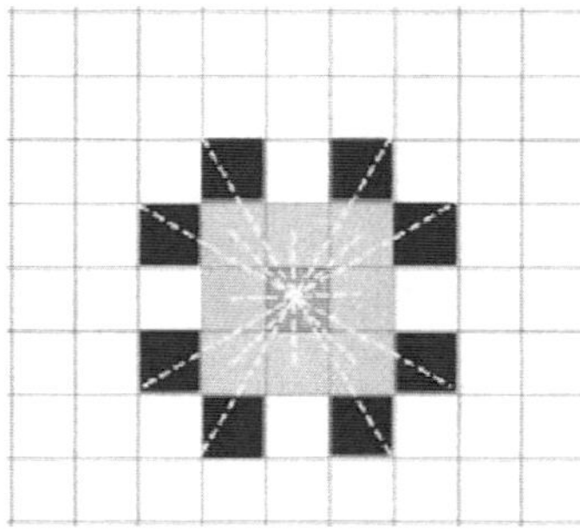

Figure 2. Exploration domain of the improved A* algorithm

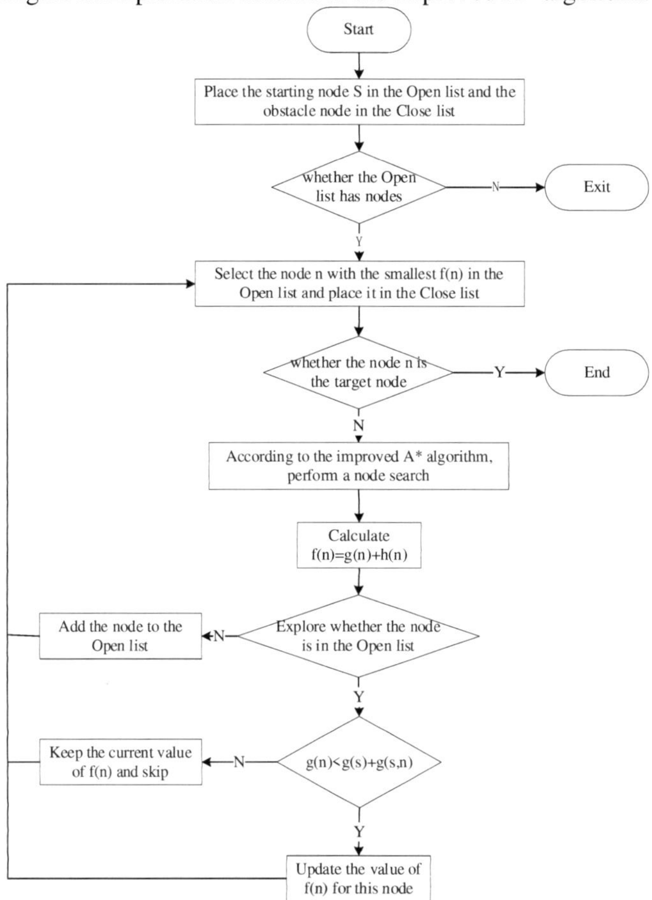

Figure 3. Flowchart of the improved A* algorithm

The improved A* algorithm has the following work process as shown in figure 3.

1) Explore the surrounding 16 grid nodes, and determine whether there is a node position beyond the map, if there is a node beyond the map, discard the node.

2) Determine whether there is an obstacle node or in the Close list (CL) among the eight nodes in the inner layer. If there is an obstacle node or in the CL, the node will not be processed. Then judge whether the node exists in the Open list (OL). If not, add the node to the OL and update its f(n) value.

3) Determine whether there is an obstacle node among the eight nodes in the outer layer. If there are obstacle nodes, the exploration of the outer layer 8 nodes is abandoned. If it is not an obstacle node, we should judge whether there is an obstacle node on the path from the node to its parent node. If exits obstacle nodes, the node will be abandoned directly. Otherwise, add the node to the OL, and calculate its f(n) value.

4) All nodes' f(n) values in the OL were compared in this round of exploration, and the node with the smallest f(n) value was added to the CL and deleted from the OL.

In our improved A* algorithm, the exploration area is increased by increasing the number of exploration domains from 8 to 16, the exploration efficiency is improved, and the best path is faster and more comprehensively in path planning.

3. TEB local path planning algorithm

3.1. TEB algorithm

TEB algorithm is optimized from elastic band (EB) algorithm. However, the implementation of EB algorithm requires the introduction of N positions and poses of Automated Vehicles in the process of movement, which are expressed as a sequence composed of $S_i=[x_i,y_i,\beta_i]$, Where (x_i,y_i) represents the coordinates of the AV, β_i represents the current posture of the AV. The sequence of positions and poses is represented as:

$$Q=\{S_i\}_{i=0\ldots n} \tag{4}$$

TEB introduces the required time interval of the path from start to end point, which is expressed as ΔT_i, the sequence of time is represented as:

$$\tau=\{\Delta T_i\}_{i=0\ldots n-1} \tag{5}$$

When the AV moves from the current position and attitude to the next in the Q sequence, combined with the movement time, the motion sequence of the AV represented as

$$B:=(Q,\tau) \tag{6}$$

In addition to introducing time, the TEB algorithm also adds the concept of multi-objective constraint on the basis of the EB algorithm. By setting different weights on multiple objective constraints, the motion sequence of the AV can be dynamically adjusted:

$$f(B)=\sum_k \gamma_k f_k(B) \tag{7}$$

$$B^*=\arg\min f(B) \tag{8}$$

As shown in Equations (7) and (8), f(B) represents the constraint weighted to B, γ_k represents the weight of constraint conditions, and B* represents the motion sequence of the optimized AV. There are two main constraints ways of the TEB algorithm. The first is based on the penalty function, which constrains the velocity and acceleration. The second takes the indexes related to path optimization as constraints, such as the shortest time, shortest path, or distance from obstacles [10].

4. Experiment and evaluation

4.1. Experiment of improved A based on PyCharm*

The improved A* algorithm is simulated by PyCHarm shown in figure 4. The blue circle is the starting point, the red circle is the endpoint, the green square is the node in the CL, and the red square is the node in the OL, and the black square is the obstacle node.

We compared the path length of the two algorithms in Table 1. It can be seen that the improved A* algorithm has a shorter planned path and fewer nodes. In Figure 4, it can be found that the path obtained by the optimized A* algorithm (blue curve) is shorter, smoother, and less angular than the traditional algorithm (red curve), which can effectively improve the efficiency of the path.

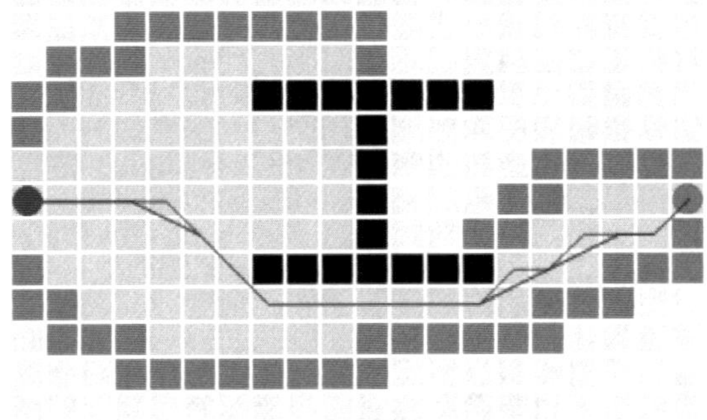

Figure 4 . Simulation results

Table 1 Path length comparison

	Linear distance	Path planning distance	Numbers of path planning nodes
Traditional A* algorithm	20	25.89	20
Improved A* algorithm	20	24.91	16

4.2. *Experiment of improved TEB based on PyCharm*

The TEB algorithm is simulated on the Rviz simulation platform. After modifying the configuration file related to the TEB algorithm and setting the min_obstacle_dist parameter to 0.7, simulation verification is performed in the scenario with moving obstacles.

The simulation diagram of the TEB algorithm before the obstacle block movement is shown in Figure 5. The red square is the movable obstacle block, and the red arrow represents the route planned by the TEB algorithm.

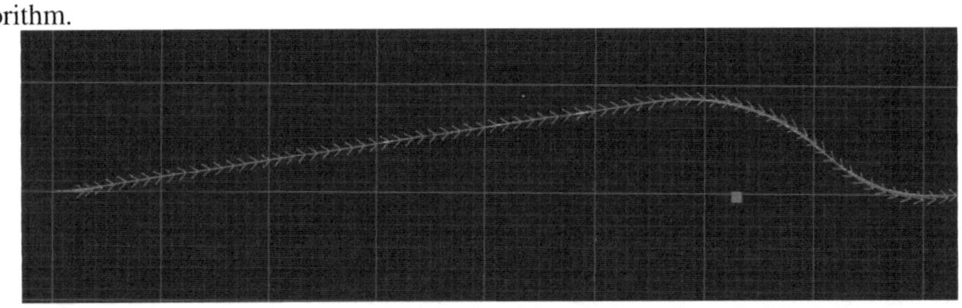

Figure 5. Simulation diagram of TEB algorithm before obstacle block movement

The simulation diagram of the TEB algorithm after obstacle block movement is shown in figure 6. When the red obstacle block moves to the left, the planned path is also changed to bypass the obstacles. This show the local path planning function of the TEB algorithm.

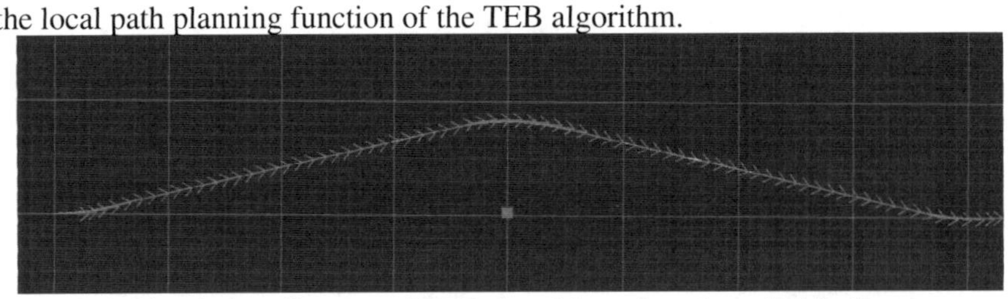

Figure 6. Simulation diagram of TEB algorithm after obstacle block movement

4.3. Experiment based on ROS

We use a mobile robot equipped with ROS system to do the experiment. The mobile robot adopts the Ackermann motion model is adopted to fit the current motion model of the car. To realize autonomous navigation, mobile robots need to hybrid path planning combined with global and local path planning [11]. We use the improved A* algorithm as the global path planning, and the TEB algorithm as the local path planning. We select several key points on the path of the global path planning as the target points of the local path planning. The robot move to these point according to the actual situation during the movement. Figure 7 shows the flow chart of hybrid path planning.

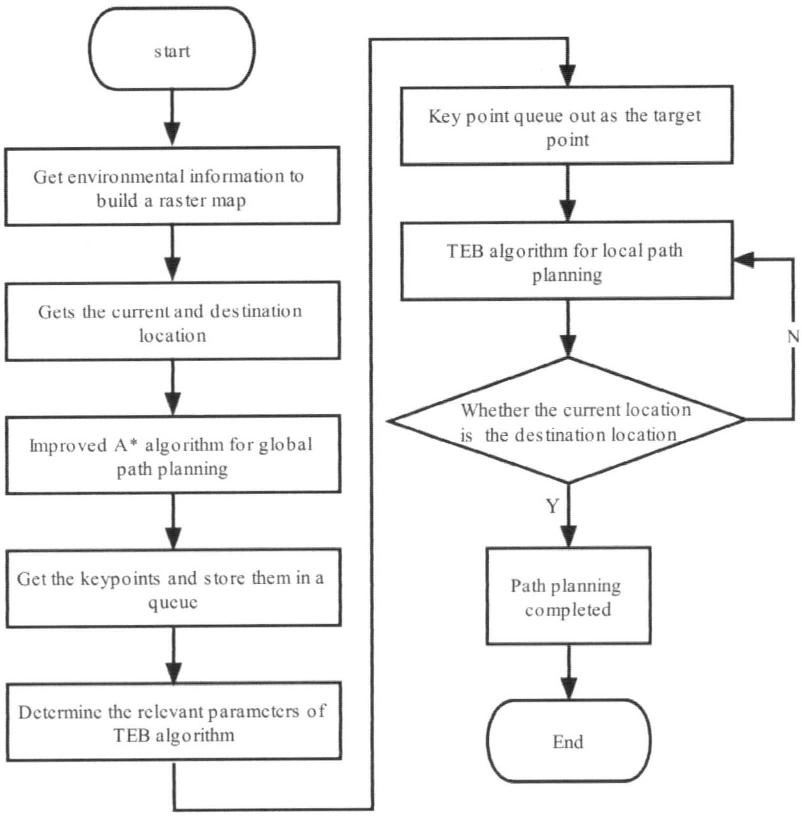

Figure 7. Flowchart of hybrid path planning

To reduce the interference brought by environmental factors, we choose to build the site indoors to simulate the operation of Automated Vehicles in the real environment under the road scenes of right-angle turns and intersections.

SLAM mapping of the surrounding environment is realized through the environmental data collected by the laser radar carried by the mobile robot as shown in Figure 8. The red pixels are the environmental data detected by laser radar, the black pixels are the infeasible area, the gray pixels are the feasible area, and the grey-green pixels are the unknown area. A complete map of the experimental site is obtained after the robot runs around the experimental site by remote control. Figure 9 is the map obtained by SLAM mapping of the robot.

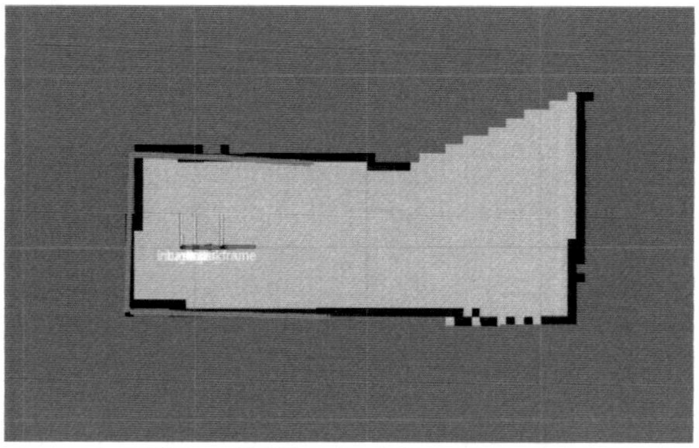

Figure 8. The mapping process

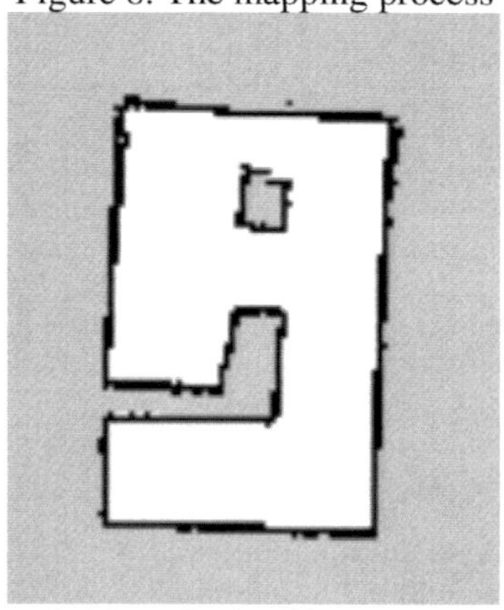

Figure 9. Map obtained by slam mapping

After obtaining the complete map of the experimental site, the experiment of hybrid path planning was done in figure 10, by combining the traditional A* algorithm and the proposed A* algorithm as global path planning respectively with TEB as local path planning.

Experiment results are shown in Figure 10. Figures (a) and (b) are the hybrid path planning with the traditional A* algorithm and improved A* algorithm as the global path planning respectively. The path planned by the traditional A* algorithm in figure (a) has many turns, and the turning angle in the path is large, so the overall path planned is longer. The improved A* algorithm in figure (b) has fewer turns, a smaller turning angle in the path, and a shorter overall path. The simulation results and the experimental results in the real environment show that the proposed algorithm has a smoother path. This is conducive to the AV in the process of driving more stable and running more smoothly.

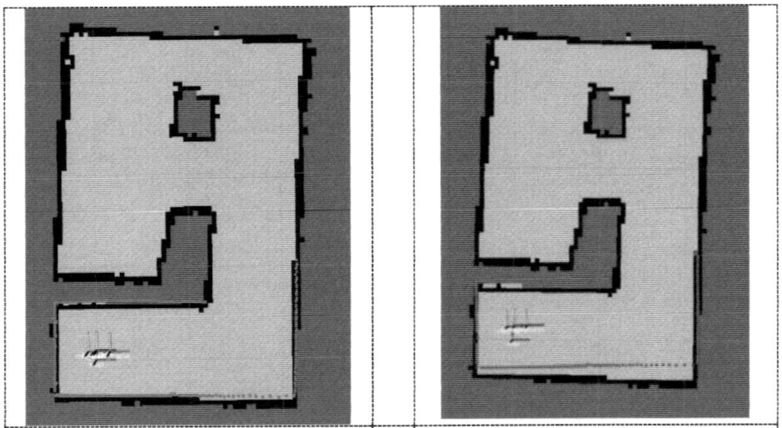

(a) Traditional A* algorithm (b) Improved A* algorithm

Figure 10. Hybrid path planning experiment of mobile robot

5. Conclusion

Based on the idea of adding an exploration domain in traditional A*, a path planning algorithm is proposed in this paper, combined with the TEB algorithm to local path plan and avoid the obstacle. which improves the search range. The raster map is constructed and the performance is simulated. The mobile robot is also used to carry out the path planning experiment in a real indoor environment. Both the simulation and experimental results show that the path planned by the improved A* algorithm is smoother and shorter, and it can realize the autonomous path planning and real-time obstacle avoidance functions of the mobile robot. We will consider how to improve the performance of AV path planning by machine learning in our future work.

Acknowledgments

This work of is supported by Education and scientific research project for young and middle-aged teachers in Fujian Province (JAT210674), and supported by Industry-school Cooperative Education Program of Ministry of Education (202102357022).The work was partially supported by the Natural Science Foundation of Fujian Province (No. 2021J01865), and the Jimei University National Fund cultivation program(No. F0107).

References

[1] Yuan J. 2020 Simulation analysis of path planning of AV. *South agricultural machinery*, 51(17):68-69.

[2] Raviteja, T, & Vedaraj, I R 2020 An introduction of autonomous vehicles and a brief survey. *J. Crit. Rev*, 7(13), 196-202.

[3] Jo K, Kim J, Kim D, Jang C, Sunwoo M 2015 Development of autonomous car—Part II: A case study on the implementation of an autonomous driving system based on a distributed architecture. *IEEE Transactions on Industrial Electronics*, *62*(8), 5119-5132.

[4] Zhang H Y, Lin W M, Chen A X 2018 Path planning for the mobile robot: A review. *Symmetry*, *10*(10), 450.

[5] Pradhan S, Mandava R K, Vundavilli P R 2021 Development of path planning algorithm for biped robot using combined multi-point RRT and visibility graph. *International Journal of Information Technology*, *13*(4), 1513-1519.

[6] Tran N, Nguyen D T, Vu D L, Truong N V 2013 Global path planning for autonomous robots using modified visibility-graph. In *2013 International Conference on Control, Automation and Information Sciences (ICCAIS)*, 317-321.
[7] Long Y, Zuo Z, Su Y, Li J, Zhang H. 2020 An A*-based Bacterial Foraging Optimisation Algorithm for Global Path Planning of Unmanned Surface Vehicles. *The Journal of Navigation*,73(6):1247-62.
[8] Patle B K, Pandey A, Parhi D R K, Jagadeesh A 2019 A review: On path planning strategies for navigation of mobile robot. *Defence Technology*, *15*(4), 582-606.
[9] Yu L, Kong D, Yan X. 2018 A Driving Behavior Planning and Trajectory Generation Method for Autonomous Electric Bus. *Future Internet*,10(6).
[10] Chen J T, Guo C and Liu Y. 2021 Path optimization method for mobile robot based on time elastic band [J]. *Science technology and engineering*, 21(26):11212-11219.
[11] Zhang X, Lai J, Xu D, Li H, Fu M 2020 2D LiDAR-based SLAM and path planning for indoor rescue using mobile robots. *Journal of Advanced Transportation, 2020.*

Smart meter life prediction method based on health index and random forest

Jun Wang[1*], Ruyi Li[1], Hao Wang[1], Lijun Lu[1], Mingzhe Li[1], Aibeiduo Li[2]

[1]Henan Xj Metering Co., Ltd. Xuchang, China
[2]Tianjin University, Tianjin, China

[*]tianyumingyue@163.com

Abstract—Smart electricity meter is the key instrument to realize the accurate measurement of electricity consumption. However, with the increase of the service life of electricity meter, its measurement accuracy will deteriorate. Aiming at the problems that a single parameter cannot accurately characterize the performance degradation process of the electric meter, and the traditional intelligent learning model is difficult to accurately fit the degradation model of the smart electric meter, a new method is proposed to fuse the data to construct the electric meter health index, and combine the multi-model similarity matching and the integrated model to calculate the residual energy of the electric meter. Methods of life prediction. Hierarchical clustering and silhouette coefficients are used to filter parameters and fuse them into a meter health index. Multi-model similarity matching is used for regression model prediction to optimize the prediction results of the model. Select a meter measurement data set to verify the effectiveness of the proposed method. The results show that the method has the advantages of fast prediction speed, and the fusion of health index and multi-model similarity matching greatly improves the prediction accuracy of the remaining life of the electric meter.

1. Introduction

As the terminal equipment of the smart grid, the reliable operation of the smart meter plays a vital role in the stability and economic operation of the smart grid [1][2]. Since the promotion and application of smart meters in my country in 2009, the number of energy meters running on the grid has reached 570 million. The State Grid Corporation requires the design life of smart meters to be more than 10 years, while the national standard JJG 596-2012 "Electronic AC Electric Energy Meter Verification Regulations" stipulates that the verification period is 8 years [3], which means that most of the measurement accuracy is qualified and the function Normal smart meters can only be scrapped in advance, which will cost tens of billions of yuan every year. Therefore, it is of great economic value and important social significance to carry out the reliability assessment and life prediction of smart meters based on the basic errors of smart meters, providing a theoretical basis for scientific rotation.

Scholars at home and abroad have conducted a lot of research on the remaining useful life (RUL) prediction of instruments and equipment. At present, remaining life methods are mainly divided into two categories: mechanism modeling and data-driven methods. Early product life is short, with the help of life test and accelerated life test, enough product failure data can be obtained, and life prediction and reliability evaluation can be carried out accordingly [4]. However, the reliability evaluation results of these models for electronic components cannot meet the user's demand for high product reliability, and accelerated stress models such as the Eyring model and Arrhenius are not suitable for accelerated tests above the component level [5]. With the development of sensor technology and signal processing technology, data-driven life prediction methods have gradually become the mainstream[6][7]. The data-driven method generally refers to the method of early warning and management of equipment based on equipment-related monitoring data without precise mechanism. It mainly uses artificial intelligence algorithms to model equipment degradation data. Commonly used time series analysis algorithms include autoregressive sliding models, support vector machines, artificial neural networks and their combination methods, but these models require a large amount of data to train models, and their results What is obtained is the size of the RUL rather than the distribution, which cannot reflect the uncertainty of the prediction results [8]. The degradation model based on stochastic process is widely used to describe the performance degradation modeling and can give the interval distribution result of RUL, which has received more and more attention and recognition [9]. The most commonly used stochastic process modeling method today is the Wiener process. It obtains the life distribution of performance degradation through the inverse Gaussian process. For example, in the literature [10], the RUL calculation is carried out on the smart meter through the accelerated test data and the field data, but the Wiener process is computationally complex and the realization cost is high.

Therefore, this paper proposes a method to fuse data to construct an engine health index, and combine multi-model similarity matching and ensemble models to predict the remaining life of the engine. First of all, combined with hierarchical clustering and silhouette coefficient, the measurement parameters of the meter are selected, and the meter health index is used to characterize the performance degradation process. This solves the problem that a single parameter cannot accurately characterize the performance degradation process of the meter, and reduces the use of all measurement data. The high-dimensional operation brought by the fusion; secondly, the performance degradation model is trained by random forest, and the remaining life prediction results are optimized by combining the multi-model similarity matching, and the prediction accuracy is improved; finally, the effectiveness of the proposed method is verified according to the field measurement data.

2. Health Index Construction

Due to the different performance degradation modes of smart meters, a single parameter cannot accurately characterize the performance degradation process of the meter. In this paper, the fusion data is used to construct the engine health index, and the selected characteristic parameters are processed by Kalman filtering, smoothing and normalization.

Hierarchical clustering is a type of clustering algorithm that creates a hierarchical nested clustering tree by calculating the similarity between data points of different categories. In a dendrogram, the raw data points of different categories are the lowest level of the tree, and the top level of the tree is the root node of a cluster. There are two methods of creating a clustering tree: bottom-up merging and top-down splitting. For hierarchical clustering of data, Euclidean distance is usually used as a metric. Before performing hierarchical clustering, all feature parameters should be normalized to eliminate the influence of different dimensions. The Euclidean distance is used as a metric in this paper, and its formula is as follows:

$$d_{ij} = \sqrt{\left(x_{i1} - x_{j1}\right)^2 + \cdots + \left(x_{ip} - x_{jp}\right)^2} \tag{1}$$

where d_{ij} is the Euclidean distance between a point $\left(x_{i1}, x_{i2}, \cdots, x_{ip}\right)$ and a point $\left(x_{j1}, x_{j2}, \cdots, x_{jp}\right)$ in the p-dimensional space. Hierarchical clustering does not need to pre-set the number of data classifications. It can optimally classify data according to different distances. Compared with other clustering algorithms, it is more suitable for engine data set classification and feature selection.

After the selected meter measurement data is filtered and smoothed, the data is processed into incremental Δf by making a difference between the real-time data and the failure data $\Delta f_N = f_{fualt} - f_t$.

After normalizing the data, the multiple measurement data are merged into a one-dimensional health index of the smart meter by the following formula:

$$\mathrm{HI}_K = \sum_{i=1}^{N} Q_i \Delta f_i \tag{2}$$

 2384 (2022) 012058 doi:10.1088/1742-6596/2384/1/012058

Where K is the time series length of the health index; Q_i is the weight value occupied by each feature, and the weight value of the fusion is the total variance occupied by a single feature parameter.

3. Data Management Model Based on Random Forest

In the remaining life modeling of smart meters, most researchers only use one model to train the performance degradation process of the electric meter, but such a model is difficult to accurately fit the degradation process of the electric meter. In view of the fact that the electric meter is composed of many different parts, the fault parts corresponding to different failure modes are different, and even the same type of electric meter has different performance degradation modes, so this paper uses the ensemble model random forest to train the performance degradation process of the electric meter.

Random forest is an ensemble learning model that combines bagging with random subspace methods, and introduces randomness on the basis of bagging, which can more accurately characterize the differences between electricity meters. As an ensemble learning model, random forest uses two ensemble strategies: bagging strategy and changing input feature strategy. The bagging strategy uses the bootstrap sampling method to integrate each underlying g-model. For each basic learning g model, the training samples are obtained by sampling with replacement. Due to the randomness of sampling, the training sample sets of each basic learning machine are not the same. Each basic learning model is trained at the same time, and finally all basic learning machines are integrated as the output of the ensemble learning model. Therefore, each basic learning machine is in a parallel relationship in the bagging strategy, and its process is shown in Figure 1.

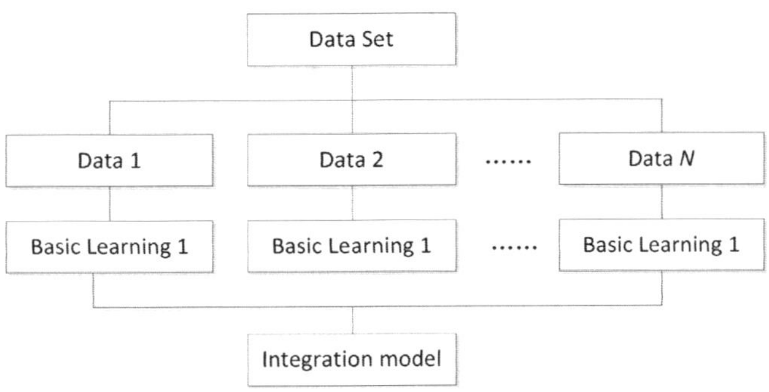

Figure 1. Bagging integration strategy

The basic learning model of the random forest model is a decision tree, or a classification and regression tree (CART). When predicting the remaining life of a smart meter, the current meter health index is input into n regression trees at the same time, and n prediction values of remaining life are obtained, that is, each regression tree has an impact on the remaining life of the current meter. After estimation, the average value of n output values is taken as the final output value of the random forest model, and the estimation of the remaining life of the current meter by the random forest model is obtained.

Below, we analyze the error rate of ensemble learning, consider the binary classification problem $y=\{0, 1\}$, the real function is $f(x)$; the simple voting method is used to integrate T basic learner $h_i(x)$, and the basic learning algorithm $H(x)$ is obtained; the error rate of each basic learner is ε, that is, $P\{h_i(x) \neq f(x)\} = \varepsilon$, and the base learner error rates are mutually independent. The error rate of the derivation base learning algorithm is as follows:

$$P\{H(x)=f(x)\}=\begin{cases} P\left(\sum_{i=1}^{T} h_i(x) \leq \frac{T}{2}\right) & \text{if } f(x)=1 \\ P\left(\sum_{i=1}^{T} h_i(x) > \frac{T}{2}\right) & \text{if } f(x)=0 \end{cases} \tag{3}$$

It is also assumed that the independent and identical distributions $x_1, x_2, \ldots, x_n$ satisfy the Bernoulli 0-1 distribution, the correct probability is p, and $S(x)=\sum_{i=1}^{n} x_i$, $\varepsilon>0$, the Hoeffding's inequality is as follows:

$$P\{S(x) \geq n(p+\varepsilon)\} \leq \exp\left\{-2\varepsilon^2 n\right\} \tag{4}$$

$$P\{S(x) \leq n(p-\varepsilon)\} \leq \exp\left\{-2\varepsilon^2 n\right\} \tag{5}$$

It can be seen from Formula (6) that as the number of individual learners T increases, the error of the ensemble learners tends to zero, so compared with using a single decision tree, the random forest can obtain higher prediction accuracy.

A random forest is essentially a classifier composed of multiple decision trees. It was an effective algorithm that was first perfected and proved by Breiman. In essence, it is a combined classifier $\left\{\bar{h}\left(X, \beta_k\right), k=1, \ldots n\right\}$. Among them, X is each feature variable, that is, the input variable. β_k is an independent and identically distributed random vector, and n represents the number of decision trees included in the random forest. Random forest is based on the majority vote of each decision tree classifier to determine the classification result.

The random forest algorithm combines bagging techniques and random tree node splitting techniques by training multiple decision tree classifiers on multiple randomly sampled data sets and then deciding the category of unknown samples through majority voting.

We denote the variable importance score as *IM* and the Gini index as *GI*. The dataset in this paper has n features $X_1, X_2, \cdots\cdots, X_n$, and now we want to calculate the Gini index score IM_n^{GI} of each feature X_n, which is also the average change in the impurity of the nth feature splitting nodes in all decision trees of random forest.

The formula for calculating the Gini index is:

$$GI_m = \sum_{k=1}^{K} \sum_{k \neq k} p_{mk} p_{mk} = 1 - \sum_{k=1}^{K} p_{mk}^2 \tag{6}$$

Among them, K represents the number of categories and p_{mk} represents the proportion of category k in node m.

The importance of feature X_n at node m, that is, the change in Gini index before and after the branch of node m is:

$$IM_n^{GI} = GI_m - GI_1 - GI_r \tag{7}$$

Among them, GI_1 and GI_r respectively represent the Gini indices of the two new nodes after the branch. If the node of feature X_n in decision tree i is set M, then the importance of X_n in the *ith* tree is:

$$IM_{in}^{GI} = \sum_{M}^{n} IM_{im}^{GI} \tag{8}$$

Finally, normalize all the obtained importance scores to get:

$$IM_n = \frac{IM_n}{\sum_{i=1} IM_{in}^{GI}} \tag{9}$$

Using the random forest model, the importance of a single feature can be calculated, the impact indicators of smart meters can be screened, and the correlation between a single impact factor and the final result can be obtained. However, because the random model has a certain degree of randomness, this model may assign different importance weights to each feature each time. The random model should be trained multiple times, and the features with higher importance are retained as the final result.

Traditional methods for predicting the remaining life of electric meters usually directly substitute the data into the regression model, or calculate through the similarity formula. In this paper, the method of combining multi-model similarity matching and regression modeling is used to predict the remaining life, and the multi-model similarity matching is used to optimize the prediction results of the regression model. Several models that best match the current performance degradation, predict the current remaining life. Usually, most researchers use similarity to find some electric meters that are most similar to the current electric meters in historical samples through similarity for instance reuse before using similarity for regression modeling. However, this method often results in that the running cycle of the current prediction exceeds the total historical cycle, reducing the accuracy of model matching. Using similarity for regression modeling can avoid the above situation and improve the accuracy of meter remaining life prediction. The remaining life prediction model of the meter is shown in Figure 2.

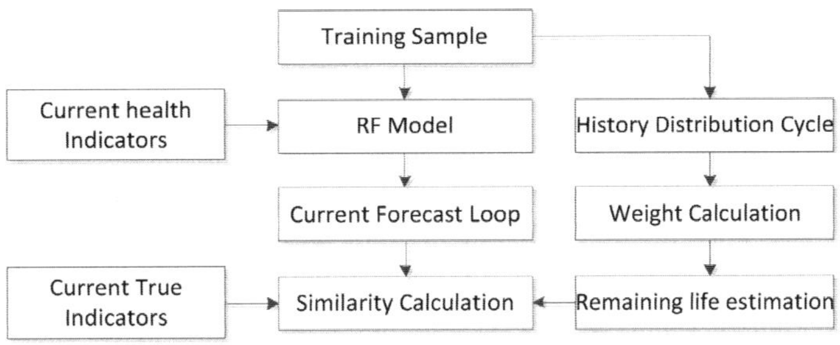

Figure 2. Prediction of smart meter lifespan

The specific steps of prediction are as follows: First, use the RF algorithm to train the performance degradation model on the historical meter data; then, select several matching models according to the error between the current cycle predicted value and the actual value, that is, the similarity with small error is large, and the similarity with large error is small.

The smart meter data used by the author is provided by a Metrology Center. There are 22 pieces of basic information for analysis of the meter, including: "manufacturer", "arrival batch", "meter mode", "access mode", "pulse constant" and "Loss of Voltage/Loss of Current Judgment". The number of electric meters used by the author is more than 150,000, and 80% of the data is used as training data and 20% is used as prediction data to test the model.

Since there are few basic information items of the electric meter extracted, the accuracy of the prediction model will be inversely proportional to the number of fault categories; therefore, the author only considers the correlation between the basic attributes of these 22 electric meters and the first three frequent faults. At the same time, the service life of the faulty electric meter can be obtained by counting the installation time and removal time of the electric meter. When predicting the failure of the electric meter, the service life is also used as one of the input features of the model, which is combined with the 22 basic attribute information and transformed into an input feature quantity consisting of 23 numbers.

4. Analysis of Results

In order to verify the validity of the proposed prediction model, experiments were carried out on the meter data provided by the metering center. These data include sorting data and in-transit data for the meters. The platform for building the random forest model is Matlab.

In this paper, from the 23 historical data of the electric meter, the variance filter is used to select 14 measurement parameters with changes, and some parameters are selected for fusion among these 14 parameters with changes. First, hierarchical clustering is performed on 14 parameters with changes.

In the case of clustering the data into 2~13 clusters, the corresponding silhouette coefficients are calculated respectively. By comparison, it is found that when the data is divided into 5 clusters, the mean value of the silhouette coefficient is 0.86, the silhouette coefficient of each cluster feature is very close to the mean value, and the clustering effect is the best.

When predicting the service life of the electric meters, the author firstly counted the usage time of more than 150,000 electric meters that were dismantled after failure. The statistical results are shown in Figure 3.

Figure 3. The service life of the fault removal meter

By calculating the service life of the existing fault removal table, the service life is divided into 4 time periods: the use time is within 2 years, the use time is 2 to 4 years, the use time is 4 to 6 years, and the use time is 6 years. above. The input is the basic information of 22 electricity meters, and the prediction accuracy and the used prediction time are shown in Table 3. Among them, the data used for training the prediction model is 113,062 electricity meters, and the data used for testing is 28,265 electricity meters.

Table 1. 4 types of life expectancy prediction results

n-tree	**Prediction accuracy/%**	**Prediction time/s**
10	74.9	5.1
30	75.3	9.1
40	75.6	11.6
50	75.4	14
80	75.5	22

It can be seen from the table that the prediction accuracy of dividing the life of the smart meter into four time periods is about 75%. At the same time, the model will give the score of each meter in each life segment. For example, the life prediction result of a random meter is selected. The score of the first class is 0.0113, the score of the second class is 0.2860, the score of class 4 is 0.7002, and the score of class 4 is 0.0026, then the probability of the meter's life being 4~6 years is 70%, the probability of its life being within 2~4 years is 28.6%, and the probability of being below 2 years is 1.13 %, the probability of more than 6 years is 0.26%, it can be judged that the life of the meter is most likely to be between 4 and 6 years.

In order to improve the prediction accuracy of the life prediction model, the service life is divided into three time periods: the use time is less than 3 years, the use time is 3-6 years, and the use time is more than 6 years. Also, the basic information of 22 electricity meters is input, and the prediction accuracy and the used prediction time are shown in Table 4. The training data is the same as the test data and the experimental data in Table 3.

Table 2. 3 types of life expectancy prediction results

n-tree	**Prediction accuracy/%**	**Prediction time/s**
10	89.4	4.6
30	89.5	8.7
40	89.6	11.4
50	89.6	13.8
80	89.7	21

 2384 (2022) 012058 doi:10.1088/1742-6596/2384/1/012058

It can be seen from the table that the prediction accuracy of dividing the life of the smart meter into three time periods is about 89%. Compared with dividing into 4 time periods, the prediction accuracy has been significantly improved. Therefore, the forecast period of the model can be divided into different time periods according to specific needs, so as to provide a better reference for the detection and rotation cycle of electricity meters.

5. Conclusion

Based on the health index and random forest algorithm, this article proposes a life prediction model of smart meters. Through the data of more than 150,000 meters provided by the Ningxia Metrology Center, the experimental test is carried out, and the following conclusions can be obtained:

1) In this paper, methods such as hierarchical clustering are used to screen out the parameters that need to be fused, so as to achieve efficient and comprehensive utilization of data. Then, the ratio of a single feature to the total variance is used as the weight, and the selected feature parameters are fused to effectively construct a health index, which overcomes the problem that a single parameter cannot accurately characterize the performance degradation of the meter.

2) The integrated model random forest is used to train the smart meter performance degradation model, which can better fit the actual degradation process; the multi-model similarity matching is combined with the RF model to optimize the prediction results of the regression model, which greatly improves the precision of the remaining life prediction.

3) Change the value of the only parameter *n-tree* of the model, the prediction accuracy will not change greatly, but the prediction time will change accordingly; therefore, it is better to set the value of the *n-tree* smaller.

References

[1] Yang T, Cai ST, Yan P, et al, Saturation Defense Method of a Power Cyber-physical System Based on Active Cut Set, IEEE Transactions on Smart Grid, 2022, DOI 10.1109/TSG.2022.3154839.
[2] Yang T, Zhao LY, Li W, et al. Dynamic energy dispatch strategy for integrated energy system based on improved deep reinforcement learning. Energy, 2021, 235: 121377.
[3] National Electromagnetic Metrology Technical Committee. Electronic AC energy meter: JJG596-2012[S]. Beijing: China Quality Inspection Press, 2013:2.
[4] LIU T Y. Research on product reliability evaluation and remaining life prediction methods under complex degradation process[D]. Changsha: National University of Defense Technology, 2016.
[5] LI X Y, JIANG T M. Summary of multi-stress acceleration models in accelerated life testing[J]. System Engineering and Electronic Technology, 2007(5): 828-831.
[6] Yang T, Zhao LY, Li W, et al. Towards healthy and cost-effective indoor environment management in smart homes: A deep reinforcement learning approach. Applied Energy, 2021, 300: 117335..
[7] Yang T, Zhang Y, Li W, et al. Decentralized networked load frequency control in interconnected power systems based on stochastic jump system theory[J]. IEEE Transactions on Smart Grid, 2020, 11(5): 4427-4439.
[8] HU CH H, FAN H D. Estimation of equipment remaining life and optimal maintenance decision [M]. Beijing: National Defense Industry Press, 2018.
[9] DU D B, SI X SH, HU CH H, et al. Life prediction method of shared load system based on stochastic degradation modeling[J]. Chinese Journal of Scientific Instrument, 2018, 39(8): 53-62.
[10] LI H L, YU H B, HE J L. Prediction of remaining operating life of smart electric energy meters based on data fusion methods [J]. Electrical Measurement and Instrumentation, 2019, 56(18):126-133.

AUTHOR INDEX

Institute of Physics
Dirac House, Temple Back
Bristol BS1 6BE UK

ISSN: 1742-6588
ISBN 978-1-7138-6490-5

CURRAN ASSOCIATES INC.
proceedings.com